Transmission Electron Microscopy

A Textbook for Materials Science

Transmission Electron Microscopy

A Textbook for Materials Science

David B. Williams
C. Barry Carter

David B. Williams
The University of Alabama in Huntsville
Huntsville AL, USA
david.williams@uah.edu

C. Barry Carter
University of Connecticut
Storrs, CT, USA
cbcarter@engr.uconn.edu

ISBN 978-0-387-76502-0 softcover (This is a four-volume set. The volumes are not sold individually.)
ISBN 978-0-387-76501-3 (eBook)

Library of Congress Control Number: 2008941103

© Springer Science+Business Media, LLC 1996, 2009
All rights reserved. This work may not be translated or copied in whole or in part without the written permission of the publisher (Springer Science+Business Media, LLC, 233 Spring Street, New York, NY 10013, USA), except for brief excerpts in connection with reviews or scholarly analysis. Use in connection with any form of information storage and retrieval, electronic adaptation, computer software, or by similar or dissimilar methodology now known or hereafter developed is forbidden.
The use in this publication of trade names, trademarks, service marks, and similar terms, even if they are not identified as such, is not to be taken as an expression of opinion as to whether or not they are subject to proprietary rights.

Printed on acid-free paper

springer.com

To our parents

Walter Dennis and Mary Isabel Carter
and
Joseph Edward and Catherine Williams,
who made everything possible.

About the Authors

David B. Williams

David B. Williams became the fifth President of the University of Alabama in Huntsville in July 2007. Before that he spent more than 30 years at Lehigh University where he was the Harold Chambers Senior Professor Emeritus of Materials Science and Engineering (MS&E). He obtained his BA (1970), MA (1974), PhD (1974) and ScD (2001) from Cambridge University, where he also earned four Blues in rugby and athletics. In 1976 he moved to Lehigh as Assistant Professor, becoming Associate Professor (1979) and Professor (1983). He directed the Electron Optical Laboratory (1980–1998) and led Lehigh's Microscopy School for over 20 years. He was Chair of the MS&E Department from 1992 to 2000 and Vice Provost for Research from 2000 to 2006, and has held visiting-scientist positions at the University of New South Wales, the University of Sydney, Chalmers University (Gothenburg), Los Alamos National

Laboratory, the Max Planck Institut für Metallforschung (Stuttgart), the Office National d'Etudes et Recherches Aérospatiales (Paris) and Harbin Institute of Technology.

He has co-authored and edited 11 textbooks and conference proceedings, published more than 220 refereed journal papers and 200 abstracts/conference proceedings, and given 275 invited presentations at universities, conferences and research laboratories in 28 countries.

Among numerous awards, he has received the Burton Medal of the Electron Microscopy Society of America (1984), the Heinrich Medal of the US Microbeam Analysis Society (MAS) (1988), the MAS Presidential Science Award (1997) and was the first recipient of the Duncumb award for excellence in microanalysis (2007). From Lehigh, he received the Robinson Award (1979), the Libsch Award (1993) and was the Founders Day commencement speaker (1995). He has organized many national and international microscopy and analysis meetings including the 2nd International MAS conference (2000), and was co-chair of the scientific program for the 12th International Conference on Electron Microscopy (1990). He was an Editor of *Acta Materialia* (2001–2007) and the *Journal of Microscopy* (1989–1995) and was President of MAS (1991–1992) and the International Union of Microbeam Analysis Societies (1994–2000). He is a Fellow of The Minerals Metals and Materials Society (TMS), the American Society for Materials (ASM) International, The Institute of Materials (UK) (1985–1996) and the Royal Microscopical Society (UK).

C. Barry Carter

C. Barry Carter became the Head of the Department of Chemical, Materials & Biomolecular Engineering at the University of Connecticut in Storrs in July 2007. Before that he spent 12 years (1979–1991) on the Faculty at Cornell University in the Department of Materials Science and Engineering (MS&E) and 16 years as the 3 M

Heltzer Multidisciplinary Chair in the Department of Chemical Engineering and Materials Science (CEMS) at the University of Minnesota. He obtained his BA (1970), MA (1974) and ScD (2001) from Cambridge University, his MSc (1971) and DIC from Imperial College, London and his DPhil (1976) from Oxford University. After a postdoc in Oxford with his thesis advisor, Peter Hirsch, in 1977 he moved to Cornell initially as a postdoctoral fellow, becoming an Assistant Professor (1979), Associate Professor (1983) and Professor (1988) and directing the Electron Microscopy Facility (1987–1991). At Minnesota, he was the Founding Director of the High-Resolution Microscopy Center and then the Associate Director of the Center for Interfacial Engineering; he created the Characterization Facility as a unified facility including many forms of microscopy and diffraction in one physical location. He has held numerous visiting scientist positions: in the United States at the Sandia National Laboratories, Los Alamos National Laboratory and Xerox PARC; in Sweden at Chalmers University (Gothenburg); in Germany at the Max Planck Institut für Metallforschung (Stuttgart), the Forschungszentrum Jülich, Hannover University and IFW (Dresden); in France at ONERA (Chatillon); in the UK at Bristol University and at Cambridge University (Peterhouse); and in Japan at the ICYS at NIMS (Tsukuba).

He is the co-author of two textbooks (the other is *Ceramic Materials; Science & Engineering* with Grant Norton) and co-editor of six conference proceedings, and has published more than 275 refereed journal papers and more than 400 extended abstracts/conference proceedings. Since 1990 he has given more than 120 invited presentations at universities, conferences and research laboratories. Among numerous awards, he has received the Simon Guggenheim Award (1985–1986), the Berndt Matthias Scholar Award (1997/1998) and the Alexander von Humboldt Senior Award (1997). He organized the 16th International Symposium on the Reactivity of Solids (ISRS-16 in 2007). He was an Editor of the *Journal of Microscopy* (1995–1999) and of *Microscopy and Microanalysis* (2000–2004), and became (co-)Editor-in-Chief of the *Journal of Materials Science* in 2004. He was the 1997 President of MSA, and served on the Executive Board of the International Federation of Societies for Electron Microscopy (IFSEM; 1999–2002). He is now the General Secretary of the International Federation of Societies for Microscopy (IFSM; 2003–2010). He is a Fellow of the American Ceramics Society (1996) the Royal Microscopical Society (UK), the Materials Research Society (2009) and the Microscopy Society of America (2009).

Preface

How is this book different from the many other TEM books? It has several unique features but what we think distinguishes it from all other such books is that it is truly a *textbook*. We wrote it to be read by, and taught to, senior undergraduates and starting graduate students, rather than studied in a research laboratory. We wrote it using the same style and sentence construction that we have used in countless classroom lectures, rather than how we have written our countless (and much-less read) formal scientific papers. In this respect particularly, we have been deliberate in *not* referencing the sources of every experimental fact or theoretical concept (although we do include some hints and clues in the chapters). However, at the end of each chapter we have included groups of references that should lead you to the best sources in the literature and help you go into more depth as you become more confident about what you are looking for. We are great believers in the value of history as the basis for understanding the present and so the history of the techniques and key historical references are threaded throughout the book. Just because a reference is dated in the previous century (or even the antepenultimate century) doesn't mean it isn't useful! Likewise, with the numerous figures drawn from across the fields of materials science and engineering and nanotechnology, we do not reference the source in each caption. But at the very end of the book each of our many generous colleagues whose work we have used is clearly acknowledged.

The book consists of 40 relatively small chapters (with a few notable Carter exceptions!). The contents of most of the chapters can be covered in a typical lecture of 50-75 minutes (especially if you talk as fast as Williams). Furthermore, each of the four softbound volumes is flexible enough to be usable at the TEM console so you can check what you are seeing against what you should be seeing. Most importantly perhaps, the softbound version is cheap enough for all serious students to buy. So we hope you won't have to try and work out the meaning of the many complex color diagrams from secondhand B&W copies that you acquired from a former student. We have deliberately used color where it is useful rather than simply for its own sake (since all electron signals are colorless anyhow). There are numerous boxes throughout the text, drawing your attention to key information (green), warnings about mistakes you might easily make (amber), and dangerous practices or common errors (red).

Our approach throughout this text is to answer two fundamental questions:

Why should we use a particular TEM technique?
How do we put the technique into practice?

In answering the first question we attempt to establish a sound theoretical basis where necessary although not always giving all the details. We use this knowledge to answer the second question by explaining operational details in a generic sense and showing many illustrative figures. In contrast, other TEM books tend to be either strongly theoretical or predominantly descriptive (often covering more than just TEM). We view our approach as a compromise between the two extremes, covering enough theory to be reasonably rigorous without incurring the wrath of electron physicists yet containing sufficient hands-on instructions and practical examples to be useful to the materials engineer/nanotechnologist who wants an answer to a materials problem rather than just a set of glorious images, spectra, and diffraction patterns. We acknowledge that, in attempting to seek this compromise, we often gloss over the details of much of the physics and math behind the many techniques but contend that the content is usually approximately right (even if on occasions, it might be precisely incorrect!).

Since this text covers the whole field of TEM we incorporate, to varying degrees, *all* the capabilities of the various kinds of current TEMs and we attempt to create a coherent view of the many aspects of these instruments. For instance, rather than separating out the broad-beam techniques of a traditional TEM from the focused-beam techniques of an analytical TEM, we treat these two approaches as different sides of the same coin. There is no reason to regard 'conventional' bright-field imaging in a parallel-beam TEM as being more fundamental (although it is certainly a more-established technique) than annular dark-field imaging in a focused-beam STEM. Convergent beam, scanning beam, and selected-area diffraction are likewise integral parts of the whole of TEM diffraction.

However, in the decade and more since the first edition was published, there has been a significant increase in the number of TEM and related techniques, greater sophistication in the microscope's experimental capabilities, astonishing improvements in computer control of the instrument, and new hardware designs and amazing developments in software to model the gigabytes of data generated by these almost-completely digital instruments. Much of this explosion of information has coincided with the worldwide drive to explore the nanoworld, and the still-ongoing effects of Moore's law. It is not possible to include all of this new knowledge in the second edition without transforming the already doorstop sized text into something capable of halting a large projectile in its tracks. It is still essential that this second edition teaches you to understand the essence of the TEM before you attempt to master the latest advances. But we personally cannot hope to understand fully all the new techniques, especially as we both descend into more administrative positions in our professional lives. Therefore, we have prevailed on almost 20 of our close friends and colleagues to put together with us a companion text (TEM; a companion text, Williams and Carter (Eds.) Springer 2010) to which we will refer throughout this second edition. The companion text is just as it says—it's a friend whose advice you should seek when the main text isn't enough. The companion is not necessarily more advanced but is certainly more detailed in dealing with key recent developments as well as some more traditional aspects of TEM that have seen a resurgence of interest. We have taken our colleagues' contributions and rewritten them in a similar conversational vein to this main text and we hope that this approach, combined with the in-depth cross-referencing between the two texts will guide you as you start down the rewarding path to becoming a transmission microscopist.

We each bring more than 35 years of teaching and research in all aspects of TEM. Our research into different materials includes metals, alloys, ceramics, semiconductors, glasses, composites, nano and other particles, atomic-level planar interfaces, and other crystal defects. (The lack of polymeric and biological materials in our own research is evident in their relative absence in this book.) We have contributed to the training of a generation of (we hope) skilled microscopists, several of whom have followed us as professors and researchers in the EM field. These students represent our legacy to our beloved research field and we are overtly proud of their accomplishments. But we also expect some combination of these (still relatively young) men and women to write the third edition. We know that they, like us, will find that writing such a text broadens their knowledge considerably and will also be the source of much joy, frustration, and enduring friendship. We hope you have as much fun reading this book as we had writing it, but we hope also that it takes you much less time. Lastly, we encourage you to send us any comments, both positive and negative. We can both be reached by e-mail: david.williams@uah.edu and cbcarter@engr.uconn.edu.

Foreword to First Edition

Electron microscopy has revolutionized our understanding of materials by completing the *processing-structure-properties* links down to atomistic levels. It is now even possible to tailor the microstructure (and mesostructure) of materials to achieve specific sets of properties; the extraordinary abilities of modern transmission electron microscopy—TEM—instruments to provide almost all the structural, phase, and crystallographic data allow us to accomplish this feat. Therefore, it is obvious that any curriculum in modern materials education must include suitable courses in electron microscopy. It is also essential that suitable texts be available for the preparation of the students and researchers who must carry out electron microscopy properly and quantitatively.

The 40 chapters of this new text by Barry Carter and David Williams (like many of us, well schooled in microscopy at Cambridge and Oxford) do just that. If you want to learn about electron microscopy from specimen preparation (the ultimate limitation); or via the instrument; or how to use the TEM correctly to perform imaging, diffraction, and spectroscopy—it's all there! This, to my knowledge, is the only complete text now available that includes all the remarkable advances made in the field of TEM in the past 30 to 40 years. The timing for this book is just right and, personally, it is exciting to have been part of the development it covers—developments that have impacted so heavily on materials science.

In case there are people out there who still think TEM is just taking pretty pictures to fill up one's bibliography, please stop, pause, take a look at this book, and digest the extraordinary intellectual demands required of the microscopist in order to do the job properly: crystallography, diffraction, image contrast, inelastic scattering events, and spectroscopy. Remember, these used to be fields in themselves. Today, one has to understand the fundamentals of *all* these areas before one can hope to tackle significant problems in materials science. TEM is a technique of characterizing materials down to the atomic limits. It must be used with care and attention, in many cases involving teams of experts from different venues. The fundamentals are, of course, based in physics, so aspiring materials scientists would be well advised to have prior exposure to, for example, solid-state physics, crystallography, and crystal defects, as well as a basic understanding of materials science, for without the latter, how can a person see where TEM can (or may) be put to best use?

So much for the philosophy. This fine new book definitely fills a gap. It provides a sound basis for research workers and graduate students interested in exploring those aspects of structure, especially defects, that control properties. Even undergraduates are now expected (and rightly) to know the basis for electron microscopy, and this book, or appropriate parts of it, can also be utilized for undergraduate curricula in science and engineering.

The authors can be proud of an enormous task, very well done.

G. Thomas
Berkeley, California

Foreword to Second Edition

This book is an exciting entry into the world of atomic structure and characterization in materials science, with very practical instruction on how you can see it and measure it, using an electron microscope. You will learn an immense amount from it, and probably want to keep it for the rest of your life (particularly if the problems cost you some effort!).

Is nanoscience "the next industrial revolution"? Perhaps that will be some combination of energy, environmental and nanoscience. Whatever it is, the new methods which now allow control of materials synthesis at the atomic level will be a large part of it, from the manufacture of jet engine turbine-blades to that of catalysts, polymers, ceramics and semiconductors. As an exercise, work out how much reduction would result in the transatlantic airfare if aircraft turbine blade temperatures could be increased by 200°C. Now calculate the reduction in CO_2 emission, and increased efficiency (reduced coal use for the same amount of electricity) resulting from this temperature increase for a coal-fired electrical generating turbine. Perhaps you will be the person to invent these urgently needed things! The US Department of Energy's Grand Challenge report on the web lists the remarkable advances in exotic nanomaterials useful for energy research, from separation media in fuel cells, to photovoltaics and nano-catalysts which might someday electrolyze water under sunlight alone. Beyond these functional and structural materials, we are now also starting to see for the first time the intentional fabrication of atomic structures in which atoms can be addressed individually, for example, as quantum computers based perhaps on quantum dots. 'Quantum control' has been demonstrated, and we have seen fluorescent nanodots which can be used to label proteins.

Increasingly, in order to find out exactly what new material we have made, and how perfect it is (and so to improve the synthesis), these new synthesis methods must be accompanied by atomic scale compositional and structural analysis. The transmission electron microscope (TEM) has emerged as the perfect tool for this purpose. It can now give us atomic-resolution images of materials and their defects, together with spectroscopic data and diffraction patterns from sub-nanometer regions. The field-emission electron gun it uses is still the brightest particle source in all of physics, so that electron microdiffraction produces the most intense signal from the smallest volume of matter in all of science. For the TEM electron beam probe, we have magnetic lenses (now aberration corrected) which are extremely difficult for our X-ray and neutron competitors to produce (even with much more limited performance) and, perhaps most important of all, our energy-loss spectroscopy provides unrivalled spatial resolution combined with parallel detection (not possible with X-ray absorption spectroscopy, where absorbed X-rays disappear, rather than losing some energy and continuing to the detector).

Much of the advance in synthesis is the legacy of half a century of research in the semiconductor industry, as we attempt to synthesize and fabricate with other materials what is now so easily done with silicon. Exotic oxides, for example, can now be laid down layer by layer to form artificial crystal structures with new, useful properties. But it is also a result of the spectacular advances in materials characterization, and our ability to see structures at the atomic level. Perhaps the best example of this is the discovery of the carbon nanotube, which was first identified by using an electron microscope. Any curious and observant electron microscopist can now discover new nanostructures just because they look interesting at the atomic scale. The important point is that if this is done in an environmental microscope, he or she will know how to make them, since the thermodynamic conditions will be recorded when using such a 'lab in a microscope'. There are efforts at materials discovery by just such combinatorial trial-and-error methods, which could perhaps be incorporated into our electron microscopes. This is needed because there are often just 'too many possibilities' in nature to explore in the computer — the number of possible structures rises very rapidly with the number of distinct types of atoms.

It was Richard Feynman who said that, "if, in some catastrophe, all scientific knowledge was lost, and only one sentence could be preserved, then the statement to be passed on, which contained the most information in the fewest words, would be that matter consists of atoms." But confidence that matter consists of atoms developed surprisingly recently and as late as 1900 many (including Kelvin) were unconvinced, despite Avagadro's work and Faraday's on electrodeposition. Einstein's Brownian motion paper of 1905 finally persuaded most, as did Rutherford's experiments. Muller was first to see atoms (in his field-ion microscope in the early 1950s), and Albert Crewe two decades later in Chicago, with his invention of the field-emission gun for his scanning transmission electron microscope (STEM). The Greek Atomists first suggested that a stone, cut repeatedly, would eventually lead to an indivisible smallest fragment, and indeed Democritus believed that "nothing exists except vacuum and atoms. All else is opinion." Marco Polo remarks on the use of spectacles by the Chinese, but it was van Leeuwenhoek (1632-1723) whose series of papers in *Phil. Trans.* brought the microworld to the general scientific community for the first time using his much improved optical microscope. Robert Hooke's 1665 Micrographica sketches what he saw through his new compound microscope, including fascinating images of facetted crystallites, whose facet angles he explained with drawings of piles of cannon balls. Perhaps this was the first resurrection of the atomistic theory of matter since the Greeks. Zernike's phase-plate in the 1930s brought phase contrast to previously invisible ultra-thin biological 'phase objects', and so is the forerunner for the corresponding theory in high-resolution electron microscopy.

The past fifty years has been a wonderfully exciting time for electron microscopists in materials science, with continuous rapid advances in all of its many modes and detectors. From the development of the theory of Bragg diffraction contrast and the column approximation, which enables us to understand TEM images of crystals and their defects, to the theory of high-resolution microscopy useful for atomic-scale imaging, and on into the theory of all the powerful analytic modes and associated detectors, such as X-rays, cathodoluminescence and energy-loss spectroscopy, we have seen steady advances. And we have always known that defect structure in most cases controls properties — the most common (first-order) phase transitions are initiated at special sites, and in the electronic oxides a whole zoo of charge-density excitations and defects waits to be fully understood by electron microscopy. The theory of phase-transformation toughening of ceramics, for example, is a wonderful story which combines TEM observations with theory, as does that of precipitate hardening in alloys, or the early stages of semiconductor-crystal growth. The study of diffuse scattering from defects as a function of temperature at phase transitions is in its infancy, yet we have a far stronger signal there than in competing X-ray methods. The mapping of strain-fields at the nanoscale in devices, by quantitative convergent-beam electron diffraction, was developed just in time to solve a problem listed on the Semiconductor Roadmap (the speed of your laptop depends on strain-induced mobility enhancement). In biology, where the quantification of TEM data is taken more seriously, we have seen three-dimensional image reconstructions of many large proteins, including the ribosome (the factory which makes proteins according to DNA instructions). Their work should be a model to the materials science community in the constant effort toward better quantification of data.

Like all the best textbooks, this one was distilled from lecture notes, debugged over many years and generations of students. The authors have extracted the heart from many difficult theory papers and a huge literature, to explain to you in the simplest, clearest manner (with many examples) the most important concepts and practices of modern transmission electron microscopy. This is a great service to the field and to its teaching worldwide. Your love affair with atoms begins!

J.C.H. Spence
Regent's Professor of Physics
Arizona State University and Lawrence
Berkeley National Laboratory

Acknowledgments

We have spent over 20 years conceiving and writing this text and the preceding first edition and such an endeavor can't be accomplished in isolation. Our first acknowledgment must be to our respective wives and children: Margie, Matthew, Bryn, and Stephen and Bryony, Ben, Adam, and Emily. Our families have borne the brunt of our absences from home (and occasionally the brunt of our presence). Neither edition would have been possible without the encouragement, advice, and persistence of (and the fine wines served by) Amelia McNamara, our first editor at Plenum Press, then Kluwer, and Springer.

We have both been fortunate to work in our respective universities with many more talented colleagues, post-doctoral associates, and graduate students, all of whom have taught us much and contributed significantly to the examples in both editions. We would like to thank a few of these colleagues directly: Dave Ackland, Faisal Alamgir, Arzu Altay, Ian Anderson, Ilke Arslan, Joysurya Basu, Steve Baumann, Charlie Betz, John Bruley, Derrick Carpenter, Helen Chan, Steve Claves, Dov Cohen, Ray Coles, Vinayak Dravid, Alwyn Eades, Shelley Gillis, Jeff Farrer, Joe Goldstein, Pradyumna Gupta, Brian Hebert, Jason Hefflefinger, John Hunt, Yasuo Ito, Matt Johnson, Vicki Keast, Chris Kiely, Paul Kotula, Chunfei Li, Ron Liu, Charlie Lyman, Mike Mallamaci, Stuart McKernan, Joe Michael, Julia Nowak, Grant Norton, Adam Papworth, Chris Perrey, Sundar Ramamurthy, René Rasmussen, Ravi Ravishankar, Kathy Repa, Kathy Reuter, Al Romig, Jag Sankar, David A. Smith, Kamal Soni, Changmo Sung, Caroline Swanson, Ken Vecchio, Masashi Watanabe, Jonathan Winterstein, Janet Wood, and Mike Zemyan.

In addition, many other colleagues and friends in the field of microscopy and analysis have helped with the book (even if they weren't aware of it). These include Ron Anderson, Raghavan Ayer, Jim Bentley, Gracie Burke, Jeff Campbell, Graham Cliff, David Cockayne, Peter Doig, the late Chuck Fiori, Peter Goodhew, Brendan Griffin, Ron Gronsky, Peter Hawkes, Tom Huber, Gilles Hug, David Joy, Mike Kersker, Roar Kilaas, Sasha Krajnikov, the late Riccardo Levi-Setti, Gordon Lorimer, Harald Müllejans, Dale Newbury, Mike O'Keefe, Peter Rez, Manfred Rühle, John-Henry Scott, John Steeds, Peter Swann, Gareth Thomas, Patrick Veyssière, Peter Williams, Nestor Zaluzec, and Elmar Zeitler. Many of these (and other) colleagues provided the figures that we acknowledge individually at the end of the book.

We have received financial support for our microscopy studies through several different federal agencies; without this support none of the research that underpins the contents of this book would have been accomplished. In particular, DBW wishes to thank the National Science Foundation, Division of Materials Research for over 30 years of continuous funding, NASA, Division of Planetary Science (with Joe Goldstein) and The Department of Energy, Basic Energy Sciences (with Mike Notis and Himanshu Jain), Bettis Laboratories, Pittsburgh, and Sandia National Laboratories, Albuquerque. While this edition was finalized at the University of Alabama in Huntsville, both editions were written while DBW was in the Center for Advanced Materials and Nanotechnology at Lehigh University, which supports that outstanding electron microscopy laboratory. Portions of both editions were written while DBW was on sabbatical or during extended visits to various microscopy labs: Chalmers University, Göteborg, with Gordon Dunlop and Hans Nordén; The Max Planck

Institut für Metallforschung, Stuttgart, with Manfred Rühle; Los Alamos National Laboratory with Terry Mitchell; Dartmouth College, Thayer School of Engineering, with Erland Schulson; and the Electron Microscope Unit at Sydney University with Simon Ringer. CBC wishes to acknowledge the Department of Energy, Basic Energy Sciences, the National Science Foundation, Division of Materials Research, the Center for Interfacial Engineering at the University of Minnesota, The Materials Science Center at Cornell University, and the SHaRE program at Oak Ridge National Laboratories. The first edition was started while CBC was with the Department of Materials Science and Engineering at Cornell University. This edition was started at the Department of Chemical Engineering and Materials Science at the University of Minnesota where the first edition was finished and was finalized while CBC was at the University of Connecticut. The second edition was partly written while CBC was on Sabbatical Leave at Chalmers University with Eva Olssen (thanks also to Anders Tholen at Chalmers), at NIMS in Tsukuba with Yoshio Bando (thanks also to Dmitri Golberg and Kazuo Furuya at NIMS at Yuichi Ikuhara at the University of Tokyo) and at Cambridge University with Paul Midgley. CBC also thanks the Master and Fellows of Peterhouse for their hospitality during the latter period.

CBC would also like to thank the team at the Ernst Ruska Center for their repeated generous hospitality (special thanks to Knut Urban, Markus Lenzen, Andreas Thust, Martina Luysberg, Karsten Tillmann, Chunlin Jia and Lothar Houben)

Despite our common scientific beginnings as undergraduates in Christ's College Cambridge, we learned our trade under different microscopists: DBW with Jeff Edington in Cambridge and CBC with Sir Peter Hirsch and Mike Whelan in Oxford. Not surprisingly, the classic texts by these renowned microscopists are referred to throughout this book. They influenced our own views of TEM tremendously, contributing to the undoubted bias in our opinions, notation, and approach to the whole subject.

List of Initials and Acronyms

The field of TEM is a rich source of initials and acronyms (these are words formed by the initials), behind which we hide both simple and esoteric concepts. While the generation of new initials and acronyms can be a source of original thinking (e.g., see ALCHEMI), it undoubtedly makes for easier communication in many cases and certainly reduces the length of voluminous textbooks. You have to master this strange language before being accepted into the community of microscopists, so we present a comprehensive listing that you should memorize.

ACF absorption-correction factor
ACT automated crystallography for TEM
A/D analog to digital (converter)
ADF annular dark field
AEM analytical electron microscope/microscopy
AES Auger electron spectrometer/spectroscopy
AFF aberration-free focus
AFM atomic force microscope/microscopy
ALCHEMI atom location by channeling-enhanced micro-analysis
ANL Argonne National Laboratory
APB anti-phase domain boundary
APFIM atom-probe field ion microscope/microscopy
APW augmented plane wave
ASW augmented spherical wave
ATW atmospheric thin window

BF bright field
BFP back-focal plane
BSE backscattered electron
BZB Brillouin-zone boundary

C1,2 condenser 1, 2, etc. lens
CASTEP electronic-potential calculation software
CAT computerized axial tomography

CB coherent bremsstrahlung
CBED convergent-beam electron diffraction
CBIM convergent beam imaging
CCD charge-coupled device
CCF cross-correlation function
CCM charge-collection microscopy
CDF centered dark field
CF coherent Fresnel/Foucault
CFE cold field emission
CL cathodoluminescence
cps counts per second
CRT cathode-ray tube
CS crystallographic shear
CSL coincident-site lattice
CVD chemical vapor deposition

DADF displaced-aperture dark field
DDF diffuse dark field
DF dark field
DFT density-functional theory
DOS density of states
DP diffraction pattern
DQE detection quantum efficiency
DSTEM dedicated scanning transmission electron microscope/microscopy
DTSA desktop spectrum analyzer

EBIC   electron beam-induced current/conductivity
EBSD   electron-backscatter diffraction
EELS   electron energy-loss spectrometer/spectrometry
EFI   energy-filtered imaging
EFTEM   energy-filtered transmission electron microscope
ELNES   energy-loss near-edge structure
ELP^{TM} energy-loss program (Gatan)
EMMA   electron microscope microanalyzer
EMS   electron microscopy image simulation
(E)MSA   (Electron) Microscopy Society of America
EPMA   electron-probe microanalyzer
ESCA   electron spectroscopy for chemical analysis
ESI   electron-spectroscopic imaging
EXAFS   extended X-ray-absorption fine structure
EXELFS   extended energy-loss fine structure

FEFF   ab-initio multiple-scattering software
FEG   field-emission gun
FET   field-effect transistor
FFP   front-focal plane
FFT   fast Fourier transform
FIB   focused ion beam
FLAPW   full-potential linearized augmented plane wave
FOLZ   first-order Laue zone
FTP   file-transfer protocol
FWHM   full width at half maximum
FWTM   full width at tenth maximum

GB   grain boundary
GIF   Gatan image filter^{TM}
GIGO   garbage in garbage out
GOS   generalized oscillator strength

HAADF   high-angle annular dark field
HOLZ   higher-order Laue zone
HPGe   high-purity germanium
HREELS   high-resolution electron energy-loss spectrometer/spectrometry
HRTEM   high-resolution transmission electron microscope/microscopy
HV   high vacuum
HVEM   high-voltage electron microscope/microscopy

ICC   incomplete charge collection
ICDD   International Center for Diffraction Data
ID   identification (of peaks in spectrum)
IDB   inversion domain boundary
IEEE   International Electronics and Electrical Engineering
IG   intrinsic Ge
IVEM   intermediate-voltage electron microscope/microscopy

K-M   Kossel-Möllenstedt

LACBED   large-angle convergent-beam electron diffraction
LCAO   linear combination of atomic orbitals
LCD   liquid-crystal display
LDA   local-density approximation
LEED   low-energy electron diffraction
LKKR   layered Korringa-Kohn-Rostoker

MAS   Microbeam Analysis Society
MBE   molecular-beam epitaxy
MC   minimum contrast
MCA   multichannel analyzer
MDM   minimum detectable mass
MLS   multiple least-squares
MMF   minimum mass fraction
MO   molecular orbital
MRS   Materials Research Society
MS   multiple scattering
MSA   multivariate statistical analysis
MSDS   material safety data sheets
MT   muffin tin
MV   megavolt

NCEMSS   National Center for Electron Microscopy simulation system
NIH   National Institutes of Health
NIST   National Institute of Standards and Technology
NPL   National Physical Laboratory

OIM   orientation-imaging microscopy
OR   orientation relationship

PARODI   parallel recording of dark-field images
PB   phase boundary
P/B   peak-to-background ratio
PEELS   parallel electron energy-loss spectrometer/spectrometry
PIPS   Precision Ion-Polishing System^{TM}
PIXE   proton-induced X-ray emission
PM   photomultiplier
POA   phase-object approximation
ppb/m   parts per billion/million
PDA   photo-diode array
PSF   point-spread function
PTS   position-tagged spectrometry

QHRTEM   quantitative high-resolution transmission electron microscopy

RB   translation boundary (yes, it does!)
RDF   radial distribution function
REM   reflection electron microscope/microscopy
RHEED   reflection high-energy electron diffraction

SACT   small-angle cleaving technique
SAD(P) selected-area diffraction (pattern)
SCF   self-consistent field
SDD   silicon-drift detector
SE   secondary electron
SEELS   serial electron energy-loss spectrometer/spectrometry
SEM   scanning electron microscope/microscopy
SESAMe sub-eV sub-Å microscope
SF   stacking fault
SHRLI   simulated high-resolution lattice images
SI   spectrum imaging
SI   Système Internationale
SIGMAK K-edge quantification software
SIGMAL L-edge quantification software
SIMS   secondary-ion mass spectrometry
S/N   signal-to-noise ratio
SOLZ   second-order Laue zone
SRM   standard reference material
STEM   scanning transmission electron microscope/microscopy
STM   scanning tunneling microscope/microscopy

TB   twin boundary
TEM   transmission electron microscope/microscopy
TFE   thermal field emission
TMBA   too many bloody acronyms

UHV   ultrahigh vacuum
URL   uniform resource locator
UTW   ultra-thin window

V/F   voltage to frequency (converter)
VLM   visible-light microscope/microscopy
VUV   vacuum ultra violet

WB   weak beam
WBDF   weak-beam dark field
WDS   wavelength-dispersive spectrometer/spectrometry
WP   whole pattern
WPOA   weak-phase object approximation
WWW   World Wide Web

XANES   X-ray absorption near-edge structure
XEDS   X-ray energy-dispersive spectrometer/spectrometry
XPS   X-ray photoelectron spectrometer/spectrometry
XRD/F   X-ray diffraction/fluorescence

YAG   yttrium-aluminum garnet
YBCO   yttrium-barium-copper oxide
YSZ   yttria-stabilized zirconia

ZAF   atomic number/absorption/fluorescence correction
ZAP   zone-axis pattern
ZLP   zero-loss peak
ZOLZ   zero-order Laue zone

List of Symbols

We use a large number of symbols. Because we are constrained by the limits of our own and the Greek alphabets, we often use the same symbol for different terms, which can confuse the unwary. We have tried to be consistent where possible but undoubtedly we have not always succeeded. The following (not totally inclusive) list may help if you remain confused after reading the text.

a	interatomic spacing	**B**	beam direction
a	relative transition probability	**B**	magnetic field strength
a	width of diffraction disk	B	background intensity
a_0	Bohr radius	$B(\mathbf{u})$	aberration function
a_0	lattice parameter		
a, b, c	lattice vectors	c	centi
a*, b*, c*	reciprocal-lattice vectors	c	velocity of light
A	absorption-correction factor	C	composition
A	active area of X-ray detector	C	contrast
A_0	amplitude	C	coulomb
A	amplitude of scattered beam	C_a	astigmatism-aberration coefficient
A	amperes	C_c	chromatic-aberration coefficient
A	atomic weight	$C_\mathbf{g}$	**g** component of Bloch wave
A	Richardson's constant	C_s	spherical-aberration coefficient
Å	Ångstrom	C_X	fraction of X atoms on specific sites
$\mathcal{A}$	Bloch wave amplitude	C_0	amplitude of direct beam
$A(\mathbf{u})$	aperture function	C_ε	combination of the elastic constants
A, B	fitting parameters for energy-loss background subtraction	$(C_s\lambda)^{1/2}$	scherzer
		$(C_s\lambda^3)^{1/4}$	glaser
b	beam-broadening parameter	c/o	condenser/objective
b	separation of diffraction disks	d	beam (probe) diameter
$\mathbf{b}_e$	edge component of the Burgers vector	d	diameter of spectrometer entrance aperture
$\mathbf{b}_p$	Burgers vector of partial dislocation		
$\mathbf{b}_T$	Burgers vector of total dislocation	d	interplanar spacing

d	spacing of moire fringes	$E_D(\mathbf{u})$	envelope function for the detector
d_c	effective source size	$E_s(\mathbf{u})$	envelope function for the source
d_d	diffraction-limited beam diameter	$E_v(\mathbf{u})$	envelope function for specimen vibration
d_{eff}	effective entrance-aperture diameter at recording plane	f	focal length
d_g	Gaussian beam diameter	$f(\mathbf{r})$	strength of object at point (x,y)
d_{hkl}	hkl interplanar spacing	$f(\theta)$	atomic-scattering factor
d_i	image distance	$f(\mathbf{k})$	atomic-scattering amplitude
d_{im}	smallest resolvable image distance	f_x	scattering factor for X-rays
d_o	object distance	$f_i(x)$	residual of least-squares fit
d_{ob}	smallest resolvable object distance	F	Fano factor
d_s	spherical-aberration limited beam diameter	F	fluorescence-correction factor
d_t	total beam diameter	$\mathbf{F}$	Lorentz force
dz	thickness of a diffracting slice	F	relativistic-correction factor
$d\sigma/d\Omega$	differential cross section of one atom	F	Fourier transform
D	aperture diameter	F'	Fourier transform of edge intensity
D	change in focus	F_B	fraction of alloying element B
D	dimension (as in 1D, 2D...)	F_g	special value of $F(\theta)$ when θ is the Bragg angle
D	distance from projector crossover to recording plane	$F(P)$	Fourier transform of plasmon intensity
D	electron dose	$F(\mathbf{u})$	Fourier transform of $f(\mathbf{r})$
D_A	distance from beam crossover to spectrometer entrance aperture	$F(0)$	Fourier transform of elastic intensity
D_{im}	depth of focus	$F(1)$	Fourier transform of single-scattering intensity
D_{ob}	depth of field	$F(\theta)$	structure factor
D_1, D_2	tie-line points on dispersion surfaces in presence of defect		
		$\mathbf{g}/\bar{\mathbf{g}}$	diffraction vector (magnitude of +/- $\mathbf{K}$ at the Bragg angle)
e	charge on the electron	$\mathbf{g}_{hkl}$	diffraction vector for hkl plane
E	energy	g	gram
$\mathbf{E}$	electric-field strength	$g(\mathbf{r})$	intensity of image at point (x,y)
E	Young's modulus	G	Bragg reflection
E	total energy	G	radius of a HOLZ ring
$\mathcal{E}$	energy loss	G	giga
E_a	spatial-coherence envelope	$G(\mathbf{u})$	Fourier transform of $g(\mathbf{r})$
E_c	chromatic-coherence envelope	Gy	gray (radiation unit)
E_c	critical ionization energy		
E_d	displacement energy	h	Planck's constant
E_F	Fermi energy/level	h	distance from specimen to the aperture
$E_{h/l}$	high/low energy for background-subtraction window	$h(\mathbf{r})$	contrast-transfer function
$E_{K/L/M}$	ionization energy for K/L/M-shell electron	(hkl)	Miller indices of a crystal plane
$E_{K/L/M}$	energy of K/L/M X-ray	hkl	indices of diffraction spots from hkl plane
$\mathcal{E}_m$	average energy loss	H	spacing of the reciprocal-lattice planes parallel to beam
E_P	plasmon energy	$H(\mathbf{u})$	Fourier transform of $h(\mathbf{r})$
$\mathcal{E}_P$	plasmon energy loss		
E_s	sputtering-threshold energy	i	beam current
E_t	threshold energy	i	imaginary number
E_0	beam energy	i	number of atoms in unit cell
$E(\mathbf{u})$	envelope function	I	intensity
$E_c(\mathbf{u})$	envelope function for chromatic aberration	I	intrinsic line width of the XEDS detector
$E_d(\mathbf{u})$	envelope function for specimen drift	i_e	emission current

i_f	filament-heating current	n	number of scattered electrons
I_g	intensity in the diffracted beam	n_0	number of incident electrons
$I_{K/L/M}$	K/L/M-shell intensity above background	n	nano
$I(\mathbf{k})$	kinematical intensity	n	principal quantum number
$I(1)$	single-scattering intensity	$\mathbf{n}$	vector normal to the surface
I_P	intensity in the first plasmon peak	n_s	number of electrons in the ionized sub-shell
I_T	total transmitted intensity		
I_0	intensity in the zero-loss peak	N	$h + k + l$
I_0	intensity in the direct beam	N	newton
$I(t)$	low-loss spectrum intensity	N	noise
		N	number of counts in ionization edge
J	current density	N	number of atoms/unit area
J	joule	N_V	number of atoms/unit volume
J	sum of spin and angular quantum numbers	$N(E)$	number of bremsstrahlung photons of energy E
k	magnitude of the wave vector	N_0	Avogadro's number
k	Boltzmann's constant		
k	kilo	O	direct beam
$\mathbf{k}_I$	**k**-vector of the incident wave		
$\mathbf{k}_D$	**k**-vector of the diffracted wave	p	integer
k_{AB}	Cliff-Lorimer factor/sensitivity factor	$\mathbf{p}$	momentum
K	bulk modulus	p	pico
K	Kelvin	P	probability of scattering
K	Kramers' constant	P	peak intensity
K	sensitivity factor	P	FWHM of a randomized electronic-pulse generator
$K/L/M$	inner-shell/characteristic X-ray/ ionization edge		
		Pa	pascal
$\mathbf{K}$	change in **k** due to diffraction	$P_{K/L/M}$	probability of K/L/M-shell ionization
$\mathbf{K}_B$	magnitude of **K** at the Bragg angle	$P(z)$	scattering matrix for a slice of thickness z
K_o	kernel		
		q	charge
l	angular quantum number	Q	cross section
L	camera length		
L	lattice spacing in beam direction	r	radius
L	length of magnetic field	r	distance a wave propagates
L_0	length of magnetic field along optic axis	r	distance between contamination spots
L	path difference	r	minimum resolvable distance/resolution
L	width of composition line-profile	r	power term to fit background in EEL spectrum
m	meters	r_M	image-translation distance
m	milli	$\mathbf{r}_n$	lattice vector
m	mirror plane	$\mathbf{r}^*$	reciprocal-lattice vector
m	number of focal increments	r_{ast}	radius of astigmatism disk
m_0	rest mass of the electron	r_{chr}	radius of chromatic-aberration disk
M	magnification	r_{sph}	radius of spherical-aberration disk
M	mega	r_{min}	minimum disk radius
M_A	angular magnification	r_{th}	theoretical disk radius
M_T	transverse magnification	$\mathbf{r}'_n$	lattice vector in strained crystal
M_1, M_2	tie-line points on dispersion surfaces	r_0	maximum radius of DP in focal plane of spectrometer
n	integer	R	ALCHEMI intensity ratio
n	free-electron density	R	count rate
n	number of counts	$\mathbf{R}$	crystal-lattice vector

R	distance on screen between diffraction spots	w	$s\xi_g$ (excitation error multiplied by extinction distance)
R	radius of curvature of EEL spectrometer	w	projected width of planar defect
R	resolution of XEDS detector	w	width
R	spatial resolution	x	distance
R	reduction in partial cross section with increasing α	$\times$	times (magnification)
R_{MAX}	diameter of beam emerging from specimen	x, y, z	atom coordinates
		X	FWHM due to XEDS detector
$\mathbf{R}_n$	lattice-displacement vector	X	rotation axis
$\mathbf{R(r)}$	displacement	y	displacement at the specimen
$\mathbf{s}$	excitation error/deviation parameter	y	number of counts in channel
s	second	y	parallax shift in the image
s	spin quantum number	z	distance within a specimen
$\mathbf{s_R}$	excitation error due to defect	z	distance along optic axis
$\mathbf{s_z(s_g)}$	excitation error	z	specimen height
s_{eff}	effective excitation error	Z	atomic number/atomic-number correction factor
S	distance from specimen to detector		
S	signal	**Greek symbols**	
S	standard deviation for n measurements	α	phase shift due to defect
sr	steradians	α	semi-angle of incidence/convergence
$\mathbf{t}$	shift vector between the ZOLZ and HOLZ	α	X-ray take-off angle
t	student (t) distribution	α_0	beam divergence semi-angle at gun crossover
t	thickness	α_{opt}	optimum convergence semi-angle
t'	absorption path length	β	brightness
t_0	thickness at zero tilt	β	ratio of electron velocity to light velocity
T	absolute temperature	β	semi-angle of collection
T	tesla	β_{opt}	optimum collection semi-angle
T_c	period of rotation	γ	degree of spatial coherence
$T(\mathbf{u})$	objective-lens transfer function	γ	phase of direct beam
$T_{\text{eff}}(\mathbf{u})$	effective transfer function	γ	relativistic-correction factor
$\mathbf{u}$	reciprocal lattice vector	γ	specimen tilt angle
$\mathbf{u}$	unit vector along the dislocation line	Δ	change/difference
$\mathbf{u}^*$	vector normal to the ZOLZ	Δ	width of energy window
U	overvoltage	Δd	change in lattice parameter
U_g	Fourier component of the perfect-crystal potential	$\Delta\phi$	phase difference
$[UVW]$	indices of a crystal direction	$\Delta\theta_i$	angle between Kossel–Möllenstedt fringes
UVW	indices of beam direction	Δ_{AB}	difference in mass-absorption coefficients
v	velocity	ΔE	energy width /spread
V	accelerating voltage	ΔE_P	plasmon-line width/change in plasmon energy
$\mathscr{V}$	potential energy	Δf	maximum difference in focus
V_c/V	volume of the unit cell	Δf	defocus error due to chromatic aberration
V_c	inner potential of cavity	Δf_{AFF}	aberration-free (de)focus
V_t	projected potential through specimen thickness	Δf_{MC}	minimum contrast defocus
		Δf_{opt}	optimum defocus
$V(\mathbf{r})$	crystal inner potential	Δ	change (in height)

Δh	relative depth in specimen	χ^2	goodness of fit (between standard and experimental spectra)
ΔI	change in intensity		
Δp	parallax shift	$\chi(\mathbf{u})$	phase-distortion function
ΔV	change in the inner potential	$\chi(\mathbf{k})$	momentum transfer
Δx	path difference/image shift		
Δx	half-width of image of undissociated screw dislocation	κ	thermal conductivity
Δx_{res}	resolution at Scherzer defocus	ξ_g	extinction distance for the diffracted beam
Δf_{sch}	Scherzer defocus	$\xi_{g'}$	absorption parameter
		ξ_0	extinction distance for the direct beam
δ	angle between XEDS detector normal and line from detector to specimen	ξ_{eff}	effective extinction distance ($s \neq 0$)
		ξ_g^{abs}	absorption-modified ξ_g
δ	angle between beam and plane of defect		
δ	diameter of disk image	λ_c	coherence length
δ	diffuseness of interface	λ	mean-free path
δ	precipitate/matrix misfit	λ	wavelength
δ	small increment	$\lambda_{K/L/M}$	mean-free path for K/L/M-shell ionization
δ	smallest resolvable distance (resolution)		
		λ_P	plasmon mean-free path
ε	deflection angle	λ_R	relativistic wavelength
ε	detector efficiency	λ^{-1}	radius of Ewald sphere
ε	energy to create an electron-hole pair		
ε	specimen-tilt angle	μ	micro
ε	strain	μ	refractive index
ε_0	permittivity of free space (dielectric constant)	μ/ρ	mass-absorption coefficient
		$\mu^{(j)}(\mathbf{r})$	Bloch function
η	phase change		
η	angle between excess Kikuchi lines at $\mathbf{s}=0$ and $\mathbf{s}>0$	ν	frequency
		ν	Poisson's ratio
$\eta(\theta)$	phase of the atomic-scattering factor		
		ψ	amplitude of a wave
Φ	phase shift accompanying scattering	ψ	the wave function
Φ	work function	ψ_{sph}	amplitude of spherical wave
ϕ	rotation angle between image and diffraction pattern	ψ_{tot}	total wave function
		ψ_0	amplitude
ϕ	angle between Kikuchi line and diffraction spot		
		ρ	angle between two directions
ϕ	angle between two Kikuchi-line pairs	ρ	density
ϕ	angle between two planes	$\rho_{c/s}$	information limit due to chromatic/spherical aberration
ϕ	angle between two plane normals		
ϕ	angle of tilt between stereo images	$\rho(\mathbf{r})$	radial distribution function
ϕ	phase of a wave	ρt	mass thickness
ϕ^*	complex conjugate of ϕ	ρ_i^2	area of a pixel
ϕ_g	amplitude of the diffracted beam		
ϕ_0	amplitude of the direct beam	σ	scattering cross section of one atom
ϕ_x	angle of deflection of the beam	σ	standard deviation
$\phi(\rho t)$	depth distribution of X-ray production	σ	stress
		$\sigma_{K/L/M}$	ionization cross section for K/L/M-shell electron
χ	wave vector outside the specimen		
χ_G	wave vector terminating on the point G in reciprocal space	σ_T	total ionization cross section
		$\sigma_{K/L/M}(\beta\Delta)$	partial ionization cross section for K/L/M-shell electron
χ_O	wave vector terminating on the point O in reciprocal space		
		θ	scattering semi-angle

LIST OF SYMBOLS

θ_B	Bragg angle	ω_p	plasmon frequency
θ_C	cut-off semi-angle		
θ_E	characteristic scattering semi-angle	Ω	filter for energy loss
θ_0	screening parameter	Ω	solid angle of collection of XEDS
τ	XEDS detector time constant	Ω	volume of unit cell
τ	dwell time		
τ	analysis time	ζ	zeta factor
ω	fluorescence yield		
ω_c	cyclotron frequency	$\otimes$	convolution (multiply and integrate)

About the Companion Volume

As described in our Preface, the many years since the publication of the first edition have seen a significant increase in the number of TEM (and related) techniques and the sophistication of the microscope's experimental capabilities, as well as new hardware designs, astonishing improvements in computer control of the instrument and amazing developments in software to handle and model the gigabytes of data generated by these (now almost completely digital) instruments. Much of this explosion of information has coincided with the world-wide drive to explore the nanoworld, and the still-ongoing effects of Moore's law. It is not possible to include all of this new knowledge in a textbook, and the primary objective of the second edition is still to teach you to understand the essence of the TEM before you attempt to master the latest advances. We also personally cannot hope to comprehend fully all of the new techniques, especially as we both descend into more administrative positions in our professional lives.

Therefore, we have prevailed on almost 20 of our close friends and colleagues to put together with us a companion applications text (Carter and Williams, eds., Springer, 2010) to which we will refer throughout this second edition. The companion text is just as it says; it's a friend whose advice you should seek when the main text is not enough. The companion is not intended to be more advanced but it certainly provides much more detail on key recent developments and some more traditional aspects of TEM that have seen a resurgence of interest. We have taken our colleagues' contributions and worked with them to produce chapters that are in a similar conversational vein to this main text. While *Transmission Electron Microscopy, Second Edition*, is a completely stand-alone textbook, we think that you will find the cross-referencing between the two texts to be of great value as you continue along the rewarding path of becoming a transmission microscopist.

Contents

About the Authors vii

Preface xi

Foreword to First Edition xiii

Foreword to Second Edition xv

Acknowledgments xix

List of Initials and Acronyms xxi

List of Symbols xxv

About the Companion Volume xxxi

Figure Credits xlix

PART 1 BASICS 1

1 The Transmission Electron Microscope 3

    Chapter Preview 3
    1.1 What Materials Should We Study in the TEM? 3
    1.2 Why Use Electrons? 4
        1.2.A An Extremely Brief History 4
        1.2.B Microscopy and the Concept of Resolution 5
        1.2.C Interaction of Electrons with Matter 7
        1.2.D Depth of Field and Depth of focus 8
        1.2.E Diffraction 8
    1.3 Limitations of the TEM 9
        1.3.A Sampling 9
        1.3.B Interpreting Transmission Images 9
        1.3.C Electron Beam Damage and Safety 10
        1.3.D Specimen Preparation 11
    1.4 Different Kinds of TEMs 11
    1.5 Some Fundamental Properties of Electrons 11
    1.6 Microscopy on the Internet/World Wide Web 15
        1.6.A Microscopy and Analysis-Related Web Sites 15
        1.6.B Microscopy and Analysis Software 15
    Chapter Summary 17

2 Scattering and Diffraction ... 23

Chapter Preview ... 23
2.1 Why Are We Interested in Electron Scattering? ... 23
2.2 Terminology of Scattering and Diffraction ... 25
2.3 The Angle of Scattering ... 26
2.4 The Interaction Cross Section and Its Differential ... 27
    2.4.A Scattering from an Isolated Atom ... 27
    2.4.B Scattering from the Specimen ... 28
    2.4.C Some Numbers ... 28
2.5 The Mean Free Path ... 28
2.6 How We Use Scattering in the TEM ... 29
2.7 Comparison to X-ray Diffraction ... 30
2.8 Fraunhofer and Fresnel Diffraction ... 30
2.9 Diffraction of Light from Slits and Holes ... 31
2.10 Constructive Interference ... 33
2.11 A Word About Angles ... 34
2.12 Electron-Diffraction Patterns ... 34
Chapter Summary ... 36

3 Elastic Scattering ... 39

Chapter Preview ... 39
3.1 Particles and Waves ... 39
3.2 Mechanisms of Elastic Scattering ... 40
3.3 Elastic Scattering from Isolated Atoms ... 41
3.4 The Rutherford Cross Section ... 41
3.5 Modifications to the Rutherford Cross Section ... 42
3.6 Coherency of the Rutherford-Scattered Electrons ... 43
3.7 The Atomic-Scattering Factor ... 44
3.8 The Origin of $f(\theta)$ ... 45
3.9 The Structure Factor $F(\theta)$ ... 46
3.10 Simple Diffraction Concepts ... 47
    3.10.A Interference of Electron Waves; Creation of the Direct and Diffracted Beams ... 47
    3.10.B Diffraction Equations ... 48
Chapter Summary ... 49

4 Inelastic Scattering and Beam Damage ... 53

Chapter Preview ... 53
4.1 Which Inelastic Processes Occur in the TEM? ... 53
4.2 X-ray Emission ... 55
    4.2.A Characteristic X-rays ... 55
    4.2.B Bremsstrahlung X-rays ... 60
4.3 Secondary-Electron Emission ... 60
    4.3.A Secondary Electrons ... 60
    4.3.B Auger Electrons ... 61
4.4 Electron-Hole Pairs and Cathodoluminescence (CL) ... 62
4.5 Plasmons and Phonons ... 63
4.6 Beam Damage ... 64
    4.6.A Electron Dose ... 65
    4.6.B Specimen Heating ... 65
    4.6.C Beam Damage in Polymers ... 66
    4.6.D Beam Damage in Covalent and Ionic Crystals ... 66
    4.6.E Beam Damage in Metals ... 66
    4.6.F Sputtering ... 68
Chapter Summary ... 68

5 Electron Sources 73

Chapter Preview 73
5.1 The Physics of Different Electron Sources 73
    5.1.A Thermionic Emission 74
    5.1.B Field Emission 74
5.2 The Characteristics of the Electron Beam 75
    5.2.A Brightness 75
    5.2.B Temporal Coherency and Energy Spread 76
    5.2.C Spatial Coherency and Source Size 77
    5.2.D Stability 77
5.3 Electron Guns 77
    5.3.A Thermionic Guns 77
    5.3.B Field-Emission Guns (FEGs) 80
5.4 Comparison of Guns 81
5.5 Measuring Your Gun Characteristics 82
    5.5.A Beam Current 82
    5.5.B Convergence Angle 83
    5.5.C Calculating the Beam Diameter 83
    5.5.D Measuring the Beam Diameter 85
    5.5.E Energy Spread 85
    5.5.F Spatial Coherency 86
5.6 What kV should You Use? 86
Chapter Summary 87

6 Lenses, Apertures, and Resolution 91

Chapter Preview 91
6.1 Why Learn About Lenses? 91
6.2 Light Optics and Electron Optics 92
    6.2.A How to Draw a Ray Diagram 92
    6.2.B The Principal Optical Elements 94
    6.2.C The Lens Equation 94
    6.2.D Magnification, Demagnification, and Focus 95
6.3 Electron Lenses 96
    6.3.A Polepieces and Coils 96
    6.3.B Different Kinds of Lenses 97
    6.3.C Electron Ray Paths Through Magnetic Fields 99
    6.3.D Image Rotation and the Eucentric Plane 100
    6.3.E Deflecting the Beam 101
6.4 Apertures and Diaphragms 101
6.5 Real Lenses and their Problems 102
    6.5.A Spherical Aberration 103
    6.5.B Chromatic Aberration 104
    6.5.C Astigmatism 106
6.6 The Resolution of the Electron Lens (and Ultimately of the TEM) 106
    6.6.A Theoretical Resolution (Diffraction-Limited Resolution) 107
    6.6.B The Practical Resolution Due to Spherical Aberration 108
    6.6.C Specimen-Limited Resolution Due to Chromatic Aberration 109
    6.6.D Confusion in the Definitions of Resolution 109
6.7 Depth of Focus and Depth of Field 110
Chapter Summary 111

7 How to 'See' Electrons **115**

Chapter Preview 115
7.1 Electron Detection and Display 115
7.2 Viewing Screens 116
7.3 Electron Detectors 117
    7.3.A Semiconductor Detectors 117
    7.3.B Scintillator-Photomultiplier Detectors/TV Cameras 118
    7.3.C Charge-Coupled Device (CCD) Detectors 120
    7.3.D Faraday Cup 121
7.4 Which Detector Do We Use for which Signal? 122
7.5 Image Recording 122
    7.5.A Photographic Emulsions 122
    7.5.B Other Image-Recording Methods 124
7.6 Comparison of Scanning Images and Static Images 124
Chapter Summary 125

8 Pumps and Holders **127**

Chapter Preview 127
8.1 The Vacuum 127
8.2 Roughing Pumps 128
8.3 High/Ultra High Vacuum Pumps 129
    8.3.A Diffusion Pumps 129
    8.3.B Turbomolecular Pumps 129
    8.3.C Ion Pumps 130
    8.3.D Cryogenic (Adsorption) Pumps 130
8.4 The Whole System 130
8.5 Leak Detection 131
8.6 Contamination: Hydrocarbons and Water Vapor 132
8.7 Specimen Holders and Stages 132
8.8 Side-Entry Holders 133
8.9 Top-entry Holders 134
8.10 Tilt and Rotate Holders 134
8.11 In-Situ Holders 135
8.12 Plasma Cleaners 138
Chapter Summary 138

9 The Instrument **141**

Chapter Preview 141
9.1 The Illumination System 142
    9.1.A TEM Operation Using a Parallel Beam 142
    9.1.B Convergent-Beam (S)TEM Mode 143
    9.1.C The Condenser-Objective Lens 145
    9.1.D Translating and Tilting the Beam 147
    9.1.E Alignment of the C2 Aperture 147
    9.1.F Condenser-Lens Defects 148
    9.1.G Calibration 149
9.2 The Objective Lens and Stage 150
9.3 Forming DPs and Images: The TEM Imaging System ... 152
    9.3.A Selected-Area Diffraction 152
    9.3.B Bright-Field and Dark-Field Imaging 155
    9.3.C Centered Dark-Field Operation 155
    9.3.D Hollow-Cone Diffraction and Dark-Field Imaging ... 157
9.4 Forming DPs and Images: The STEM Imaging System .. 158

		9.4.A	Bright-Field STEM Images	159
		9.4.B	Dark-Field STEM Images	161
		9.4.C	Annular Dark-Field Images	161
		9.4.D	Magnification in STEM	161
	9.5	Alignment and Stigmation		161
		9.5.A	Lens Rotation Centers	161
		9.5.B	Correction of Astigmatism in the Imaging Lenses	162
	9.6	Calibrating the Imaging System		164
		9.6.A	Magnification Calibration	164
		9.6.B	Camera-Length Calibration	165
		9.6.C	Rotation of the Image Relative to the DP	167
		9.6.D	Spatial Relationship Between Images and DPs	168
	9.7	Other Calibrations		168
	Chapter Summary			169

10 Specimen Preparation — 173

- Chapter Preview — 173
- 10.1 Safety — 173
- 10.2 Self-Supporting Disk or Use a Grid? — 174
- 10.3 Preparing a Self-Supporting Disk for Final Thinning — 175
  - 10.3.A Forming a Thin Slice from the Bulk Sample — 176
  - 10.3.B Cutting the Disk — 176
  - 10.3.C Prethinning the Disk — 177
- 10.4 Final Thinning of the Disks — 178
  - 10.4.A Electropolishing — 178
  - 10.4.B Ion Milling — 178
- 10.5 Cross-Section Specimens — 182
- 10.6 Specimens on Grids/Washers — 183
  - 10.6.A Electropolishing—The Window Method for Metals and Alloys — 183
  - 10.6.B Ultramicrotomy — 183
  - 10.6.C Grinding and Crushing — 184
  - 10.6.D Replication and Extraction — 184
  - 10.6.E Cleaving and the SACT — 186
  - 10.6.F The 90° Wedge — 186
  - 10.6.G Lithography — 187
  - 10.6.H Preferential Chemical Etching — 187
- 10.7 FIB — 188
- 10.8 Storing Specimens — 189
- 10.9 Some Rules — 189
- Chapter Summary — 191

PART 2 DIFFRACTION — 195

11 Diffraction in TEM — 197

- Chapter Preview — 197
- 11.1 Why Use Diffraction in the TEM? — 197
- 11.2 The TEM, Diffraction Cameras, and the TV — 198
- 11.3 Scattering from a Plane of Atoms — 199
- 11.4 Scattering from a Crystal — 200
- 11.5 Meaning of n in Bragg's Law — 202
- 11.6 A Pictorial Introduction to Dynamical Effects — 203
- 11.7 Use of Indices in Diffraction Patterns — 204
- 11.8 Practical Aspects of Diffraction-Pattern Formation — 204
- 11.9 More on Selected-Area Diffraction Patterns — 204
- Chapter Summary — 208

12 Thinking in Reciprocal Space ... 211

Chapter Preview ... 211
12.1 Why Introduce Another Lattice? ... 211
12.2 Mathematical Definition of the Reciprocal Lattice ... 212
12.3 The Vector **g** ... 212
12.4 The Laue Equations and their Relation to Bragg's Law ... 213
12.5 The Ewald Sphere of Reflection ... 214
12.6 The Excitation Error ... 216
12.7 Thin-Foil Effect and the Effect of Accelerating Voltage ... 217
Chapter Summary ... 218

13 Diffracted Beams ... 221

Chapter Preview ... 221
13.1 Why Calculate Intensities? ... 221
13.2 The Approach ... 222
13.3 The Amplitude of a Diffracted Beam ... 223
13.4 The Characteristic Length ξ_g ... 223
13.5 The Howie-Whelan Equations ... 224
13.6 Reformulating the Howie-Whelan Equations ... 225
13.7 Solving the Howie-Whelan Equations ... 226
13.8 The Importance of $\gamma^{(1)}$ and $\gamma^{(2)}$ ... 226
13.9 The Total Wave Amplitude ... 227
13.10 The Effective Excitation Error ... 228
13.11 The Column Approximation ... 229
13.12 The Approximations and Simplifications ... 230
13.13 The Coupled Harmonic Oscillator Analog ... 231
Chapter Summary ... 231

14 Bloch Waves ... 235

Chapter Preview ... 235
14.1 Wave Equation in TEM ... 235
14.2 The Crystal ... 236
14.3 Bloch Functions ... 237
14.4 Schrödinger's Equation for Bloch Waves ... 238
14.5 The Plane-Wave Amplitudes ... 239
14.6 Absorption of Bloch Waves ... 241
Chapter Summary ... 242

15 Dispersion Surfaces ... 245

Chapter Preview ... 245
15.1 Introduction ... 245
15.2 The Dispersion Diagram When $U_g = 0$ ... 246
15.3 The Dispersion Diagram When $U_g \neq 0$ ... 247
15.4 Relating Dispersion Surfaces and Diffraction Patterns ... 247
15.5 The Relation Between U_g, ξ_g, and s_g ... 250
15.6 The Amplitudes of Bloch Waves ... 252
15.7 Extending to More Beams ... 253
15.8 Dispersion Surfaces and Defects ... 254
Chapter Summary ... 254

16 Diffraction from Crystals ... 257

Chapter Preview ... 257
16.1 Review of Diffraction from a Primitive Lattice ... 257
16.2 Structure Factors: The Idea ... 258

16.3	Some Important Structures: BCC, FCC and HCP	259
16.4	Extending fcc and hcp to Include a Basis	261
16.5	Applying the bcc and fcc Analysis to Simple Cubic	262
16.6	Extending hcp to TiAl	262
16.7	Superlattice Reflections and Imaging	262
16.8	Diffraction from Long-Period Superlattices	264
16.9	Forbidden Reflections	265
16.10	Using the International Tables	265
	Chapter Summary	267

17 Diffraction from Small Volumes — 271

	Chapter Preview	271
17.1	Introduction	271
	17.1.A The Summation Approach	272
	17.1.B The Integration Approach	273
17.2	The Thin-Foil Effect	273
17.3	Diffraction from Wedge-Shaped Specimens	274
17.4	Diffraction from Planar Defects	275
17.5	Diffraction from Particles	277
17.6	Diffraction from Dislocations, Individually and Collectively	278
17.7	Diffraction and the Dispersion Surface	279
	Chapter Summary	281

18 Obtaining and Indexing Parallel-Beam Diffraction Patterns — 283

	Chapter Preview	283
18.1	Choosing Your Technique	284
18.2	Experimental SAD Techniques	284
18.3	The Stereographic Projection	286
18.4	Indexing Single-Crystal DPs	287
18.5	Ring Patterns from Polycrystalline Materials	290
18.6	Ring Patterns from Hollow-Cone Diffraction	291
18.7	Ring Patterns from Amorphous Materials	293
18.8	Precession Diffraction	295
18.9	Double Diffraction	296
18.10	Orientation of the Specimen	298
18.11	Orientation Relationships	302
18.12	Computer Analysis	303
18.13	Automated Orientation Determination and Orientation Mapping	305
	Chapter Summary	305

19 Kikuchi Diffraction — 311

	Chapter Preview	311
19.1	The Origin of Kikuchi Lines	311
19.2	Kikuchi Lines and Bragg Scattering	312
19.3	Constructing Kikuchi Maps	313
19.4	Crystal Orientation and Kikuchi Maps	317
19.5	Setting the Value of S_g	318
19.6	Intensities	319
	Chapter Summary	320

20 Obtaining CBED Patterns — 323

	Chapter Preview	323
20.1	Why Use a Convergent Beam?	323

	20.2	Obtaining CBED Patterns	324
		20.2.A Comparing SAD and CBED	325
		20.2.B CBED in TEM Mode	326
		20.2.C CBED in STEM Mode	326
	20.3	Experimental Variables	327
		20.3.A Choosing the C2 Aperture	327
		20.3.B Selecting the Camera Length	328
		20.3.C Choice of Beam Size	329
		20.3.D Effect of Specimen Thickness	329
	20.4	Focused and Defocused CBED Patterns	329
		20.4.A Focusing a CBED Pattern	330
		20.4.B Large-Angle (Defocused) CBED Patterns	330
		20.4.C Final Adjustment	332
	20.5	Energy Filtering	334
	20.6	Zero-Order and High-Order Laue-Zone Diffraction	335
		20.6.A ZOLZ Patterns	335
		20.6.B HOLZ Patterns	336
	20.7	Kikuchi and Bragg Lines in CBED Patterns	338
	20.8	HOLZ Lines	339
		20.8.A The Relationship Between HOLZ Lines and Kikuchi Lines	339
		20.8.B Acquiring HOLZ Lines	341
	20.9	Hollow-Cone/Precession CBED	342
	Chapter Summary		343

21 Using Convergent-Beam Techniques — 347

Chapter Preview — 347

	21.1	Indexing CBED Patterns	348
		21.1.A Indexing ZOLZ and HOLZ Patterns	348
		21.1.B Indexing HOLZ Lines	351
	21.2	Thickness Determination	352
	21.3	Unit-Cell Determination	354
		21.3.A Experimental Considerations	354
		21.3.B The Importance of the HOLZ-Ring Radius	355
		21.3.C Determining the Lattice Centering	356
	21.4	Basics of Symmetry Determination	357
		21.4.A Reminder of Symmetry Concepts	357
		21.4.B Friedel's Law	358
		21.4.C Looking for Symmetry in Your Patterns	358
	21.5	Lattice-Strain Measurement	361
	21.6	Determination of Enantiomorphism	363
	21.7	Structure Factor and Charge-Density Determination	364
	21.8	Other Methods	365
		21.8.A Scanning Methods	365
		21.8.B Nanodiffraction	366
	Chapter Summary		366

PART 3 IMAGING — 369

22 Amplitude Contrast — 371

Chapter Preview — 371

	22.1	What Is Contrast?	371
	22.2	Amplitude contrast	372
		22.2.A Images and Diffraction Patterns	372
		22.2.B Use of the Objective Aperture or the STEM Detector: BF and DF Images	372

	22.3	Mass-Thickness Contrast	373
		22.3.A Mechanism of Mass-Thickness Contrast	373
		22.3.B TEM Images	374
		22.3.C STEM Images	376
		22.3.D Specimens Showing Mass-Thickness Contrast	377
		22.3.E Quantitative Mass-Thickness Contrast	378
	22.4	Z-Contrast	379
	22.5	TEM Diffraction Contrast	381
		22.5.A Two-Beam Conditions	381
		22.5.B Setting the Deviation Parameter, s	382
		22.5.C Setting Up a Two-Beam CDF Image	382
		22.5.D Relationship Between the Image and the Diffraction Pattern	384
	22.6	STEM Diffraction Contrast	384
	Chapter Summary		386

23 Phase-Contrast Images — 389

	Chapter Preview		389
	23.1	Introduction	389
	23.2	The Origin of Lattice Fringes	389
	23.3	Some Practical Aspects of Lattice Fringes	390
		23.3.A If s = 0	390
		23.3.B If s ≠ 0	390
	23.4	On-Axis Lattice-Fringe Imaging	391
	23.5	Moiré Patterns	392
		23.5.A Translational Moiré Fringes	393
		23.5.B Rotational Moiré Fringes	393
		23.5.C General Moiré Fringes	393
	23.6	Experimental Observations of Moiré Fringes	393
		23.6.A Translational Moiré Patterns	394
		23.6.B Rotational Moiré Patterns	394
		23.6.C Dislocations and Moiré Fringes	394
		23.6.D Complex Moiré Fringes	396
	23.7	Fresnel Contrast	397
		23.7.A The Fresnel Biprism	397
		23.7.B Magnetic-Domain Walls	398
	23.8	Fresnel Contrast from Voids or Gas Bubbles	399
	23.9	Fresnel Contrast from Lattice Defects	400
		23.9.A Grain Boundaries	402
		23.9.B End-On Dislocations	402
	Chapter Summary		402

24 Thickness and Bending Effects — 407

	Chapter Preview		407
	24.1	The Fundamental Ideas	407
	24.2	Thickness Fringes	408
	24.3	Thickness Fringes and the DP	410
	24.4	Bend Contours (Annoying Artifact, Useful Tool, Invaluable Insight)	411
	24.5	ZAPs and Real-Space Crystallography	412
	24.6	Hillocks, Dents, or Saddles	413
	24.7	Absorption Effects	413
	24.8	Computer Simulation of Thickness Fringes	414
	24.9	Thickness-Fringe/Bend-Contour Interactions	414
	24.10	Other Effects of Bending	415
	Chapter Summary		416

25 Planar Defects 419
Chapter Preview 419
25.1 Translations and Rotations 419
25.2 Why Do Translations Produce Contrast? 421
25.3 The Scattering Matrix 422
25.4 Using the Scattering Matrix 423
25.5 Stacking Faults in fcc Materials 424
    25.5.A Why fcc Materials? 424
    25.5.B Some Rules 425
    25.5.C Intensity Calculations 426
    25.5.D Overlapping Faults 426
25.6 Other Translations: π and δ Fringes 427
25.7 Phase Boundaries 429
25.8 Rotation Boundaries 430
25.9 Diffraction Patterns and Dispersion Surfaces 430
25.10 Bloch Waves and BF/DF Image Pairs 431
25.11 Computer Modeling 432
25.12 The Generalized Cross Section 433
25.13 Quantitative Imaging 434
    25.13.A Theoretical Basis and Parameters 434
    25.13.B Apparent Extinction Distance 435
    25.13.C Avoiding the Column Approximation 435
    25.13.D The User Interface 436
Chapter Summary 436

26 Imaging Strain Fields 441
Chapter Preview 441
26.1 Why Image Strain Fields? 441
26.2 Howie-Whelan Equations 442
26.3 Contrast from a Single Dislocation 444
26.4 Displacement Fields and Ewald's Sphere 447
26.5 Dislocation Nodes and Networks 448
26.6 Dislocation Loops and Dipoles 448
26.7 Dislocation Pairs, Arrays, and Tangles 450
26.8 Surface Effects 451
26.9 Dislocations and Interfaces 452
26.10 Volume Defects and Particles 456
26.11 Simulating Images 457
    26.11.A The Defect Geometry 457
    26.11.B Crystal Defects and Calculating the Displacement Field 458
    26.11.C The Parameters 458
Chapter Summary 459

27 Weak-Beam Dark-Field Microscopy 463
Chapter Preview 463
27.1 Intensity in WBDF Images 463
27.2 Setting S_g Using the Kikuchi Pattern 464
27.3 How to Do WBDF 466
27.4 Thickness Fringes in Weak-Beam Images 467
27.5 Imaging Strain Fields 468
27.6 Predicting Dislocation Peak Positions 469
27.7 Phasor Diagrams 470
27.8 Weak-Beam Images of Dissociated Dislocations 473
27.9 Other Thoughts 477

	27.9.A	Thinking of Weak-Beam Diffraction as a Coupled Pendulum	477
	27.9.B	Bloch Waves	478
	27.9.C	If Other Reflections are Present	478
	27.9.D	The Future Is Now	478
	Chapter Summary		479

28 High-Resolution TEM — 483

Chapter Preview — 483
- 28.1 The Role of an Optical System — 483
- 28.2 The Radio Analogy — 484
- 28.3 The Specimen — 485
- 28.4 Applying the WPOA to the TEM — 487
- 28.5 The Transfer Function — 487
- 28.6 More on $\chi(u)$, $\sin\chi(u)$, and $\cos\chi(u)$ — 488
- 28.7 Scherzer Defocus — 490
- 28.8 Envelope Damping Functions — 491
- 28.9 Imaging Using Passbands — 492
- 28.10 Experimental Considerations — 493
- 28.11 The Future for HRTEM — 494
- 28.12 The TEM as a Linear System — 494
- 28.13 FEG TEMs and the Information Limit — 495
- 28.14 Some Difficulties in Using an FEG — 498
- 28.15 Selectively Imaging Sublattices — 500
- 28.16 Interfaces and Surfaces — 502
- 28.17 Incommensurate Structures — 503
- 28.18 Quasicrystals — 504
- 28.19 Single Atoms — 505
- Chapter Summary — 506

29 Other Imaging Techniques — 511

Chapter Preview — 511
- 29.1 Stereo Microscopy and Tomography — 511
- 29.2 $2\tfrac{1}{2}$D Microscopy — 512
- 29.3 Magnetic Specimens — 514
  - 29.3.A The Magnetic Correction — 514
  - 29.3.B Lorentz Microscopy — 515
- 29.4 Chemically Sensitive Images — 517
- 29.5 Imaging with Diffusely Scattered Electrons — 517
- 29.6 Surface Imaging — 519
  - 29.6.A Reflection Electron Microscopy — 519
  - 29.6.B Topographic Contrast — 521
- 29.7 High-Order BF Imaging — 521
- 29.8 Secondary-Electron Imaging — 522
- 29.9 Backscattered-Electron Imaging — 523
- 29.10 Charge-Collection Microscopy and Cathodoluminescence — 523
- 29.11 Electron Holography — 524
- 29.12 In Situ TEM: Dynamic Experiments — 526
- 29.13 Fluctuation Microscopy — 528
- 29.14 Other Variations Possible in a STEM — 528
- Chapter Summary — 529

30 Image Simulation — 533

Chapter Preview — 533
- 30.1 Simulating images — 533
- 30.2 The Multislice Method — 533

	30.3	The Reciprocal-Space Approach	534
	30.4	The FFT Approach	536
	30.5	The Real-Space approach	536
	30.6	Bloch Waves and HRTEM Simulation	536
	30.7	The Ewald Sphere Is Curved	537
	30.8	Choosing the Thickness of the Slice	537
	30.9	Beam Convergence	538
	30.10	Modeling the Structure	540
	30.11	Surface Grooves and Simulating Fresnel Contrast	540
	30.12	Calculating Images of Defects	542
	30.13	Simulating Quasicrystals	543
	30.14	Bonding in Crystals	544
	30.15	Simulating Z-Contrast	545
	30.16	Software for Phase-Contrast HRTEM	545
	Chapter Summary		545

31 Processing and Quantifying Images — 549

	Chapter Preview		549
	31.1	What Is Image Processing?	549
	31.2	Processing and Quantifying Images	550
	31.3	A Cautionary Note	550
	31.4	Image Input	550
	31.5	Processing Techniques	551
		31.5.A Fourier Filtering and Reconstruction	551
		31.5.B Analyzing Diffractograms	552
		31.5.C Averaging Images and Other Techniques	554
		31.5.D Kernels	556
	31.6	Applications	556
		31.6.A Beam-Sensitive Materials	556
		31.6.B Periodic Images	557
		31.6.C Correcting Drift	557
		31.6.D Reconstructing the Phase	557
		31.6.E Diffraction Patterns	558
		31.6.F Tilted-Beam Series	559
	31.7	Automated Alignment	560
	31.8	Quantitative Methods of Image Analysis	561
	31.9	Pattern Recognition in HRTEM	562
	31.10	Parameterizing the Image Using QUANTITEM	563
		31.10.A The Example of a Specimen with Uniform Composition	563
		31.10.B Calibrating the Path of R	565
		31.10.C Noise Analysis	565
	31.11	Quantitative Chemical Lattice Imaging	567
	31.12	Methods of Measuring Fit	568
	31.13	Quantitative Comparison of Simulated and Experimental HRTEM Images	570
	31.14	A Fourier Technique for Quantitative Analysis	571
	31.15	Real or Reciprocal Space?	572
	31.16	Software	573
	31.17	The Optical Bench—A Little History	573
	Chapter Summary		575

PART 4 SPECTROMETRY — 579

32 X-ray Spectrometry — 581

Chapter Preview 581

	32.1	X-ray Analysis: Why Bother?	581
	32.2	Basic Operational Mode	584
	32.3	The Energy-Dispersive Spectrometer	584
	32.4	Semiconductor Detectors	585
		32.4.A How Does an XEDS Work?	585
		32.4.B Cool Detectors	586
		32.4.C Different Kinds of Windows	586
		32.4.D Intrinsic-Germanium Detectors	587
		32.4.E Silicon-Drift Detectors	588
	32.5	Detectors with High-Energy Resolution	589
	32.6	Wavelength-Dispersive Spectrometers	589
		32.6.A Crystal WDS	589
		32.6.B CCD-Based WDS	590
		32.6.C Bolometers/Microcalorimeters	590
	32.7	Turning X-rays into Spectra	591
	32.8	Energy Resolution	593
	32.9	What You Should Know about Your XEDS	594
		32.9.A Detector Characteristics	594
		32.9.B Processing Variables	596
	32.10	The XEDS-AEM Interface	598
		32.10.A Collection Angle	598
		32.10.B Take-Off Angle	599
		32.10.C Orientation of the Detector to the Specimen	599
	32.11	Protecting the Detector from Intense Radiation	600
	Chapter Summary		601
33	**X-ray Spectra and Images**		**605**
	Chapter Preview		605
	33.1	The Ideal Spectrum	605
		33.1.A The Characteristic Peaks	605
		33.1.B The Continuum Bremsstrahlung Background	606
	33.2	Artifacts Common to Si(Li) XEDS Systems	606
	33.3	The Real Spectrum	608
		33.3.A Pre-Specimen Effects	608
		33.3.B Post-Specimen Scatter	611
		33.3.C Coherent Bremsstrahlung	613
	33.4	Measuring the Quality of the XEDS-AEM Interface	614
		33.4.A Peak-to-Background Ratio	614
		33.4.B Efficiency of the XEDS System	614
	33.5	Acquiring X-ray Spectra	615
		33.5.A Spot Mode	615
		33.5.B Spectrum-Line Profiles	616
	33.6	Acquiring X-ray Images	616
		33.6.A Analog Dot Mapping	617
		33.6.B Digital Mapping	618
		33.6.C Spectrum Imaging (SI)	619
		33.6.D Position-Tagged Spectrometry (PTS)	620
	Chapter Summary		620
34	**Qualitative X-ray Analysis and Imaging**		**625**
	Chapter Preview		625
	34.1	Microscope and Specimen Variables	625
	34.2	Basic Acquisition Requirements: Counts, Counts, and More Caffeine	626

	34.3	Peak Identification...............................	627
	34.4	Peak Deconvolution................................	630
	34.5	Peak Visibility....................................	632
	34.6	Common Errors....................................	634
	34.7	Qualitative X-ray Imaging: Principles and Practice.....	634
	Chapter Summary...		636

35 Quantitative X-ray Analysis................................ 639

	Chapter Preview..	639
35.1	Historical Perspective..............................	639
35.2	The Cliff-Lorimer Ratio Technique.................	640
35.3	Practical Steps for Quantification..................	641
	35.3.A Background Subtraction.................	641
	35.3.B Peak Integration........................	644
35.4	Determining k-Factors............................	646
	35.4.A Experimental Determination of k_{AB}.........	646
	35.4.B Errors in Quantification: The Statistics.......	647
	35.4.C Calculating k_{AB}.........................	648
35.5	The Zeta-Factor Method...........................	652
35.6	Absorption Correction.............................	654
35.7	The Zeta-Factor Absorption Correction.............	656
35.8	The Fluorescence Correction.......................	656
35.9	ALCHEMI..	657
35.10	Quantitative X-ray Mapping......................	658
	Chapter Summary..	660

36 Spatial Resolution and Minimum Detection............... 663

	Chapter Preview..	663
36.1	Why Is Spatial Resolution Important?..............	663
36.2	Definition and Measurement of Spatial Resolution.....	664
	36.2.A Beam Spreading.........................	665
	36.2.B The Spatial-Resolution Equation...........	666
	36.2.C Measurement of Spatial Resolution.........	667
36.3	Thickness Measurement............................	668
	36.3.A TEM Methods...........................	669
	36.3.B Contamination-Spot Separation Method......	670
	36.3.C Convergent-Beam Diffraction Method.......	671
	36.3.D Electron Energy-Loss Spectrometry Methods..	671
	36.3.E X-ray Spectrometry Method................	671
36.4	Minimum Detection................................	672
	36.4.A Experimental Factors Affecting the MMF.....	673
	36.4.B Statistical Criterion for the MMF...........	673
	36.4.C Comparison with Other Definitions.........	674
	36.4.D Minimum-Detectable Mass.................	674
	Chapter Summary..	675

37 Electron Energy-Loss Spectrometers and Filters.......... 679

	Chapter Preview..	679
37.1	Why Do EELS?....................................	679
	37.1.A Pros and Cons of Inelastic Scattering........	679
	37.1.B The Energy-Loss Spectrum................	680
37.2	EELS Instrumentation..............................	681
37.3	The Magnetic Prism: A Spectrometer and a Lens.....	681
	37.3.A Focusing the Spectrometer.................	682
	37.3.B Spectrometer Dispersion..................	683

		37.3.C	Spectrometer Resolution	683
		37.3.D	Calibrating the Spectrometer	684
	37.4	Acquiring a Spectrum		684
		37.4.A	Image and Diffraction Modes	685
		37.4.B	Spectrometer-Collection Angle	685
		37.4.C	Spatial Selection	688
	37.5	Problems with PEELS		688
		37.5.A	Point-Spread Function	688
		37.5.B	PEELS Artifacts	689
	37.6	Imaging Filters		690
		37.6.A	The Omega Filter	691
		37.6.B	The GIF	692
	37.7	Monochromators		693
	37.8	Using Your Spectrometer and Filter		694
	Chapter Summary			696

38 Low-Loss and No-Loss Spectra and Images — 699

	Chapter Preview		699
38.1	A Few Basic Concepts		699
38.2	The Zero-Loss Peak (ZLP)		701
	38.2.A	Why the ZLP Really Isn't	701
	38.2.B	Removing the Tail of the ZLP	701
	38.2.C	Zero-Loss Images and Diffraction Patterns	702
38.3	The Low-Loss Spectrum		703
	38.3.A	Chemical Fingerprinting	704
	38.3.B	Dielectric-Constant Determination	705
	38.3.C	Plasmons	705
	38.3.D	Plasmon-Loss Analysis	707
	38.3.E	Single-Electron Excitations	709
	38.3.F	The Band Gap	709
38.4	Modeling The Low-Loss Spectrum		710
Chapter Summary			711

39 High Energy-Loss Spectra and Images — 715

	Chapter Preview		715
39.1	The High-Loss Spectrum		715
	39.1.A	Inner-Shell Ionization	715
	39.1.B	Ionization-Edge Characteristics	717
39.2	Acquiring a High-Loss Spectrum		721
39.3	Qualitative Analysis		723
39.4	Quantitative Analysis		723
	39.4.A	Derivation of the Equations for Quantification	724
	39.4.B	Background Subtraction	726
	39.4.C	Edge Integration	728
	39.4.D	The Partial Ionization Cross Section	728
39.5	Measuring Thickness from the Core-Loss Spectrum		730
39.6	Deconvolution		731
39.7	Correction for Convergence of the Incident Beam		733
39.8	The Effect of the Specimen Orientation		733
39.9	EFTEM Imaging with Ionization Edges		733
	39.9.A	Qualitative Imaging	734
	39.9.B	Quantitative Imaging	734
39.10	Spatial Resolution: Atomic-Column EELS		735

		39.11	Detection Limits	736
	Chapter Summary			737

40 Fine Structure and Finer Details ... 741

Chapter Preview ... 741
- 40.1 Why Does Fine Structure Occur? ... 741
- 40.2 ELNES Physics ... 742
  - 40.2.A Principles ... 742
  - 40.2.B White Lines ... 744
  - 40.2.C Quantum Aspects ... 744
- 40.3 Applications of ELNES ... 745
- 40.4 ELNES Fingerprinting ... 746
- 40.5 ELNES Calculations ... 747
  - 40.5.A The Potential Choice ... 748
  - 40.5.B Core-Holes and Excitons ... 749
  - 40.5.C Comparison of ELNES Calculations and Experiments ... 750
- 40.6 Chemical Shifts in the Edge Onset ... 750
- 40.7 EXELFS ... 751
  - 40.7.A RDF via EXELFS ... 752
  - 40.7.B RDF via Energy-Filtered Diffraction ... 753
  - 40.7.C A Final Thought Experiment ... 753
- 40.8 Angle-Resolved EELS ... 755
- 40.9 EELS Tomography ... 755

Chapter Summary ... 757

Index ... I-1

Figure Credits

TEM is a visual science, and any TEM text is heavily dependent on figures, halftones, and (more recently) full color images to transmit its message. We have been fortunate to work with many colleagues over the years who have generously given us fine examples of the art and science of TEM; we would like to acknowledge them here. We have also used our own work, and the work of others, whose permission has been sought as listed below.

Chapter 1

Figure 1.1: From Ruska, E (1980) *The Early History of the Electron Microscope*, Fig. 6 reproduced by permission of S. Herzel Verlag GmbH & Co.

Figure 1.2B,C: Specimen courtesy of Y Ikuhara and T Yamamoto, University of Tokyo, reproduced by permission of JEOL Ltd.

Figure 1.4: Courtesy of M Watanabe.

Figure 1.6: Courtesy of KS Vecchio.

Figure 1.7: Courtesy of T Hayes, from Hayes, T (1980) in O Johari Ed. SEM-1980 1 1, Fig. 8 reproduced by permission of Scanning Microscopy International.

Figure 1.9A: Courtesy of M Kersker, reproduced by permission of JEOL USA Inc.

Figure 1.9B: Courtesy of E Essers, reproduced by permission of Carl Zeiss SMT.

Figure 1.9C: Courtesy of K Jarausch, reproduced by permission of Hitachi High Technologies.

Figure 1.9D: Courtesy of M Kersker, reproduced by permission of JEOL USA Inc.

Figure 1.9E: Courtesy of OL Krivanek, reproduced by permission of NION Inc.

Figure 1.9F: Courtesy of JS Fahy, reproduced by permission of FEI Co.

Chapter 2

Figure 2.4: Courtesy of J Bruley and VJ Keast.

Figure 2.11: Modified from Hecht, E (1988) Optics, Fig. 10.21 Addison-Wesley.

Figure 2.13A,D: Courtesy of KS Vecchio.

Figure 2.13C: Courtesy of DW Ackland.

Chapter 3

Figure 3.3: Courtesy of DE Newbury, modified from Newbury, DE (1986) in DC Joy *et al.* Eds. *Principles of Analytical Electron Microscopy* p 6, Fig. 2 original reproduced by permission of Plenum Press.

Figure 3.4: Courtesy of DE Newbury, modified from data in Newbury, DE (1986) in DC Joy *et al.* Eds. *Principles of Analytical Electron Microscopy* p 8, Table II reproduced by permission of Plenum Press.

Chapter 4

Figure 4.1: Courtesy of DE Newbury, modified from Newbury, DE (1986) in DC Joy *et al.* Eds. *Principles of Analytical Electron Microscopy* p 20, Fig. 4 original reproduced by permission of Plenum Press.

Figure 4.3: Modified from Woldseth, R (1973) *X-ray Energy Spectrometry*, Fig. 3 original reproduced by permission of Kevex Instruments.

Figure 4.4: Modified from Williams, DB (1987) *Practical Analytical Electron Microscopy in Materials Science,* 2nd Edition, Fig. 4.3 reproduced by permission of Philips Electron Optics.

Figure 4.11: Courtesy of LW Hobbs, modified from Hobbs, LW (1979) in JJ Hren *et al.* Eds. *Introduction to Analytical Electron Microscopy*, Fig. 17.2 original reproduced by permission of Plenum Press.

Figure 4.12: Courtesy of LW Hobbs, modified from Hobbs, LW (1979) in JJ Hren *et al.* Eds. *Introduction to Analytical Electron Microscopy*, Fig. 17.4 original reproduced by permission of Plenum Press.

Table 4.1: Courtesy of JI Goldstein, from Goldstein, JI *et al.* (1992) *Scanning Electron Microscopy and X-ray Microanalysis,* 2nd Edition, Table 3.11 reproduced by permission of Plenum Press.

Table 4.2: Data obtained from National Physical Laboratory, Teddington, UK, web site. http://www.kayelaby.npl.co.uk/atomic_and_nuclear_physics/4_2/4_2_1.html

Table 4.3: Courtesy of NJ Zaluzec and JF Mansfield, from Zaluzec, NJ and Mansfield, JF (1987) in K Rajan Ed. *Intermediate Voltage Electron Microscopy and Its Application to Materials Science* p 29, Table 1 reproduced by permission of Philips Electron Optics.

Chapter 5

Figure 5.1: Modified from Hall, CE (1966) *Introduction to Electron Microscopy*, Fig. 7.8 McGraw-Hill.

Figure 5.4B: Courtesy of JI Goldstein, modified from Goldstein, JI et al. (1992) *Scanning Electron Microscopy and X-ray Microanalysis,* 2nd Edition, Fig. 2.7 original reproduced by permission of Plenum Press.

Figure 5.5: Courtesy of DW Ackland.

Figure 5.6A: Modified from Crewe, AV et al. (1969) Rev. Sci. Instrum. **40** 241, Fig. 2.

Figure 5.6B: Courtesy of DW Ackland.

Figure 5.7: Courtesy of M Watanabe, modified from Watanabe, M and Williams, DB (2006) J. Microsc. **221** 89, Fig. 14.

Figure 5.10: Courtesy of JR Michael, modified from Michael, JR and Williams, DB (1987) J. Microsc. **147** 289, Fig. 3 original reproduced by permission of the Royal Microscopical Society.

Figure 5.11: Modified from Williams, DB (1987) *Practical Analytical Electron Microscopy in Materials Science,* 2nd Edition, Fig. 2.12B Philips Electron Optics.

Figure 5.12: Courtesy of JR Michael, from Michael, JR and Williams, DB (1987) J. Microsc. **147** 289, Fig. 2 original reproduced by permission of the Royal Microscopical Society.

Figure 5.13A: Courtesy of DW Ackland.

Figure 5.13B: Reproduced by permission of NSA Hitachi Scientific Instruments Ltd.

Chapter 6

Figure 6.7: Courtesy of DW Ackland.

Figure 6.8A: Reproduced by permission of Philips Electronic Instruments Inc.

Figure 6.8B: Reproduced by permission of Kratos Ltd.

Figure 6.8C: From Mulvey, T (1974) Electron Microscopy-1974 17, Fig. 1 reproduced by permission of the Australian Academy of Science.

Figure 6.8D: From Reimer, L (1993) *Transmission Electron Microscopy,* 3rd Edition, Fig. 2.12 reproduced by permission of Springer Verlag.

Figure 6.9: Modified from Reimer, L (1993) Transmission Electron Microscopy, 3rd Edition, Fig. 2.3 Springer Verlag.

Figure 6.10B: Courtesy of AO Benscoter.

Figure 6.11: Modified from Reimer, L (1993) *Transmission Electron Microscopy,* 3rd Edition, Fig. 2.13 Springer Verlag.

Figure 6.12A: Courtesy of OL Krivanek, reproduced by permission of NION Inc.

Figure 6.12A: Courtesy of M Haider, reproduced by permission of CEOS GmbH.

Figure 6.15: Modified from Reimer, L (1993) *Transmission Electron Microscopy,* 3rd Edition, Fig. 4.23 Springer Verlag.

Chapter 7

Figure 7.1: Modified from Stephen, J et al. (1975) J. Phys. E 8 607, Fig. 2.

Figure 7.5: Modified from Williams, DB (1987) *Practical Analytical Electron Microscopy in Materials Science,* 2nd Edition, Fig. 1.2 Philips Electron Optics.

Figure 7.6: Modified from Berger, SD et al. (1985) *Electron Microscopy and Analysis* p 137, Fig. 1 original by permission of The Institute of Physics Publishing.

Chapter 8

Figure 8.1: Courtesy of WC Bigelow, modified from Bigelow, WC (1994) *Vacuum Methods in Electron Microscopy*, Fig. 4.1 original by permission of Portland Press Ltd.

Figure 8.2: Courtesy of WC Bigelow, modified from Bigelow, WC (1994) *Vacuum Methods in Electron Microscopy*, Fig. 5.1 original by permission of Portland Press Ltd.

Figure 8.3: Reproduced by permission of Leybold Vacuum Products Inc.

Figure 8.4: Courtesy of WC Bigelow, modified from Bigelow, WC (1994) *Vacuum Methods in Electron Microscopy*, Fig. 7.1 original by permission of Portland Press Ltd.

Figure 8.6: Reproduced by permission of Gatan Inc.

Figure 8.7: Modified from Valdrè, U and Goringe, MJ (1971) in U Valdrè Ed. *Electron Microscopy in Materials Science* p 217, Fig. 6 original by permission of Academic Press Inc.

Figure 8.8: Courtesy of NSA Hitachi Scientific Instruments Ltd.

Figure 8.9A,B: Reproduced by permission of Gatan Inc.

Figure 8.10A: Reproduced by permission of Gatan Inc.

Figure 8.11: Reproduced by permission of Gatan Inc.

Figure 8.12: Modified from Komatsu, M et al. (1994) J. Amer. Ceram. Soc. **77** 839, Fig. 1 original by permission of The American Ceramic Society.

Figure 8.13: Original by permission of NSA Hitachi Scientific Instruments Ltd.

Figure 8.14A: Courtesy PE Fischione, reproduced by permission of EA Fichione Instruments Inc.

Figure 8.14B: Courtesy PE Fischione, reproduced by permission of EA Fichione Instruments Inc.

Figure 8.15A,B: Courtesy NJ Zaluzec.

Chapter 9

Figure 9.5: Courtesy of J Rodenburg, modified from original diagram on web site.

Figure 9.6: Modified from Reimer, L (1993) *Transmission Electron Microscopy*, 3rd Edition, Fig. 4.14 Springer Verlag.

Figure 9.10B: Courtesy of M Watanabe, modified from Watanabe, M *et al.* (2006) Microsc. Microanal. **12** 515, Fig. 6

Figure 9.15: Courtesy of R Ristau.

Figure 9.17: Modified from Williams, DB (1987) *Practical Analytical Electron Microscopy in Materials Science*, 2nd Edition, Fig. 1.7 original reproduced by permission of Philips Electron Optics.

Figure 9.19B-D: Courtesy of DW Ackland.

Figure 9.20: Modified from Edington, JW (1976) Practical Electron Microscopy in Materials Science, Fig. 1.5 original reproduced by permission of Philips Electron Optics.

Figure 9.21: Courtesy of S Ramamurthy.

Figure 9.22: Courtesy of DW Ackland.

Figure 9.24: Courtesy of DW Ackland.

Figure 9.25: Courtesy of DW Ackland.

Figure 9.26: Courtesy of S Ramamurthy.

Table 9.1: From Williams, DB (1987) *Practical Analytical Electron Microscopy in Materials Science*, 2nd Edition, Table 2.4 reproduced by permission of Philips Electron Optics.

Table 9.2: From Williams, DB (1987) *Practical Analytical Electron Microscopy in Materials Science*, 2nd Edition, Table 2.2 reproduced by permission of Philips Electron Optics.

Chapter 10

Figure 10.1: Modified from Médard, L *et al.* (1949) Rev. Met. **46** 549, Fig. 5.

Figure 10.3: Reproduced by permission of SPI Inc.

Figure 10.4: Reproduced by permission of South Bay Technology.

Figure 10.5A: Reproduced by permission of Electron Microscopy Sciences.

Figure 10.5B: Reproduced by permission of VCR Inc.

Figure 10.8A: Modified from Thompson-Russell, KC and Edington, JW (1977) *Electron Microscope Specimen Preparation Techniques in Materials Science*, Fig. 9 original reproduced by permission of Philips Electron Optics.

Fig. 10.8B: Modified from Thompson-Russell, KC and Edington, JW (1977) *Electron Microscope Specimen Preparation Techniques in Materials Science*, Fig. 7 original reproduced by permission of Philips Electron Optics.

Figure 10.9A: Modified from Thompson-Russell, KC and Edington, JW (1977) *Electron Microscope Specimen Preparation Techniques in Materials Science*, Fig. 12 original reproduced by permission of Philips Electron Optics.

Figure 10.9B: Courtesy PE Fischione, reproduced by permission of EA Fichione Instruments Inc.

Figure 10.10: Modified from Thompson-Russell, KC and Edington, JW (1977) *Electron Microscope Specimen Preparation Techniques in Materials Science*, Fig. 11 Philips Electron Optics.

Figure 10.11: Courtesy of R Alani, reproduced by permission of Gatan Inc.

Figure 10.12: Courtesy of AG Cullis, from Cullis, AG *et al.* (1985) Ultramicrosc. **17** 203, Figs. 1A, 3 reproduced by permission of Elsevier Science BV.

Figure 10.13: Modified from van Hellemont, J *et al.* (1988) in J Bravman *et al.* Eds. *Specimen Preparation for Transmission Electron Microscopy of Materials* Mater. Res. Soc. Symp. Proc. **115** 247, Fig. 1 original by permission of MRS.

Figure 10.16A,B: Modified from Thompson-Russell, KC and Edington, JW (1977) *Electron Microscope Specimen Preparation Techniques in Materials Science*, Figs. 20, 21 original reproduced by permission of Philips Electron Optics.

Figure 10.17A: Modified from Thompson-Russell, KC and Edington, JW (1977) *Electron Microscope Specimen Preparation Techniques in Materials Science*, Fig. 25 original reproduced by permission of Philips Electron Optics.

Figure 10.17B: Courtesy of M Aindow.

Figure 10.19A-F: Courtesy of SD Walck.

Figure 10.20: Modified from Hetherington, CJD (1988) in J Bravman *et al.* Eds. *Specimen Preparation for Transmission Electron Microscopy of Materials* Mater. Res. Soc. Symp. Proc. **115** 143, Fig. 1 original reproduced by permission of MRS.

Figure 10.21: Modified from Dobisz, EA et al. (1986) J. Vac. Sci. Technol. B 4 850, Fig. 1 original reproduced by permission of MRS.

Figure 10.22A,B: After Fernandez, A (1988) in J Bravman *et al.* Eds. *Specimen Preparation for Transmission Electron Microscopy of Materials*. Mater. Res. Soc. Symp. Proc. **115** 119, Fig. 1.

Figure 10.22C,D: Courtesy of J Basu.

Figure 10.23: Reproduced by permission of FEI Inc.

Figure 10.24A-F: Courtesy of L Giannuzzi.

Figure 10.25A,B: Thanks to JR Michael.

Figure 10.26: Modified from Goodhew, PJ (1988) in J Bravman *et al.* Eds. *Specimen Preparation for Transmission Electron Microscopy of Materials*, Mater. Res. Soc. Symp. Proc. **115** 52.

Table 10.1: Courtesy of T Malis.

Chapter 11

Table 11.1: Modified from Hirsch, PB *et al.* (1977) *Electron Microscopy of Thin Crystals,* 2nd Edition p 19, Krieger, NY.

Chapter 13

Table 13.2: Modified from Reimer, L (1993) *Transmission Electron Microscopy,* 3rd Edition Table 7.2 p 296 Springer Verlag.

Chapter 14

Figure 14.2: Modified from Hashimoto, H *et al.* (1962) Proc. Roy. Soc. (London) **A269** 80, Fig. 2.
Table 14.2: Modified from Reimer, L (1993) *Transmission Electron Microscopy,* 3rd Edition Table 3.2 p 58, Springer Verlag.

Chapter 16

Figure 16.5: Courtesy of ML Jenkins, from Jenkins, ML *et al.* (1976) Philos. Mag. **34** 1141, Fig. 2 reproduced by permission of Taylor and Francis.
Figure 16.6: Courtesy of BC De Cooman.
Figure 16.7: From Dodsworth, J *et al.* (1983) Adv. Ceram. **6** 102, Fig. 3 reproduced by permission of the American Ceramic Society.
Figure 16.8: Courtesy of BC De Cooman.
Figure 16.9: Courtesy of M Gajdardziska-Josifovska, from Gajdardziska-Josifovska M *et al.* (1995) Ultramicrosc. **58** 65, Fig. 1 reproduced by permission of Elsevier Science BV.
Figure 16.10: Courtesy of S McKernan.
Figure 16.11: Modified from Hahn, T (Ed.) *International Tables for Crystallography A* pp 538–539, No. 164 original by permission of The International Union of Crystallography.
Table 16.1: Modified from Edington, JW (1976) *Practical Electron Microscopy in Materials Science,* Appendix 8 Van Nostrand Reinhold.

Chapter 17

Figure 17.2: Modified from Edington, JW (1976) *Practical Electron Microscopy in Materials Science,* Fig. 2.16 original reproduced by permission of Philips Electron Optics.
Figure 17.7: From Carter, CB *et al.* (1981) Philos. Mag. **A43** 441, Fig. 5C reproduced by permission of Taylor and Francis.
Figure 17.9: Modified from Hirsch, PB *et al.* (1977) *Electron Microscopy of Thin Crystals,* 2nd Edition, Fig. 4.11, Krieger.
Figure 17.10: From Driver, JH *et al.* (1972) Phil Mag. **26** 1227, Fig. 3 reproduced by permission of Taylor and Francis.
Figure 17.11A-C: From Lewis, MH and Billingham, J (1972) JEOL News 10e(l) 8, Fig. 3 reproduced by permission of JEOL USA Inc.
Figure 17.11D: Modified from Sauvage, M and Parthè, E (1972) Acta Cryst. **A28** 607, Fig. 2.
Figure 17.12: Modified from Carter, CB *et al.* (1981) Philos. Mag. **A43** 441, Fig. 5A,B.
Figure 17.13: Modified from Carter, CB *et al.* (1980) J. Electron Microsc. **63** 623, Fig. 8.
Figure 17.14: Modified from Carter, CB (1984) Philos. Mag. **A50** 133, Figs. 1–3.

Chapter 18

Figure 18.2: Modified from Edington, JW (1976) *Practical Electron Microscopy in Materials Science,* Fig. A1.7 original reproduced by permission of Philips Electron Optics.
Figure 18.7: Courtesy of S Ramamurthy.
Figure 18.9: Courtesy of S McKernan.
Figure 18.10A,C: Courtesy of S McKernan.
Figure 18.10B,D: Modified from Vainshtein, BK *et al.* (1992) in JM Cowley Ed. *Electron Diffraction Techniques* **1**, Fig. 6.13 original reproduced by permission of Oxford University Press.
Figure 18.10E: From Vainshtein, BK *et al.* (1992) in JM Cowley Ed. *Electron Diffraction Techniques* **1**, Fig. 6.13 reproduced by permission of Oxford University Press.
Figure 18.11: Modified from James, RW (1965) in L Bragg Ed. *The Optical Principles of the Diffraction of X-rays* The Crystalline State **II**, Figs. 170, 184 Cornell University Press.
Figure 18.12: Courtesy of DJH Cockayne, modified from Sproul, A *et al.* (1986) Philos. Mag. **B54** 113, Fig. 1 original by permission of Taylor and Francis.
Figure 18.13: From Graczyk, JF and Chaudhari, P (1973) Phys. stat. sol. b **58** 163, Fig. l0A reproduced by permission of Akademie Verlag GmbH.
Figure 18.14: Courtesy of A Howie, from Howie, A (1988) in PR Buseck *et al.* Eds. *High-Resolution Transmission Microscopy and Associated Techniques* p 60, Fig. 14.12 reproduced by permission of Oxford University Press.
Figure 18.15: Courtesy of LD Marks and CS Own, modified from Own, CS and Marks, LD (2005) Rev. Sci. Instrum. **76** 033703, Fig. 1.
Figure 18.16: Courtesy of J-P Morniroli.

Figure 18.17: From Tietz, LA *et al.* (1995) Ultramicrosc. **60** 241, Figs. 2–4 reproduced by permission of Elsevier Science BV.

Figure 18.18: Modified from Tietz, LA *et al.* (1995) Ultramicrosc. **60** 241, Fig. 5 original by permission of Elsevier Science BV.

Figure 18.19: Modified from Andrews, KW *et al.* (1971) *Interpretation of Electron Diffraction Patterns*, 2nd Edition, Fig. 41 original reproduced by permission of Plenum Press.

Figure 18.20: Modified from Andrews, KW *et al.* (1971) *Interpretation of Electron Diffraction Patterns*, 2nd Edition, Fig. 41 original reproduced by permission of Plenum Press.

Figure 18.21: Modified from Andrews, KW *et al.* (1971) *Interpretation of Electron Diffraction Patterns*, 2nd Edition, Fig. 41 original reproduced by permission of Plenum Press.

Figure 18.22: Modified from Edington, JW (1976) *Practical Electron Microscopy in Materials Science*, Fig. 2.20 original reproduced by permission of Philips Electron Optics.

Figure 18.24: Modified from Li, C and Williams, DB (2003) Interface Science **11** 461–472, Figs. 2A, 4. Courtesy of C Li.

Chapter 19

Figure 19.6A: Courtesy of G. Thomas, modified from Levine, E *et al.* (1966) J. Appl. Phys. **37** 2141, Fig. 1A original reproduced by permission of the American Institute of Physics.

Figure 19.7: Modified from Okamoto, PR *et al.* (1967) J. Appl. Phys. **38** 289, Fig. 5.

Figure 19.8: Courtesy of S Ramamurthy.

Figure 19.9A: Courtesy of G. Thomas, modified from Thomas, G and Goringe, MJ (1979) *Transmission Electron Microscopy of Metals*, Fig. 2.30 John Wiley & Sons Inc.

Figure 19.9B: Modified from Edington. JW (1976) *Practical Electron Microscopy in Materials Science*, Fig. 2.27 Van Nostrand Reinhold.

Figure 19.11: Modified from Thomas, G and Goringe, MJ (1979) *Transmission Electron Microscopy of Metals*, Fig. 2.29 John Wiley & Sons Inc. Thanks to G Thomas.

Chapter 20

Figure 20.2A: Courtesy of KS Vecchio, from Williams, DB *et al.* Eds. (1992) *Images of Materials*, Fig. 6.5 reproduced by permission of Oxford University Press.

Figure 20.2B: Courtesy of KS Vecchio, from Williams, DB *et al.* Eds. (1992) *Images of Materials*, Fig. 6.17 reproduced by permission of Oxford University Press,

Figure 20.3: Modified from Williams, DB (1987) *Practical Analytical Electron Microscopy in Materials Science*, 2nd Edition, Fig. 6.6 Philips Electron Optics.

Figure 20.5: Courtesy of JF Mansfield, from Mansfield, JF (1984) *Convergent Beam Diffraction of Alloy Phases*, Fig. 5.3 reproduced by permission of Institute of Physics Publishing.

Figure 20.6: From Lyman, CE *et al.* Eds. (1990) *Scanning Electron Microscopy, X-ray Microanalysis and Analytical Electron Microscopy—A Laboratory Workbook*, Fig. A 27.2 reproduced by permission of Plenum Press.

Figure 20.7A,B: Courtesy of J-P Morniroli, from Morniroli, J-P (2002) *Large-Angle Convergent-Beam Electron Diffraction*, Figs. V.8, V.12B (Thanks to SF Paris).

Figure 20.8A: Courtesy of J-P Morniroli, modified from Morniroli, J-P (2002) *Large-Angle Convergent-Beam Electron Diffraction*, Fig. VI.I.

Figure 20.8B-D: Courtesy of J-P Morniroli, from Morniroli, J-P (2002) *Large-Angle Convergent-Beam Electron Diffraction*, Fig. VI.2A-C (Thanks to SF Paris).

Figure 20.9: Courtesy of KS Vecchio, from Williams, DB *et al.* Eds. (1992) *Images of Materials*, Fig. 6.21 reproduced by permission of Oxford University Press.

Figure 20.10: Courtesy of JA Hunt, reproduced by permission of Gatan Inc.

Figure 20.11D: Courtesy of R Ayer.

Figure 20.12: Modified from Ayer, R (1989) J. Electron Microscopy Tech. **13** 3, Fig. 3.

Figure 20.13: Courtesy of WAT Clark from Heilman, P *et al.* (1983) Acta Metall. **31** 1293, Fig. 4 reproduced by permission of Elsevier Science BV.

Figure 20.14: Modified from Williams, DB (1987) *Practical Analytical Electron Microscopy in Materials Science*, 2nd Edition, Fig. 6.9 original by permission of Philips Electron Optics.

Figure 20.15: Modified from Williams, DB (1987) *Practical Analytical Electron Microscopy in Materials Science*, 2nd Edition, Fig. 6.16 original by permission of Philips Electron Optics.

Figure 20.16: Courtesy of CM Sung.

Figure 20.17: Courtesy of M Terauchi.

Figure 20.18: Courtesy of CS Own and LD Marks.

Chapter 21

Figure 21.1: Courtesy of ZL Wang, after Wang, ZL *et al.* (2003) Phys. Rev. Lett. **91** 185502, Fig. 2.

Figure 21.2: Modified from Williams, DB (1987) *Practical Analytical Electron Microscopy in Materials*

Science, 2nd Edition, Fig. 6.13 original by permission of Philips Electron Optics.

Figure 21.3: Modified from Williams, DB 1987 *Practical Analytical Electron Microscopy in Materials Science,* 2nd Edition, Fig. 6.14 original by permission of Philips Electron Optics.

Figure 21.4: Courtesy of JM Zuo, simulation from WebEMAPS.

Figure 21.5: Courtesy of B Ralph, modified from Williams, DB (1987) *Practical Analytical Electron Microscopy in Materials Science,* 2nd Edition, Fig. 6.18 original by permission of Philips Electron Optics.

Figure 21.6: From Williams, DB (1987) *Practical Analytical Electron Microscopy in Materials Science,* 2nd Edition, Fig. 4.29A reproduced by permission of Philips Electron Optics.

Figure 21.8: Modified from Williams, DB (1987) *Practical Analytical Electron Microscopy in Materials Science,* 2nd Edition, Fig. 4.29B,C original by permission of Philips Electron Optics.

Figure 21.9: Courtesy of R Ayer, from Raghavan, M et al. (1984) Metall. Trans. **15A** 783, Fig. 6 reproduced by permission of ASM International.

Figure 21.10A: Courtesy of KS Vecchio, modified from Williams, DB et al. Eds. (1992) *Images of Materials,* Fig. 6.23 original by permission of Oxford University Press.

Figure 21.10B: Courtesy of R Ayer, modified from Ayer, R (1989) J. Electron Microsc. Tech. **13** 3, Fig. 7 original by permission of John Wiley & Sons Inc.

Figure 21.12: Courtesy of KS Vecchio, modified from Williams. DB et al. Eds. (1992) *Images of Materials,* Fig. 6.19 original by permission of Oxford University Press.

Figure 21.13: Courtesy of JM Zuo, modified from Kim, M et al. (2004) Appl. Phys. Lett. 84 2181, Fig. 1.

Figure 21.14: Modified from Johnson, A (2007) Acta Cryst. **B63** 511, Fig.7 reproduced by permission of The International Union of Crystallography. Courtesy A Johnson.

Figure 21.15: Modified from Zuo, JM et al. (1999) Nature **401** 49, Fig. 3A reproduced by permission of Macmillan Magazines Ltd. Courtesy JCH Spence.

Figure 21.16: Courtesy of R McConville, from Williams, DB et al. Eds. (1992) *Images of Materials,* Fig. 6.33 reproduced by permission of Oxford University Press.

Figure 21.17: Courtesy of JM Cowley, from Liu, M and Cowley, JM (1994) Ultramicrosc. **53** 333, Figs. 1, 2 reproduced by permission of Elsevier Science BV.

Table 21.1: Data from Williams, DB (1987) *Practical Analytical Electron Microscopy in Materials Science,* 2nd Edition p 79, reproduced by permission of Philips Electron Optics.

Table 21.2: Data from Williams, DB (1987) *Practical Analytical Electron Microscopy in Materials Science,* 2nd Edition p 79, reproduced by permission of Philips Electron Optics.

Chapter 22

Figure 22.5: Courtesy of KA Repa.

Figure 22.6: From Williams, DB (1987) *Practical Analytical Electron Microscopy in Materials Science,* 2nd Edition, Fig. 3.7D reproduced by permission of Philips Electron Optics.

Figure 22.7: Courtesy of KB Reuter.

Figure 22.8: From Williams, DB (1987) *Practical Analytical Electron Microscopy in Materials Science,* 2nd Edition, Fig. 3.7C reproduced by permission of Philips Electron Optics.

Figure 22.9A,B: Courtesy of HL Tsai, from Williams, DB (1987) *Practical Analytical Electron Microscopy in Materials Science,* 2nd Edition, Fig. 1.19A, B reproduced by permission of Philips Electron Optics.

Figure 22.9C: Courtesy of K-R Peters.

Figure 22.10: Modified from Williams, DB (1983) in Krakow, W et al. (Eds.) *Electron Microscopy of Materials* Mater. Res. Soc. Symp. Proc. **31** 11, Fig. 3A,B.

Figure 22.11A,B: Courtesy of IM Watt, from Watt, I (1996) *The Principles and Practice of Electron Microscopy,* 2nd Edition, Fig. 5.5A,B reproduced by permission of Cambridge University Press.

Figure 22.12: Courtesy of MMJ Treacy, from Williams, DB (1987) *Practical Analytical Electron Microscopy in Materials Science,* 2nd Edition, Fig. 5.26B reproduced by permission of Philips Electron Optics.

Figure 22.14: Courtesy of SJ Pennycook, from Pennycook, SJ et al. (1986) J. Microsc. **144** 229, Fig. 8 reproduced by permission of the Royal Microscopical Society.

Figure 22.15A,B: Courtesy of SJ Pennycook, from Lyman, CE (1992) *Microscopy: The Key Research Tool,* special publication of the EMSA Bulletin **22** 7, Fig. 7 reproduced by permission of MSA.

Figure 22.15C: Courtesy of SJ Pennycook, from Browning, ND et al. (1995) Interface Science **2** 397, Fig. 4D reproduced by permission of Kluwer.

Figure 22.16A: From Edington, JW (1976) *Practical Electron Microscopy in Materials Science,* Fig. 2.34 reproduced by permission of Philips Electron Optics.

Figure 22.17: Courtesy of D Cohen.

Chapter 23

Figure 23.3A: From Izui, KJ et al. (1977) J. Electron Microsc. **26** 129, Fig. 1 reproduced by permission of the Japanese Society of Electron Microscopy.

Figure 23.3C: Courtesy of JCH Spence, from Spence, JCH *Experimental High-Resolution Electron Microscopy,* Fig. 5.15 reproduced by permission of Oxford University Press.

Figure 23.4B: Courtesy of JL Hutchison, from Hutchison, JL et al. (1991) in J Heydenreich and W Neumann Eds. *High-Resolution Electron Microscopy—Fundamentals and Applications* p 205, Fig. 3 reproduced by permission of Halle/Saale.
Figure 23.4C: Courtesy of S McKernan.
Figure 23.4D: From Carter, CB et al. (1989) Philos. Mag. **A63** 279, Fig. 3 reproduced by permission of Taylor and Francis.
Figure 23.8: From Tietz, LA et al. (1992) Philos. Mag. **A65** 439, Figs. 3A, 12A,C reproduced by permission of Taylor and Francis.
Figure 23.10: Courtesy of J Zhu.
Figure 23.12: Modified from Vincent, R (1969) Philos. Mag. **19** 1127, Fig. 4.
Figure 23.13: Modified from Norton, MG and Carter, CB (1995) J. Mater. Sci. 30, Fig. 6.
Figure 23.14: Courtesy of U Dahmen, from Hetherington, CJD and Dahmen, U (1992) in PW Hawkes Ed. Signal and Image Processing in Microscopy and Micro-analysis *Scanning Microscopy* Supplement **6** 405, Fig. 9 reproduced by permission of Scanning Microscopy International.
Figure 23.15: From Heidenreich, RD (1964) *Fundamentals of Transmission Electron Microscopy*, Figs. 5.4, 5.6 reproduced by permission of John Wiley & Sons Inc.
Figure 23.16A: Modified from Heidenreich, RD (1964) *Fundamentals of Transmission Electron Microscopy*, Fig. 11.2 original by permission of John Wiley & Sons Inc.
Figure 23.16B: From Boersch, H et al. (1962) Z. Phys. **167** 72, Fig. 4 reproduced by permission of Springer Verlag.
Figure 23.17: Courtesy of M Rühle.
Figure 23.18: Modified from Kouh, YM et al. (1986) J. Mater. Sci. **21** 2689, Fig, 9.
Figure 23.19: Courtesy of M Rühle, from Rühle, M and Sass, SL (1984) Philos. Mag. **A49** 759, Fig. 2 reproduced by permission of Taylor and Francis.
Figure 23.20B,E: From Carter, CB et al. (1986) Philos. Mag. **A55** 21, Fig. 11 reproduced by permission of Taylor and Francis.

Chapter 24

Figure 24.1: Courtesy of S Ramamurthy.
Figure 24.2: Redrawn after Edington, JW (1976) *Practical Electron Microscopy in Materials Science*, Fig. 3.2A.
Figure 24.3B: Courtesy of D Cohen.
Figure 24.3C: Courtesy of S King when not busy founding Cricinfo.
Figure 24.5: Courtesy of D Susnitzky.
Figure 24.7: Redrawn after Edington, JW (1976) *Practical Electron Microscopy in Materials Science*, Fig. 3.3.
Figure 24.8: Redrawn after Edington, JW (1976) *Practical Electron Microscopy in Materials Science*, Figs. 3.4B,D. Images reproduced by permission of Philips Electron Optics.
Figure 24.9: Courtesy of S Ramamurthy.
Figure 24.10: From Hashimoto, H et al. (1962) Proc. Roy. Soc. (London) **A269** 80, Fig. 11 reproduced by permission of The Royal Society.
Figure 24.11A: Courtesy of NSA Hitachi Scientific Instruments Ltd.
Figure 24.11B,C: Courtesy of D Cohen.
Figure 24.12: From Edington, JW (1976) *Practical Electron Microscopy in Materials Science*, Fig. 3.3D reproduced by permission of Philips Electron Optics.
Figure 24.13B,C: From De Cooman, BC et al. (1987) in JD Dow and IK Schuller Eds. *Interfaces, Superlattices, and Thin Films*, Mater. Res. Soc. Symp. Proc. **77** 187, Fig. 1 reproduced by permission of MRS.

Chapter 25

Figure 25.4A-D: Courtesy of D Cohen.
Figure 25.4E,F: Modified from Gevers, R et al. (1963) Phys. stat. sol. **3** 1563, Table 3.
Figure 25.5: From Föll, H et al. (1980) Phys. stat. sol. (a) **58** 393, Fig. 6A,C reproduced by permission of Akademie Verlag GmbH.
Figure 25.7A,B: From Lewis, MH (1966) Philos. Mag. **14** 1003, Fig. 9 reproduced by permission of Taylor and Francis.
Figure 25.7C,D: Courtesy of S Amelinckx, from Amelinckx, S and Van Landuyt, J (1978) in S Amelinckx et al. Eds. *Diffraction and Imaging Techniques in Material Science* **I** 107, Figs. 3, 18 North-Holland.
Figure 25.8: From Rasmussen, DR et al. (1991) Phys. Rev. Lett. **66** (20) 262, Fig. 2 reproduced by permission of The American Physical Society.
Figure 25.9: Courtesy of S Summerfelt.
Figure 25.13: Modified from Metherell, AJ (1975) in U Valdrè and E Ruedl Eds. *Electron Microscopy in Materials Science* **II** 397, Fig. 13 Commission of the European Communities.
Figure 25.14: From Hashimoto, H et al. (1962) Proc. Roy. Soc. (London) **A269** 80, Fig. 15 original by permission of The Royal Society.
Figure 25.16: Modified from Rasmussen, R et al. (1991) Philos. Mag. **63** 1299, Fig. 4.

Chapter 26

Figure 26.2B: Modified from Amelinckx, S (1964) Solid State Physics Suppl. **6**, Fig. 76.
Figure 26.6A-C: Modified from Carter, CB (1980) Phys. stat. sol. (a) **62** 139, Fig. 4.

Figure 26.6F: From Van Landuyt, J *et al.* (1970) Phys. stat. sol. **41** 271, Fig. 1 reproduced by permission of Akademie Verlag GmbH.

Figure 26.6G-H: Courtesy of BC De Cooman.

Figure 26.7: Modified from Hirsch, PB *et al.* (1977) *Electron Microscopy of Thin Crystals,* 2nd Edition, Fig. 7.8 Krieger.

Figure 26.8: From Delavignette, P and Amelinckx, S (1962) J. Nucl. Mater. **5** 17, Fig. 7 reproduced by permission of Elsevier Science BV.

Figure 26.10: From Urban, K (1971) in S Koda Ed. *The World Through the Electron Microscope* Metallurgy **V** 26, reproduced by permission of JEOL USA Inc.

Figure 26.11: Courtesy of A Howie, from Howie, A and Whelan, MJ (1962) Proc. Roy. Soc. (London) **A267** 206, Fig. 14 reproduced by permission of The Royal Society.

Figure 26.12: Modified from M Wilkens (1978) in S Amelinckx *et al.* Eds. *Diffraction and Imaging Techniques in Material Science* **I** 185, Fig. 4 North-Holland.

Figure 26.14: From Dupouy, G and Perrier, F (1971) in S Koda Ed. *The World Through the Electron Microscope* Metallurgy **V** 100, reproduced by permission of JEOL USA Inc.

Figure 26.15A: From Modeer, B and Lagneborg, R (1971) in S Koda Ed. *The World Through the Electron Microscope* Metallurgy **V** 44, reproduced by permission of JEOL USA Inc.

Figure 26.15B: Courtesy of DA Hughes, from Hansen, N and Hughes, DA (1995) Phys. stat. sol. (a) **149** 155, Fig. 5 reproduced by permission of Akademie Verlag GmbH.

Figure 26.16A: From Siems, F *et al.* (1962) Phys. stat. sol. **2** 421, Fig. 5A reproduced by permission of Akademie Verlag GmbH.

Figure 26.16C: From Siems, F *et al.* (1962) Phys. stat. sol. **2** 421, Fig. 15A reproduced by permission of Akademie Verlag GmbH.

Figure 26.17A: Modified from Whelan, MJ (1958–1959) J. Inst. Met. **87** 392, Fig. 25A.

Figure 26.17B: Courtesy of K Ostyn.

Figure 26.18: From Takayanagi, L (1988) Surface Science **205** 637, Fig. 5 reproduced by permission of Elsevier Science BV.

Figure 26.19A: From Tunstall, WJ *et al.* (1964) Philos. Mag. **9** 99, Fig. 9 reproduced by permission of Taylor and Francis.

Figure 26.19B: From Amelinckx, S in PG Merli and VM Anti-sari Eds. *Electron Microscopy in Materials Science* p 128, Fig. 45 reproduced by permission of World Scientific.

Figure 26.20: Courtesy of W Skrotski.

Figure 26.21: Courtesy of W Skrotski.

Figure 26.22: From Carter, CB *et al.* (1986) Philos. Mag. **A55** 21, Fig. 2 reproduced by permission of Taylor and Francis.

Figure 26.23: From Carter, CB *et al.* (1981) Philos. Mag. **A43** 441, Fig. 3 reproduced by permission of Taylor and Francis.

Figure 26.24: Courtesy of K Ostyn.

Figure 26.25: Courtesy of L Tietz.

Figure 26.26A: Courtesy of LM Brown, from Ashby, MF and Brown, LM (1963) Philos. Mag. **8** 1083, Fig. 10 reproduced by permission of Taylor and Francis.

Figure 26.26B: Modified from Whelan, MJ (1978) in S Amelinckx *et al.* Eds. *Diffraction and Imaging Techniques in Material Science* **I** 43, Fig. 36 North-Holland.

Figure 26.26C: Courtesy of LM Brown, from Ashby, MF and Brown, LM (1963) Philos. Mag. **8** 1083, Fig. 12 reproduced by permission of Taylor and Francis.

Figure 26.27: From Rasmussen, DR and Carter, CB (1991) J. Electron Microsc. Technique **18** 429, Fig. 2 reproduced by permission of John Wiley & Sons Inc.

Chapter 27

Figure 27.7: Courtesy of S King.

Figure 27.10: Courtesy of DJH Cockayne, from Cockayne, DJH (1972) Z. Naturforschung **27a** 452, Fig. 6 original by permission of Verlag der Zeitschrift für Naturforschung, Tübingen.

Figure 27.13: Modified from Carter, CB *et al.* (1986) Philos. Mag. **A55** 1, Fig. 9.

Figure 27.15: Modified from Föll, H *et al.* (1980) Phys. stat. sol. (a) **58** 393, Fig. 6B,C.

Figure 27.17: From Heidenreich. RD (1964) *Fundamentals of Transmission Electron Microscopy*, Fig. 9.20 reproduced by permission of John Wiley & Sons Inc.

Figure 27.18: Courtesy of DJH Cockayne, from Ray, ILF and Cockayne, DJH (1971) Proc. Roy. Soc. (London) **A325** 543, Fig. 10 reproduced by permission of The Royal Society.

Figure 27.23: Modified from Carter, CB (1979) J. Phys. (A) **54** (1) 395, Fig. 8A.

Chapter 28

Figure 28.4: Courtesy of R Gronsky, from Gronsky, R (1992) in DB Williams *et al.* Eds. *Images of Materials*, Fig. 7.6 original by permission of Oxford University Press.

Figure 28.5: Courtesy of S McKernan.

Figure 28.6: Courtesy of S McKernan.

Figure 28.7: Modified from Cowley, JM (1988) in PR Buseck *et al.* Eds. *High-Resolution Electron Microscopy and Associated Techniques*, Fig. 1.9 Oxford University Press.

Figure 28.8: Courtesy of JCH Spence, from Spence, JCH (1988) *Experimental High-Resolution Electron*

Microscopy, 2nd Edition, Fig. 4.3 original by permission of Oxford University Press.

Figure 28.10: From de Jong, AF and Van Dyck, D (1993) Ultramicrosc. **49** 66, Fig. 1 original by permission of Elsevier Science BV.

Figure 28.11: Courtesy of MT Otten, from Otten, MT and Coene, WMJ (1993) Ultramicrosc. **48** 77, Fig. 8 reproduced by permission of Elsevier Science BV.

Figure 28.12: Courtesy of MT Otten, from Otten, MT and Coene, WMJ (1993) Ultramicrosc. **48** 77, Fig. 11 reproduced by permission of Elsevier Science BV.

Figure 28.13: Courtesy of MT Otten, from Otten, MT and Coene, WMJ (1993) Ultramicrosc. **8** 77, Fig. 10 reproduced by permission of Elsevier Science BV.

Figure 28.14A,B: From Amelinckx, S *et al.* (1993) Ultramicrosc. **51** 90, Fig. 2 original by permission of Elsevier Science BV.

Figure 28.15: From Amelinckx, S *et al.* (1993) Ultramicrosc. **51** 90, Fig. 3 reproduced by permission of Elsevier Science BV.

Figure 28.16: From Rasmussen, DR *et al.* (1995) J. Microsc. **179** 77, Fig. 2C,D reproduced by permission of the Royal Microscopical Society.

Figure 28.18A: Courtesy of S McKernan.

Figure 28.18B: From Berger, A *et al.* (1994) Ultramicrosc. **55** 101, Fig. 4B reproduced by permission of Elsevier Science BV.

Figure 28.18C: Courtesy of S Summerfelt.

Figure 28.18D: Courtesy of S McKernan.

Figure 28.19: Courtesy of DJ Smith.

Figure 28.21A: From Van Landuyt, J *et al.* (1991) in J Heydenreich and W Neumann Eds. *High-Resolution Electron Microscopy—Fundamentals and Applications,* p 254, Fig. 6 reproduced by permission of Halle/Saale.

Figure 28.21D: From Van Landuyt, J *et al.* (1991) in J Heydenreich and W Neumann Eds. *High-Resolution Electron Microscopy—Fundamentals and Applications,* p 254, Fig. 8 reproduced by permission of Halle/Saale.

Figure 28.22: From Nissen, H-U and Beeli, C (1991) in J Heydenreich and W Neumann Eds. *High-Resolution Electron Microscopy—Fundamentals and Applications,* p 272, Fig. 4 reproduced by permission of Halle/Saale.

Figure 28.23: From Nissen, H-U and Beeli, C (1991) in J Heydenreich and W Neumann Eds. *High-Resolution Electron Microscopy—Fundamentals and Applications,* p 272, Fig. 2 reproduced by permission of Halle/Saale.

Figure 28.24: From Parsons, JR *et al.* (1973) Philos. Mag. **29** 1359, Fig. 2 reproduced by permission of Taylor and Francis.

Table 28.1: Modified from de Jong, AF and Van Dyck, D (1993) Ultramicrosc. **49** 66, Table 1.

Chapter 29

Figure 29.2: Courtesy of R Sinclair, from Sinclair, R *et al.* (1981) Met. Trans. 12A, 1503, Figs. 13, 14 reproduced by permission of ASM International.

Figure 29.4A,B: From Marcinkowksi, MJ and Poliak, RM (1963) Philos. Mag. 8, 1023, Fig. 15a,b reproduced by permission of Taylor and Francis.

Figure 29.4C,D: Courtesy of J Silcox, from Silcox, J (1963) Philos. Mag. 8, 7, Fig. 7 reproduced by permission of Taylor and Francis.

Figure 29.5: Courtesy of AJ Craven, from Buggy, TW *et al.* (1981) *Analytical Electron Microscopy-1981,* p 231, Fig. 5 reproduced by permission of San Francisco Press.

Figure 29.6D,E: Courtesy of NSA Hitachi Scientific Instruments Ltd. and S McKernan.

Figure 29.7: Courtesy of R Sinclair.

Figure 29.8: From Kuesters, K-H *et al.* (1985) J. Cryst. Growth **71**, 514, Fig. 4 reproduced by permission of Elsevier Science BV.

Figure 29.9: Courtesy of M. Mallamaci.

Figure 29.10A: Modified from De Cooman, BC *et al.* (1985) J. Electron Microsc. Tech. **2**, 533, Fig. 1.

Figure 29.10B: Courtesy of SM Zemyan.

Figure 29.10C-E: Courtesy of BC De Cooman.

Figure 29.12: Courtesy of G Thomas, from Bell, WL and Thomas, G (1972) in G Thomas *et al.* Eds. *Electron Microscopy and Structure of Materials,* p 53, Fig. 28 reproduced by permission of University of California Press.

Figure 29.13: Courtesy of K-R Peters, from Peters, K-R (1984) in DF Kyser *et al.* Eds. *Electron Beam Interactions with Solids for Microscopy, Microanalysis and Lithography,* p 363, Fig. 1 original by permission of Scanning Microscopy International.

Figure 29.14: Courtesy of R McConville, from Williams, DB (1987) *Practical Analytical Electron Microscopy in Materials Science,* 2nd Edition, Fig. 3.11 reproduced by permission of Philips Electron Optics.

Figure 29.15: Courtesy of Philips Electronic Instruments, from Williams, DB (1987) *Practical Analytical Electron Microscopy in Materials Science,* 2nd Edition, Fig. 3.10 reproduced by permission of Philips Electron Optics.

Figure 29.16: Courtesy of H Lichte, from Lichte, H (1992) Scanning Microscopy, p 433, Fig. 1 reproduced by permission of Scanning Microscopy International.

Figure 29.17: Modified from Lichte, H (1992) Ultramicrosc. 47, 223, Fig. 1.

Figure 29.18: Modified from Tonomura, A. Courtesy of NSA Hitachi Scientific Instruments Ltd.

Figure 29.19A-C: From Tonomura, A (1992) Adv. Phys. 41, 59, Fig. 29 reproduced by permission of Taylor and Francis.

Figure 29.19D: From Tonomura, A (1987) Rev. Mod. Phys. 59, 639, Fig. 41 reproduced by permission of The American Physical Society.

Figure 29.20A: From Tonomura, A (1992) Adv. Phys. 41, 59, Fig. 38, reproduced by permission of Taylor and Francis.

Figure 29.20B: From Tonomura, A (1992) Adv. Phys. 41, 59, Fig. 42 reproduced by permission of Taylor and Francis.

Figure 29.20C: From Tonomura, A (1992) Adv. Phys. 41, 59, Fig. 44 reproduced by permission of Taylor and Francis.

Figure 29.21: Courtesy of R Sinclair, from Sinclair, R *et al.* (1994) Ultramicrosc. 56, 225, Fig. 5 reproduced by permission of Elsevier Science BV.

Figure 29.22: Courtesy of M Treacy.

Chapter 30

Figure 30.2A,B: Courtesy of MA O'Keefe, from O'Keefe, MA and Kilaas, R (1988) in PW Hawkes *et al.* Eds. Image and Signal Processing in Electron Microscopy, Scanning Microscopy Supplement 2 p 225, Fig. 1 original by permission of Scanning Microscopy International.

Figure 30.3: From Kambe, K (1982) Ultramicrosc **10** 223, Fig. 1A-D reproduced by permission of Elsevier Science BV.

Figure 30.4: Courtesy of MA O' Keefe, from O'Keefe, MA and Kilaas, R (1988) in PW Hawkes *et al.* Eds. Image and Signal Processing in Electron Microscopy Scanning Microscopy Supplement 2 p 225, Fig. 4 reproduced by permission of Scanning Microscopy International.

Figure 30.5: Modified from Rasmussen, DR and Carter, CB (1990) Ultramicrosc. **32** 337, Figs. 1, 2.

Figure 30.8: From Beeli, C and Horiuchi, S (1994) Philos. Mag. **B70** 215, Fig. 6A-D reproduced by permission of Taylor and Francis.

Figure 30.9: From Beeli, C and Horiuchi, S (1994) Philos. Mag. **B70** 215, Fig. 7A-D reproduced by permission of Taylor and Francis.

Figure 30.10: From Beeli, C and Horiuchi, S (1994) Philos. Mag. **B70** 215, Fig. 8 reproduced by permission of Taylor and Francis.

Figure 30.11: From Jiang, J *et al.* (1995) Phil. Mag. Lett. **71** 123, Fig. 4 reproduced by permission of Taylor and Francis.

Chapter 31

Figure 31.1: Courtesy of J. Heffelfinger.

Figure 31.2: From Rasmussen, DR *et al.* (1995) J. Microsc. 179, 77, Fig. 1b original by permission of The Royal Microscopical Society.

Figure 31.3: From Rasmussen, DR *et al.* (1995) J. Microsc. 179, 77, Fig. 5 reproduced by permission of The Royal Microscopical Society.

Figure 31.4: Courtesy of OL Krivanek, from Krivanek, OL (1988) in PR Buseck *et al.* Eds. *High-Resolution Electron Microscopy and Associated Techniques,* Fig. 12.6 reproduced by permission of Oxford University Press.

Figure 31.5: Courtesy of OL Krivanek, Krivanek, OL (1988) in PR Buseck *et al.* Eds. *High-Resolution Electron Microscopy and Associated Techniques,* Fig. 12.7 reproduced by permission of Oxford University Press.

Figure 31.6A: Courtesy of JCH Spence, from Spence, JCH and Zuo, JM (1992) *Electron Microdiffraction,* Fig. A1.3 reproduced by permission of Plenum Press.

Figure 31.6B: Courtesy of OL Krivanek, from Krivanek, OL (1988) in PR Buseck *et al.* Eds. *High-Resolution Electron Microscopy and Associated Techniques,* Fig. 12.8 reproduced by permission of Oxford University Press.

Figure 31.7: Courtesy of S McKernan.

Figure 31.8: Courtesy of ZL Wang. Wwang, ZL *et al.* (2007) MRS Bulletin, 109–116. Reproduced by permission of MRS.

Figure 31.9: Courtesy of ZC Lin, from Lin, ZC (1993) Ph.D. dissertation, Fig. 4.15, University of Minnesota.

Figure 31.10: Courtesy of O Saxton, from Kirkland, AI (1992) in PW Hawkes Ed. Signal and Image Processing in Microscopy and Microanalysis, *Scanning Microscopy* Supplement 6, 139, Figs. 1–3 reproduced by permission of Scanning Microscopy International.

Figure 31.11: From Zou, XD and Hovmöller, S (1993) Ultramicrosc. 49, 147, Fig. 1 reproduced by permission of Elsevier Science BV.

Figure 31.12A: From Kirkland, AI *et al.* (1995) Ultramicrosc. 57, 355, Fig. 1 reproduced by permission of Elsevier Science BV.

Figure 31.12B; From Kirkland. AI *et al.* (1995) Ultramicrosc. 57, 355, Fig. 3 reproduced by permission of Elsevier Science BV.

Figure 31.13A-C: From Kirkland, AI *et al.* (1995) Ultramicrosc. 57, 355, Fig. 8 reproduced by permission of Elsevier Science BV.

Figure 31.14: Courtesy of OL Krivanek, from Krivanek, OL and Fan, GY (1992) in PW Hawkes (Ed.) Signal and Image Processing in Microscopy and Microanalysis, Scanning Microscopy Supplement 6, p 105, Fig. 4 reproduced by permission of Scanning Microscopy International.

Figure 31.15: Courtesy of OL Krivanek, Krivanek, OL and Fan, GY (1992) in PW Hawkes (Ed.) Signal and Image Processing in Microscopy and Microanalysis,

Scanning Microscopy Supplement 6, p 105, Fig. 5 reproduced by permission of Scanning Microscopy International.

Figure 31.16: After U Dahmen, from Paciornik, S *et al.* (1996) Ultramicrosc. 62, 15, Fig. 1.

Figure 31.17: Courtesy of U Dahmen, from Paciornik, S *et al.* (1996) Ultramicrosc. 62, 15, Fig. 5 reproduced by permission of Elsevier Science BV.

Figure 31.18: Courtesy of A Ourmazd, from Kisielowski, C *et al.* (1995) Ultramicrosc. 58, 131, Figs. 2–4 reproduced by permission of Elsevier Science BV.

Figure 31.19: Courtesy of A Ourmazd, from Kisielowski, C *et al.* (1995) Ultramicrosc. 58, 131, Figs. 8, 10, 12 reproduced by permission of Elsevier Science BV.

Figure 31.20A,B: Data from Ourmazd, A *et al.* (1990) Ultramicrosc. 34, 237, Fig. 1.

Figure 31.20C,D: From Ourmazd, A *et al.* (1990) Ultramicrosc. 34, 237, Figs. 2, 5 reproduced by permission of Elsevier Science BV. Courtesy of A Ourmazd.

Figure 31.21A-F: From Kisielowski, C *et al.* (1995) Ultramicrosc. 34, 237, Fig. I5 reproduced by permission of Elsevier Science BV. Courtesy of A Ourmazd.

Figure 31.22: Courtesy of U Dahmen, from Paciornik, S *et al.* (1996) *Ultramicroscopy,* in press, Fig. 2 reproduced by permission of Elsevier Science BV.

Figure 31.23: From King, WE and Campbell, GH (1994) Ultramicrosc. 56, 46, Fig. 1 reproduced by permission of Elsevier Science BV.

Figure 31.24: From King, WE and Campbell, GH (1994) Ultramicrosc. 56, 46, Fig. 6 reproduced by permission of Elsevier Science BV.

Figure 31.25: Courtesy of M Rühle, from Möbus, G *et al.* (1993) Ultramicrosc. 49, 46, Fig. 6 reproduced by permission of Elsevier Science BV.

Figure 31.26: From Thon, F (1970) in U Valdrè Ed. *Electron Microscopy in Materials Science,* p 571, Fig. 36 reproduced by permission of Academic Press.

Figure 31.27: Courtesy of J Heffelfinger.

Chapter 32

Figure 32.1: Courtesy of JE Yehoda, from Messier, R and Yehoda, JE (1985) J. Appl. Phys. **58** 3739, Fig. 1 reproduced by permission of the American Institute of Physics.

Figure 32.2: Courtesy of M Watanabe.

Figure 32.3B,C: Courtesy of JH Scott.

Figure 32.4A: Courtesy of JH Scott.

Figure 32.4B: Courtesy of P Statham, reproduced by permission of Oxford Instruments.

Figure 32.5: Courtesy of N Rowlands, reproduced by permission of Oxford Instruments.

Figures 32.6,7: Courtesy of SM Zemyan.

Figure 32.8A-C: Courtesy of JH Scott, modified from figure originally supplied by Photon Detector Technologies.

Figure 32.8D: Courtesy of DE Newbury.

Figure 32.9A-C: Courtesy of M Terauchi.

Figure 32.9D: Courtesy of D Wollman, SW Nam and DE Newbury.

Figure 32.11: Courtesy of SM Zemyan, modified from Zemyan, S and Williams, DB (1995) in DB Williams *et al.* Eds. *X-ray Spectrometry in Electron Beam Instruments,* Fig. 12.9 original by permission of Plenum Press.

Figure 32.12A: Courtesy of SM Zemyan, modified from Zemyan, SM and Williams, DB (1995) in DB Williams *et al.* Eds. *X-ray Spectrometry in Electron Beam Instruments,* Fig. 12.10 original by permission of Plenum Press.

Figure 32.12B: Courtesy of JH Scott, modified from diagram originally supplied by Photon Detector Technologies .

Figure 32.13: Courtesy of JJ Friel, modified from Mott, RB and Friel, JJ (1995) in DB Williams *et al.* Eds. *X-ray Spectrometry in Electron Beam Instruments,* Fig. 9.8 original reproduced by permission of Plenum Press.

Figure 32.14: Modified from Williams, DB (1987) *Practical Analytical Electron Microscopy in Materials Science,* 2nd Edition, Fig. 4.5A original reproduced by permission of Philips Electron Optics.

Figure 32.15: Modified from Williams, DB (1987) *Practical Analytical Electron Microscopy in Materials Science,* 2nd Edition, Fig. 4.30 original reproduced by permission of Philips Electron Optics.

Chapter 33

Figure 33.1: Courtesy of SM Zemyan.

Figure 33.2: Courtesy of SM Zemyan.

Figure 33.3: Courtesy of DE Newbury, from Newbury, DE (1995) in DB Williams *et al.* Eds. *X-ray Spectrometry in Electron Beam Instruments,* Fig. 11.18 reproduced by permission of Plenum Press.

Figure 33.4: Courtesy of SM Zemyan.

Figure 33.5A,B: Courtesy of G Cliff, modified from Cliff, G and Kenway, PB (1982) Microbeam Analysis-1982 p 107, Figs. 5, 4 original reproduced by permission of San Francisco Press.

Figure 33.6: Modified from Williams, DB and Goldstein, JI (1981) in KFJ Heinrich *et al.* Eds. *Energy-Dispersive X-ray Spectrometry* p 346, Fig. 7A NBS.

Figure 33.7: Courtesy of SM Zemyan.

Figure 33.8: Courtesy of SM Zemyan.

Figure 33.9A: Courtesy of SM Zemyan.

Figure 33.9B: Courtesy of KS Vecchio, modified from Vecchio, KS and Williams, DB (1987) J. Microsc. **147** 15, Fig. 1 original by permission of the Royal Microscopical Society.

Figure 33.10A: Courtesy of SM Zemyan.

Figure 33.10B: Courtesy of SM Zemyan, modified from Zemyan, SM and Williams, DB (1994) J. Microsc. **174** 1, Fig. 6.

Figure 33.10C: Courtesy of SM Zemyan, modified from Zemyan, SM and Williams, DB (1995) in DB Williams *et al.* Eds. *X-ray Spectrometry in Electron Beam Instruments*, Fig. 12.7.

Figure 33.11: Courtesy of M Watanabe.

Figure 33.12: Courtesy of M Watanabe.

Figure 33.13A-C: Courtesy of CE Lyman, from Lyman, CE (1992) in CE Lyman *et al.* Eds. *Compositional Imaging in the Electron Microscope: An Overview, Microscopy: The Key Research Tool* p 1, Fig. 2.

Figure 33.14A: Modified frontispiece image by Hunneyball, PD *et al.* (1981) in GW Lorimer *et al.* Eds. *Quantitative Microanalysis with High Spatial Resolution* The Institute of Metals, London.

Figure 33.14B,C: Courtesy of DT Carpenter.

Figure 33.15: Courtesy of M Watanabe.

Chapter 34

Figure 34.1: Courtesy of M Watanabe.

Figure 34.2: Courtesy of SM Zemyan and M Watanabe.

Figure 34.3: Courtesy of SM Zemyan and M Watanabe.

Figure 34.4A-D: Courtesy M Watanabe, from Watanabe, M and Williams, DB (2003) *Microsc. Microanal.* **9** Suppl. 2 p 124, Figs. 1–4 reproduced by permission of the Microscopy Society of America.

Figures 34.5–7: Courtesy of SM Zemyan and M Watanabe.

Figure 34.8: Courtesy of CH Kiely. Full report given by Enache, DI *et al.* (2006) Science **311** 362.

Chapter 35

Figures 35.1–4: Courtesy of SM Zemyan and M Watanabe.

Figure 35.5A: Modified from Williams, DB (1987) *Practical Analytical Electron Microscopy in Materials Science,* 2nd Edition, Fig. 4.20 original by permission of Philips Electron Optics.

Figure 35.5B,C: Courtesy of SM Zemyan.

Figure 35.6: Courtesy of SM Zemyan.

Figure 35.7: Modified from Wood, JE *et al.* (1984) J. Microsc. **133** 255, Fig. 2.8 original by permission of the Royal Microscopical Society.

Figure 35.8: Modified from Bender, BA *et al.* (1980) J. Amer. Ceram. Soc. **63** 149, Fig. 1 original by permission of the American Ceramic Society.

Figure 35.10B: Courtesy of JA Eades, from Christenson, KK and Eades, JA (1986) Proc. 44th EMSA Meeting p 622, Fig. 2 original by permission of the Electron Microscopy Society of America.

Figure 35.11A-C: Courtesy of M Watanabe and MG Burke.

Figure 35.11D: Courtesy of M Watanabe, modified from Watanabe, M *et al.* (2006) Microsc. Microanal. **12** 515, Figs. 10, 11 reproduced by permission of the Microscopy Society of America.

Tables 35.1, 2: From Williams, DB (1987) *Practical Analytical Electron Microscopy in Materials Science,* 2nd Edition, Table 4.2A,B reproduced by permission of Philips Electron Optics.

Table 35.3A,B: From Wood, JE *et al.* (1984) J. Microsc. **133** 255, Tables 9, 11 reproduced by permission of the Royal Microscopical Society.

Table 35.4: Courtesy of M Watanabe.

Chapter 36

Figure 36.1: Courtesy of M Watanabe.

Figure 36.2: Courtesy of M Watanabe.

Figure 36.3: Courtesy of JR Michael, modified from Williams, DB *et al.* (1992) Ultramicrosc. **47** 121, Fig. 1.

Figure 36.4: Courtesy of JR Michael, modified from Williams, DB *et al.* (1992) Ultramicrosc. **47** 121, Fig. 2 original by permission of Elsevier Science BV.

Figure 36.5A-C: Courtesy of M Watanabe, modified from Watanabe, M *et al.* (2006) Microsc. Microanal. **12** Suppl. 2, 1568, Figs. 2–4 reproduced by permission of the Microscopy Society of America.

Figure 36.8A,B: From Williams, DB (1987) *Practical Analytical Electron Microscopy in Materials Science,* 2nd Edition, Fig. 4.27 reproduced by permission of Philips Electron Optics.

Figure 36.9: After Williams, DB *et al.* (2002) J. Electron Microscopy **51** (Suppl), S113, Figure courtesy of M Watanabe.

Figure 36.10: Courtesy of CE Lyman, modified from Lyman, CE (1987) in J Kirschner *et al.* Eds. *Physical Aspects of Microscopic Characterization of Materials* p 123, Fig. 1, original reproduced by permission of Scanning Microscopy International.

Figure 36.11: Courtesy of M Watanabe, modified from Watanabe, M *et al.* (2006) Microsc. Microanal. **12** 515, Fig. 13; in turn modified from Lyman, CE (1987) in J Kirschner *et al.* Eds. *Physical Aspects of Microscopic Characterization of Materials* p 123, Fig. 7.

Figure 36.12: Courtesy of M Watanabe, modified from Watanabe, M *et al.* (2006) Microsc. Microanal. **12** 515, Fig. 5 reproduced by permission of the Microscopy Society of America.

Figure 36.13: Courtesy of M Watanabe.

Chapter 37

Figure 37.1: Courtesy of J Bruley.

Figure 37.2A Courtesy of JA Hunt, original by permission of Gatan Inc.

Figure 37.2B,C: Courtesy of RF Egerton, modified from Egerton, RF (1996) *Electron Energy-Loss Spectroscopy in the Electron Microscope,* 2nd edition,

Fig. 2.2 original reproduced by permission of Plenum Press.
Figure 37.3: Courtesy of K Scudder, reproduced by permission of Gatan Inc.
Figure 37.4A: Courtesy of JA Hunt, reproduced by permission of Gatan Inc.
Figure 37.4B: Courtesy of M Watanabe.
Figure 37.5: Courtesy of JA Hunt, reproduced by permission of Gatan Inc.
Figure 37.9A: Courtesy of JA Hunt, modified from Hunt, JA and Williams, DB (1994) Acta Microsc. **3** 1, Fig. 7 original by permission of the Venezuelan Society for Electron Microscopy.
Figure 37.9B,C: Courtesy of C Colliex.
Figure 37.10: Courtesy of JA Hunt, from Hunt, JA and Williams, DB (1994) Acta Microsc. **3** 1, Fig. 5 reproduced by permission of the Venezuelan Society for Electron Microscopy.
Figure 37.11: Courtesy of JA Hunt, from Hunt, JA and Williams, DB (1994) Acta Microsc. **3** 1, Fig. 4 reproduced by permission of the Venezuelan Society for Electron Microscopy.
Figure 37.12: Courtesy of J Bruley, from Hunt, JA and Williams, DB (1994) Acta Microsc. **3** 1, Fig. 6 reproduced by permission of the Venezuelan Society for Electron Microscopy.
Figure 37.13A: Courtesy of OL Krivanek, modified from Krivanek, OL *et al.* (1991) Microsc. Microanal. Microstruct. **2** 315, Fig. 8.
Figure 37.13B: Courtesy of JA Hunt, reproduced by permission of Gatan Inc.
Figure 37.14: Courtesy of M Watanabe.
Figure 37.15A,B: Courtesy of JA Hunt, reproduced by permission of Gatan Inc.
Figure 37.16A,B: Courtesy of F Hofer.
Figure 37.17A: Courtesy of V Dravid.
Figure 37.17B: Courtesy of C Colliex.
Figure 37.17C,D: Courtesy of JA Hunt, reproduced by permission of Gatan Inc.

Chapter 38

Figure 38.1: Courtesy of J Bruley.
Figures 38.2–4: Courtesy of JA Hunt, reproduced by permission of Gatan Inc.
Figure 38.5A: Courtesy of J Bruley.
Figure 38.5B: Courtesy of JA Hunt, reproduced by permission of Gatan Inc.
Figure 38.6A,B: Courtesy of RH French, modified from van Benthem, K, Elsässer, C and French, RH (2001) J. Appl. Phys. **90** 6156, Figs.1, 7.
Figure 38.7A: Courtesy of J Bruley.
Figure 38.7B: Courtesy of JA Hunt, reproduced by permission of Gatan Inc.
Figure 38.8A,B: From Williams, DB and Edington, JW (1976) Acta Metall. **24** 323, Fig. 7 reproduced by permission of Elsevier Science BV.
Figure 38.8C: Courtesy of AJ Strutt.
Figure 38.9A: Courtesy of JA Hunt.
Figure 38.9B: Courtesy of M Libera, modified from Kim *et al.* (2006) J. Am. Chem. Soc. **128** 6570, Figs.1, 2.
Figure 38.10A,B: Courtesy of JA Hunt.
Figure 38.11: Courtesy of VJ Keast.
Table 38.1: Courtesy of RF Egerton, from Egerton, RF (1996) *Electron Energy-Loss Spectroscopy in the Electron Microscopy,* 2nd edition p 157, Table 3.2.

Chapter 39

Figure 39.1: Courtesy of OL Krivanek, modified from Ahn, CC and Krivanek, OL (1983) *EELS Atlas* p iv, original by permission of Gatan Inc.
Figure 39.2: Courtesy of J Bruley, modified from Joy, DC (1986) in DC Joy *et al.* Eds. *Principles of Analytical Electron Microscopy* p 249, Fig. 8 Plenum Press.
Figure 39.3: Courtesy of J Bruley.
Figure 39.4: Courtesy of CE Lyman, modified from Lyman, CE (1987) in J Kirschner *et al.* Eds. *Physical Aspects of Microscopic Characterization of Materials* p 123, Fig. 2 original by permission of Scanning Microscopy International.
Figure 39.5: Courtesy of DC Joy, modified from Joy, DC (1979) in JJ Hren *et al.* Eds. *Introduction to Analytical Electron Microscopy* p 235, Fig. 7.6 original by permission of Plenum Press.
Figure 39.6: Courtesy of M Kundmann.
Figure 39.7: Courtesy of J Bruley.
Figure 39.8: Courtesy of K Sato and Y Ishiguro, modified from Sato, K and Ishiguro, Y (1996) Materials Transactions, **37** 643, Figs. 1, 7 Japan Institute of Metals.
Figure 39.9: Courtesy of J Bruley.
Figure 39.10: Courtesy of JA Hunt, from Hunt, JA and Williams, DB (1994) Acta Microsc. **3** 1, Fig. 14 original by permission of the Venezuelan Society for Electron Microscopy.
Figure 39.11: Courtesy of JA Hunt, from Williams, DB and Goldstein, JI (1992) Microbeam Analysis **1** 29, Fig. 11 reproduced by permission of VCH.
Figures 39.12,13: Courtesy of JA Hunt, from Hunt, JA and Williams, DB (1994) Acta Microsc. **3** 1, Fig. 17A,B reproduced by permission of the Venezuelan Society for Electron Microscopy.
Figure 39.14: Courtesy of RF Egerton, modified from Egerton, RF (1993) Ultramicrosc. **50** 13, Fig. 6.
Figure 39.15–18: Courtesy of J Bruley.
Figure 39.19A,B: Courtesy of JA Hunt, from Hunt, JA and Williams, DB (1994) Acta Microsc. **3** 1, Fig. 16

reproduced by permission of the Venezuelan Society for Electron Microscopy.

Figure 39.20: Courtesy of F Hofer.

Figure 39.21: Modified from Egerton, RF (1996) *Electron Energy-Loss Spectroscopy in the Electron Microscope,* 2nd edition, Fig. 1.11 original by permission of Plenum Press. Courtesy of RF Egerton.

Figure 39.22: Courtesy of M Varela.

Figure 39.23: Courtesy of M Varela.

Chapter 40

Figure 40.4: Modified from Zaluzec, NJ (1982) Ultramicrosc. **9** 319, Fig. 3. Courtesy of NJ Zaluzec.

Figure 40.5B: Courtesy of J Bruley.

Figure 40.6: Courtesy of PE Batson, from Batson, PE (1993) Nature **366** 727, Fig. 1 reproduced by permission of Macmillan Journals Ltd.

Figures 40.7A,B: Courtesy of VJ Keast, modified from Keast, VJ *et al.* (1998) Acta Mater. **46** 481.

Figure 40.8. Courtesy of RF Brydson, modified from Garvie, LAJ *et al.* (1994) Amer. Mineralogist **79** 411, Fig. 4.

Figure 40.11: Courtesy of J Bruley and J Mayer.

Figure 40.12: Courtesy of J Bruley.

Figure 40.13 Modified from Alamgir, FM *et al.* (2000) Microscopy and Microanalysis Suppl. **2** 194. Courtesy FM Almagir.

Figure 40.14 Modified from Botton, GA (2005) J. Electr. Spect. Rel. Phen. **143** 129, Fig. 5. Courtesy of GA Botton.

Figure 40.15: Courtesy of M Aronova and RD Leapman, modified from Leapman, RD and Aronova, MA (2006) in *Cellular Electron Microscopy* Ed. JR McIntosh.

Part 4

Spectrometry

32

X-ray Spectrometry

CHAPTER PREVIEW

To make use of the X-rays generated when the beam strikes the specimen, we have to detect them and identify from which element they originated. This is accomplished by X-ray spectrometry, which is one way to transform the TEM into a far more powerful instrument, called an analytical electron microscope (AEM). Currently, the only commercial spectrometer that we use on the TEM is an X-ray energy-dispersive spectrometer (XEDS), which uses a Si semiconductor detector or sometimes a Ge detector. New detector technologies are emerging, which we'll describe briefly. While some of these may render the Si detector obsolete, we'll nevertheless emphasize this particular detector.

The XEDS is a sophisticated instrument that utilizes the fast processing speeds made possible by modern semiconductors. The detector generates voltage pulses that are proportional to the X-ray energy. Electronic processing of the pulses translates the X-ray energy into a signal in a specific channel in a computer-controlled storage system. The counts in the energy channels are then displayed as a spectrum or, more usefully, transformed into a quantitative compositional profile or, better still, a compositional image or 'map.'

> **COUNTS**
> We'll see over and over again that maximizing the number of X-ray counts is paramount.

The Si detector is compact enough to fit in the confined region of the TEM stage and, in one form or another, is sensitive enough to detect all the elements above Li in the periodic table. We'll start with the basic physics you need to understand how the detectors work and give you a brief overview of the processing electronics. We then describe a few simple tests you can perform to confirm that your XEDS is working correctly and the choices you have to make due to the way the XEDS is interfaced to the AEM column.

It is really most important from a practical point of view that you know the limitations of your XEDS and understand the spectrum. Therefore, we'll describe these limitations in detail, especially the unavoidable artifacts (Chapter 33). In Chapter 34, we'll show how the spectra can easily give a qualitative elemental analysis of any chosen feature in your image and, in just a little more time (Chapter 35), a full quantitative analysis. In Chapter 36, we'll show that this information can be obtained with a spatial resolution approaching a nanometer or below and offers detection limits close to a single atom. So '*micro*analysis' is not a good term; '*nano*analysis' is more accurate but sounds worse. 'Analysis' is how we'll describe it.

32.1 X-RAY ANALYSIS: WHY BOTHER?

The limitations of only using TEM imaging should, by now, be obvious to you. Our eyes are accustomed to the interpretation of 3D, reflected-light images. However, as we have seen in great detail in Part 3, the TEM gives 2D projected images of thin 3D specimens and you, the operator, need substantial experience to interpret these images correctly. For example, Figure 32.1 shows six images, taken with light and electron microscopes (can you distinguish which images are from which kind of microscope?). The scale of the images varies over 6 orders of magnitude from nanometers to millimeters and yet they all appear similar. Without any prior knowledge it would not be possible, even for an experienced microscopist, to identify the nature of these specimens simply from the images.

Now if you look at Figure 32.2, you can see six X-ray spectra, one from each of the specimens in Figure 32.1. The spectra are plots of X-ray *counts* (imprecisely

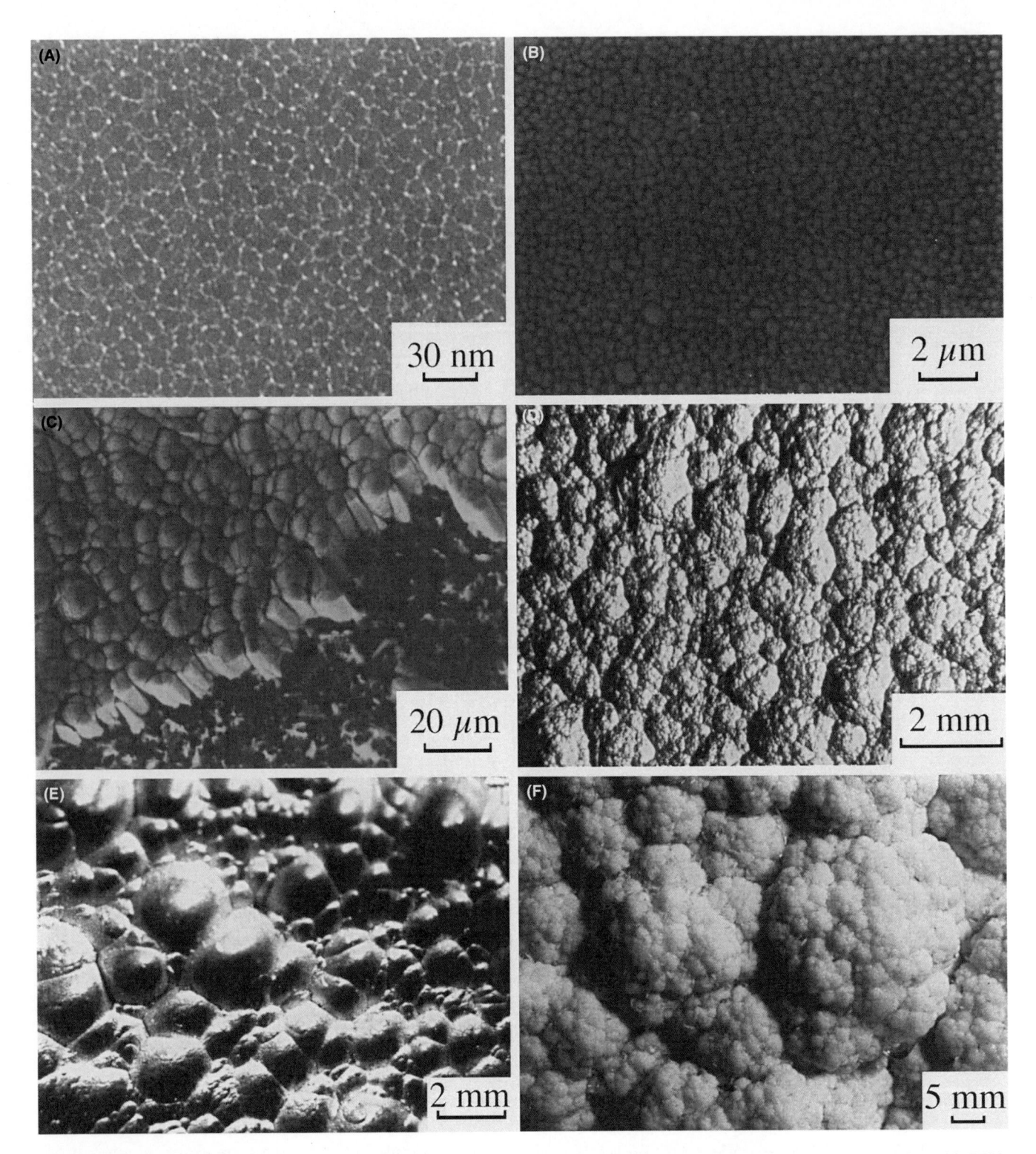

FIGURE 32.1. Six images of various specimens, spanning the dimensional range from nanometers to millimeters. The images were taken with TEMs, SEMs, and light microscopes, but the characteristic structures are very similar, and it is not possible, without prior knowledge, to identify the specimens.

termed 'intensity') versus X-ray *energy* and basically consist of Gaussian-shaped peaks on a slowly changing background. From Chapter 4 you already know that the peaks are characteristic of the elements in the specimen and the background is also called the bremsstrahlung.

But even with no knowledge of XEDS, you can easily see that each specimen gives a different spectrum.

Different characteristic peaks mean different elemental constituents; it is possible to obtain this information in a matter of minutes or even seconds.

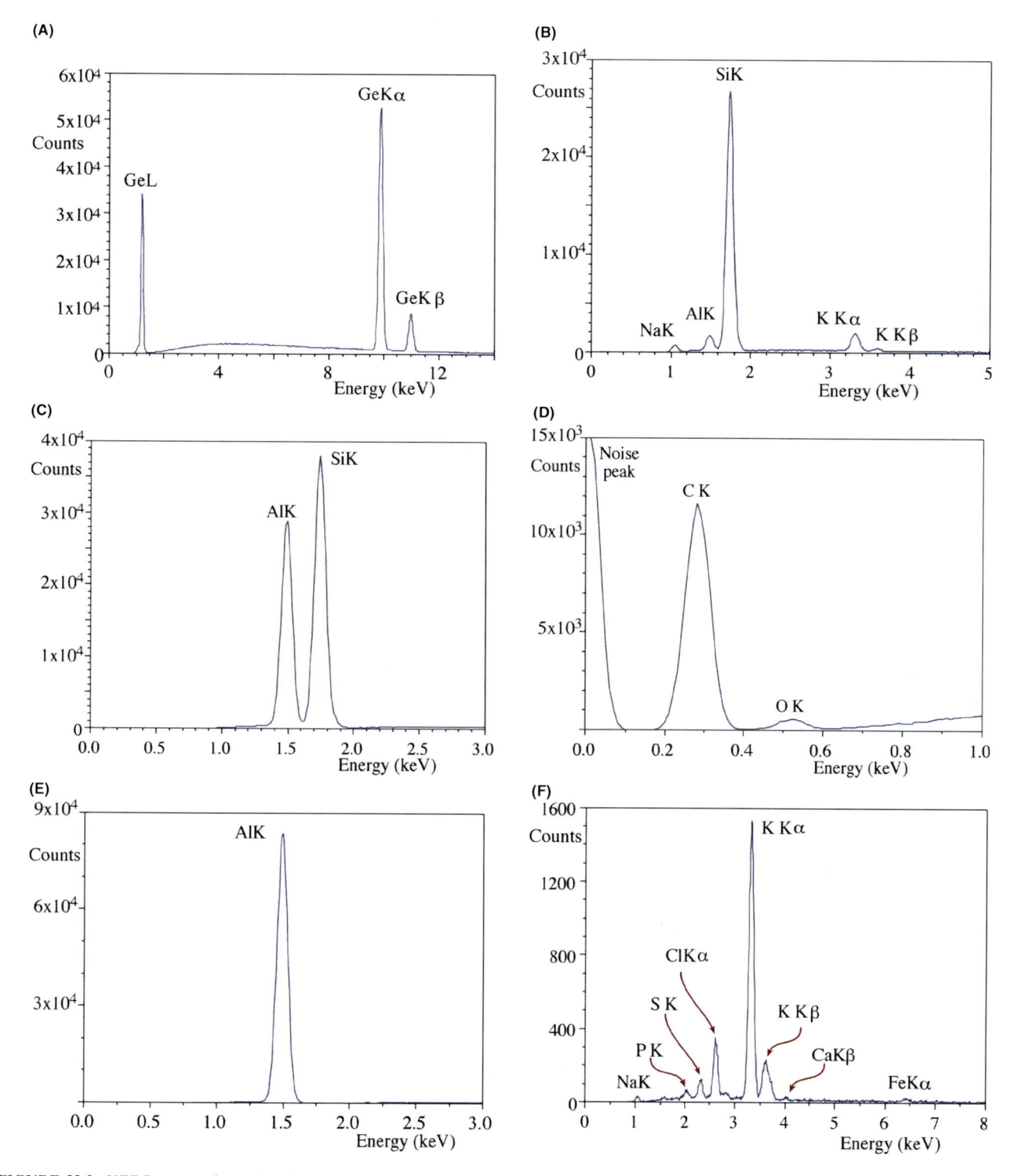

FIGURE 32.2. XEDS spectra from the six specimens in Figure 32.1. Each spectrum is clearly different from the others, and helps to identify the specimens as (A) pure Ge, (B) silica glass, (C) Al evaporated on a Si substrate, (D) pyrolitic graphite, (E) pure Al, and (F) a cauliflower.

When you have such elemental information, any subsequent image and/or diffraction analysis is greatly facilitated. For your interest, the identity of each specimen is given in the caption to Figure 32.2. While Figure 32.2A–E is from common inorganic materials, Figure 32.2F is from a cauliflower which, once you get it into the electron microscope, provides a very distinctive spectrum, albeit from a somewhat carbonized relic of the original vegetable. The familiar morphology of this specimen, now obvious in Figure 32.1F, also accounts for the generic term 'cauliflower structure' which is given to these and similar microstructures.

32.1 X-RAY ANALYSIS: WHY BOTHER? 583

The main message you should get from this illustration is that the combination of imaging and spectroscopy transforms a TEM into the much more powerful AEM.

32.2 BASIC OPERATIONAL MODE

To produce spectra such as those in Figure 32.2, you first obtain a TEM or STEM image of the area you wish to analyze. In TEM mode, you then have to condense the beam to an appropriate size for analysis. This means exciting the C1 lens more strongly, decreasing the C2 aperture size and adjusting the C2 lens strength. These steps will misalign the illumination system and it can be tedious to move between TEM-image and focused-spot analysis modes, unless you are driving a fully computer-controlled (S)TEM. So, we recommend that you operate in STEM mode. First, create your STEM image as we described back in Section 9.4. Then simply stop the scanning probe and position it on the feature you wish to analyze and switch on the XEDS. In STEM mode, digital software can also check for specimen drift during your analysis.

> **STEM MODE**
> Use STEM for AEM. It makes it easier to change from image to analysis mode, easy to form compositional images, and easier to compensate for drift.

In this 'spot' mode you can simply move the beam around the specimen and get a sense of the elemental chemistry of different features you select. However, this approach is very limited from a statistical sampling standpoint and highly biased toward what you think looks interesting in the image. It is now feasible to gather not merely a spectrum from a feature in your specimen, as in Figure 32.2, but a spectrum at *every pixel* in a digital STEM image. From such 'spectrum images' we can extract maps showing the distribution of each element in the specimen and its relationship to the features in the electron image, thus adding another dimension (literally) to the power of the AEM (go back and check the X-ray map in Figure 1.4). We'll talk more about this in Chapter 33 and discuss both qualitative and quantitative maps in Chapters 34 and 35.

For reasons that we'll describe in detail later, you should *always* perform XEDS with your specimen in a low-background (Be) holder. Unless you have an UHV AEM, the holder should be cooled to liquid-N_2 temperature to minimize contamination, and we recommend a double-tilt version, so you can simultaneously carry out diffraction and/or imaging along with your analysis.

32.3 THE ENERGY-DISPERSIVE SPECTROMETER

The XEDS was developed in the late 1960s and by the mid-1970s was an option on many TEMs and even more widespread on the SEM. This rapid spread testifies to the fact that the XEDS is really quite a remarkable instrument, embodying many of the most advanced features of semiconductor technology. It is compact, stable, robust, easy to use, and you can quickly interpret the readout. Several books have been devoted to XEDS on electron-beam instruments and these are listed in the general references. Figure 32.3A shows a schematic diagram of the complete XEDS system and we'll deal with each of the major components as we go through this chapter.

The computer controls all three parts. First, it controls whether the detector is on or off. Ideally, we only want to process one incoming X-ray photon at one time. So the detector is switched off when an X-ray photon is detected and switched on again after that signal is processed (notice we use the particle description of an X-ray here; other detectors work in ways that assume the X-ray is a wave). Second, the computer controls the processing electronics, assigning the signal to the correct energy channel in the storage system. Third, the computer calibrates the spectrum display and tells you the conditions under which you acquired the spectrum, the peak identity, the number of X-rays in a specific channel

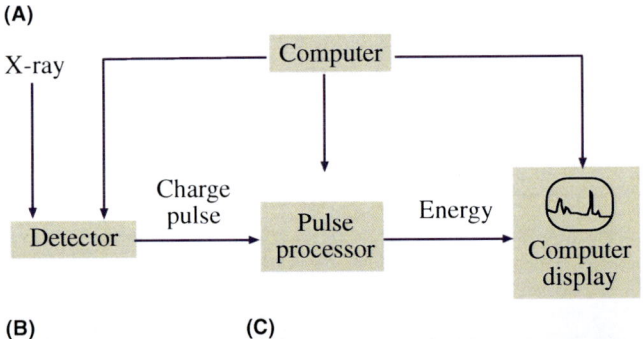

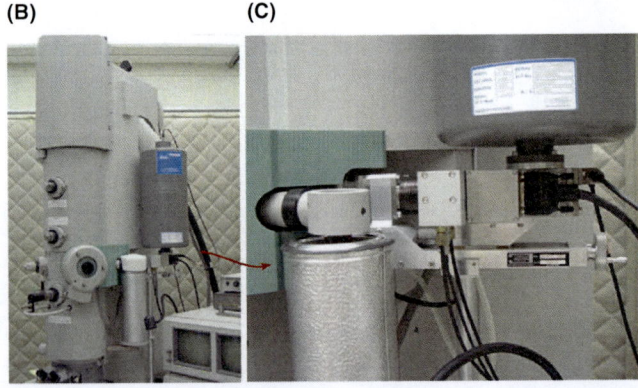

FIGURE 32.3. (A) Schematic diagram of the principle of XEDS; the computer controls the detector, the processing electronics and the display. (B) An XEDS system interfaced to the stage of an AEM. Even in close-up (inset), all that is visible is the large liquid-N_2 dewar attached to the side of the column.

or 'window' of several channels, etc. Any subsequent data processing is also carried out using the computer.

> **3 COMPONENTS**
> The three main parts of an XEDS system are
> (i) the detector
> (ii) the processing electronics
> (iii) the computer

We can summarize the working of the XEDS as follows

- The detector generates a charge pulse proportional to the X-ray energy.
- This pulse is first converted to a voltage.
- The voltage is amplified through a field-effect transistor (FET), isolated from other pulses, further amplified, then identified electronically as resulting from an X-ray of specific energy.
- A digitized signal is stored in the channel assigned to that energy in the computer display.

The speed of this process is such that the spectrum appears to be generated in parallel with the full range of X-ray energies detected simultaneously, but the process actually involves very rapid serial processing of individual X-ray signals. Thus, the XEDS both detects X-rays and separates (*disperses*) them into a spectrum according to their *energy*; hence the name of the spectrometer.

Figure 32.3B shows an XEDS interfaced to an AEM. In fact, you can't see the processing electronics, the display, or even the detector itself because it sits close to the specimen within the column. The only feature that you can see is the dewar containing liquid-N_2 to cool the detector and even this is disappearing from the latest detectors.

32.4 SEMICONDUCTOR DETECTORS

The Si detector in an XEDS is a reverse-biased p-i-n diode and since this is still, by far, the most common detector, we will take this as our model. Later in this section we'll discuss the role of other semiconductor detectors, such as intrinsic-Ge (IG) and Si-drift detectors (SDDs).

32.4.A How Does an XEDS Work?

While you don't need to know precisely how the detector works in order to use it, a basic understanding will help you optimize your system and it will also become obvious why certain experimental procedures and precautions are necessary.

When X-rays deposit energy in a semiconductor, electrons are transferred from the valence band to the conduction band, creating electron-hole pairs, as we saw back in Section 4.4. The energy required for this transfer in Si is ~3.8 eV at liquid-N_2 temperature. (This energy is a statistical quantity, so don't try to link it directly to the band gap.) Since characteristic X-rays typically have energies well above 1 keV, thousands of electron-hole pairs can be generated by a single X-ray. The number of electrons or holes created is directly proportional to the energy of the X-ray photon. Even though all the X-ray energy is not, in fact, converted to electron-hole pairs, enough are created for us to collect sufficient signal to distinguish most elements in the periodic table, with good statistical precision. Figure 32.4 is a schematic

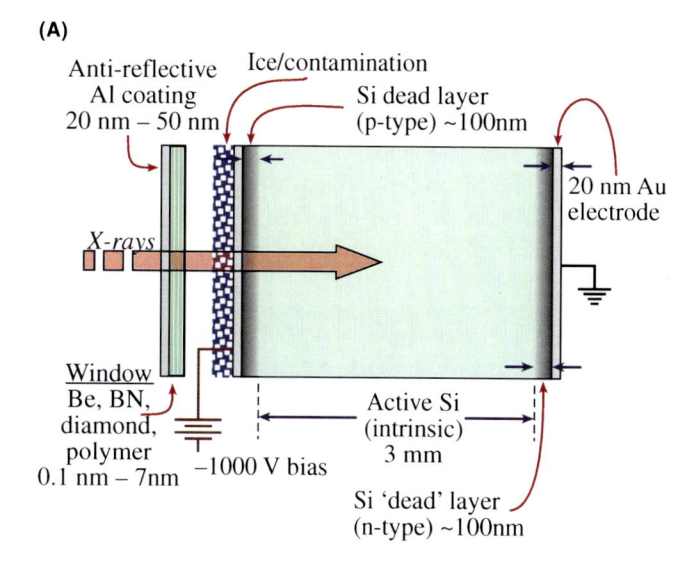

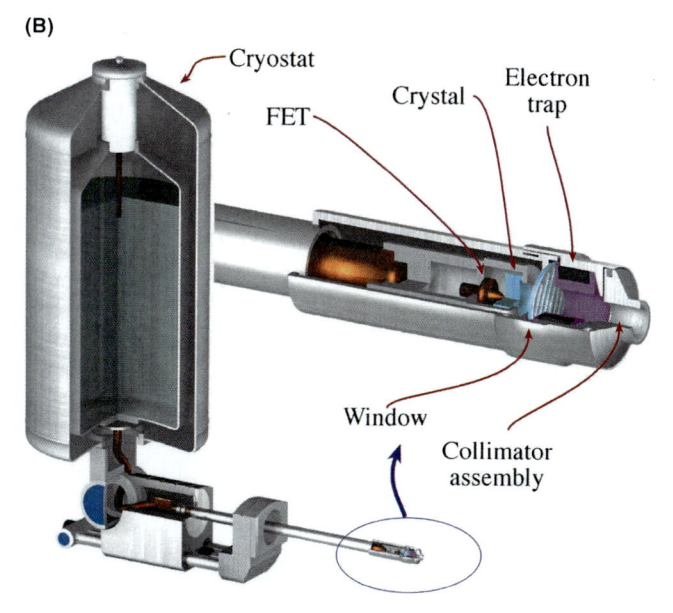

FIGURE 32.4. (A) Cross section of a Si(Li) detector with dimensions indicated (not to scale). In the intrinsic Si region the incoming X-rays generate electron-hole pairs which are separated by an applied bias. A positive bias attracts the electrons to the rear ohmic contact and this charge pulse is amplified by an FET. (B) Exploded diagram of how the individual parts fit together.

diagram of a Si detector and it is similar to the semiconductor electron detectors we discussed back in Chapter 7.

Electron detectors separate the electrons and holes by an internal reverse bias across a very narrow p-n junction, but we need a much thicker detector for X-rays to generate electron-hole pairs since X-rays penetrate matter much more easily than electrons.

Even the purest commercial Si contains acceptor impurities and exhibits p-type behavior. So we compensate for the impurities, which would aid recombination of electron-hole pairs by 'filling' any recombination sites with Li, thus creating intrinsic Si in which the electrons and holes can be separated. Henceforth, we'll refer to Si(Li) (often pronounced "silly") detectors.

The thousands of electrons and holes generated by an X-ray still constitute a very small charge pulse ($\sim 10^{-16}$ C), and so we apply a 0.5–1 keV bias between evaporated Au or Ni ohmic contacts to separate most of the charge. The metal film on the front face creates a p-type region and the back of the crystal is doped to produce n-type Si under a thicker rear contact. So the whole crystal is now a p-i-n device, with shallow junctions on either side of an intrinsic region.

> **THE ENERGY OF A PULSE**
> Remember that the magnitude of the charge pulse is proportional to the energy of the X-ray that generated the electron-hole pairs.

When a reverse bias is applied (i.e., a negative charge is placed on the p-type region and a positive charge on the n-type), the electrons and holes are separated and an electron pulse can be measured at the rear contact.

In the p and n regions at either end of the detector, the Li compensation is not completely effective. These regions are effectively unresponsive to the X-ray because most of the electron-hole pairs recombine, and don't contribute to the pulse. These so-called 'dead layers' are an inevitable result of the fabrication process and reduce the detector efficiency. In practice, it is the p-type dead layer at the entrance surface that is most important since the X-rays must traverse it to be detected and we will refer to this as *the* dead layer.

> **DEAD ACTIVE**
> The p and n regions are called 'dead layers'; the intrinsic region between them is referred to as the 'active layer.'

The dead layer has become thinner as the detector technology has improved and its effects on the spectrum continue to be reduced (although not to zero, as we shall see).

32.4.B Cool Detectors

Why do we have to cool the detector? Well, if the detector were at room temperature, three highly undesirable effects would occur

- Thermal energy would activate electron-hole pairs, giving a noise level that would swamp the X-ray signals we want to detect.
- The Li atoms would diffuse under the bias, destroying the intrinsic nature of the detector.
- The noise level in the FET would mask signals from low-energy X-rays.

So the detector and the FET are usually cooled with liquid N_2, hence the characteristic dewar shown in Figure 32.3B. The weight of the dewar and the need for constant filling with liquid N_2 are major drawbacks. While the majority of XEDS systems on AEMs still use liquid-N_2 cooled Si(Li) detectors, alternatives are available, such as compact dewars (which use much less N_2), cryo-cooling, compressor-based devices which attain liquid-N_2 temperature mechanically, non-compressive technologies and Peltier-cooled systems, which cool sufficiently (and very rapidly) to deliver reasonable energy resolution with fewer problems. Liquid-N_2 cooling has other drawbacks. Residual hydrocarbons and water vapor in the column form carbon contamination or ice films on the cold detector surface, causing absorption of low-energy X-rays. There are obvious solutions to this problem. We can either isolate the detector from the vacuum, or remove hydrocarbons and water vapor from the column. The latter is a more desirable solution but the former is far easier and much less expensive.

> **THE XEDS DEWAR**
> The cylinder hanging on the side of the AEM is the dewar holding liquid N_2 that quietly boils away.

32.4.C Different Kinds of Windows

Liquid-N_2 cooled detectors are usually isolated from the AEM stage in a pre-pumped tube with a sealed 'window' which allows most X-rays through into the detector. There are three kinds of detector; those with a Be window, those with an ultra-thin window and those without a protective window.

Let's examine the pros and cons of each window; a good review has been given by Lund.

Beryllium-window detectors use a thin Be sheet. The best foil is ~ 7 μm which is transparent to most X-rays, and can withstand atmospheric pressure when the stage is vented to air. But 7 μm Be is expensive (~$3 M/pound!), rare, and slightly porous, so a thicker sheet (~12–25 μm) is more commonly used. Rolling such a thin Be sheet is a remarkable metallurgical achievement but it still absorbs X-rays with energies < ~1 keV. Therefore, we cannot detect K_α X-rays from elements below about Na ($Z = 11$) in the periodic table, preventing analysis of B, C, N, and O, which are important in the materials, biological, and geological sciences. Other factors such as the low fluorescence yield and increased absorption within the specimen make light-element X-ray analysis somewhat of a challenge, and EELS is often preferable (see Chapters 38 and 39).

> **KLM TIME**
>
> Remind yourself now of the energies involved for K, L, and M for different elements.

Ultra-thin window (UTW) detectors use <100 nm polymer films, diamond, boron nitride, or silicon nitride, all of which can withstand atmospheric pressure while transmitting 192-eV boron K X-rays and the best UTW windows can even analyze Be K X-rays (110 eV). Early polymer UTWs would break if you accidentally vented the column to air without withdrawing the detector and window behind a valve. This problem was overcome by strengthening the polymers with Al films and this accounts for the term 'atmospheric thin window' (ATW) which you may hear. You should remember that different window materials absorb light-element X-rays differently, so you need to know the characteristics of your window. For example, carbon-containing windows absorb nitrogen K_α X-rays very strongly, nitrogen absorbs oxygen, etc.

Windowless detectors only make sense in UHV AEMs such as old VG instruments and Nion dedicated STEMs which minimize hydrocarbons and keep the partial pressure of water vapor by operating with a stage vacuum < ~10^{-8} Pa. Windowless systems routinely detect Be K X-rays as shown in Figure 32.5, which is a remarkable feat of electronics technology.

The relative performance of the various windows is summarized in Figure 32.6. Here we plot the detector efficiency as a function of X-ray energy. You can clearly see the rapid drop in efficiency at low energies and the improved performance of windowless/UTW systems. In fact, Si(Li) detectors absorb (i.e., detect) X-rays with almost 100% efficiency over the range from ~2 to 20 keV, as shown in Figure 32.7. Within this range are X-rays from all the elements in the periodic table above P. This uniform high efficiency is a major advantage of the XEDS detector. Table 32.1 is a concise summary of the pros and cons of each kind of window.

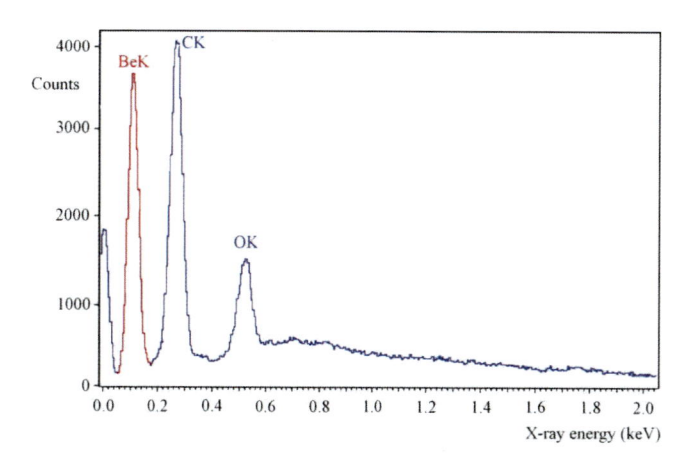

FIGURE 32.5. XEDS spectrum showing the detection of Be in an oxidized Be foil in an SEM at 10 keV. The Be K_α line is not quite resolved from the noise peak.

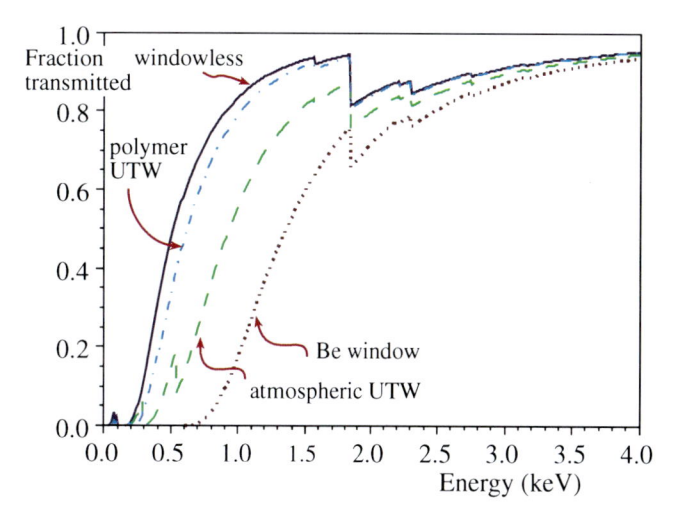

FIGURE 32.6. Low-energy efficiency calculated for a windowless detector, an UTW (1-μm Mylar coated with 20 nm of Al) detector, an ATW detector, and a 13-μm Be-window detector. Note that the efficiency is measured in terms of the percentage of X-rays *transmitted* by the window.

> **3.8 eV**
>
> It takes ~3.8 eV to generate an electron-hole pair in Si, so a Be K_α X-ray will create at most ~29 electron-hole pairs, giving a charge pulse of ~5×10^{-18} C!

32.4.D Intrinsic-Germanium Detectors

You can also see in Figure 32.7 that Si(Li) detectors show a drop in efficiency >~20 keV. This is because such high-energy X-rays can pass through the detector without creating electron-hole pairs. This effect limits the use of Si(Li) in 300–400 keV AEMS in

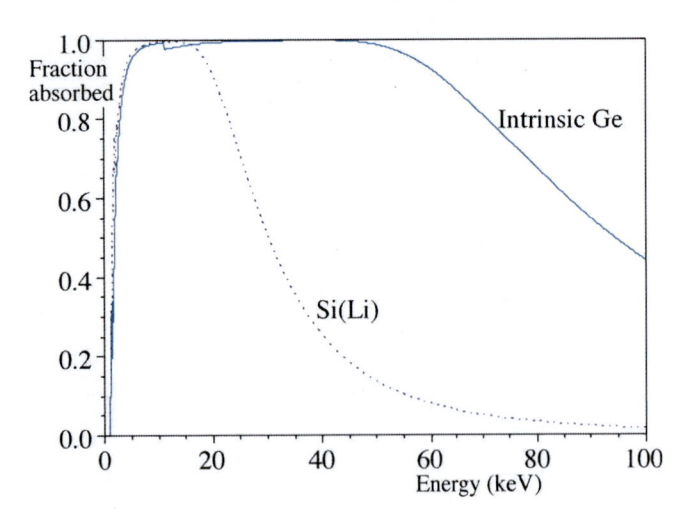

FIGURE 32.7. High-energy efficiency up to 100-keV X-ray energy calculated for Si(Li) and IG detectors, assuming a detector thickness of 3 mm in each case. Note the effect of the Ge absorption edge at about 11 keV. In contrast to Figure 32.6, the efficiency is measured by the percentage of X-rays *absorbed* within the detector.

which we can generate K_α X-rays from all the high-Z elements, e.g., 75 keV Pb K_α X-rays are easily formed at 300 keV. As we'll see in Chapter 35, there are advantages to using the K lines for quantification rather than the lower-energy L or M lines and with a Si(Li) detector, the K lines from elements above Ag ($Z = 47$) are barely detectable. One possible solution is to use an intrinsic Ge (IG) detector which more strongly absorbs high-energy X-rays, as detailed by Sareen.

We can manufacture Ge of higher purity than Si and such Ge is inherently intrinsic, so, Li compensation is not necessary. IGs are more robust and can be warmed up repeatedly which, as we'll see, can solve certain detector problems.

> **PROTECT YOUR DETECTOR**
>
> The intense doses of high-energy electrons or X-rays which can easily be generated in an AEM (e.g., when the beam hits a grid bar) can destroy the Li compensation in a Si(Li) detector, but there is no such problem in an IG crystal.

Furthermore, the intrinsic region can easily be made ~5 μm thick, giving 100% detection of Pb K_α X-rays. Figure 32.7 compares the efficiency of Si(Li) and IG detectors up to 100 keV. There is an even more fundamental advantage to IG detectors. Since it takes only ~2.9 eV to create an electron-hole pair in Ge, compared with 3.8 eV in Si, a given X-ray produces more electron-hole pairs in Ge, and so the energy resolution and signal to noise are better. The only drawback is that the ionization cross sections for high-energy K X-ray excitations are very small for 300–400 keV electrons, so the peak intensities are low. So, why aren't we all using IG detectors? Well, there's no good technical answer but the facts that Si(Li) detectors are easier to manufacture and have a long history of dependable operation are sound commercial reasons why IG detectors have never seriously penetrated the market.

32.4.E. Silicon-Drift Detectors

Si-drift detectors (SDD) may eventually displace traditional Si(Li) detectors; although relatively new, they are already the detector of choice in the much larger SEM and XRF markets. The SDD is basically a CCD (go back and look at Figure 7.3) consisting of concentric rings of p-doped Si implanted on a single crystal of n-Si across which a high voltage is applied to pick up the electrons generated as X-rays enter the side opposite the p-doped rings (Figure 32.8A–C). Applying a voltage from inside to outside the detector (rather than front to back as in a Si(Li)) permits collection of the electrons generated in the n-Si with a 4× lower voltage. Because the central anode in the middle of the p-doped rings has a much smaller capacitance than the large anode at the rear face of a Si(Li) detector (Figure 32.3A–C), a very high throughput of counts is possible, peaking at output rates of many hundred of kcps (Figure 32.8D and look ahead to Figure 32.12B)). If we could in fact generate such enormous count rates in an AEM, we could reduce quantification errors (see Chapter 35), increase analytical sensitivity (see Chapter 36), and seriously improve the statistics of X-ray mapping (see Chapters 33 and 35). In addition to a high throughput, the SDD can operate with no cooling or minimal thermoelectric (Peltier) cooling, while maintaining energy resolutions competitive with Si(Li) detectors (see Section 32.8). This is possible because modern Si-processing technology has reduced thermal-electron generation to extremely low values.

TABLE 32.1. Comparison of Windows

Type	Name	Thickness	Material	Advantage	Disadvantage
Be	Beryllium	~7 μm	Be	Robust	Absorption
UTW	Ultra-thin window	300 nm	Polymer	Low absorption	Breaks easily
ATW	Atmospheric thin window	300 nm	Polymer on grid	Low absorption, robust	Less effective area
None	Windowless	0 nm	None	No absorption	Contamination, light transmitted, need UHV

The only problem here is that, as we discuss at length throughout the subsequent chapters, the use of small probes and thin specimens means that total X-ray count rates in AEMs are usually small, negating a principal advantage of SDDs. However, all is not lost and with the advent of intermediate-voltage FEGs and, more recently, C_s correctors, we can create electron probes of < 0.2 nm with > 1 nA of current. If you are prepared to sacrifice spatial resolution by increasing the size of the probe-limiting aperture, then several nA can be generated in probes of a few nm. If you are prepared to sacrifice spatial resolution still further by using thicker specimens, then it is easy to reach the current signal-processing limits (see next section) of Si(Li) electronics (>50 kcps), in which case SDDs might become attractive alternatives. But the jury is still out.

Because an SDD is made up of arrays of individual cells, each one like Figure 32.8A, it is also feasible to consider designing specially shaped SDDs that conform to the inside of the AEM stage. Thus, we could increase the collection angle well beyond typical flat Si(Li) collection angles of a few tens of mrads, and overcome the count-rate limitation (see Section 32.9.A). Count rates are, of course, no problem in SEM-based, bulk-specimen, X-ray microanalysis, which explains the rapid increase of SDDs on those instruments. If you want to learn more, Newbury gives a thorough analysis of the pros and cons of SDDs for mapping in the SEM.

32.5 DETECTORS WITH HIGH-ENERGY RESOLUTION

The poor energy resolution (typically ~135±10 eV) is a fundamental limitation of any of the semiconductor detectors. This limitation arises because, as we discuss in Section 32.7, the detection and processing steps are statistical processes. This poor resolution gives rise to significant peak overlaps (see Section 34.4) and fundamentally limits the sensitivity (detection limits) of analyses (see Section 36.4). However, there are X-ray spectrometers available with significantly better resolution than EDS (<1–10 eV), which may provide better options in the future, so it is worth noting these potential technologies.

32.6 WAVELENGTH-DISPERSIVE SPECTROMETERS

32.6.A Crystal WDS

Before the invention of the XEDS, the wavelength-dispersive (WDS) or crystal spectrometer was widely used. The WDS uses crystals of known interplanar

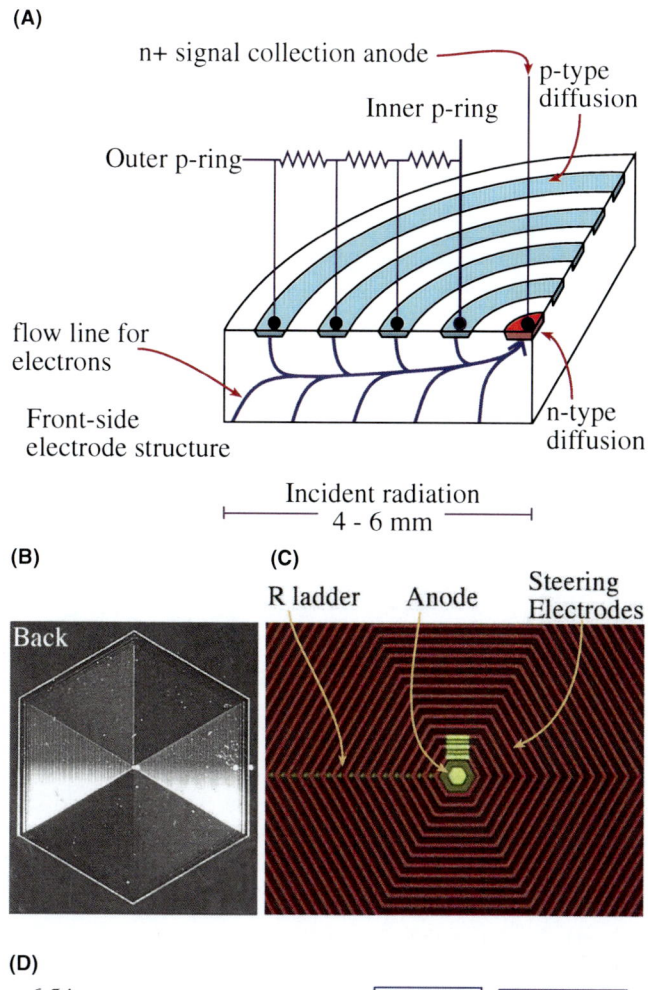

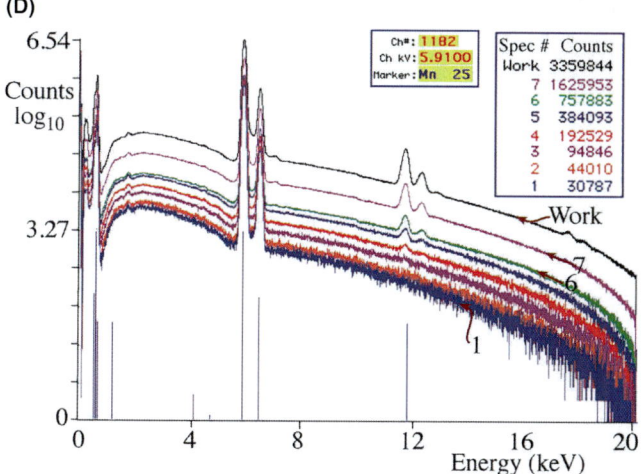

FIGURE 32.8. (A) Schematic diagram showing the back of a quadrant of an SDD consisting of concentric rings of p-type Si on a single crystal of n-type Si. The FET is integrated onto the back of the detector and the bias is applied between the outside p-type ring and the anode inside the inner ring giving electron paths inside the detector as shown in blue. (B) Low-magnification and (C) high-magnification image of the electrode structure on the back of the SDD. (D) SDD spectra from a bulk Mn sample showing no degradation in resolution with increasing output count rates. Be careful: the counts are displayed on a log scale which distorts the usual linear vertical counts scale. The maximum counts are in the Mn Kα peak at 5.91 keV. The black spectrum has over 3.3×10^6 counts in a single channel at that energy. The colored spectra have less counts down to the blue one with 30×10^3 counts in the same channel. But all spectra display the same shape. A Si(Li) spectrometer cannot handle such count rates.

spacing which (see Part 2) disperse X-rays of wavelength (λ) through different angles (θ) according to Bragg's law ($n\lambda = 2d \sin\theta$). So a WDS treats electrons as waves and this gives considerable advantages over XEDS

- Better energy resolution (~5–10 eV) to minimize peak overlaps.
- Higher peak to background (*P/B*) ratio to improve detection limits.
- Better detection of light elements (minimum $Z = 4$, Be) by careful choice of analyzing crystal rather than solely through a dependence upon electronics, as in the XEDS.
- No artifacts in the spectrum from the detection and signal processing, except for higher-order lines from fundamental reflections (when $n \geq 2$ in the Bragg equation).
- Higher throughput count rate using a gas-flow proportional counter.

So why don't we have WDS systems on our AEMs? Well, as you'll see in Chapter 35, the forerunner of the AEM, back in the 1960s and early 1970s, was the electron microscope micro-analyzer (EMMA), which did indeed use WDS. However, the WDS was a large and inefficient addition to the TEM, and never attained general acceptance by TEM users for two reasons

- The crystal has to be moved to a precise angle where it collects a tiny fraction of the X-rays from the specimen, whereas the XEDS detector subtends a relatively large solid angle.
- The WDS collects a single λ at any time while the XEDS detects X-rays over a large range of energies. WDS is a very slow, serial collector; XEDS is a fast, effectively parallel, collector.

The geometrical advantage of XEDS (remember, we need to maximize X-ray counts) combined with rapid detection over a wide energy range, without the mechanical motion of the WDS, accounts for the dominance of Si(Li) XEDS systems in all AEMs. Two alternative approaches to overcoming the poor energy resolution of XEDS are being explored. One is a development of the traditional WDS and the other is a totally new approach to X-ray detection.

32.6.B CCD-Based WDS

In an attempt to detect ultra-soft (i.e., very low energy) X-rays, Terauchi and Kawana have designed a WDS using an aberration-corrected, concave, diffraction grating (instead of the usual bent crystal) with a CCD detector in place of the proportional counter. The

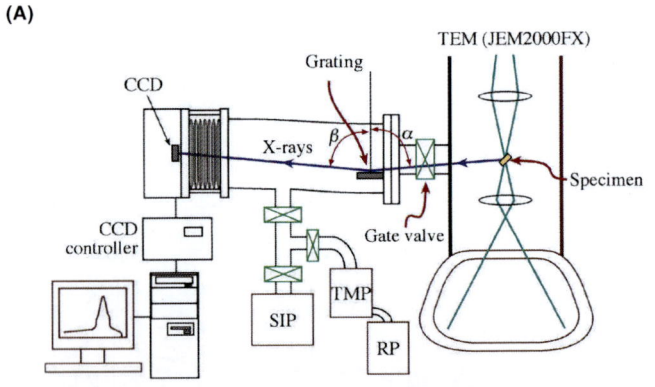

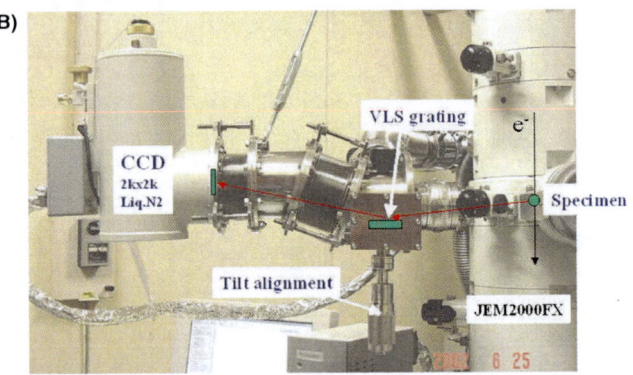

FIGURE 32.9. High-energy resolution X-ray spectrometry; (A) schematic diagram and (B) image of the diffraction-grating WDS system fixed to a TEM column. (C) High-resolution X-ray spectra from hexagonal, cubic, and wurtzite forms of BN showing differences in the B K_α peak shapes due to differences in bonding. (D) Comparison of Si(Li) and bolometer spectra obtained in an SEM illustrating the tremendous difference in energy resolution.

detector is much more compact than a standard WDS, as shown in Figure 32.9A and B, and delivers an energy resolution of 0.6 eV, which is 200× better than a typical EDS resolution and sufficient to resolve intensity variations in the characteristic peaks that arise due to changes in the density of states (DOS) of the valence band (see Figure 32.9C). Study of the DOS and related bonding effects is usually the role of EELS (see Chapter 40), so CCD-WDS is a real breakthrough for X-ray spectrometry. The energy resolution would improve further if the CCD pixel size were decreased, and the grating dispersion increased. Unfortunately, the count rate is very low and to get the spectra shown in Figure 32.9C, a probe size of ~1 μm was used. A larger CCD would permit smaller probes and these will become available via digital-camera technology. Also, recent developments in capillary optics may increase the WDS collection angle, so the technology is worth watching.

32.6.C Bolometers/Microcalorimeters

A totally different approach to detecting X-rays is to measure the heat emitted when an X-ray is absorbed. At first sight this might appear ridiculous, but the

while offering the effectively parallel-collection aspects of XEDS (see Figure 32.9D). Bolometers have been installed on SEMs where their small area is offset by the large count rate. Despite initial enthusiasm, bolometers are not commercially available because they have to be cooled to a few mK with liquid He, which increases the cost close to that of a new SEM and their size to the point where attaching one to a TEM is a major mechanical-engineering feat. The technology to create bolometer arrays is available but until compact, large collection angle WDS or cheap bolometer systems are available, we'll have to live with the poor resolution and other limitations of Si(Li) detectors, to which we'll devote the rest of the chapter.

Table 32.2 below summarizes much of the preceding and following discussion.

32.7 TURNING X-RAYS INTO SPECTRA

The electronics attached to a Si(Li) or SDD convert the charge pulse created by the incoming X-ray into a voltage pulse, which is stored in the appropriate energy channel of the computer display (which used to be called a multi-channel analyzer or MCA). The pulse-processing electronics must maintain good energy resolution across the spectrum without peak shift or distortion, even at high count rates. To accomplish this, all the electronic components beyond the detector crystal must have low-noise characteristics and employ some means of handling pulses that arrive in rapid succession. This whole process used to rely on analog pulse processing, but many of the problems inherent in the analog process have been solved by digital techniques (Mott and Friel), and all current XEDS systems process the charge pulses digitally.

Let's consider first of all what happens if a single, isolated X-ray enters the detector and the electron-hole pairs are separated and captured to create a charge pulse.

- The charge enters the FET preamplifier and is converted into a voltage pulse.
- The pulse is digitized and the X-ray energy that generated the pulse is computed.
- The computer assigns the signal to the appropriate energy channel on the display.

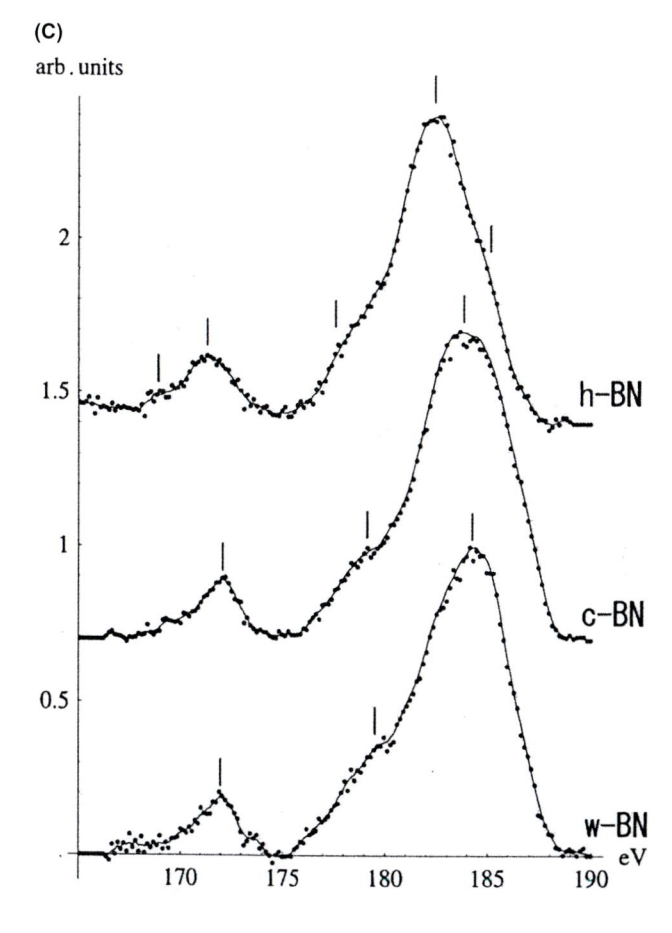

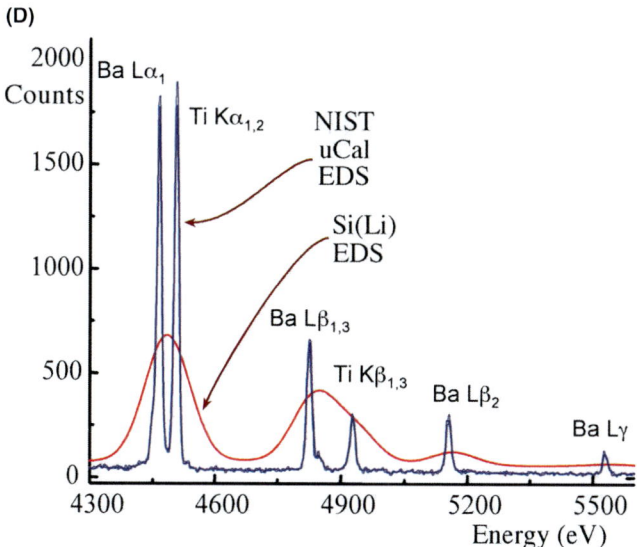

FIGURE 32.9. (Continued).

concept has been demonstrated (Wollman et al.) by creating a microcalorimeter or bolometer (i.e., a sensitive thermometer). As with WDS, the bolometer is limited by its small area, so the solid angle is small and the count rate correspondingly low, but it delivers a resolution comparable with WDS (~5–10 eV),

The accumulation of pulses or counts entering each channel at various rates produces a histogram of counts versus energy that is a digital representation of the X-ray spectrum. The computer display offers multiples of 1024 channels in which to display the spectrum and various energy ranges can be

TABLE 32.2. Comparison of X-ray Spectrometers

Characteristic	IG	Si(Li)	SDD	WDS	Bolometer
Energy resolution (typical/on column)	135 eV	150 eV	140 eV	10 eV	10 eV
Energy resolution (best)	114 eV	128 eV	127 eV	5 eV	5 eV
Energy to form electron-hole pairs (77 K)	2.9 eV	3.8 eV	3.8 eV	n.a.	n.a.
Band gap energy (indirect)	0.67 eV	1.1 eV	1.1 eV	n.a.	n.a.
Cooling required	LN$_2$ or thermoelectric	LN$_2$/thermoelectric	None/thermoelectric	None	100 mK
Detector active area	10–$\geq$50 mm^2	10–$\geq$50 mm^2	$\geq$50 mm^2	n.a.	1 mm^2
Detector arrays available	No	No	Yes	No	Yes
Typical output rates	5–10 kcps	5–20 kcps	1000 kcps	50 kcps	1 kcps
Time to collect full spectrum	~1 min	~1 min	few secs	~30 min	~30 min
Collection angle (sr)	0.03–0.20	0.03–0.30	0.3	10^{-4}–10^{-3}	10^{-4}–10^{-3}
Take-off angle	0°/20°/72°	0°/20°/72°	20°	40°–60°	40°–60°
Artifacts	Escape, sum peaks Ge K/L peaks	Escape, sum peaks Si K peak	Multiple sum peaks	High-order lines	

Data in this table come from the Web sites of the leading XEDS manufactures. For the latest information, check the URLs listed in the reference section.

assigned to these channels. For example, a 10-, 20-, or 40-keV energy range can be used (or even 80 keV for an IG detector on an intermediate-voltage AEM). The display resolution chosen depends on the number of channels available.

A typical energy range that you might select for a Si(Li) and SDD detector is 10 or 20 keV and in 2048 channels this gives you a display resolution of 5 or 10 eV per channel.

> **DISPLAY RESOLUTION**
> You should keep the resolution at 10 eV per channel or better. Smaller values use more memory, but memory is cheap. Larger values mean fewer channels for each characteristic peak, giving the peak a serrated step-like appearance rather than a smooth Gaussian shape.

Details of the pulse-processing electronics are not important except for two variables over which you have control. These are the time constant and the dead time. The time constant (τ) is important only if you have an old analog system; it is the time (~5–100 μs) allowed for the analog processor to evaluate the magnitude of the charge pulse.

- Choosing the shortest τ (typically, a few μs) will allow more counts per second (cps) to be processed but with a greater error in the assignment of a specific energy to the pulse, and so the energy resolution (see Section 32.8 below) will be poorer.
- Choosing a longer τ will give you better resolution but the count rate will be lower.

> **LONG OR SHORT τ?**
> For most routine thin-foil analyses, you should maximize the count rate (shortest τ), unless there is a specific reason why you want to get the best possible energy resolution (longest τ).

With an analog system you can't have a high count rate and good resolution, so for most routine thin-foil analyses you should maximize the count rate (shortest τ), unless there is a specific reason why you want to get the best possible energy resolution (longest τ). This recommendation is based on a detailed argument presented by Statham.

If you have a digital system, the individual pulses are monitored and τ is varied for each pulse, depending on how close they are together. We call this 'adaptive pulse processing'; it gives a continuous variation of output count rate with input rate rather than a discrete range for each value of τ.

Now, in reality, there are many X-rays entering the detector but, because of the speed of modern electronics, the system can usually discriminate between the arrival of two, almost simultaneous, X-rays. You can find details of the electronics in Goldstein et al. When the electronic circuitry detects the arrival of a pulse, it takes less than a μs before the detector is effectively switched off for a period of time called the dead time, while the pulse processor analyzes that pulse. The dead time is clearly closely related to τ, and is so short that you should expect your XEDS system to process outputs up to 10 kcps quite easily (if it is a later analog system) and 30 kcps or more for a digital system. Even higher outputs (up to 70–100 kcps) need to be handled with an SDD (but in AEMs the beam current and/or specimen thickness is rarely great enough). The dead time increases as more X-rays enter your detector, because it shuts down more often. The dead time can be defined in several ways. If you take the ratio of the output count rate (R_out) to the input count rate (R_in), which you can usually measure, then

$$\text{Dead time in \%} = \left(1 - \frac{R_\text{out}}{R_\text{in}}\right) \times 100\% \quad (32.1\text{a})$$

An alternative definition is

$$\text{Dead time in \%} = \frac{(\text{clock time} - \text{live time})}{\text{clock time}} \times 100\% \quad (32.1b)$$

These different 'times' can be confusing, but it helps to think of equation 32.1b as follows: if you ask the computer to collect a spectrum for a live time of 100s, then the detector must be live and receiving X-rays for this amount of time. If the detector is actually dead for 20s because it is processing X-rays, it will actually take 120s of 'clock time' to accumulate a spectrum, so the dead time (from equation 32.1b) will be 20/120 = 16.7%. As the input count rate increases, the output count rate will drop and the clock time will increase accordingly. Dead times in excess of 50–60% (or as little as 30% in very old systems) mean that your detector is saturated with X-rays and collection becomes increasingly inefficient. Then you should turn down the beam current or move to a thinner area of the specimen to lower the count rate; but this is a rare situation for a thin-foil analyst to face. Just remember these times.

- Dead time is when the detector is not counting X-rays but processing the previous photon.
- Live time is when the detector is ready to detect an X-ray and not processing any signal.
- Clock time is what it says.

32.8 ENERGY RESOLUTION

The natural line width of the emitted X-rays is only a few eV but the measured widths are usually >> 100 eV. The electronic noise in the XEDS system is a major source of the difference between the practical and theoretical energy resolutions and the width of the electronic noise is described as the 'point-spread function' of the detector. Since the poor energy resolution of XEDS is a major limitation, we need to examine this concept more closely.

We can define the energy resolution R of the detector as

$$R^2 = P^2 + I^2 + X^2 \quad (32.2)$$

where P is a measure of the quality of the processing electronics, defined as the full width at half maximum (FWHM) of a randomized electronic-pulse generator. X is the FWHM-equivalent attributable to detector leakage current and incomplete charge collection (see Section 32.9.A). I is the intrinsic line width of the detector which is controlled by fluctuations in the numbers of electron-hole pairs created by a given X-ray and is given by

$$I = 2.35(F\varepsilon E)^{1/2} \quad (32.3)$$

Here F is the Fano factor of the distribution of X-ray counts from Poisson statistics, ε is the energy to create an electron-hole pair in the detector, and E is the energy of the X-ray line. Because of these two factors, the experimental resolution can only be defined under standard analysis conditions defined by the IEEE.

THE IEEE STANDARD FOR R
This is the FWHM of the Mn K_α peak, generated (not in the microscope) by an Fe^{55} source which produces 10^3 cps with an 8 μs pulse-processor time constant.

Rather than using IEEE-required radioactive Fe^{55} we recommend measuring the R on your AEM column! Now since Mn is not a common specimen to have lying around, you will find it useful to keep a thin NiO specimen (Egerton and Cheng) to check the resolution when the detector is on the column. You can also use the O K peak to measure the low-energy resolution of your detector. Suitable NiO films < 50 nm thick are available from commercial companies that provide supplies for EM laboratories. Ni is close enough to Mn in the periodic table that you can get good measure of resolution (although it will be slightly worse than that at Mn since resolution degrades with increasing X-ray energy). Others have used thin Cr films instead of NiO. You should be more concerned with changes in R over time than the absolute value since changes indicate that the detector is responding differently to the X-ray flux.

COUNT-RATE EFFECT
All detectors lose resolution as their temperature increases and the count rate increases, although digital electronics handle higher count rates better.

Your computer system will have software that calculates R rather than directly measuring the FWHM of the Mn or Ni peak. It's good to measure the FWHM yourself and you do this by determining the energy width between the channels either side of the peak that contain half the maximum counts in the central (peak) channel, as shown in Figure 32.10.

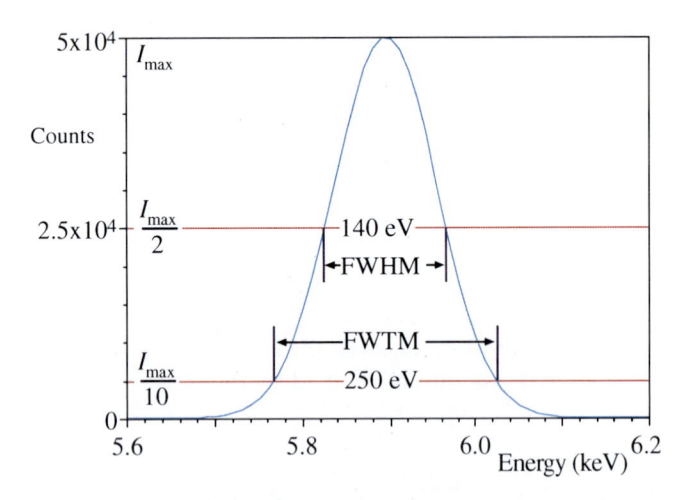

FIGURE 32.10. Measurement of the energy resolution of an XEDS detector by determining the number of channels that encompass the FWHM of the Mn K$_\alpha$ peak. The number of channels multiplied by the eV/channel gives the resolution which, typically, should be about 130–140 eV on the column. You can also measure the FWTM to give an indication of the degree of the ICC which distorts the low-energy side of the peak. The FWTM should be ~1.83× the FWHM if ICC is insignificant.

> **DETECTOR RESOLUTION**
> Typically, Si(Li) detectors have a resolution of ~140 eV at Mn K$_\alpha$ with the best being <130 eV. The best reported IG resolution is 114 eV. SDDs offer about 140 eV but can get down to ~130 eV with Peltier cooling.

Because the value of ϵ is lower for Ge (2.9 eV) than for Si (3.8 eV), IG detectors have higher R than Si(Li). The resolution is also a function of the detector area, and the best values we just gave are for 10 mm^2 detectors. The 30 mm^2 or 50 mm^2 detectors, which are typically installed on AEMs to increase the count rate, have resolutions ~5–10 eV worse than the figures just mentioned. So you should be aware that, when you measure R on your AEM, there will be a further degradation in resolution. It is rare to find a 30 mm^2 Si(Li) detector delivering a resolution <~140 eV on the AEM column, even though quoted values are typically ~10 eV less.

How close are XEDS detectors to their theoretical resolution limit? If we assume that there is no leakage and the electronics produce no noise, then $P = X = 0$ in equation 32.2, so $R = I$. For Si, $F = 0.1$, $\epsilon = 3.8$ eV, and the Mn K$_\alpha$ line occurs at 5.9 keV, which gives $R = 111$ eV. So it seems that there is not much more room for improvement. The resolution of semiconductor detectors won't approach that of crystal spectrometers or bolometers (< 1 to ~10 eV). However, because of the dependence of I on X-ray energy, light-element K lines do have FWHMs well below 100 eV. We'll see in Section 34.5 that there are signal-processing methods to improve the resolution.

32.9 WHAT YOU SHOULD KNOW ABOUT YOUR XEDS

There are several fundamental parameters which you can specify, measure, and monitor to ensure that your XEDS is performing acceptably. Many of these tests are standard procedures (e.g., see the XEDS laboratories described by Lyman et al.) and are summarized by Zemyan and Williams. In an SEM, which is relatively well behaved, Si(Li) detectors have been known to last 10 years or more before requiring service or replacement. In contrast, an AEM is a hostile environment and the life of a detector can be considerably shortened unless there is a protective shutter (see Section 32.11), which should *always* be closed unless you are acquiring a spectrum. If you analyze a thick portion of your specimen (a waste of time), traverse a grid bar across the field of view (easy to do), or accidentally insert the objective diaphragm while the shutter is open (in which case you should be sentenced to memorize this chapter) you can 'flood' the detector and close down the electronics. If this happens, ask for help, but be prepared to wait a while because the system takes time to recover. If it happens too often, you can permanently damage your detector.

> **DAMAGING THE DETECTOR**
> High X-ray or electron fluxes can damage the detector; it is particularly important to monitor the detector performance on your AEM, so that quantitative analyses you make at different times may be compared in a valid manner.

You need to know the operating specifications for your own XEDS, and how to measure them. We can break these specifications down into detector variables and signal-processing variables.

For all of the tests/actions that we describe below, you must discuss the procedure with the instrument technician/laboratory manager **before** doing anything because it is easy to damage the detector if you don't do it right.

32.9.A Detector Characteristics

The *detector resolution* that we just defined may degrade for a variety of reasons. Two are particularly common

- Damage to the intrinsic region by high-energy fluxes of radiation.
- Bubbling in the liquid-N$_2$ dewar due to ice crystals building up.

We've just told you how to avoid the first problem. If part of your responsibility in the lab is to top up the liquid N₂ then it should be filtered before putting it into the dewar and *never* re-cycled. If the N₂ is bubbling, you must ask that the problem be corrected and you can suggest warming up the detector.

Or you can suggest that the AEM needs a detector cooled by means other than liquid N₂!

> **A WARNING ON WARMING**
> Warming the detector to room temperature should only happen after consultation with the manufacturer, and after turning off the applied bias. (Think what happens to the Li otherwise.)

Incomplete-charge collection (ICC): because of the dead layer, the X-ray peak will not be a perfect Gaussian shape. Usually the peak will have a low-energy tail, because some X-ray energy will be deposited in the dead layer and will not create electron-hole pairs in the intrinsic region. You can measure this ICC effect from the ratio of the full width at tenth maximum (FWTM) to the FWHM of the displayed peak, as shown schematically in Figure 32.10.

In Si(Li) detectors, the phosphorus K_α peak shows the worst ICC effects because this X-ray fluoresces Si very efficiently. ICC will also occur if the Si has a large number of defects arising, for example, from damage via a high flux of backscattered electrons. The crystal defects act as recombination sites, but they can be annealed by warming the detector, as we just described. An IG detector should meet the same FWTM/FWHM criterion as a Si(Li) detector. If the ratio is higher than 2 for the Ni K_α peak there is something seriously wrong with the detector and it needs to be replaced. An SDD has an extremely thin dead layer, so ICC should be minimal.

Detector contamination: over a period of time, even in a UHV STEM, ice and/or hydrocarbons will eventually build up on the detector surface, or the window. When contamination occurs, it reduces the detection efficiency for low-energy X-rays. Ultimately, any detector will contaminate because of residual water vapor in the detector vacuum or because the window may be slightly porous. In all cases, the problem is insidious because the effects develop over time and you might not notice the degradation of your spectrum quality until differences in light-element quantification occur in the same specimen analyzed at different times. Therefore, you should regularly monitor the quality of the low-energy spectrum. The ratio of the Ni K_α/Ni L_α can be used (as described by Michael) as a signal to warn of icing/contamination.

> **IDEAL GAUSSIAN**
> An ideal Gaussian shape gives a ratio FWTM/FWHM of 1.82 (Mn K_α or Ni K_α) but this will be larger for lower-energy X-rays that are more strongly absorbed by the detector.

The K/L ratio will differ for different dead layers, different UTWs or ATWs, and different specimen thicknesses, so we can't define an acceptable figure of merit. The best you can do is to measure the ratio (ideally when you first use your detector) and monitor any changes, being aware that, as the ratio increases, quantification of low-energy X-ray lines becomes increasingly unreliable. When the ratio become unacceptable, the detector should be warmed up. Automatic, in situ, heating devices which raise the detector temperature sufficiently to sublime off the ice, without warming the dewar up to ambient temperature, make this process routine. If your detector doesn't have such a device then you should again ask that it be warmed. An SDD doesn't have icing problems if operated at ambient temperatures and, even if Peltier cooling is used, it is still nowhere as cold as liquid N₂.

> **ICE ON THE DETECTOR**
> If contamination or ice builds up on the detector, the K_α/L_α ratio rises; the ice selectively absorbs the lower-energy L line.

In summary, you should measure and continually monitor changes in

- *The detector resolution* on the column using the Mn or Ni K_α line (typically 150 eV for Si(Li) and 140 eV for an IG or SDD).
- *The ICC* defined by the FWTM/FWHM ratio of the Ni K_α line (ideally 1.82).
- *The ice/contamination build-up* reflected in the Ni K_α/L_α ratio.

If any of these figures of merit get significantly larger than *your* baseline values, then warming up the detector, if necessary to room temperature, may help.

In summary, you must be very careful with your XEDS.

- DO NOT generate high fluxes of X-rays or backscattered electrons unless your detector is shuttered.
- DO NOT ever warm up the detector yourself; get help. But make sure that the bias is turned off and the manufacturer is consulted.
- DO NOT use unfiltered or re-cycled liquid N₂.

All these DO NOTs make liquid-N$_2$-free and minimal dead-layer SDDs rather attractive.

32.9.B Processing Variables

You need to ensure that the output counts (i.e., the spectrum) reflect the input X-ray counts. There are three things you can check to make sure the pulse-processing electronics are working properly

- First, check the calibration of the energy range of the spectrum.
- Second, check the dead-time correction circuitry.
- Third, check the maximum output count rate.

The energy resolution should not change significantly from day to day, unless you change the (analog) time constant. Electronic-circuit stability has improved to a level where such checks need only be done a couple of times a year. Will you do the check or rely on someone else?

Calibration of the energy-display range: this process is quite simple; collect a spectrum from a specimen which generates a pair of X-ray lines separated by about the width of the display range (e.g., Cu for 0–10 keV, Mo for 0–20 keV). Some systems use an internal electronic strobe to define zero, and in this case you only need a specimen with a dominant line at the high end of the energy range. Having gathered a spectrum, see if the computer markers are correctly positioned at the peak centroid (e.g., Cu L$_\alpha$ at 0.932 keV and Cu K$_\alpha$ at 8.04 keV). (You'll learn more about the specific energies in Chapter 34 and you can also go back to Chapter 4 to remind yourself of the relationship between E and Z.)

> **RECALIBRATE**
> If the peak and the marker are >1 channel (10 eV) apart, then you should recalibrate your display using the commercial software.

Checking the dead-time correction circuit: if the dead-time correction electronics are working properly, the electronics will give an increase in output counts directly proportional to the increase in input counts, for a fixed live time. This behavior is absolutely essential for valid quantification.

- Choose a pure element, say our favorite NiO foil which we know will give a strong K$_\alpha$ peak.
- Choose a live time, say 50s, and a beam current to give a dead-time readout of about 10%.
- Measure the total Ni K$_\alpha$ counts that accumulate in about 30–60s (longer is better).

Then repeat the experiment with higher input count rates (e.g., dead times of 30, 50, 70%).

> **MUST BE LINEAR**
> The processing electronics must show linear behavior for valid quantification.

To increase the count rate, increase the beam current by choosing a larger diameter beam or larger C2 aperture. The dead time should increase as the input count rate goes up, but the live time remains fixed (by your choice). If you plot the number of output counts against the beam current, measured with a Faraday cup, or a calibrated exposure-meter reading, then it should be linear, as shown in Figure 32.11. But you will see when you do the experiment that it takes an increasingly longer (clock) time to attain the preset live time as the dead time increases. If you don't have a Faraday cup, you can use the input count rate as a measure of the current; remember that the Faraday cup is useful for many other functions, such as characterizing the performance of the electron source, as we saw in Chapter 5. Remember

Determination of the maximum output count rate: again the procedure is simple

- Gather a spectrum for a fixed clock time, say 10–30s, with a given dead time, say 10%.
- Increase the dead time by increasing the beam current, C2 aperture, or specimen thickness.
- See how many counts accumulate in the Ni K$_\alpha$ peak in the same fixed clock time.

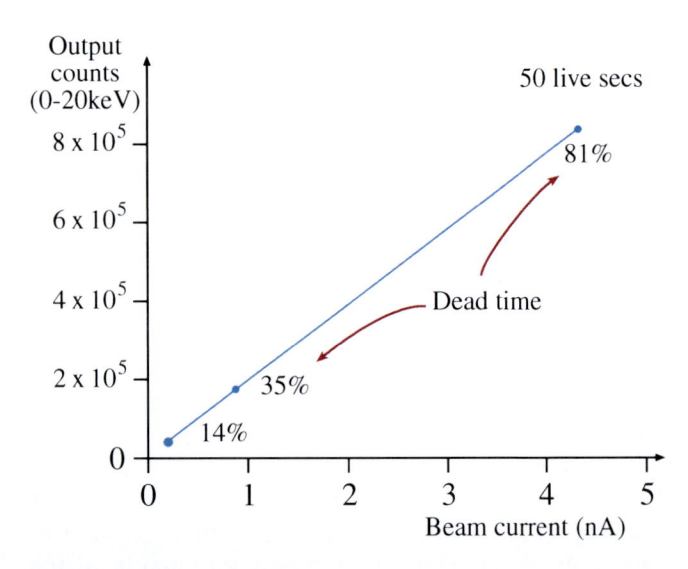

FIGURE 32.11. A plot of the output counts in a fixed live time as a function of increasing beam current showing good linear behavior over a range of dead times, demonstrating that the dead-time correction circuitry is operating correctly.

The number of counts should rise to a maximum and then drop off above a certain dead time, which depends on the system electronics. Beyond the maximum, the detector will be dead more than it is live and so the counts in a given clock time will decrease. In Figure 32.12A, this maximum is at about 60% dead time, typical of modern systems, although in older XEDS units this peak can occur at as little as 30%. If you have analog electronics, you can repeat this experiment for different time constants, τ, and the counts should increase as τ is lowered (at the expense of energy resolution), as also shown in Figure 32.12A. Clearly, if you operate at the maximum in such a curve (if you can generate enough input counts) then you will be getting the maximum possible counts in the shortest possible time. As we've already said, it is almost always better to have more counts than to have the best energy resolution, so select the shortest τ unless you have a peak overlap problem. Digital processing is easier than analog for this reason.

If your specimen is too thin it might not be possible to generate sufficient X-ray counts to reach dead times in excess of 50%, so the curve may not reach a maximum, particularly if τ is very short or you have digital processing. In this case, just use a thicker specimen.

While it is rare that a good thin foil produces enough X-rays to overload the detector electronics, there are situations (e.g., maximizing analytical sensitivity—see Chapter 36) when you need to generate as many counts as possible. However, if you use a thick specimen and high beam current, you may produce too many counts for analog processing, or even digital, systems. If high count rate is your primary mode of operation, you might want to consider an SDD. Figure 32.12B shows the prospects for high output count rates with an SDD. The problem in TEM, as we've noted, is to generate sufficient counts to make use of this extraordinary processing capability.

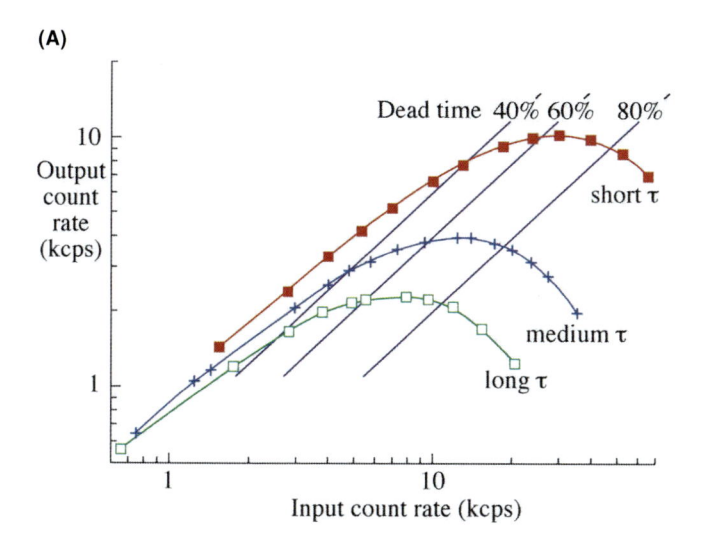

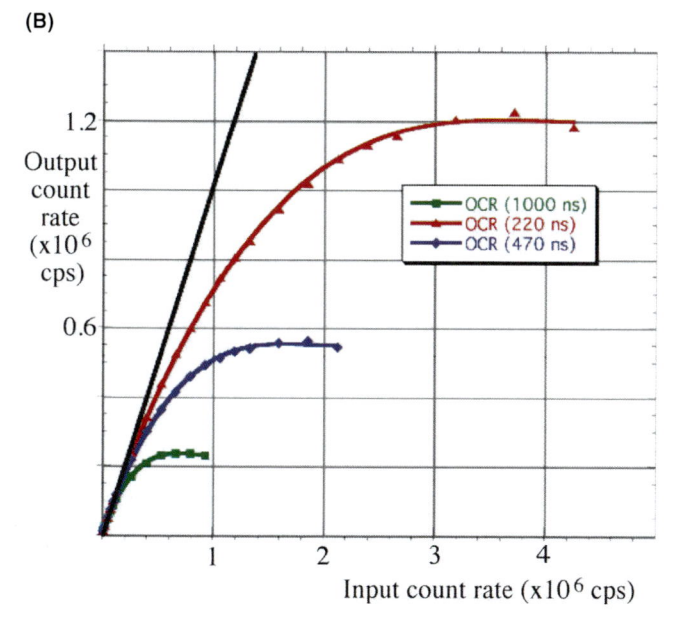

FIGURE 32.12. (A) The output count rate in a given clock time as a function of dead time. The maximum processing efficiency is reached at ~60% dead time. It is very inefficient to use the system above the maximum output rate because of the very long (clock) times needed to gather a spectrum. Increasing the (analog) time constant results in fewer counts being processed and a drop in the output count rate. (B) Data from an SDD on an SEM at three different time constants: note the enormous increase in output count rate up to 1.2 million cps.

> **TOO THIN**
> In XEDS, your specimen may be too thin: you need enough counts.

As shown in Figure 32.13, digital processing permits a higher throughput over a continuous range of energy resolution than the fixed ranges available from each specific (in this case six) τ for an analog system. We've already mentioned that C_s correction permits larger apertures to be used and therefore, significantly more current can be put into small probes, but it's unlikely that if you have a C_s-corrected AEM it will be still interfaced to an old analog XEDS system.

In summary, you should occasionally

- Check the energy calibration of the computer display.
- Check the dead-time circuitry by the linearity of the output count rate versus beam current (Figure 32.11).
- Check the counts in a fixed clock time as a function of beam current to determine the maximum output count rate (Figure 32.12).

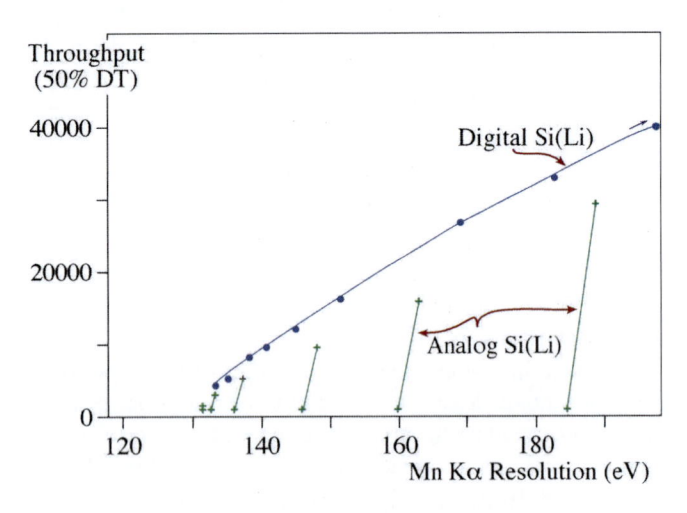

FIGURE 32.13. Digital pulse processing gives a continuous range (blue line) of X-ray throughput at 50% dead time, compared with a set of fixed throughput ranges (green lines) for specific (analog) time constants.

32.10 THE XEDS-AEM INTERFACE

In your TEM, an intense beam of high-energy electrons bombards your specimen, which scatters many electrons. The specimen *and any other part of the TEM* that is hit by these electrons emit both characteristic and bremsstrahlung X-rays (which have energies up to that of the electron beam). X-rays of several tens or hundreds of keV can penetrate long distances into the material and fluoresce characteristic X-rays from anything that they hit. Ideally, the XEDS should only 'see' the X-rays from the beam-specimen interaction volume. However, as shown in Figure 32.14, it is not possible to prevent radiation from the stage and other areas of the specimen from entering the detector. As you can see from Figure 32.14, the XEDS has a collimator in front of the detector crystal. This collimator is the last line of defense against the entry of undesired radiation from the stage region of the microscope.

It's worth checking if you have an ideal collimator, constructed of a high-Z material, such as W, Ta, or Pb, coated externally and internally with a low-Z material, such as Al, C, or Be. The low-Z coating minimizes the production of X-rays from backscattered electrons that happen to spiral into the collimator. The high-Z material absorbs high-energy bremsstrahlung. The inside of the collimator should be baffled to prevent any backscattered electrons from generating X-rays that then penetrate the detector. No collimator is entirely successful and we'll describe the contribution of system-produced X-rays in the next chapter.

> **THE COLLIMATOR**
> The collimator defines both the (desired) collection angle of the detector and the average take-off angle of X-rays entering the detector.

32.10.A Collection Angle

The detector collection angle (Ω) is the solid angle subtended at the analysis point on the specimen by the active area of the front face of the detector. The collection angle is shown in Figure 32.14 and is defined as

$$\Omega = \frac{A \cos \delta}{S^2} \qquad (32.4)$$

where A is the active area of the detector (usually 30 mm^2), S is the distance from the analysis point to the detector face, and δ is the angle between the normal to the detector face and a line from the detector to the specimen. In many XEDS systems, the detector crystal is tilted toward the specimen so $\delta = 0$, then $\Omega = A/S^2$. It is clear that to maximize Ω the detector should be placed as close to the specimen as possible.

As we'll see, in most AEM experiments, it is the low X-ray count rate that limits the accuracy of the experiment. Commercial Si(Li) crystals have A values from 10 to 30 mm^2 and 50 mm^2 is becoming more common. As a result, values of Ω dictated by the closest distance between the specimen and detector/collimator lie in the range from 0.3 down to 0.03 sr. ATW detectors invariably have lower Ω values than Be-window or windowless detectors because the polymer window has to be supported on a grid which reduces the collection angle by ~20%. IG detectors need a reflective window to prevent IR radiation from generating noise in the

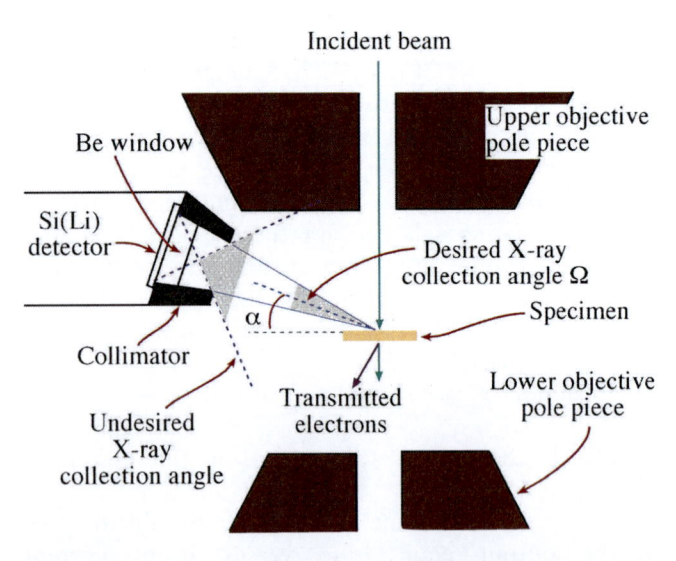

FIGURE 32.14. Schematic diagram of the interface between the XEDS and the AEM stage showing how the detector can 'see' X-rays from regions other than the beam-specimen interaction volume over the (relatively large) undesired collection angle. The (relatively small) desired collection angle Ω and take-off angle α are also shown.

detector. So, even with the largest detectors, Ω is a small fraction of the total solid angle of characteristic X-ray generation, which is, of course, 4π sr.

> **Ω**
> The value of Ω is the most important parameter in determining the quality of your X-ray analysis. You need three things for good AEM X-ray analysis: counts, counts, and more counts.

This value of Ω is calculated from the dimensions of the stage and the collimator. There is no way, however, that you can measure this critical parameter directly, although you can compare X-ray count rates between different detector systems using a standard specimen such as our thin NiO film and a known beam current. A useful figure of merit is the X-ray counts per second detected from a standard specimen, with a given beam current and a given detector collection angle (cps/nA/sr). Typically, for an AEM with a nominal Ω of 0.13 sr and a beam energy of 300 keV the figure of merit is >8000. For a beam energy of 100 keV, it is about 13,000 (Zemyan and Williams). The increase at lower keV is due to the increased ionization cross section.

The magnitude of Ω is limited because the upper polepiece of the objective lens gets in the way of the collimator, thus limiting S. To avoid this limitation, we could increase the polepiece gap but doing so would lower the maximum beam current and degrade the image resolution, both of which are highly undesirable. The advent of C_s-corrected AEMs has removed this limitation and will open up a new chapter in high count-rate X-ray analysis in AEMs (Watanabe and Williams). Arrays of SDDs have already been used to give solid angles >1 sr on PIXE systems (Doyle et al.), so the future of XEDS in the AEM has exciting possibilities.

32.10.B Take-Off Angle

The take-off angle α is the angle between the specimen surface (at 0° tilt) and a line to the center of the detector, as shown in Figure 32.14. Sometimes, it is also defined as the angle between the transmitted beam and the line to the detector, which is simply (90° + α). In the SEM/EPMA, we keep α high to minimize X-ray absorption in bulk specimens. Unfortunately, in an AEM, if we maximize α, the price we pay is lowering Ω. A high-angle detector has to be positioned above the upper objective polepiece and will be much farther from the specimen. In the EPMA, low Ω is not a problem because there are always sufficient X-rays from a bulk specimen, but in the AEM the highest possible Ω is essential, as we've already emphasized (counts, counts...).

In those AEMs where the detector has a high take-off angle but a low Ω, the poor X-ray count rate makes quantitative analysis much more time consuming. Keeping the detector below the polepiece restricts α to a maximum of about 20°. In most cases you will find that such a small value of α is not a problem because one of the major advantages of thin-specimen AEM compared to EPMA is that absorption can usually be neglected. If absorption is a problem in your particular specimen, you can reduce the X-ray path length by tilting it toward the detector, thus increasing α (see Section 35.5). We don't recommend tilting because it increases spurious effects (see Chapter 33) and also lowers the P/B (peak to background) ratio in the spectrum.

> **TAKE-OFF ANGLE AND COUNTS**
> We would like to maximize the take-off angle *and* maximize the count rate, but we can't.

32.10.C Orientation of the Detector to the Specimen

There are two simple questions that you must be able to answer.

(a) *Is the detector pointing on axis?* The detector is inserted to a point where it is almost touching the objective polepiece, and you hope that it is 'looking' at the region of your specimen that is on the optic axis when the specimen is eucentric and at zero tilt. To find out if your system is thus aligned, take a low-magnification X-ray map from a homogeneous specimen, such as our thin NiO film. If the detector is not pointing on axis, the map will show an asymmetric intensity. Alternatively, if you can't map at a low enough magnification, simply see how the Ni K_α intensity varies from area to area on the foil with the stage traverses set at zero and different areas selected using the beam deflectors. The maximum intensity should be recorded in the middle grid square and for some distance around. It is also instructive to do the same test with the specimen moved above or below the eucentric plane. Again, the maximum intensity should be recorded at eucentricity. If the intensity is asymmetric then the detector or the collimator is not well aligned and some of your precious X-ray counts are being shadowed from the detector, probably by the collimator. So you need to ask for technical help.

(b) *Where is the detector with respect to the image?* It is best if the detector is 'looking' toward a thin region of the specimen, rather than toward a thicker region, as shown in Figure 32.15A. This alignment minimizes the X-ray path length through your specimen. If the BF image rotates when changing magnification in your TEM then the apparent detector orientation with respect to the image (on the screen) will vary with

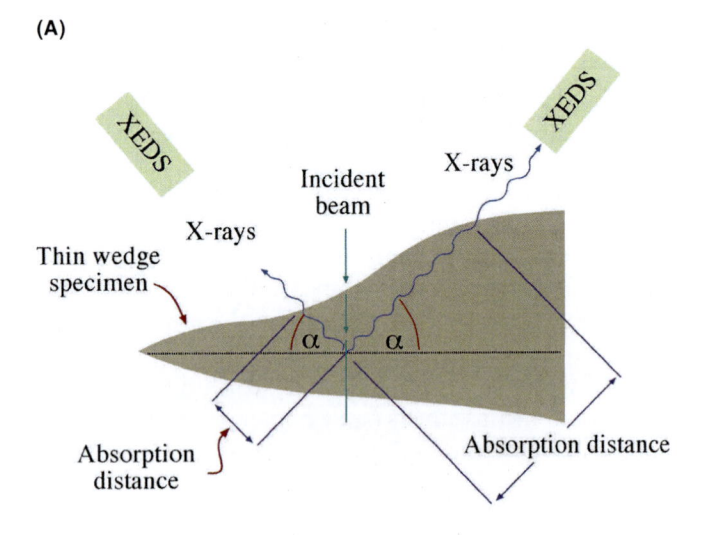

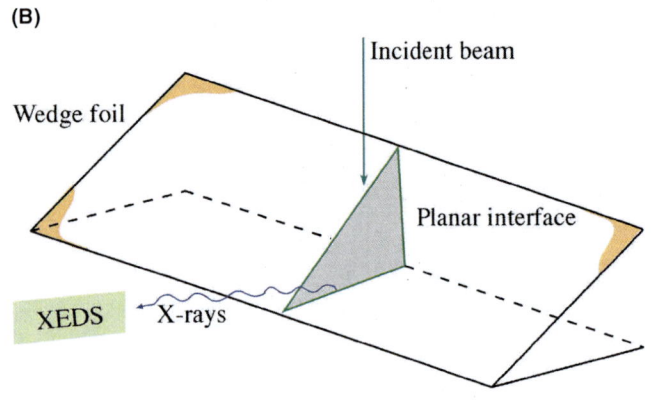

FIGURE 32.15. (A) The position of the XEDS relative to a wedge-shaped thin foil results in different X-ray path lengths. The shortest path length with the detector 'looking' at the thinnest region of the foil minimizes any X-ray absorption. (B) The preferred orientation of the XEDS when analyzing a planar defect: the interface plane is parallel to the detector axis and the incident-beam direction.

magnification. A STEM BF doesn't rotate so the relative orientation of the detector to your image is fixed. It is simple to find this orientation if the detector axis (y axis) is normal to a principal traverse axis (x) of the stage. If you press gently on the end of your holder, the image will move in the $+x$ direction. Then you can determine geometrically the direction ($+y$ or $-y$) along which the detector is 'looking' with respect to that $+x$ direction in the TEM image. In STEM, the image might be rotated 180° with respect to the TEM image so you have to take this into account. But it's all a good exercise in 3D geometry.

If you're doing analysis or mapping across a planar interface, which is a common AEM application, then you should also orient your specimen such that the interface is parallel to the detector axis and the beam, so the detected X-rays have come through regions of similar composition and don't cross the interface. A tilt-rotation holder is ideal for this but, if a low-background version is not available, then you may need to take out the holder and re-position your foil manually a few times until the interface is in the right orientation (see Figure 32.15B).

> **LOOKING AT THE SPECIMEN**
> The XEDS detector must be 'looking' at the thin edge of your specimen and aligned with any planar interface you are studying.

32.11 PROTECTING THE DETECTOR FROM INTENSE RADIATION

If you are not careful, the XEDS electronics can be temporarily saturated if high doses of electrons or X-rays hit the detector. The detector itself may also be damaged, particularly in intermediate-voltage AEMs. As we've noted, these situations occur when you inadvertently move thick areas of your specimen under the beam, or if you are traversing around a thin specimen and the support grid hits the beam. To avoid these problems, shutter systems are built into most XEDS systems which automatically protect the Si(Li) crystal if the AEM is switched to low magnification or if the pulse processor detects too high a flux of radiation.

If you don't have a shutter then you can physically retract the detector to lower Ω (if it is retracted along a line of sight to the specimen) or remove the detector from out of view of the specimen. The drawback to this approach is that constant retraction and reinsertion of the detector may cause undue wear on the sliding 'O'-ring seal. Also you may reposition your detector slightly differently each time, unless the system is designed so you can push the detector to a fixed stop, thus insuring a constant Ω and α. A shutter is highly recommended!

> **SHUT THE SHUTTER**
> To avoid reliance on the automatic system, it is best to have the shutter closed until you have decided which area you want to analyze and it is thin enough that the generated X-ray flux doesn't saturate the detector. Also, never have the objective aperture inserted.

CHAPTER SUMMARY

The XEDS (usually of the Si(Li) variety) is the only X-ray spectrometer currently used in TEMs. It is remarkably compact, efficient, and sensitive. A combination of Si(Li)/SDD and IG detectors can detect K_α lines from all the elements from Be to U. The XEDS is limited by its poor energy resolution, artifacts in the spectra, and the need for cooling, but it is simple to run and maintain. You must take care to perform certain basic procedures and refrain from certain others that can damage the detector. Sometimes, it may be too simple; beware. You need to

- Measure your detector resolution every 6 months at the Mn or Ni K_α line (at best 130–140 eV for Si(Li) or SDD and 120–130 eV for IG).
- Measure the ICC (FWTM/FWHM ratio of the Ni K_α line: ideally 1.82), every 6 months.
- Unless you have an SDD or a Peltier-cooled Si(Li), monitor ice build-up via in the Ni K_α/L_α ratio on a monthly basis.
- Check the calibration of the energy range of your computer display every 6 months especially if you have an SDD.
- Check the dead-time correction circuitry by the linearity of the output count rate versus beam current, every 6 months.
- Check the counts in a fixed clock time as a function of beam current to determine the maximum output count rate, every 6 months.
- Always operate with the shutter closed until you are ready for analysis.
- Always retract the objective diaphragm prior to analysis.
- Ensure the XEDS is pointing toward the thin edge of any wedge/disk specimen.

Interfacing your XEDS to the AEM is crucial since it determines the count rate, the need for an absorption correction, and the intrusion of spurious X-rays into your spectrum. In any decision involving XEDS in the AEM, you should always choose the option that optimizes the count rate.

For the sake of completeness, Table 32.2 shows you the relative merits of the various detectors that we have discussed in this chapter.

GENERAL TEXTS

Garratt-Reed, AJ and Bell, DC 2002 *Energy-dispersive X-ray Analysis in the Electron Microscope* Bios (Royal Microsc. Soc.) Oxford UK. Similar in scope to the XEDS chapters in this textbook.

Goldstein, JI, Newbury, DE, Echlin, P, Joy, DC, Romig, AD Jr, Lyman, C., Fiori, CE and Lifshin, E 2003 *Scanning Electron Microscopy and X-ray Microanalysis* 3rd Ed. Springer New York. In-depth treatment of all aspects of XEDS in the SEM/EPMA. Includes details of the electronics (Section 3.2.7)

Goodhew, PJ, Humphreys, FJ, and Beanland, R 2001 *Electron Microscopy and Analysis* 3rd Ed. Taylor and Francis New York. A broad introduction covering SEM, TEM, and AEM.

Jones, IP 1992 *Chemical Microanalysis Using Electron Beams* Institute of Materials London. Quantitative AEM; lots of calculations to illustrate the analytical principles; essential for the serious X-ray analyst.

Williams, DB, Goldstein, JI and Newbury, DE, Eds. 1995 *X-Ray Spectrometry in Electron Beam Instruments* Plenum Press New York. Tells all you need to know and more about X-ray detection and processing in SEM/EPMA (mainly) and TEM.

REFERENCES FOR THE CHAPTER

Doyle, BL, Walsh, DS, Kotula, PG, Rossi, P, Schülein, T and Rohde, M 2004 *An Annular Si Drift Detector μPIXE System Using AXSIA Analysis* X-Ray Spectrom. **34** 279–284. Illustrating the use of an array of SDDs.

Egerton RF and Cheng SC 1994 *The Use of NiO Test Specimens in Analytical Electron Microscopy* Ultramicrosc. **55** 43–54. As it says!.

Lund, MW 1995 *Current Trends in Si(Li) Detector Windows for Light Element Analysis* in *X-Ray Spectrometry in Electron Beam Instruments* DB Williams, JI Goldstein and DE Newbury, Eds. 21–31 Plenum Press New York. Detailed review of windows.

Lyman, CE, Newbury, DE, Goldstein, JI, Williams, DB, Romig, AD Jr, Armstrong, JT, Echlin, PE, Fiori, CE, Joy, D., Lifshin, E and Peters, KR 1990 *Scanning Electron Microscopy, X-Ray Microanalysis and*

Analytical Electron Microscopy; A Laboratory Workbook Plenum Press New York. Includes some standard tests for XEDS.

Michael, JR, 1995 Energy-Dispersive X-ray Spectrometry in Ultra-High Vacuum Environments in X-Ray Spectrometry in Electron Beam Instruments Eds. DB Williams, JI Goldstein and DE Newbury p83 Plenum Press New York. Using the Ni K_α/Ni L_α ratio.

Mott, RB and Friel, JJ, 1995 Improving EDS Performance with Digital Pulse Processing in X-Ray Spectrometry in Electron Beam Instruments Eds. DB Williams, JI Goldstein and DE Newbury 127–155 Plenum Press New York. Digital processing as we discuss in Section 32.7.

Newbury, DE 2006 The New X-ray Mapping: X-ray Spectrum Imaging Above 100 kHz Output Count Rate with the Silicon Drift Detector Microscopy and Microanalysis 12 26–35. Using SDDs for mapping.

Sareen, RA, 1995 Germanium X-ray Detectors in X-Ray Spectrometry in Electron Beam Instruments Eds. DB Williams, JI Goldstein and DE Newbury 33–51 Plenum Press New York. IG detectors.

Statham, PJ 1995 Quantifying Benefits of Resolution and Count Rate in EDX Microanalysis in X-Ray Spectrometry in Electron Beam Instruments Eds. DB Williams, JI Goldstein and DE Newbury 101–126 Plenum Press New York. Balancing count rate and resolution.

Terauchi, M and Kawana, M 2006 Soft-X-ray Emission Spectroscopy Based on TEM—Toward a Total Electronic Structure Analysis Ultramicrosc. 106 1069–1075. CCD-based WDS.

Terauchi, M, Yamamoto, H and Tanaka, M 2001 X-ray Emission Spectroscopy, Transmission Electron Microscope, DOS of the Valence Band, Soft-X-ray Spectrometer, B K-emission Spectra, Hexagonal Boron-Nitride J. Electr. Microsc. 50(2) 101–104. Development of a sub-eV resolution soft-X-ray spectrometer for a transmission electron microscope.

Watanabe, M and Williams, DB 2006 Frontiers of X-ray Analysis in Analytical Electron Microscopy: Toward Atomic-Scale Resolution and Single-Atom Sensitivity Microscopy and Microanalysis 12 515–526. C_s-corrected AEM.

Wollman, DA, Nam, SW, Hilton, GC, Irwin, KD, Bergren, NF, Rudman, DA, Martinis, JM and Newbury DE 2000 Microcalorimeter Energy-Dispersive Spectrometry Using a Low Voltage Scanning Electron Microscopy J. Microsc 199 37–44. The bolometer for XEDS.

Zemyan, SM and Williams, DB 1995 Characterizing an Energy-Dispersive Spectrometer on an Analytical Electron Microscope in X-Ray Spectrometry in Electron Beam Instruments Eds. DB Williams, JI Goldstein and DE Newbury 203–219 Plenum Press New York. Summary of standard tests for XEDS.

URLs: SELECTION OF XEDS MANUFACTURERS

1) www.bruker-axs.de/
2) www.edax.com/
3) www.oxinst.com
4) www.pgt.com
5) www.thermo.com/

SELF-ASSESSMENT QUESTIONS

Q32.1 Define: XEDS, IG, SDD, AEM.
Q32.2 Explain in four steps how X-rays from the specimen are converted into a spectrum.
Q32.3 Distinguish between dead time, live time, and clock time.
Q32.4 Distinguish between dead layer(s) and active regions of detectors.
Q32.5 List four ways that you can damage your detector while you're operating the AEM.
Q32.6 What is more important during X-ray microanalysis and why: X-ray counts, X-ray energy resolution, X-ray take-off angle, or specimen tilt?
Q32.7 What is the best accelerating voltage to use for XEDS?
Q32.8 Why must Si detectors be doped with Li while IG detectors are not doped (Hint: what does the 'I' mean?)?
Q32.9 How does this Li affect the operation of the detector? Why is there no Li in a SDD?
Q32.10 What type of detector is better for detecting (a) high-Z materials, (b) lower-Z materials and why?
Q32.11 Why is pulse processing required to translate X-ray photons into a spectrum?
Q32.12 What is a reasonable dead time and how is this affected by the vintage of your detector electronics?
Q32.13 Why is digital pulse processing preferred over analog processing? Give one exception to this preference.
Q32.14 What advantage does XEDS have over WDS?
Q32.15 What are some of the disadvantages to XEDS?
Q32.16 Why is a large collection angle, Ω, useful? What limits the value of Ω in practice?
Q32.17 Why is a high take-off angle, α, useful? What limits the value of α in practice?
Q32.18 Why is a larger Ω more important than a higher α?
Q32.19 Why is the collimator so important in front of your detector?
Q32.20 What aspects of TEM design restrict the use of large arrays of detector crystals such as SDDs?
Q32.21 Why is it not a major issue that we cannot optimize the take-off angle α?

Q32.22 What materials are used in the collimator to avoid X-ray generation within the collimator and what might cause the generation of these X-rays in the first place?

Q32.23 You switch on your XEDS detector but you don't register any X-rays coming through the system. Explain what may be causing this situation.

TEXT-SPECIFIC QUESTIONS

T32.1 Explain what factors control the shape of the characteristic peaks and the background in a typical energy-dispersive spectrum such as Figure 32.2A.

T32.2 List three reasons why we cool Si(Li) or IG detectors with liquid N_2 and three undesirable consequences. Search the Web sites of the major EDS manufacturers to see if there are alternatives to liquid-N_2 cooling. Why do we not have to cool an SDD to liquid-N_2 temperatures?

T32.3 Explain why, in comparison to a Si(Li) detector, an IG detector exhibits (a) better energy resolution, (b) more resistance to high-energy electron damage, (c) less susceptibility to accidental loss of liquid N_2 and (d) better detection of high-energy X-rays. Given these advantages, why are IG detectors not more common on AEMs?

T32.4 Explain why, in comparison to a Si(Li) detector, a wavelength-dispersive (crystal) spectrometer (WDS) offers (a) better resolution, (b) more throughput of counts, (c) fewer artifacts. Given these advantages, why is the WDS not more common on AEMs?

T32.5 Why does the output X-ray count rate rise to a maximum and then fall as the input count rate continues to rise (as in Figure 32.12A)?

T32.6 So why don't we always operate the AEM at the maximum count rate (which would, inter alia, improve counting statistics, thus reducing errors), reduce the time required to acquire the spectrum (thus reducing damage to beam-sensitive specimens)?

T32.7 Consider the curves in Figure 32.12. Explain why the statements in the caption are true. (Are they?)

T32.8 Give examples of how the limitations of the curves in Figure 32.12A might be overcome?

T32.9 Explain why SDDs can operate at much higher count rates than other XEDS detectors.

T33.10 An engineer decided to remove the collimator baffles from her AEM. What does she put up with as a consequence?

T33.11 Why would it be good to have both the largest detector take-off angle and the highest detector collection angle? Why do we have to make a compromise choice in practice? Explain why correction of the spherical aberration coefficient of the objective lens would remove this compromise.

T32.12 Figure 32.9 shows a WDS system on a TEM. Why is it not commonly used now? (A careful discussion is required.)

T32.13 Look at Figure 32.9D. Why are the EDS peaks so much broader than those from the bolometer and could this possibly be improved?

T32.14 An image produced using a backscattered-electron detector shows the presence of three regions of significantly different contrast. Where should you begin to further analyze this specimen?

T32.15 A strong background appears in the 0–10 keV spectrum of a geological specimen. What is this specimen, and how can the high background be compensated for?

T32.16 Using a windowless detector it becomes apparent by comparing past and previous spectra of the same specimen that there is likely a coating of ice or hydrocarbons on the detector. Should you just thaw out the detector or not?

T32.17 After measuring XEDS spectra from a biological or polymeric specimen, you close your AEM session but find yourself physically unable to open the chamber to remove the specimen. What is likely to be the cause?

T32.18 You have decided to switch an IG detector in for the Si(Li) already in your microscope to identify X-ray lines above 20 keV. After the switch, you notice what appears to be a large increase in the number of escape peaks in your spectra. Name one reason why this might be happening.

T32.19 Barry and David are laughing about something you did in lab last week. You were attempting to optimize the energy resolution of the XEDS spectra you obtained. What is their point?

T32.20 Are there hidden peaks obscured by the low-energy edge of several of your more intense peaks and what can you do about it if there are?

T32.21 A novice at using the XEDS on the TEM decides to employ WDS instead, with the notion in mind that it wouldn't be such a bad thing to get more resolution for the spectra. Which rude awakening is this person close to discovering?

T32.22 A geological specimen is elementally analyzed using XEDS. This turns out to contain Si, Al, O, and Fe. Out of curiosity, an X-ray map in search of Cu is performed on this same region of the specimen. Multiple locations of Cu spring up in the map. Explain.

CHAPTER SUMMARY

33

X-ray Spectra and Images

CHAPTER PREVIEW

The X-ray spectrum generated within your specimen consists of element-specific characteristic peaks with well-defined energies superimposed on a non-characteristic background. While the XEDS system is a remarkable piece of technology, we've already described its limited resolution and we will see in this chapter that it is also prone to creating small artifact peaks in the spectrum. Furthermore, the unavoidable presence of scattered electrons and X-rays within the AEM conspire to degrade the quality of the generated spectrum and increase the number of false peaks in the displayed spectrum. The AEM illumination system and specimen stage are rich sources of powerful radiation, not all of it by any means coming from the area of interest in your specimen. So you have to take precautions to ensure that the X-ray spectrum you collect comes predominantly from the area of the specimen that you want to analyze and we describe several tests you should perform to ensure that the XEDS-TEM interface is optimized. Once you understand the desirable and undesirable contents of a spectrum, we'll show you the various ways to gather your spectra, display them, and also form X-ray images from them. In particular, digital spectrum imaging is the most powerful technique for optimizing the information that you can gather from the low number of counts generated in thin foils.

33.1 THE IDEAL SPECTRUM

Back in Chapter 4 we learned that electrons generate two kinds of X-rays. When electrons ionize an atom which returns to ground state, the emitted characteristic X-ray *energy* is unique to the ionized atom (Figure 4.2). When electrons are slowed by electrostatic interaction with the nucleus they produce a continuum of bremsstrahlung X-rays and together the characteristic peaks and the bremsstrahlung background comprise the X-ray spectra we detect and display via the XEDS (as shown schematically back in Figure 4.6 and experimentally in Figure 32.2 and many others).

33.1.A The Characteristic Peaks

Some more detailed revision: a beam electron ionizes an atom in your specimen by ejecting an inner or core-shell electron, leaving a hole in the shell. The probability of this event occurring is governed by the ionization cross section. Then a cascade of electron transitions occurs, with each transition filling the hole with an electron from a more weakly bound shell (leaving a hole in that shell, and so on) and ultimately the last electron falls into a core-shell from the conduction band. Depending on the fluorescence yield, each transition results in either a characteristic X-ray or an Auger electron. The characteristic X-rays have a well-defined energy and a natural 'line width' (the FWHM of the Gaussian distribution of X-ray energies) of typically 1–5 eV. But, as you already know from the previous chapter, the XEDS degrades this width to a Gaussian-shaped peak with a FWHM of about 135 ± 10 eV. We'll use the term 'line' to denote the actual X-ray energy at the peak of the Gaussian and we'll talk about K, L, and M families of lines (and indeed the XEDS computer display includes lines superimposed on the spectra at the K, L, M, etc., peak energies). The actual number of counts (intensities) of the characteristic peaks from a given element, and the relative differences between spectra from different elements, are really quite complex and we'll go into this more in the next chapter when we discuss details of spectra and qualitative analysis. For the time being, all you need to know is that, generally speaking, lower-energy X-ray peaks are more intense than higher-energy ones and the heavier the element, the more complex the characteristic spectrum. Also X-rays with energies < ~1 keV are absorbed both within your specimen and within the detector and ultimately, the combination of absorption and low fluorescence yield means it is not possible to detect X-rays with energies < ~110 eV (the Be K line).

33.1.B The Continuum Bremsstrahlung Background

The background in the X-ray spectrum is bremsstrahlung (braking) radiation arising as beam electrons are slowed down or stopped by electrostatic interactions with nuclei in the specimen. The intensity of the bremsstrahlung is zero at the beam energy (since we can't create more energy than the beam has) and rises until it is effectively infinite at zero energy. This distribution is well understood and described mathematically by variations on Kramers' law as we also noted back in Figure 4.6. Modifications to Kramers' law account well for the absorption of the X-rays and accurately describe the experimentally detected bremsstrahlung distribution, even from thin specimens. So the net result is that X-ray spectra consist, as we have seen, for example, in Figures 1.4A and 32.2, of characteristic, Gaussian-shaped peaks superimposed on a background of bremsstrahlung X-rays, most clearly visible in Figure 32.2F. Many more spectra will appear throughout this and subsequent chapters and, as already noted, we can create elemental images from specific characteristic peaks.

Unfortunately, what makes X-ray spectrometry challenging is that the spectrum generated within your thin specimen, the spectrum detected by your Si(Li) detector, and the spectrum displayed on your computer screen are all quite different. So understanding what controls the X-ray counts that you measure in the characteristic peaks in your spectra is absolutely essential. Otherwise it will be difficult to first identify (Chapter 34) and then quantify (Chapter 35) the peak intensities in terms of the presence and the amount, respectively, of the elements in the analysis volume of your specimen.

We'll first describe how the XEDS system generates artifact (escape, sum, and internal fluorescence) peaks and then we'll show how the AEM itself also contributes unwanted X-rays, all of which can confuse the unwary operator into thinking certain elements are in their specimen when, in fact, they are not. Misinterpretation of the presence or absence of certain elemental signals could cause real problems if, for example, you are making decisions on the suitability or otherwise of a specific material for a given application, or even more seriously, making a forensic or medical decision based on the chemistry of your specimen: more about this in Chapter 34.

33.2 ARTIFACTS COMMON TO Si(Li) XEDS SYSTEMS

The XEDS system introduces its own artifacts into the spectrum. Fortunately, we understand all these artifacts, but they still occasionally mislead the unwary operator; see the review by Newbury. We can separate the artifacts into two groups

- Signal-detection artifacts: examples are escape peaks and internal fluorescence peaks.
- Signal-processing artifacts: e.g., the sum peaks.

Escape peak: Because the detector is not a perfect sink for all the X-ray energy, it is possible that a small fraction of the energy is lost and not transformed into electron-hole pairs. The easiest way for this to happen is if the incoming photon of energy E fluoresces a Si K_α X-ray (energy 1.74 keV) which escapes from the intrinsic region of the detector. The detector then registers an apparent X-ray energy of $(E - 1.74)$ keV, as shown in Figure 33.1.

> **ESCAPE PEAK**
> Si escape peaks appear in the spectrum 1.74 keV below the true characteristic peak position.

The magnitude of the escape peak depends on the design of the detector and the energy of the fluorescing X-ray. The most efficient X-ray to fluoresce Si K_α X-rays is the P K_α, but in a well-designed detector even the P escape peak will only amount to <2% of the P K_α intensity. This fact explains why you can only see escape peaks if there are major characteristic peaks in the spectrum. More escape peaks occur in IG spectra because we can fluoresce both Ge K_α (9.89 keV) and L_α (1.19 keV)

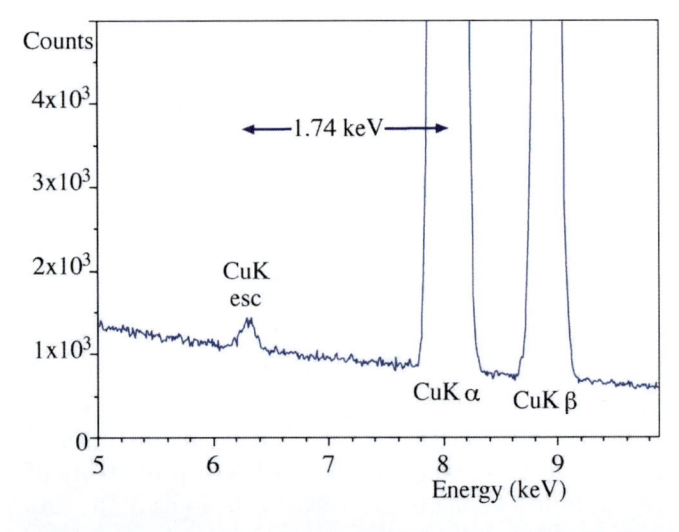

FIGURE 33.1. The escape peak in a spectrum from pure Cu, 1.74 keV below the Cu K_α peak. The intense K_α peak is truncated in the display because it is 50–100× more intense than the escape peak.

characteristic X-rays in the detector. Each of these will cause corresponding escape peaks. SDDs are thinner than typical Si(Li) crystals so there may be a slightly enhanced possibility for an escape peak. The analysis software should be able to recognize any escape peak, remove it, and add the intensity back into the characteristic peak where it belongs. Because the escape-peak intensity is so small it is rarely a problem unless you misinterpret it as coming from an element which is not present.

The internal fluorescence peak: This is a characteristic peak from the Si (or Ge) in the detector dead layer. Incoming photons can fluoresce atoms in the dead layer and the resulting Si K_α or Ge K/L X-rays enter the intrinsic region of the detector which cannot distinguish their source and, therefore, register a small peak in the spectrum. As semiconductor detector design has improved and dead layers have decreased in thickness, the internal fluorescence peak artifact has shrunk but it has not yet disappeared entirely. So beware.

Obviously if you are looking for small amounts of Si in your specimen you'll always find it in a Si(Li) spectrum after long enough counting time! Depending on the dead-layer thickness, the Si K peak intensity corresponds to ~0.1–1% of the specimen composition (see Figure 33.2) so it is hardly a major problem if you are aware of it. Similar effects are observed in IG detectors also, but the very thin dead layer in SDDs should be an advantage.

LONG COUNTING TIMES
A small Si K_α peak will occur in ALL spectra from Si(Li) detectors after long counting times.

Sum peak: As we described earlier, the processing electronics are designed to switch off the detector while each pulse is analyzed and assigned to the correct energy channel. The sum peak arises when the count rate exceeds the electronics' ability to discriminate all the individual pulses and so-called 'pulse pile-up' occurs. This is likely to occur when

- The input count rate is high.
- The dead times are $>\sim 60\%$.
- There are major characteristic peaks in the spectrum.

The electronics simply cannot be perfect. Occasionally, two photons will enter the detector at exactly the same time. The analyzer then registers an energy corresponding to the sum of the two photons. Since this coincidence event is most likely for the X-ray giving the major peak, a sum peak first appears at twice the energy of the major peak, as shown in Figure 33.3.

If you are using an SDD because you have a C_s-corrected AEM and are seeking to maximize count rates, you can generate multiple sum peaks, causing serious interpretation problems in your analyses. There is evidence (see Newbury's paper) that digital processing is, in fact, worse than analog processing in terms of generating sum peaks in SDD systems, and hybrid analog/digital systems are being considered in the SEM field where this is a bigger problem.

THE SUM PEAK
The sum peak should be invisible if you maintain a reasonable input count rate, typically <10,000 cps, which should give a dead time of <60%.

Since you can't usually generate very high count rates, unless your specimen is really thick, there is little need to worry about sum peaks in AEM spectra but, as always,

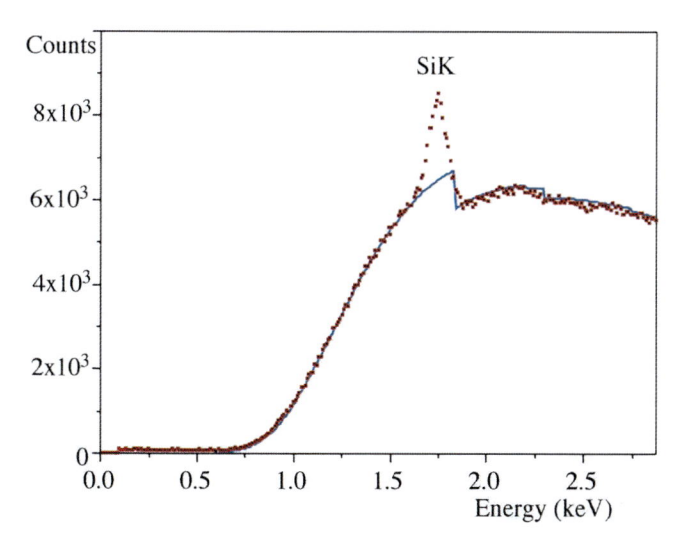

FIGURE 33.2. The Si internal fluorescence peak in a spectrum from pure carbon obtained with a Be-window Si(Li) detector. The ideal spectrum is fitted as a continuous line that only shows the Si absorption edge.

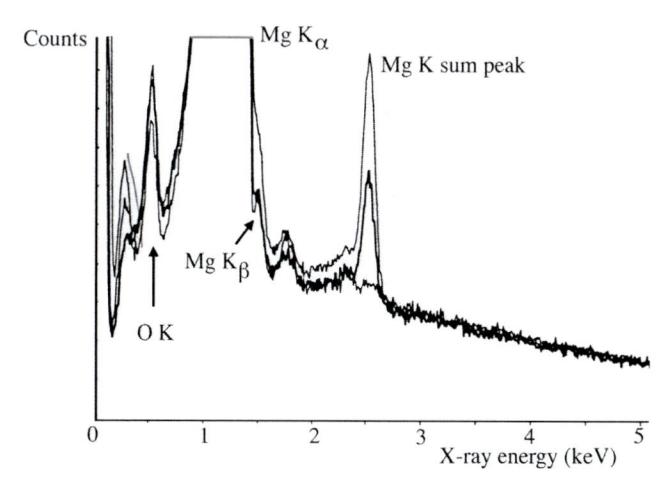

FIGURE 33.3. The Mg K sum (coincidence) peak occurs at twice the Mg K_α peak in this spectrum from a bulk specimen of (oxidized) pure Mg. The sum peak decreases change rapidly with decreasing dead times; upper trace 70%, middle trace 47%, lower trace 14% dead time. The sum-peak artifact is close to the background intensity at 14% dead time.

33.2 Artifacts Common to Si(Li) XEDS Systems 607

you should at least be aware of the danger. (For example, the Ar K_α energy is almost exactly twice the Al K_α energy and the sum peak has led some researchers to report Ar being present in Al specimens when it wasn't, and has caused others to ignore Ar, which actually was present in ion-milled specimens!) One exception to this lack of concern is if you have intense low-energy peaks where the residual noise in the electronic circuitry interferes with the pile-up rejection. So if you're analyzing elements lighter than Mg, take care to use low input count rates. Reducing the dead time to 10–20% should remove even the Mg K_α sum peak, as shown in Figure 33.3.

Much of what we have just discussed and most of what we'll cover in Chapters 34 and 35, can be observed experimentally on the AEM. But it is often just as instructive, and certainly easier, to simulate the spectra. To this end, we strongly advise you to download the public-domain simulation software 'Desktop Spectrum Analyzer' (DTSA) from NIST (Fiori and Swyt and URL #1); it is listed in the recommended software in Section 1.6 and its use is described in depth in the companion text. In fact, almost all the spectra in this and the subsequent chapter were generated by DTSA (in part, to ensure a uniform appearance). This software permits realistic simulation of XEDS data in TEMs (and SEMs) and introduces you to all the aspects of spectral processing, artifacts, modeling, etc., that are discussed in this and the next three chapters.

33.3 THE REAL SPECTRUM

In the perfect AEM, all spectra would be characteristic only of the chosen analysis region of your specimen with which the beam interacts. The analysis of bulk specimens in the SEM/EPMA approaches this ideal but, in the AEM, two factors combine to introduce false information which we call system and spurious X-rays. These X-rays introduce small errors into both qualitative and quantitative analysis, unless you are aware of the dangers and take appropriate precautions to identify and minimize the problems. The two factors that make the AEM different to the EPMA and that are responsible for these problems are

- The high-voltage electrons which generate intense doses of stray X-rays and scattered electrons in the illumination system.
- The thin specimen which scatters high-energy electrons and X-rays around the limited confines of the AEM stage.

Modern AEMs are designed to minimize some of these problems, but they have not disappeared entirely. It is important for you to understand that these problems are *not* major distortions of the X-ray spectrum, and only introduce small peaks or changes in intensity equivalent to composition changes at the 1% level or less. Nevertheless, identification and quantification of such small elemental amounts are often the *raison d'être* for an analysis (e.g., impurity segregation), so you have to be wary of these artifacts which we'll now discuss in some detail. Remember, these artifacts are in addition to the XEDS artifacts, described in the previous section.

33.3.A Pre-specimen Effects

The TEM illumination system produces high-energy bremsstrahlung X-rays and electrons scattered outside the main beam, both of which may strike the specimen, producing spurious X-rays.

In inhomogeneous specimens (which are usually just the kind that we want to analyze) the presence of significant amounts of spurious X-rays means that the quantification process could give the wrong answer. There are reviews (e.g., Williams and Goldstein or Allard and Blake), which describe in detail how to identify and minimize these artifacts from the illumination system, so we will just describe the precautions necessary to insure that the AEM is operating acceptably. Since these artifacts are primarily a result of the high-energy electrons interacting with column components, such as diaphragms and polepieces, you must take extra care when using intermediate-voltage instruments.

> **SPURIOUS X-RAYS**
> We define **spurious X-rays** as those that come from the specimen but are **not** generated by the electron probe in the chosen analysis region.

The standard way to detect stray radiation from the illumination system is to position the focused electron beam to pass through a hole in your specimen and see if you can detect an X-ray spectrum characteristic of the specimen.

If the major peak in the hole-count spectrum contains more than a few percent of the characteristic intensity in the same major peak obtained from a thin area of your specimen, under similar conditions, then the illumination system needs attention.

You can easily determine whether stray electrons or X-rays are the problem, as illustrated in Figure 33.4, which shows spectra obtained on a silver disk specimen (A) and down the hole (B). Ag has a high-energy K_α line at ~23 keV and a low-energy L_α line at ~3 keV. With the beam through the hole in the disk, stray X-rays will fluoresce the high-energy peak more efficiently and stray electrons will preferentially excite the low-energy peak. So changes in the K/L ratio from a heavy metal test specimen such as Ag are determinative experiments. The NiO specimen that we described in Chapter 32 is available on Mo support grids and the Mo K/L ratio is equally diagnostic.

THE HOLE COUNT

We cannot reduce spurious radiation sources to zero, so a spectrum, sometimes termed a 'hole count,' is **invariably** obtained in all AEMs if you count for long enough.

Let's consider first the problem of high-energy bremsstrahlung X-rays generated in and penetrating through a standard Pt C2 diaphragm (go back and look at Figure 6.10 to remind yourself why diaphragms stop beam electrons). These X-rays flood your specimen and fluoresce characteristic X-rays from a large area around the analysis region. These stray X-rays would give a high Ag (or Mo) K/L ratio in the spectrum from your test specimen (see Figure 33.4B).

These diaphragms should be a standard fixture in your AEM (check with your lab manager) but they are expensive and they cannot be flame-cleaned in the usual way. When the thick diaphragms do contaminate, you have to discard them, otherwise the contamination itself will become a source of X-rays and will also deviate electrons out of the main beam by charging.

Pt TOP HAT

The solution to this problem is to use very thick (several mm) platinum diaphragms which have a top-hat shape, and a slightly tapered bore to maintain good electron collimation (see Figure 33.4C).

Alternatively, some AEMs incorporate a small diaphragm just above the upper objective lens to shadow the thicker outer regions of the specimen from stray X-rays.

Another way you can minimize the effects of this undesirable bremsstrahlung is to use an evaporated film or window-polished flake on a Be grid rather than a self-supporting disk, but you don't always have this choice. If your specimen is thinner than the path length for fluorescence (a few tens of nm in many cases), spurious X-rays will not be generated. Of course, it may not be possible to prepare such thin specimens, or it may take a great deal of effort, while self-supporting disks are relatively easy and quick to produce; so this isn't a popular suggestion with graduate students. However, with increasing interest in the properties of nano-scale

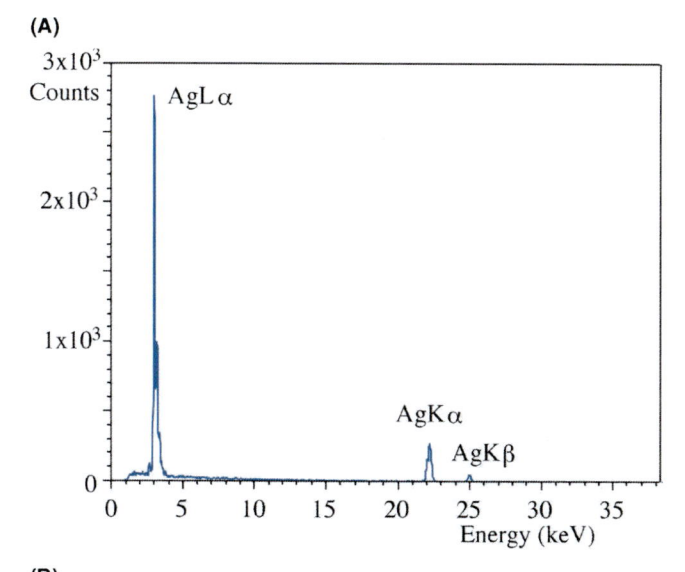

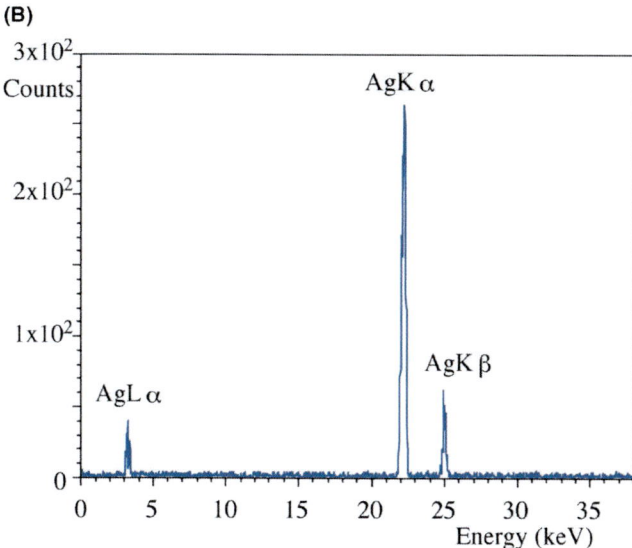

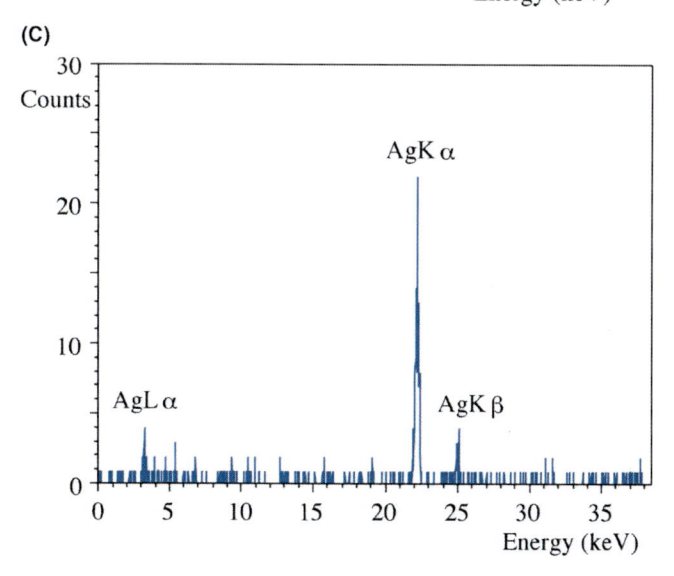

FIGURE 33.4. The 'hole count.' (A) A Ag self-supporting disk produces an electron-characteristic (high L/K ratio) spectrum when struck by the primary beam. (B) Without a thick C2 diaphragm, a reasonably intense Ag spectrum is also detected when the beam is placed down a hole in the specimen. This spectrum has a high K/L ratio which indicates high-energy

FIGURE 33.4 (Continued) bremsstrahlung fluorescence. Note that the (spurious) Ag K_α hole-count intensity (B) is ~ 50% of the (real) Ag K_α intensity (A) recorded from the specimen! (C) Use of a thick Pt C2 diaphragm reduces the intensity of the Ag K_α hole count substantially. The K_α intensity in (C) is about 30 times less than in (B). (Note the scale change.)

thin films you may be fortunate enough to be studying such ideal specimens.

For a quantitative, reproducible measure of the hole count, you should use a uniform thin specimen, such as the NiO film we have described. This film should be supported on a bulk material that has a low-energy (<~3 keV) L line and a high-energy (>~15 keV) K line. A thick Mo or Au washer or support grid is ideal. Any high-energy bremsstrahlung X-rays penetrating through the C2 diaphragm will strongly fluoresce the Mo K or Au L line while stray electrons will excite the Mo L or Au M lines preferentially.

With thick diaphragms and thin foils, the remaining stray X-rays will not influence the accuracy of quantification or introduce detectable peaks from elements not in the analysis region. For more details on this test, see Lyman and Ackland. If you don't want to go to the trouble of this test, then the least you should do is measure the in-hole spectrum on your specimen and subtract it from your experimental spectrum if the hole spectrum contains peaks with >1% of the counts in the characteristic peaks that come from your specimen.

> **RULE OF THUMB**
> The ratio of Mo K_α or Au L_α intensity detected (when the beam is down the hole) to the Ni K_α intensity obtained with the beam on the specimen should be less than 1%.

Now that we've minimized stray X-rays, let's consider the possibility that all the electrons are not confined to the beam. Stray electrons will generate a hole spectrum with a low K/L ratio (the opposite of Figure 33.4B). If your microscope has a non-beam defining spray diaphragm below the C2 diaphragm it will eliminate such stray electrons without generating unwanted X-rays. Then the main source of poorly collimated electrons is usually the 'tail' of electrons around the non-Gaussian-shaped probe that arises from spherical aberration in the C2 lens, as shown in Figure 33.5 (Cliff and Kenway). The best way to minimize this effect is to image the beam on the TEM screen under the conditions that you will use for analysis, and select the best C2 aperture size to define the probe, as we discussed way back in Section 5.5. It is a simple test to move your probe closer and closer to the edge of your specimen and see when you start generating X-rays. Do this with different size, top-hat C2 diaphragms. If you are fortunate enough to have a C_s corrector in your probe-forming lens then the probe will have much smaller tails, but this is a very expensive solution, not available to all.

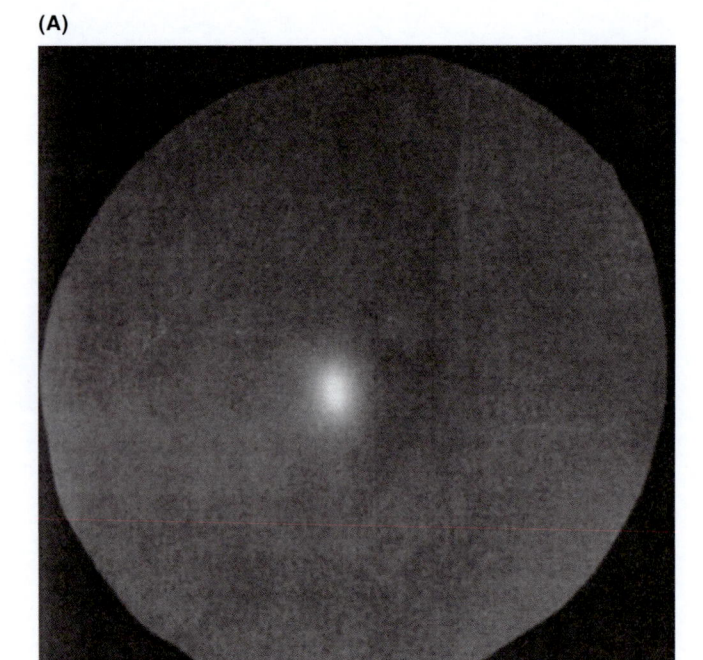

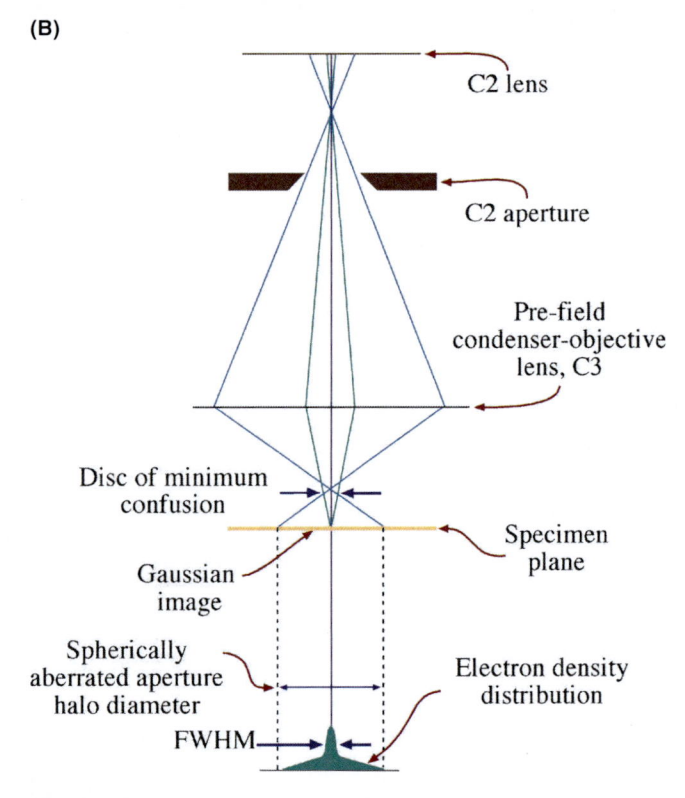

FIGURE 33.5. (A) The shadow of the C2 diaphragm defines the extent of the halo of electrons which excites X-rays well away from the chosen analysis region. (B) Ray diagram showing how the STEM probe formed with a large C2 aperture generates a broad halo around the intense Gaussian-shaped probe. Such beam tailing is the major source of uncollimated electrons and arises due to spherical aberration in the probe-forming lens. Choosing a smaller aperture will limit the tailing but cut down the probe current.

In summary

- Stray X-rays will give a high Ag (or Mo) K/L ratio.
- Stray electrons will give a low Ag (or Mo) K/L ratio.
- Always operate with clean, thick, top-hat C2 diaphragms.
- Use very thin flake specimens or uniform thin films, if possible.
- Always image the electron beam on the TEM screen prior to analysis, to insure that it is well collimated by the C2 aperture.
- Use a C_s-corrected AEM if you can find one.

Now the only significant X-ray source will be the region where you position the probe.

33.3.B Post-specimen Scatter

The electrons interact with the specimen and are scattered elastically or inelastically. It is fortunate for us that the intensity of such scatter from a thin specimen is greatest in the forward direction. So most of the scattered electrons are gathered by the field of the lower objective polepiece and proceed into the imaging system of the AEM away from the XEDS detector. Unfortunately, some electrons are scattered through high enough angles that they strike other parts of your specimen, the support grid, the holder, or the objective lens polepiece or other material in the AEM stage.

> **SYSTEM X-RAYS**
> X-rays that come from parts of the AEM other than the specimen.

It is instructive to insert the objective diaphragm during an analysis (just once!) to see the enormous increase in spurious and system X-rays that result. (Actually, you'll do this experiment by mistake anyway.) Usually the X-ray flux is so great that the pulse-processing electronics are saturated, the dead time reaches 100%, and the automatic shutter will activate. However, even when you remove the diaphragm, electrons scattered by the specimen may still create X-rays characteristic of the materials in the holder (brass), the polepiece (Fe and Cu), and the collimator (e.g., Al, W) and any of these X-rays could be picked up by the XEDS detector.

> **Cu IS EVERYWHERE**
> Remember the post-specimen scatter will still generate specimen-characteristic X-rays remote from the area of interest, even if a Cu peak is not present.

Furthermore, despite the strong field of the upper objective polepieces in probe-forming STEMs (another good reason to always operate in STEM mode) some back-scattered electrons may travel directly into the XEDS detector, generating electron-hole pairs. Other scattered electrons may hit your specimen at some point well away from the area of interest where they will still produce specimen-characteristic (and therefore) spurious, X-rays. All these possibilities are undesirable but unavoidable because, without the beam-specimen interactions that produce this scattering, we would get no information at all from the specimen. Figure 33.6 summarizes all the possible sources of spurious and system X-rays from post-specimen scatter.

In addition to electron scatter, there will be a flux of bremsstrahlung X-rays produced in the specimen. The intensity of these X-rays is also greatest in the forward direction (see gray shaded area in Figure 33.6). Since they possess a full spectrum of energy, the bremsstrahlung will fluoresce some characteristic X-rays from any material that they strike. The easiest way to discern the magnitude of this problem is to use a uniformly thin foil (such as our standard NiO film) on a Cu grid. When you position the probe on the film in the middle of a grid square, many micrometers from any grid bar, the collected spectrum will invariably show a Cu peak arising from the grid, as a

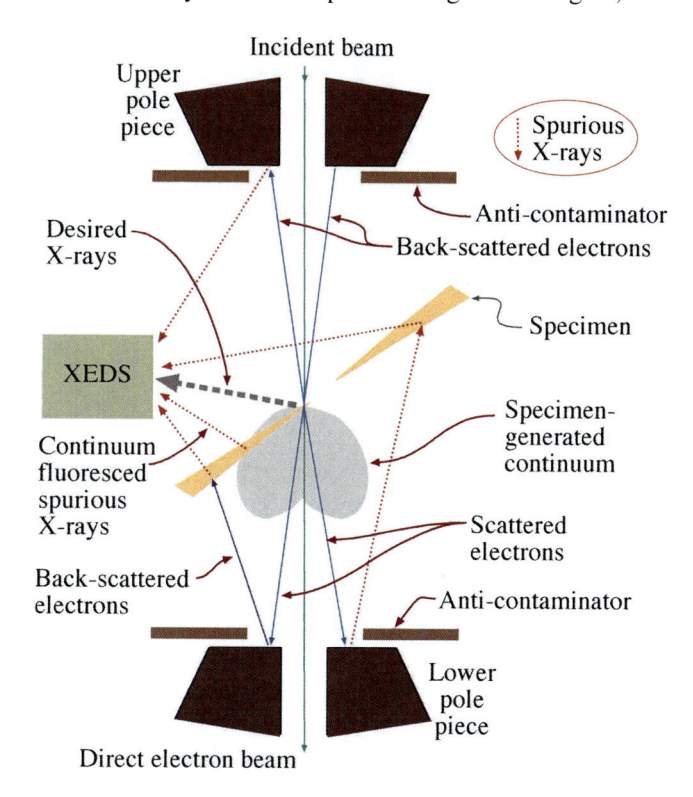

FIGURE 33.6. Sources of system and spurious X-rays generated when the incident beam (green) is scattered by the specimen. BSEs and forward-scattered electrons (blue) excite system X-rays in the stage and spurious X-rays (red dotted lines) elsewhere in the specimen. Bremsstrahlung (gray-shaded region) fluoresces the specimen away from the analysis region (also red dotted lines). The gray dotted line represents the desired X-rays from the analysis region.

result of interactions with electrons or X-rays scattered by the film. An example of this effect in a Cr film is shown in Figure 33.7. The presence of the Cu peak can be removed by using a Be grid, since Be K_α X-rays are much more difficult to detect. However, using Be grids merely removes the observable effect in the spectrum, not the cause.

To minimize the effects of the scattered radiation, you should keep your specimen close to zero tilt (i.e., normal to the beam). If you tilt $< \sim 10°$ then the background intensity is not measurably increased. Under these conditions, your specimen will undergo minimum interaction with both the forward-directed X-rays and any backscattered electrons. Both of these phenomena have only a small horizontal component of intensity. The effects of your specimen interacting with self-generated X-rays will be further reduced by using specimens, such as evaporated films or window-polished flakes rather than self-supporting disks, just as we suggested in the previous section. In self-supporting disks, the bulk regions will interact more strongly with the bremsstrahlung. It is not known what fraction of the post-specimen scatter consists of electrons and what fraction is X-rays, because this will vary with both specimen and microscope conditions. However, there is no evidence to suggest that this X-ray fluorescence limits the accuracy of quantitative analyses (which is at best $\pm 3-5\%$, as we'll see in Chapter 35).

In addition to keeping your specimen close to zero tilt, you can further reduce the effects of post-specimen scatter by surrounding the specimen with low-Z material. Use of these materials will also remove from the spectrum any characteristic peaks due to the microscope constituents. Be is the best material for this purpose and, as we said right at the beginning of this part of the book, Be specimen holders and Be support grids are essential for X-ray analysis. Ideally, all solid surfaces in the microscope stage region that could be struck by scattered radiation should also be shielded with Be. Unfortunately, such modifications are rarely available commercially.

> **Be**
> Be oxide is highly toxic if inhaled, so if you have to handle Be grids or other Be components, use gloves and tweezers and don't breathe!

The narrow polepiece gap, required to produce high probe currents, and the cold finger, used to reduce hydrocarbon contamination, both tend to increase the problems associated with post-specimen scatter. In the ideal AEM, the vacuum would be such that a cold finger would not be necessary and the polepiece gap would be chosen to optimize both the detected peak to background ratio and the probe current. When an AEM stage was substantially modified with low-Z material (e.g., by Lyman et al.) a large reduction in bremsstrahlung intensity was reported and X-ray peak to background ratios were produced that are still unmatched by most commercial AEMs. We'll discuss this more in Section 33.5.

You must note, however, that whatever precautions you take, the scattered electrons and X-rays, which are invariably present, result in a specific limitation to X-ray analysis. (See the below box on this.)

> **SMALL AND LARGE**
> If you are seeking small amounts ($<2\%$) of element A in a specific region of your specimen, and that same element A is present in large amounts, either elsewhere in your specimen or in the microscope stage, then you *cannot* unequivocally identify the presence of that element A in the specific region of your specimen! If you count for long enough, a small peak from A will *invariably* be present in all spectra, just as surely as the Si internal-fluorescence peak from your detector will be present.

Obviously then, you must determine the contributions to the X-ray spectrum from your microscope, and this is best achieved by inserting a low-Z specimen in the beam that generates mainly a bremsstrahlung spectrum, such as an amorphous-carbon film, supported on a Be grid or a pure B foil. If a spectrum is accumulated for a substantial period of time (say 10–20 minutes, or even over lunch), then in addition to the C or B peak (if your XEDS can detect them) the various instrumental

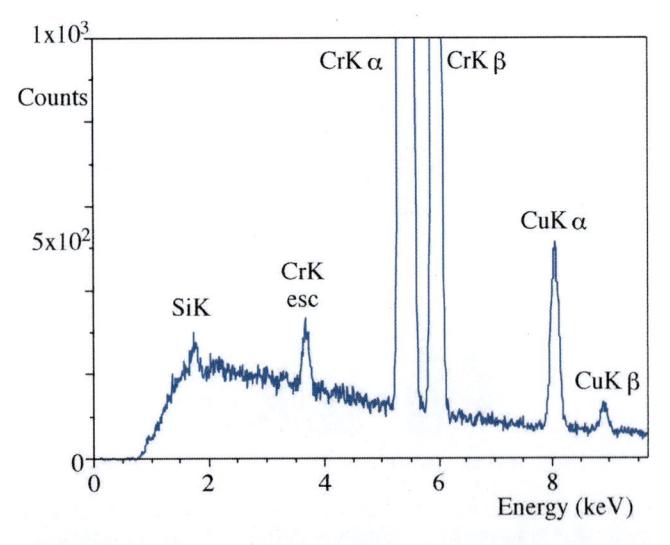

FIGURE 33.7. Cu peaks in a spectrum from a thin Cr film on a Cu grid. Although the beam was many micrometers from the grid, Cu X-rays are excited by electron scatter from within the specimen, and their intensity generally increases with specimen tilt. The Cr escape peak and the Si internal-fluorescence peaks are also visible.

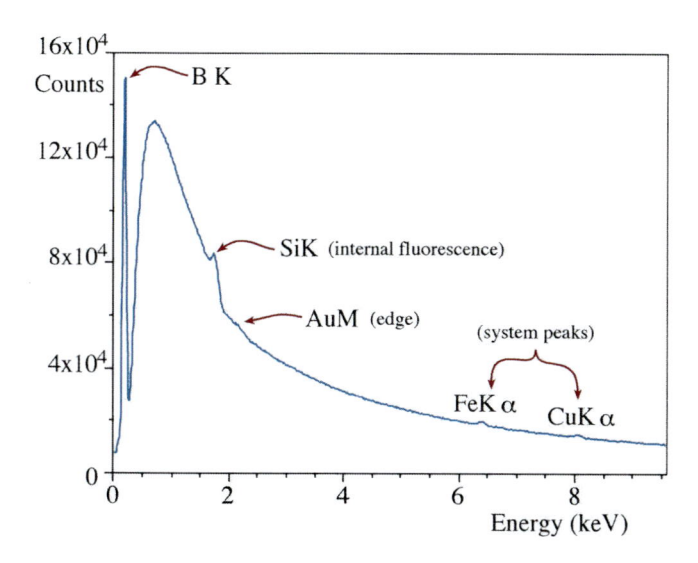

FIGURE 33.8. An XEDS spectrum from high purity boron, showing system peaks. The Si K_α peak and the Au M absorption edges are detector artifacts but the small peaks at 4.6 and 7.5 keV are system peaks from elements in the microscope stage.

experiments, but no one thought it would occur at AEM voltages until it was clearly demonstrated by Reese et al. Figure 33.9A shows a portion of an X-ray spectrum contributions to the spectrum should become visible. Such an 'instrument spectrum' (see Figure 33.8) should only exhibit the internal-fluorescence peak and possibly the Au absorption edge from the detector. Any other peaks will be from the TEM itself, assuming the specimen is pure. These peaks will tell you which elements it is *not* possible to seek in small quantities in your specimen because of their presence in your AEM.

We can summarize the methods used to minimize the effects of post-specimen scattering quite simply

- Always remove the objective diaphragm.
- Operate as close to zero tilt as possible.
- Use a Be specimen holder and Be grids.
- Use thin foils, flakes, or films rather than self-supporting disks.

Remember, that even with these precautions, you will still have to look out for artifacts in the spectrum, particularly those from the XEDS system.

33.3.C Coherent Bremsstrahlung

As we noted earlier, the bremsstrahlung spectrum is sometimes referred to as the continuum because the intensity is assumed to be a smooth, slowly varying function of energy. This assumption is perfectly reasonable when the bremsstrahlung is generated in bulk polycrystalline materials by electrons with energies $<\sim 30$ keV, such as in a SEM. However, in thin single-crystal specimens illuminated by high-energy electrons, it is possible to generate a bremsstrahlung X-ray spectrum that contains small, Gaussian-shaped peaks known as 'coherent bremsstrahlung' (CB). The phenomenon of CB is well known from high-energy physics

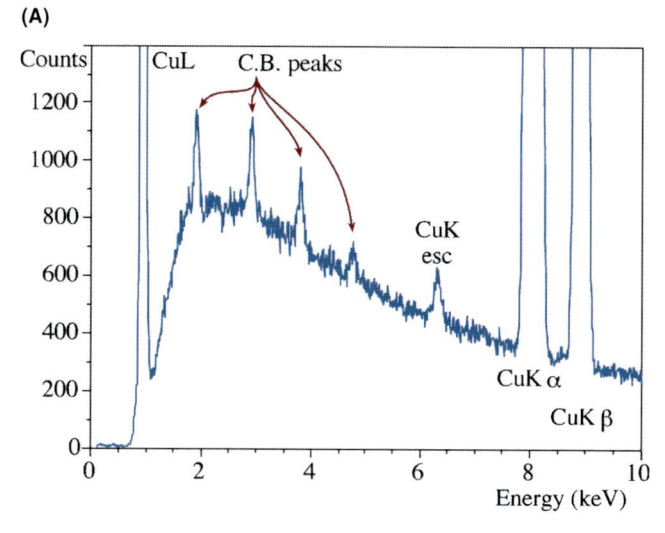

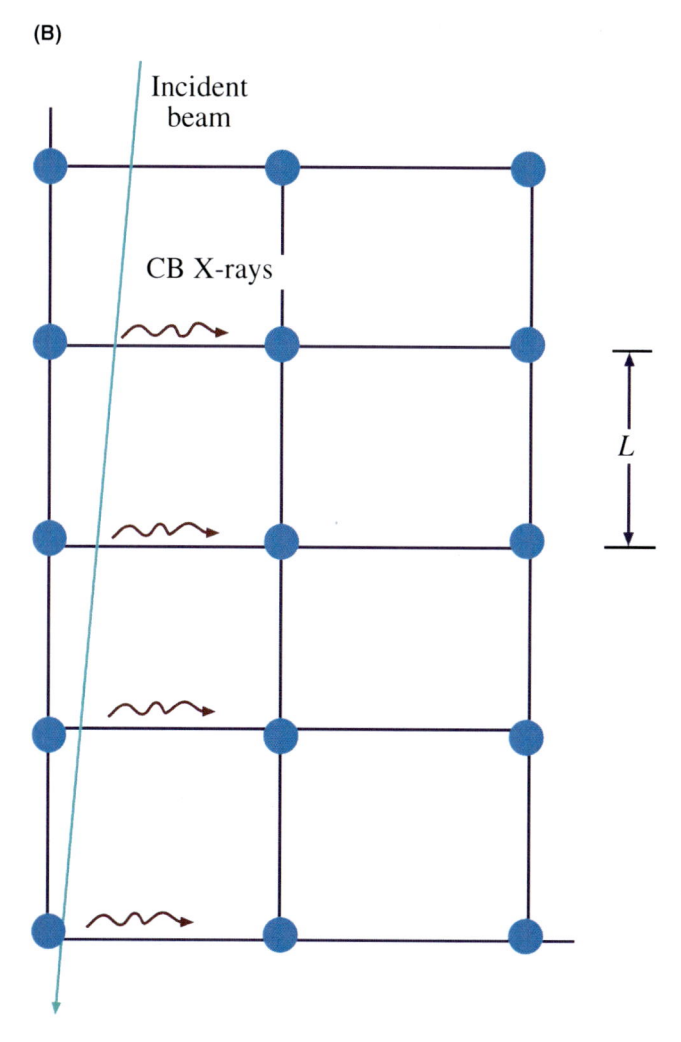

FIGURE 33.9. (A) CB peaks in a spectrum from pure Cu and (B) the schematic generation of CB when the beam passes close to a row of atoms in the specimen.

from a thin foil of pure copper taken at 120 keV. The primary peaks, as expected, are the Cu $K_{\alpha/\beta}$ and the L family of lines. In addition, the escape peak is identified. The other small peaks are the CB peaks. They arise, as shown in Figure 33.9B, by the nature of the coulomb interaction of the beam electrons with the regularly spaced nuclei in the crystal specimen. As the beam electron proceeds through the lattice, close to a row of atoms, each bremsstrahlung-producing event is similar in nature and so the resultant radiation tends to have the same energy. The regular interactions result in X-ray photons of energy E_{CB} given by

$$E_{CB} = \frac{12.4\beta}{L(1 - \beta\cos(90 + \alpha))} \quad (33.2)$$

where β is the electron velocity (v) divided by the velocity of light (c), L is the real lattice spacing in the beam direction, equal to $1/H$ in a zone-axis orientation (go back and check Section 21.3.B), and α is the detector take-off angle. More than one CB peak arises because different Laue zones give different values of L. The CB peak intensity seems greatest when the beam is close to a low-index zone axis, and these conditions should be avoided if possible. Operating with a convergent beam reduces the intensity of the CB peaks and this will be helped with a C_s-corrected AEM since larger convergence angles can be used to give more probe current without degrading the probe size.

> **CB PEAKS**
> You may mistakenly identify these CB peaks as characteristic peaks from a small amount of some element in the specimen, but fortunately, you can easily distinguish CB peaks from characteristic peaks.

Unfortunately, you can't remove the CB effects entirely, even by operating far from a major zone axis, since some residual peaks are invariably detectable if you count for long enough.

As predicted by equation 33.2, the CB peaks will move depending on both the accelerating voltage (which will alter v and hence β) and the specimen orientation, which will change the value of L. Of course, characteristic peaks show no such behavior and are dependent only on the elements present in your specimen. While CB peaks are a problem, you should only be concerned if you are seeking to detect a small amount of specific element in your specimen. More about this problem in the next chapter.

33.4 MEASURING THE QUALITY OF THE XEDS-AEM INTERFACE

In the end, what we need is some measure of how well our XEDS system is working and to be able to compare it with values from other systems. There are two ways to do this, both of which use a thin film, such as our standard NiO (although in these examples, we'll use Cr since that was how the original experiments were done. The principles are the same).

33.4.A Peak-to-Background Ratio

The first test of how well your XEDS is interfaced to your TEM is to measure the peak to background (P/B) ratio in the film.

> **FIORI P/B**
> There are several definitions of P/B ratio, but the best one, termed the Fiori definition, is shown in Figure 33.10A.

For the Ni K_α peak, you should integrate the peak intensity from 7.1 to 7.8 keV and divide this by the average background intensity in a 10 eV window (i.e., one or two channels depending on the display resolution) under the peak. In the Cr thin-film example shown in Figure 33.10A, the Cr K_α peak is summed from 5.0 to 5.7 keV. In a well-constructed AEM, the P/B ratio will increase with keV. The P/B values shown in Figure 33.10B (measured again on a Cr film rather than a NiO film (Zemyan and Williams)) should be achievable in any modern AEM. This value is an important test of the XEDS-AEM interface and the design of the stage. Higher is better!

33.4.B Efficiency of the XEDS System

The relative detector efficiency is a measure of how many X-ray counts per second (cps) are collected, detected and processed by the XEDS system. This is very important because of the overwhelming need to gather the most X-ray counts possible, given all that we've described about the inherent inefficiencies of X-ray generation and detection in XEDS of thin films.

In a fixed live time the detector efficiency will be affected by the specimen thickness, the probe current, and the solid angle of collection of the detector. So you should use the standard NiO specimen again, to fix the thickness variable and measure the probe current with a Faraday cup, as we've mentioned several times before (this gives cps/nA). Last, you need to factor in the collection angle given by the XEDS manufacturer (which is in fact calculated, not measured) to give the best figure of merit in terms of cps/nA/sr as described by Zemyan and Williams. Typical values are shown (again for Cr) in Figure 33.10C and, as you can see, the efficiency decreases with increasing kV because of the decrease in ionization cross section. But at any given kV, a higher number indicates higher efficiency.

It is possible to determine both the P/B ratio and the efficiency from direct spectral measurement and software to do this is described in the companion text.

33.5 ACQUIRING X-RAY SPECTRA

There are many commercial systems for acquiring spectra and images and many of these differ in terms of the file formats used within the computer system. Consequently, you might find it difficult to exchange spectra between different XEDS systems (e.g., for comparison of data gathered on two different AEMs). There has been a concerted attempt by the various manufacturers and professional societies to rectify this by creating the so-called 'EMSA/MAS standard file format' (Egerton et al.). You should check to ensure that your XEDS system supports this format.

So from what we've told you so far, you now know what information is likely to be in your spectra; which peaks might be real, which are most likely artifacts, and which ones you can and can't interpret as coming from the analysis region. Now we can concentrate on how best to actually gather the spectra before proceeding with analyses.

33.5.A Spot Mode

The standard way of gathering spectra, from the earliest days of AEM (and preceded by SEM/EPMA) was simply to use the beam deflectors to position the spot on a feature in your image and switch on the XEDS. We call this 'spot mode.' You could do this in a TEM by condensing the beam down with the C2 lens and adjusting C1 iteratively until the beam is small enough to interact only with the feature you wanted to analyze, such as a precipitate. In STEM spot mode you simply stop the beam from scanning and move the probe onto the feature, hoping to get it in the right position before the image fades from the STEM screen. In either case you also have to hope that both the probe position and the feature in your image stay stationary for long enough to gather a spectrum with sufficient counts to give you the composition information you need. On vintage AEMs, this method ensures that any carbon contamination buildup precisely buries the feature of interest! However, that same contamination spot would also show you if the beam or specimen had drifted during the analysis.

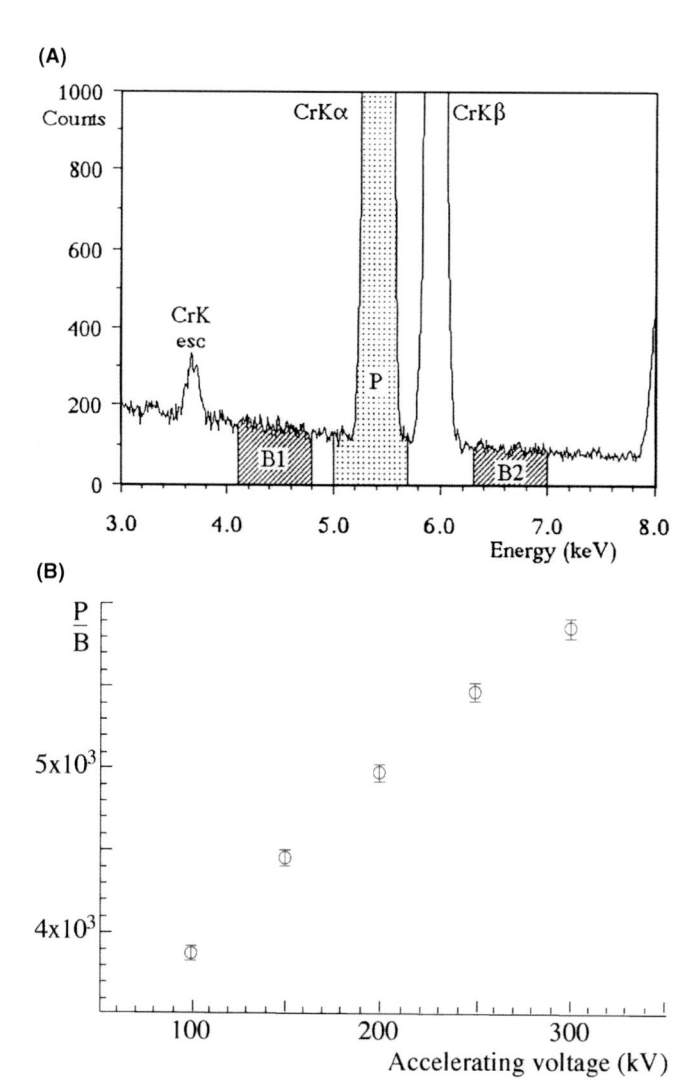

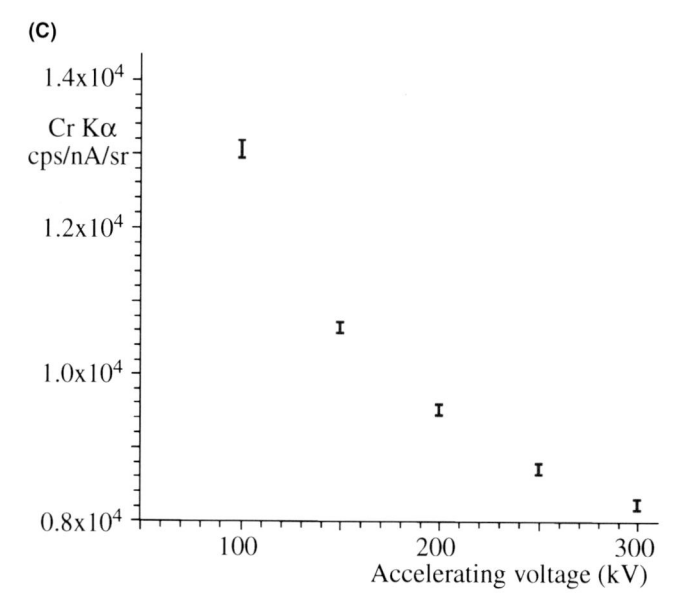

FIGURE 33.10. (A) Definition of the Fiori P/B ratio in a Cr thin film. (B) Change in the Fiori P/B as a function of kV for a high-performance 300-kV FEG AEM. (C) Decrease in the cps/nA/sr from a thin Cr film demonstrating decreasing detector collection and processing efficiency with increasing kV.

So spot mode is time consuming, the statistical confidence is appalling, and the pre-selection of which feature(s) to analyze introduces serious operator bias (and if you're not careful, all that you might analyze is an artifact of the specimen preparation rather than a key microstructural feature). Despite all this, it is still common to see publications in which this method is precisely how the analysis was performed and, in terms of giving a gross indication of local variations in chemistry, say for a complex extraction replica containing multiple phases, it certainly has some benefit, as shown in Figure 33.11. So, be careful when you build on someone else's results.

33.5.B Spectrum-Line Profiles

A variation of the spot mode of analysis is to take a series of spot analyses across a linear feature of interest in your specimen, such as a grain boundary or interphase interface, as shown in Figure 33.12A, and build up a set of spectra which, when analyzed, will reveal the composition profile across the interface. The information can be displayed as a set of superimposed spectra, termed a spectrum-line profile, as shown in Figure 33.12B, and the changes in the characteristic-peak intensities reflect significant composition changes. Since any kind of interface is a major planar defect in an engineering material, this approach at least removes some of the operator bias of the single-spot mode. However, line profiles still only reveal the composition across a single point on the boundary and many such profiles are required if you are to determine composition variations along the defect. So this approach is also tedious.

The solution to the limitations of both spot and line profiles approaches is to gather parts of spectra or preferably full spectra at every pixel in the STEM image, producing compositional images or maps.

This method is by far the best way to gather X-ray information with some semblance of statistical significance and without operator bias, so we'll now spend the rest of the chapter describing this method.

33.6 ACQUIRING X-RAY IMAGES

Mapping, or compositional imaging, used to be rarely used for analysis in the AEM because of the low X-ray count rate due to the small probe currents and collection angles. The overall inefficiency of the process meant that to gather a map with sufficient X-ray counts to be able to draw any conclusions about the variations in chemistry in the thin foil, you would have to scan the area for many hours. During this time, specimen drift, contamination or damage would occur and the resultant information would be compromised. However, with the developments of intermediate-voltage FEG sources, detectors with higher collection efficiency, drift-correction software, cleaner stage vacuums, and, most recently, C_s correctors, it is now possible, on the best AEMs, to gather quantitative X-ray maps in a matter of minutes. So it's worth looking at the various imaging options which have been developed, along with these improvements in AEM design and computer technology. In all of these approaches it will take significant

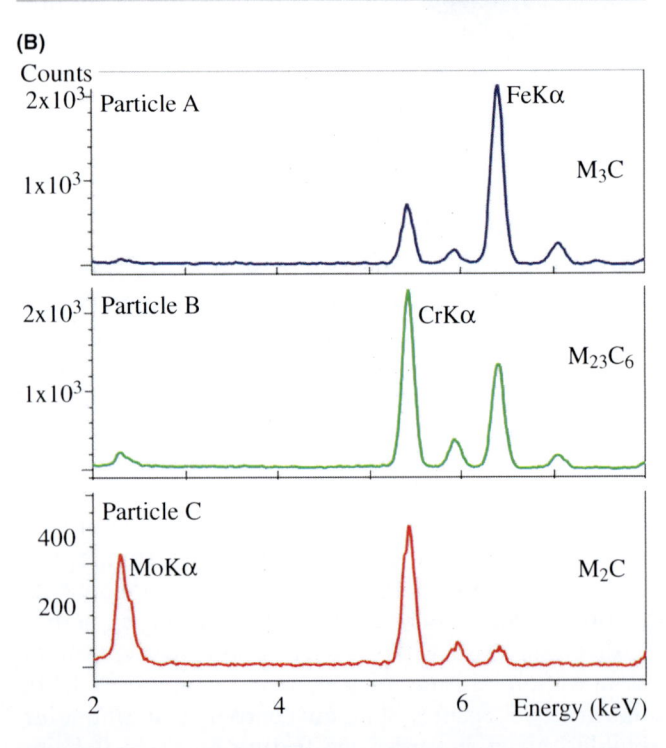

FIGURE 33.11. (A) STEM image of different carbide particles on an extraction replica from stainless steel specimen. (B) Multiple spectra taken from an array of points in the micrograph illustrating the variable chemistry of the different carbides.

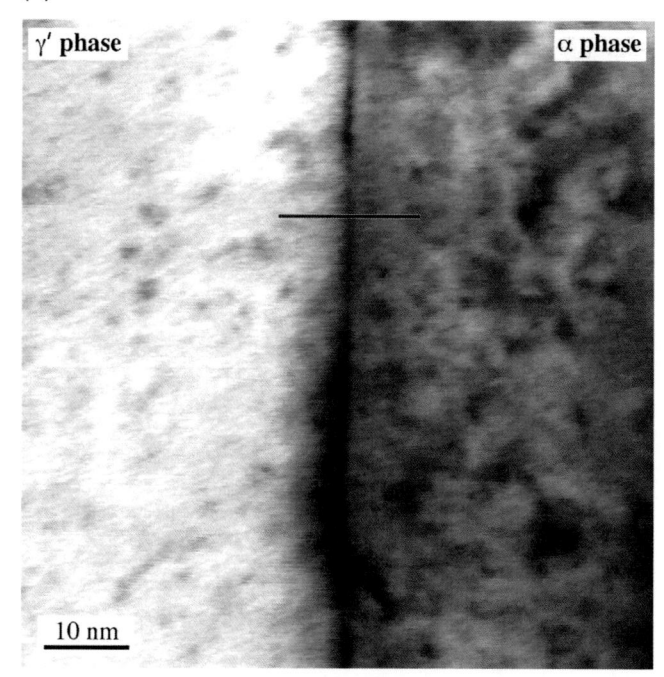

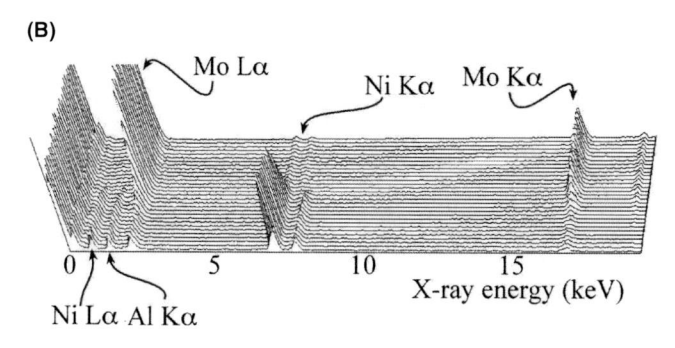

FIGURE 33.12. (A) STEM image of an α/γ′ interphase interface in a Ni-base superalloy with a line showing from where X-ray spectra were obtained. (B) A spectrum-line profile taken along the line in (A) showing clearly the change in Ni and Mo composition across the interface.

time to acquire a map and the biggest danger is that the specimen will drift during that time or the beam current will change (particularly, if you have a cold-FEG source). So learn about the drift-correction software that should be available in any standard commercial XEDS software package. Unless you can measure the probe current *on the fly* in a cold-FEG AEM, a Schottky FEG is recommended for mapping.

Analysis of bulk specimens in the SEM or EPMA is not limited by X-ray counts so the X-ray imaging techniques we'll discuss were generally pioneered on the SEM/EPMA. Likewise, as we'll see in Chapter 37 and beyond, EELS measurements have many millions or even billions of counts and so EELS mapping of thin films was used long before thin-foil X-ray analysts were able to benefit. But in this book, we'll rewrite history a little and talk first about all these techniques in reference to X-ray mapping.

33.6.A Analog Dot Mapping

Dot maps are the original method of acquiring qualitative X-ray images and, somewhat surprisingly, were first created more than half a century ago by Cosslett and Duncumb and progress since then has been reviewed by Friel and Lyman. The approach is simple: you select a specific energy (or wavelength) channel (or a window (range) of channels) in your X-ray spectrum, scan the beam across the area of interest and when the X-ray detector registers an X-ray of the selected energy (range), record and display it. So as the beam scans, the display intensity builds up and the changes in intensity reflect the changes in the number of X-rays detected. For example, if you select the Pd L_α peak channel or a window covering the peak, then regions showing lots of dots on the screen are high in Pd, as shown in Figure 33.13 (we'll go into more detail about how to do this in Section 34.7). This qualitative, analog

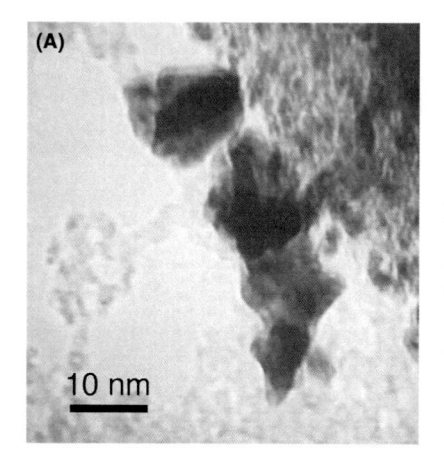

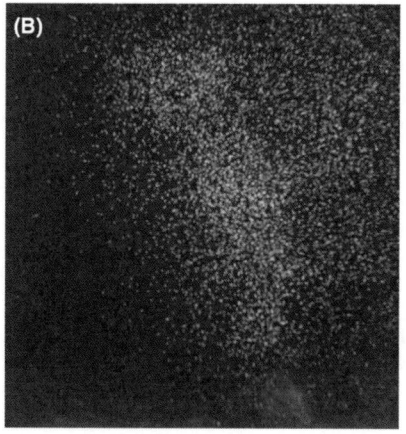

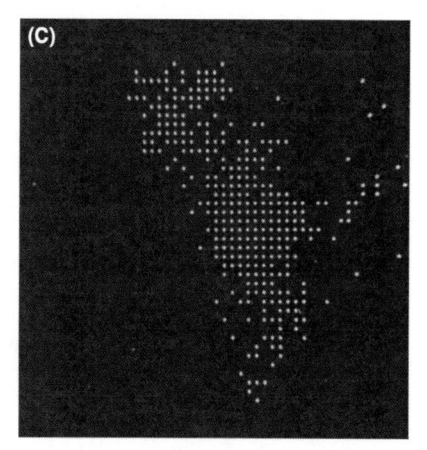

FIGURE 33.13. (A) STEM BF image of a Pd catalyst particle on a support film. (B) Analog dot map using the Pd L_α signal. Note the correlation between thicker (darker) regions of the specimen and increased Pd signal. (C) Early digital map taken with Pd L_α signal but subtracting the background signal.

approach is not directly quantifiable since the background can't be removed, unless you simultaneously gather a dot map from a bremsstrahlung window adjacent to the peak window and subtract the one from the other (more about this in Chapter 35). Furthermore, in thin foils, changes in thickness will also produce changes in intensity in both the peak and background maps and this is clear in Figure 33.13B. It is possible to refine this approach by gathering multiple X-ray maps, assigning colors to different X-rays, and overlaying the maps to give an indication of relative composition changes. Again, this process is *not* quantitative.

33.6.B Digital Mapping

As computers became more powerful, it was possible to collect X-rays from multiple channels or windows and thus acquire several maps simultaneously. If one or more of the maps was the bremsstrahlung intensity then you could produce quantitative maps, on the fly, as the data accumulates. The first success in digitizing thin-foil X-ray maps was carried out by Hunneyball et al. almost 30 years ago. It was only possible to build up 128×128 pixel images with 256 X-ray counts at each pixel using a 100-nm probe with 5 nA of probe current over 200s. But the maps were fully quantitative, removed the effects of foil-thickness variation, and revealed relatively small composition variations around GBs in aged Al alloys, as shown in Figure 33.14A. Because of the central role of the computer in the acquisition, this approach has the advantage of post-acquisition processing and comparison of different quantification routines.

Also after acquisition it was possible to use such processing techniques as false coloring, computerized image overlays, scatter diagrams, etc. (Bright and Newbury). All these advantages combine to make digital imaging a most attractive approach for displaying XEDS data. This process really came of age in the mid-1990s with the availability of faster computers, improved data storage, intermediate-voltage FEG sources, reasonable X-ray collection angles, and stable AEMs permitting long collection times without drift. An example from such an AEM is shown in Figure 33.14B which is 256×256 pixels and was acquired using a 1-nm probe with 0.9 nA of current for 5400 s. Comparing Figure 33.14A and B is instructive and

FIGURE 33.14. (A) The first quantitative digital map (128 × 128 pixels) obtained in an old thermionic-source AEM generating a low count rate. The image from a thin foil of Al-Zn shows depletion of Zn around a triple point (compare the colors in the map with the quantitative look-up 'table' on the right side of the image). (B) More recent digital map from a 300-kV FEG AEM showing enrichment of Al at GBs in an electro-migrated specimen of Al-4% Cu. The bright regions are $CuAl_2$ intermetallics. In both (A) and (B) quantification was achieved by subtracting the bremsstrahlung intensity at each pixel. (C) Quantitative Cu line profile taken across a GB indicated in (B), by arrow.

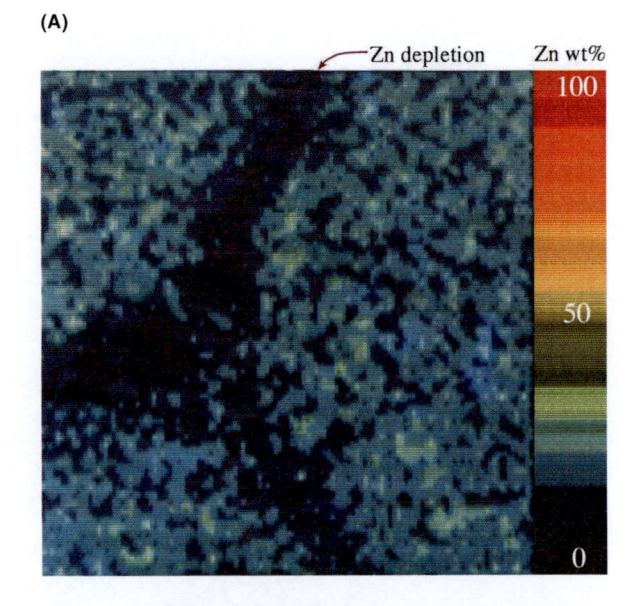

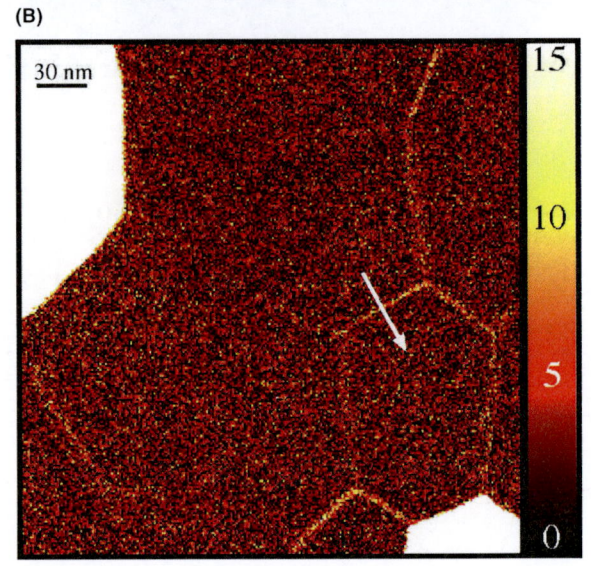

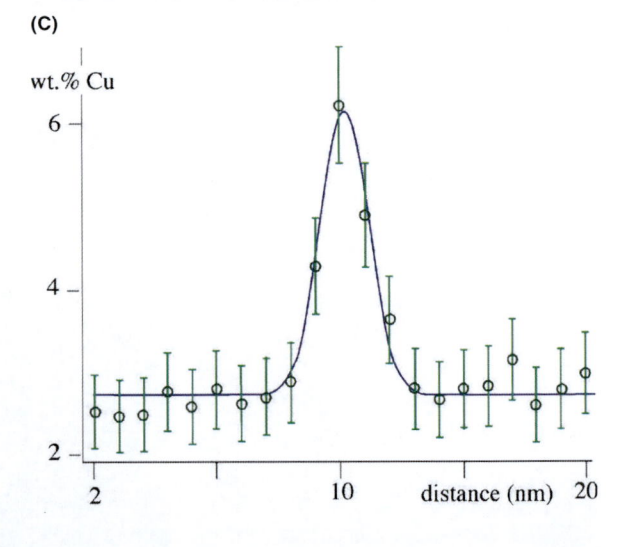

shows the enormous progress in both the quality of mapping and the spatial resolution. Once you have a digital map such as this you can go into the map and extract quantitative data from any region, e.g., the line profile taken across the GB in Figure 33.14C. However, you still have to select the window(s) or peak(s) in the spectrum that you map, thus introducing your own bias in terms of what you expect to be in the specimen. Furthermore, once the map is recorded and stored, you cannot return to re-check the data, or map another element from the same area, since all the other information in the spectrum was lost. Likewise, unless you store the image from the mapped area, it too cannot be re-examined. So you have to get everything right, collect all the X-ray images you need, as well as the background spectra to permit subsequent quantification. Gathering multiple maps was limited by the computer memory, but that is no longer a problem since memory is cheap. Now, as we discuss below, you don't even have to bother about pre-selecting which X-rays you want to map, you just gather them all and later on decide which you want to image. How do we do this? It's called 'spectrum imaging.'

> **SPECTRUM IMAGING**
> SI is *the* preferred method for X-ray (and EELS) mapping.

33.6.C Spectrum Imaging (SI)

As the term implies, SI collects a full spectrum at every pixel in the digital image (so you can only do this in STEM mode (although there are analog versions in energy-filtered TEM, as we'll see in Chapter 39)). The result of the SI process is a 3D data cube, as shown in Figure 33.15A, with the electron image constituting the *x–y* plane and all the XEDS spectra in the *z* direction. The SI term was first used for EELS in the late 1980s and we'll mention this topic again in Chapter 37. Only much later did SI became feasible for X-ray mapping although it is now common enough to be used in materials problem solving (e.g., Wittig et al.). On a historical note, you should be aware that SI methods have been practiced in other fields, such as radio astronomy for several decades and, indeed, Legge and Hammond took the output of their EDS and WDS spectrometers and synchronized the

FIGURE 33.15. (A) Schematic spectrum-image data cube showing how as the beam stops at each pixel in the *x–y* plane as it is creating a STEM image, a full X-ray spectrum is gathered at each pixel. The different colors at different energy (*z*) values indicate different signals from different elements that appear at different energies. (B) A series of X-ray maps of a GB region in a Ni-base superalloy taken at specific energies from an SI data cube. (C) An example of a map from a single channel in the X-ray spectrum (i.e., a single image plane) in the SI in (B) coinciding with the Nb K_α peak energy. (D) The application of multivariate statistical analysis and principal-component analysis to remove noise and enhance the Nb signal.

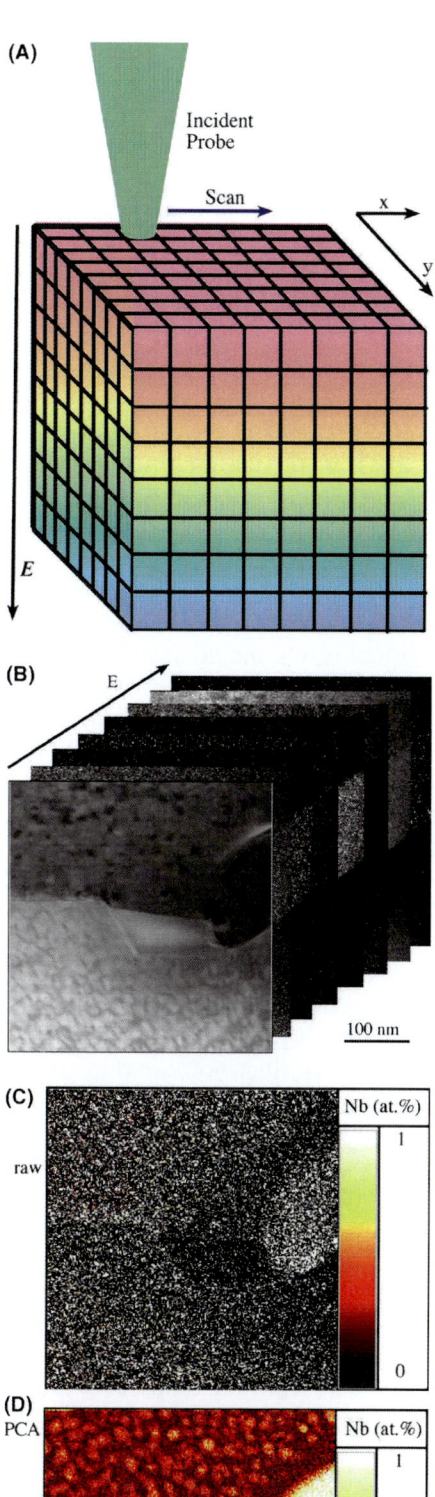

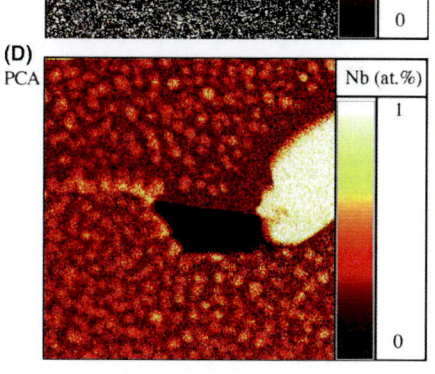

detectors' output pulses with the position of the beam 30 years ago. They collected the data on magnetic tape, and later reconstructed the data into a 3D file, so it can be argued that this was the first demonstration of the SI concept, at least in electron-beam instruments, so we aren't really revising the historical record!

The beauty of the SI technique is that, once you have this data cube stored, you can go back to it at any time, recheck the data, and re-do any analysis, search for other spectral features that you might not originally have thought present or important, look at different images at different energies, and yet always have the original image and spectra at your disposal.

If you think about the diagram in Figure 33.15A, you'll see that there are many ways to slice and dice the data cube which reproduce all the other methods of analysis that we've described. If you select a single pixel in the x–y (image) plane, then there is a full spectrum attached to it and so, if you wish, you can select a set of individual pixels and do as many spot analyses as you wish. Likewise, you can select a line of pixels in the image, effectively slicing the cube along the x–z, y–z, or some combination of these directions and thus produce spectrum-line profiles. You can (and this is where it gets really useful) slice the cube at any plane in the z (labeled E for energy) direction and, at each plane, you'll get a different image consisting of X-rays of that particular energy. You can add planes, subtract planes, and sum image pixels only in certain features (e.g., a strangely shaped precipitate or along a boundary plane). You can also envisage gathering spectra at certain pixels as a function of time (chrono-spectrometry?) and there are surely other options which you can think of. Now if you do the math and consider a 1 k × 1 k pixel image with a 2048-channel spectrum at each pixel, you'll find that the data cube is about 2 GB in size. In fact, in gathering such an enormous amount of data, the bigger challenge is to find ways both to search this enormous database efficiently and extract meaningful data. Various advanced software methods, such as multivariate statistical analysis, principal-component analysis, and maximum-pixel spectrum analysis are available to remove noise and extract rare events from the data cube. Such processes are closely related to the parameterizing of HRTEM images discussed in Chapter 31, and are discussed in far more detail in the companion text.

Just to give you a hint of what can be done, Figure 33.15B shows a series of X-ray images cutting the x–y plane at different energy values, displayed 'behind' the original STEM ($E = 0$) image. If we select a single x–y plane such as that at the Nb K_α line energy (Figure 33.15C), that image is very noisy because of the limited signal captured in that one plane. However, we can use sophisticated software to remove all the noise components in the plane and also add in all the other Nb signals from other energy planes to produce the image showing the Nb distribution (Figure 33.15D). Before you think that such an extraordinary change is unreasonable and that information is being created where it did not exist, compare the processed image (D) with the original STEM electron image (the top slice in B). Clearly the small matrix precipitates that are imaged in (A) are also mapped in (D). Extracting information like this from the SI data cube tells you how much more powerful the SI approach is compared with spot or line-profile analyses. Now it is routine to gather both XES and EELS SI simultaneously, thus completely optimizing the acquisition of analytical data.

33.6.D Position-Tagged Spectrometry (PTS)

PTS is a specific commercial version of generic SI, developed by Mott and Friel at Princeton Gamma Tech (PGT, now Bruker AXS). PTS eliminates the conflict between your wishing to view full X-ray images quickly versus the analytical advantages of having complete spectra saved at each pixel (which even on the best AEMs still takes 30–45 minutes if you want to quantify the data). In PTS, the beam is scanned rapidly relative to traditional mapping, and the X-rays are counted in the analysis computer, preserving both spatial information from the image and spectral information. Sophisticated processing software can be used to interrogate the data *during* acquisition, which is not possible in conventional SI. Alternatively, this software can be used after the complete spectrum is stored, as with conventional SI, where you acquire the full spectrum at a single pixel and then move to the next pixel and gather another spectrum. PTS also permits relatively easy monitoring of such phenomena as specimen drift, contamination, or damage during the acquisition.

CHAPTER SUMMARY

XEDS in the AEM is a challenge because the detection/processing system creates artifact peaks in the spectrum and X-rays are generated and detected from sources in the AEM other than the region of your specimen where you put the beam. Nevertheless, there are well-defined precautions you can take so that you are sure that the artifacts, the spurious and system X-rays are minimized and that your subsequent interpretation and quantification are not compromised. There are also several standard tests you can carry out to compare your AEM system performance with other instruments.

In summary, to understand your AEM and to acquire meaningful spectra, you should
- Buy your own NiO standard thin film on a Mo grid.
- Test your XEDS to determine the artifact peaks that it produces.
- Gather a spectrum down a hole in the NiO film to see what spurious X-rays the AEM illumination system produces. Use top-hat C2 apertures at all times.
- Gather an XEDS spectrum from a light-element film to see what system peaks your AEM introduces into the spectrum. Use thin foils, flakes, or films rather than self-supporting disks if possible.
- Be aware of CB if you count for a long time to detect small peaks from trace elements.
- Image the electron beam on the TEM screen to ensure that it is Gaussian.
- Always remove the objective diaphragm.
- Operate with the specimen as close to zero tilt as possible.

Check that

- The hole count is <1% of your experimental spectrum under the same operating conditions.
- The P/B ratio and the detector efficiency data are acceptable and do not change with time.
- If you're looking for characteristic peaks from trace or minor elements in your specimen, understand that such peaks are much more likely to be confused with artifacts or system peaks (much more about this in the next chapter).

Once you are sure of what's in your spectrum, decide whether you want to do quick and dirty point analyses of some features in your image or, if you are confident of the importance of the area that you want to analyze completely, gather X-ray maps or, ideally, acquire a SI data cube which will really give you the full picture of your specimen composition.

GENERAL REFERENCES

Garrett-Reed, AJ and Bell, DC 2003 *Energy-dispersive X-ray Analysis in the Electron Microscope* Bios (Royal Microsc. Soc.) Oxford UK. Different descriptions of many of the issues discussed in this chapter.

Lyman, CE (Ed.) 2006 Microscopy and Microanalysis **12** 1. A commemorative edition celebrating 50 years of X-ray mapping; great history and examples of good practice.

Williams, DB, Goldstein, JI and Newbury, DE (Eds.) 1995 *X-Ray Spectrometry in Electron Beam Instruments* Plenum Press New York. Still the best available source of background information on XEDS hardware and software, even though increasingly dated in content.

SPECIFIC REFERENCES

Allard, LF and Blake, DF 1982 *The Practice of Modifying an Analytical Electron Microscope to Produce Clean X-ray Spectra* in *Microbeam Analysis-1982* 8–20 Ed. KFJ Heinrich San Francisco Press San Francisco CA. Early review of system artifacts.

Bright, DS and Newbury, DE 1991 *Concentration Histogram Imaging* Analytical Chemistry **63** 243A–250A. Processing maps for appearance and more.

Cliff, G and Kenway, PB 1982 *The Effects of Spherical Aberration in Probe-forming Lenses on Probe Size and Image Resolution* in *Microbeam Analysis-1982* 107–110 Ed. KFJ Heinrich San Francisco Press San Francisco CA. The tail on the probe.

Cosslett, VE and Duncumb, P 1956 *Microanalysis by a Flying-spot X-ray Method* Nature **177** 1172–1173. First dot maps by XEDS—more than 50 years ago!!

Egerton, RF, Fiori, CE, Hunt, JA, Isaacson, MS, Kirkland EJ and Zaluzec, NJ 1991 *EMSA/MAS Standard File Format for Spectral Data Exchange* EMSA Bulletin **21** 35–41. The EMSA/MAS standard file format.

Fiori, CE, Swyt, CR and Ellis, JR 1982 *The Theoretical Characteristic to Continuum Ratio in Energy Dispersive Analysis in the Analytical Electron Microscope* in *Microbeam Analysis-1982*, 57–71 Ed. KFJ Heinrich San Francisco Press San Francisco CA. The Fiori P/B definition.

Fiori CE and Swyt, CR 1994 *Desk Top Spectrum Analyzer (DTSA)*, U.S. Patent 5 299 138. The original description of the essential software.

Friel JJ and Lyman CE 2006 *Tutorial Review: X-ray Mapping in Electron-Beam Instruments* Microscopy and Microanalysis **12** 2–25. A good place to start before you map.

Hunneyball, P D, Jacobs, MH and Law, TJ 1981 *Digital X-ray Mapping from Thin Foils* in *Quantitative Microanalysis with High Spatial Resolution* 195–202 Eds. GW Lorimer, MH Jacobs and P Doig, The Metals Society London. First digital XEDS maps.

Legge, GJF and Hammond, I 1979 *Total Quantitative Recording of Elemental Maps and Spectra with a Scanning Microprobe* J. Microsc. **117** 201–210. A little history.

Lyman, CE and Ackland, DW 1991 *The Standard Hole Count Test: a Progress Report* in *Microbeam Analysis-1991*, 720–721 Ed. DG Howitt San Francisco Press San Francisco CA. Measuring hole count.

Lyman, C.E, Goldstein, JI, Williams, DB, Ackland, DW, von Harrach, S, Nicholls, AW and Statham, P.J 1994 *High Performance X-ray Detection in a New Analytical Electron Microscope* J. Microsc. **176** 85–98. Modifying the stage to use only low-Z materials.

Mott RB and Friel, JJ 1999 *Saving the Photons: Mapping X-rays by Position-Tagged Spectrometry* J. Microsc. **193** 2–14. The first commercial SI software

Newbury DE 1995 *Artifacts in Energy Dispersive X-Ray Spectrometry in Electron Beam Instruments; Are Things Getting Any Better?* in *X-Ray Spectrometry in Electron Beam Instruments* DB Williams, JI Goldstein and DE Newbury (Eds.) 167–201 Plenum Press New York. Discussion of artifacts and how digital imaging changes them.

Newbury, DE 2005 *X-ray Spectrometry and Spectrum Image Mapping at Output Count Rates above 100 kHz with a Silicon Drift Detector on a Scanning Electron Microscope* Scanning **27** 227–239. SDD performance.

Reese, GM, Spence, JCH and Yamamoto, N 1984 *Coherent Bremsstrahlung from Kilovolt Electrons in Zone Axis Orientations* Phil. Mag. **A 49** 697–716. Early demonstration of CB.

Williams, DB and Goldstein, JI 1981 *Artifacts Encountered in Energy Dispersive X-ray Spectrometry in the Analytical Electron Microscopy* in *Energy Dispersive X-ray Spectrometry* 341–349 Eds. KFJ Heinrich, DE Newbury, RL Myklebust and CE Fiori NBS Special Publication 604 U.S. Department of Commerce/NBS Washington D.C. Early review of system artifacts.

Wittig, JE, Al-Sharaba, JF, Doerner, M, Bian, X, Bentley, J and Evans, ND 2003 *Influence of Microstructure on the Chemical Inhomogeneities in Nanostructured Longitudinal Magnetic Recording Media* Scripta Mater. **48** 943–948. Early example of SI.

Zemyan, SM and Williams, DB 1995 *Characterizing an Energy-Dispersive Spectrometer on an Analytical Electron Microscope* in *X-Ray Spectrometry in Electron Beam Instruments* DB Williams, JI Goldstein and DE Newbury (Eds.) 203–219 Plenum Press New York. Basic understanding of XEDS performance.

URLs

1) www.cstl.nist.gov/div837/Division/outputs/DTSA/DTSA.htm To download DTSA

SELF-ASSESSMENT QUESTIONS

Q33.1 What's the difference between spurious X-rays and system X-rays?
Q33.2 What are the best steps you can take to minimize spurious X-rays?
Q33.3 What causes incomplete charge collection (ICC) in an XEDS?
Q33.4 Name three common artifacts in the spectra generated by XEDS systems.
Q33.5 Which of the various artifacts should you be particularly aware of while operating your AEM and why?
Q33.6 Distinguish the sources of desired and undesired radiation impinging on the detector.
Q33.7 How can you find out what X-rays are generated by your AEM-EDS system rather than by your specimen?
Q33.8 Why is it important to know what your system X-rays are?
Q33.9 What's the danger of not minimizing any spurious X-rays?
Q33.10 Why would you not use a Cu grid to analyze diffusion profiles in brass using XEDS in the AEM?
Q33.11 Why is it challenging to detect trace (∼0.1 wt%) amounts of Si unambiguously in an Fe alloy using XEDS?
Q33.12 Why is a top-hat C2 aperture/diaphragm so called and why is it essential to have such apertures in your AEM?
Q33.13 Ideally how many C2 apertures should you have in an AEM?
Q33.14 How can one minimize post-specimen scattering?
Q33.15 What causes coherent bremsstrahlung (CB)?
Q33.16 If the hole count is >>1%, what is likely causing this?
Q33.17 Is STEM well suited for AEM? If so, why?
Q33.18 From the extra information in this chapter, explain why it is so important to align your X-ray detector with any planar interface.
Q33.19 List four ways by which you can minimize post-specimen scattering.

Q33.20 Why might CB be a problem that could cause you to misinterpret your analysis?
Q33.21 Give a simple way to distinguish a CB peak from a characteristic peak.
Q33.22 Why is it better to form X-ray images rather than individual spectra or line profiles?
Q33.23 Why is it much more difficult to acquire good X-ray images than individual spectra?

TEXT-SPECIFIC QUESTIONS

T33.1 A P/B ratio is measured on the Ni K_α line at 300 kV using the Fiori method and is determined to be 1000. What should you probably do?

T33.2 Why don't microscope manufacturers just coat the insides of the AEM with Be on all surfaces?

T33.3 Why coat any of the insides of the microscope at all with a low-Z material?

T33.4 To decrease the X-ray background for a particular specimen, the kV should be increased. Why?

T33.5 In an experiment, some unidentified X-ray peaks are present. Name some possible causes.

T33.6 Upon measuring a hole count for a particular specimen, you discover that the illumination system is not 'clean.' Besides being floored by this discovery, you want to know what you can do to fix it. So what do you do?

T33.7 Having become quite the expert on XEDS in the SEM, your lab partner tilts the specimen in the TEM chamber to 45° 'just like we did in the SEM.' Before smacking your lab partner repeatedly, what should you remind him of?

T33.8 Upon analyzing the XEDS spectra of your material, you cannot conclude what several peaks correspond to. You have tried sum peaks, escape peaks, and the internal fluorescence peak with no luck. Have you discovered a new element, or is there another answer that is more likely?

T33.9 David and Stuart are working at the 1210 when we arrive one morning. They are using XEDS, and all their spectra have peaks below 1 keV. What in the world are they doing?

T33.10 How would you discern if your X-ray spectrum contains a significant amount of spurious or system X-rays? How can you distinguish the spurious contributions from stray X-rays and stray electrons?

T33.11 Calculate the energies of the principal coherent bremsstrahlung peaks generated from a thin foil of Cu in a <001> orientation by 120-kV electrons when the EDS detector has a 20° take-off angle. Compare your answer with Figure 33.9 and comment on any discrepancies. (Hint: you'll have to find the lattice parameter of pure Cu.)

T33.12 How wide do you think the probe in Figure 33.5 is? Explain your reasoning.

T33.13 If the electron probe that you use to do analysis is as shown in Figure 33.5, what effect would this have, for example, on a line-profile analysis across an interface?

T33.14 Which effect(s) in Figure 33.6 is (are) causing the extra peaks in Figure 33.7?

T33.15 Are any other peaks in Figure 33.7 not caused by effects in Figure 33.6 and if so what is causing them?

T33.16 Explain the reasons for each of the directives at the end of Section 33.3.B.

T33.17 Distinguish analog and digital mapping and explain why one is so much better than the other.

T33.18 Distinguish SI and PTS.

T33.19 Make a list of the different ways to cut up the SI data cube and explain the different information that you would get from each cut. See if you find one that has not been published and make a name for yourself.

34
Qualitative X-ray Analysis and Imaging

CHAPTER PREVIEW

It is a waste of time to proceed with *quantitative* analysis of your XEDS spectrum or image without first carrying out *qualitative* analysis. Qualitative analysis requires that *every* peak in the spectrum be identified unambiguously, with statistical certainty, otherwise it should be ignored for both subsequent quantitative analysis and imaging. We emphasize this point because of the many opportunities for the misidentification of small peaks in the spectrum. In this chapter, we'll deal initially with acquisition and identification of the elemental information in spectra and images. First, we will show you how to choose the best operating conditions for your particular AEM and XEDS system. Then we'll explain the best way to obtain a spectrum for qualitative analysis. You have to acquire a spectrum with sufficient X-ray counts to allow you to draw the right conclusions with a given degree of confidence. There are a few simple rules to follow which allow you to do this.

> **A QUALITATIVE MUST!**
> Although such an approach may seem time-consuming and unnecessarily tedious, the need for initial qualitative analysis of the spectrum cannot be stressed too strongly.

Two advantages are gained from rigorous qualitative analysis. First, you may be able to solve the analytical problem at hand without needing to perform full quantification. Second, when quantification is carried out (see the next chapter), you will not spend an inordinate amount of time analyzing an element that isn't there, and you can be confident that your results are valid. We'll go over the many ways to misidentify peaks in your spectra, particularly small ones, which may, in fact, arise from important trace elements but, which might be artifacts, could be peaks from another element, or are possibly statistically insignificant. Commercial peak-identification software, while improving all the time, is not error-free. We'll end with a few words about qualitative X-ray imaging.

34.1 MICROSCOPE AND SPECIMEN VARIABLES

When you first acquire a spectrum, the operating conditions should maximize the X-ray count rate to give you sufficient intensity in the characteristic peaks in your spectrum, in the shortest time, with the minimum number of artifacts. You need sufficient counts so you can detect, unambiguously, the presence of *all* the elements in your specimen (within the limitations of your XEDS detector) *with statistical certainty*. As we'll explain, the best conditions for such qualitative analyses require that you obtain the spectrum from a reasonably thick, large area of your specimen, using a large probe and a large aperture to give the most current, but in doing this you'll compromise other desirable analysis qualities, particularly high spatial resolution. So right up front you need to know two key points

- There are only three requirements for good qualitative analyses; counts, counts and more counts
- The conditions for the best qualitative (and quantitative) analysis (which are also those that give the best analytical sensitivity) are precisely the worst for obtaining the best spatial resolution.

A more complicated factor in getting the most X-ray counts in your spectrum is choosing the right operating voltage. Remember, back in Figure 33.10.C, we showed that you get a higher detection and collection efficiency if you decrease the kV because the scattering cross section (σ) increases when the kV decreases; that was a specimen effect. Now we are talking about a gun effect;

as the kV increases, the gun brightness increases (go back and check this in Chapter 5). While the two effects counter each other somewhat, the added advantage of an increased P/B ratio (which helps with detecting small peaks) with higher kV, as well as improved spatial resolution (due to less beam spreading), tips the scales in favor of using the maximum kV at all times. Only choose a lower voltage if knock-on damage is a problem, as might be the case, for example, in a 200–400 kV AEM with a beam-sensitive specimen such as a ceramic, a mineral, most semiconductors or a low-Z (<15) metal/alloy.

Pick a portion of your specimen that is single phase in the area of interest and make sure it is tilted well away from strong diffraction conditions to minimize crystallographic effects (more on this in Chapter 35) and coherent bremsstrahlung. Ideally, you will need a probe current of several tens of nanoamps. The necessary combination of probe size and final aperture depends on the type of source in your AEM. To get several tens of nanoamps from a LaB$_6$ source, you have to select a relatively large probe size, say, a few tens of nanometers and a large C2 aperture. An FEG source will give much less total current than a thermionic source and it won't be possible to generate more than a few nanoamps at best. So, for reasonable qualitative analysis and subsequent quantitative analysis, you'll have to accumulate your spectrum for a longer time with an FEG compared to a thermionic source. A C_s-corrector on your FEG-AEM will help tremendously, since a larger probe-forming aperture can be used without compromising the quality of the probe, and tens of nanoamps can be generated in a C_s-corrected probe of a couple of nanometers dimension. So if your institution has enough money to spend on your AEM you can minimize the otherwise necessary compromise between needing lots of counts and getting high spatial resolution

You can always gather more counts in your spectrum by choosing a thicker region of the specimen. There is nothing wrong with doing this when you are carrying out *qualitative* analysis. A thick specimen degrades your spatial resolution, but we've already agreed to compromise that aspect of the analysis during this initial qualitative procedure. The only danger is that, if you are interested in finding a few weight percent of a light element, those weak X-rays may be absorbed in your specimen and so may not be detected. However, from an experimental standpoint, having carried out qualitative analysis of a relatively large, thick region, you can always do further analyses of smaller, thinner areas, under conditions that optimize spatial resolution, which we'll discuss in Chapter 36.

Remember that we have been talking about several different 'resolutions'. Don't confuse them

- spatial resolution: distances measured in nm (see Chapter 36).
- chemical resolution: analytical sensitivity/detection limits depending on P/B (see Chapter 36).
- energy resolution: identifying elements by distinguishing their spectral peaks at different energies (see this chapter and Chapter 33).

So, just in case you haven't got the message by now, good qualitative analysis (and subsequent quantification) requires a large number of X-ray counts in the spectrum (just how many we'll tell you in a while). These counts might take a long time to generate, so you run the danger of damaging, or changing the chemistry of any beam-sensitive specimens. You may also contaminate your chosen area if your AEM is not UHV and/or your specimen is not clean. So it's always good to use a plasma cleaner before analysis, unless doing so will destroy your specimen. (Go and check Chapter 10 or the companion text where we talk about specimen preparation.) To minimize damage and contamination, you should spread the beam over as large an area as possible, either by overfocusing C2 if you're in TEM mode or by rastering the beam in STEM mode, remembering that, in doing so, your analysis will be an average over the chosen area. Use a liquid-N$_2$ cooled, low-background holder, especially if contamination is still a problem.

34.2 BASIC ACQUISITION REQUIREMENTS: COUNTS, COUNTS, AND MORE CAFFEINE

The first and most important step in qualitative analysis is to acquire a spectrum across the complete X-ray energy range. Analysis can often be accomplished using X-rays with energies from ~1 to 10 keV, and this is the typical range used in the SEM. However, the TEM has a much higher accelerating voltage, and the consequent increase in available overvoltage (remind yourself what this is. Hint: see Chapter 4) means that you can easily generate and detect much higher-energy X-rays. If you are using an intermediate-voltage AEM and a windowless IG detector, we noted in Chapter 32 that all the possible K$_\alpha$ lines from all the elements above Be in the periodic table can be detected.

> **ENERGY RANGE**
> The first thing to do is to adjust your computer display to the widest possible energy range. For a Si(Li) detector or an SDD, 0–40 keV is sufficient and for an IG detector, 0–80 keV may be more useful.

Of course, if you know the specimen you are analyzing, such a step may not seem essential, but it is still a wise initial precaution since unanticipated contaminants or trace impurities may be present. The next steps are the basics for acquiring a spectrum for qualitative analysis

- Collect a spectrum over, say, 0–40 keV for several hundred seconds (take a coffee break) and ascertain the actual energy range over which all the detectable characteristic peaks occur.
- If all the peaks that you can see are present in an energy range < 40 keV, re-gather the spectrum over that reduced range (take another coffee break), thereby improving the resolution of the computer display by lowering the number of eV per channel.
- The spectrum that you finally gather for qualitative analysis must be displayed with *no more than 10 eV per channel* resolution or better. A display range of 0–20 keV should be possible under these conditions (i.e., 2048 channels in total) in all but the most ancient of XEDS computer systems.
- If you have analog processing electronics, you can increase the counts in your spectrum by reducing the detector time constant to maximize the throughput. This step degrades the energy resolution of the XEDS but, for many qualitative analyses, this is not important. A digital system will automatically optimize the throughput of counts.
- Watch the dead-time readout while acquiring the spectrum to make sure you haven't chosen a combination of probe current and specimen thickness that overloads the detector electronics. Remember that you want to keep the dead time below about 50–60%, and an output count rate of around 10 kcps (analog) to 30 kcps (digital) is about the best that can be handled by current detector electronics under these conditions. This will rarely be a problem in thin-foil analysis!
- The total counts in this qualitative spectrum should exceed 1,000,000 *over the full energy range*. While this may seem a lot, at 3 kcps, it will take you just over 15 minutes to accumulate this number of counts (now it's probably time for a bathroom break, anyhow). So adjust the probe size/current, the size of the C2 diaphragm and (if possible) the specimen thickness until the count rate is sufficient.

When you've got a good high-count spectrum over a suitable energy range, there is a well-defined sequence of steps developed for analysis of spectra from the SEM (see Goldstein et al. as usual) that should be followed to ensure that you correctly identify each peak in the spectrum and disregard those peaks that are artifacts or are not statistically significant and we'll describe a modified form of this procedure for thin-foil specimens in the next section.

Figure 34.1 shows the effects of increasing acquisition time (i.e., increasing counts) on the visibility of small peaks. The longer the time, the better the quality of the spectrum. So coffee breaks can be very beneficial.

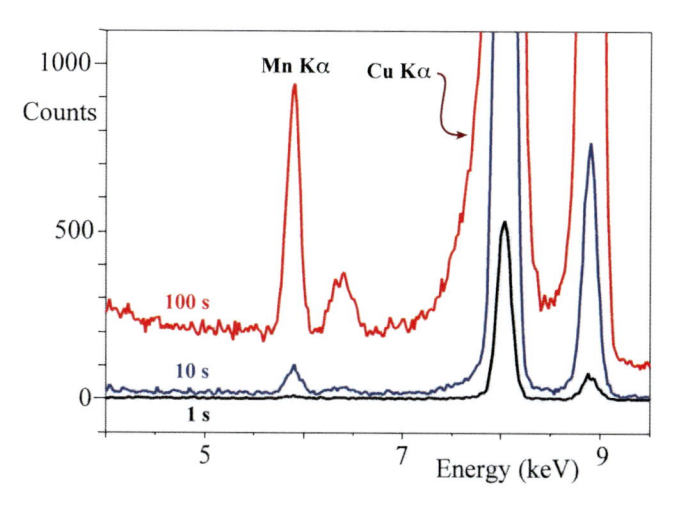

FIGURE 34.1. The improvement in the quality of a spectrum from a Cu-1% Mn thin foil with increasing acquisition time. After 1 s (black) the small Mn $K_{\alpha/\beta}$ peaks are not visible; after 10 s (blue) the MnK_α is detectable but the K_β is barely visible. After 100 s (red) all peaks are clearly visible. With increasing acquisition time it is generally much simpler to discern peaks from the background and specimen peaks from artifact peaks, so peak identification will proceed a lot more easily.

34.3 PEAK IDENTIFICATION

There are four key steps to running a successful qualitative analysis.

First: let's assume that you have read, understood, and applied the contents of Chapters 32 and 33. So you know what artifacts are likely to arise from your XEDS system and what system peaks occur in your AEM and you're aware of CB, etc. Now, ensure that the computer display is calibrated to be as accurate as the display resolution, over the energy range you've selected. If your spectrum is displayed at 10 eV per channel, the characteristic peak centroids must all be within ±10 eV of their true position on the energy scale.

Second: the computer system can be used to run an automatic identification check on the peaks in the spectrum, assuming the energy display is well calibrated. If the spectrum is simple, containing a few well-separated peaks, this automatic step may be all that is required. However, as we'll discuss in Section 34.5, misidentification occasionally occurs during such an 'auto-search' or 'peak ID' software routine. The more complex the spectrum, the more likely this is to happen, e.g., if the spectrum contains many peaks, particularly if peak overlap is occurring (e.g., Zn L_α confused with Na K_α) or if the spectrum contains complex peak families from heavier and/or rarer elements (e.g., Ta M_α confused with Si K_α). This problem is exacerbated if you don't understand the complexities of X-ray families or don't follow all the precautions that we'll go through below. In addition, even in the best software, small peaks may sometimes be missed and phenomena such as CB are often not taken into account.

THE KEY	LABEL THE PEAKS
For good qualitative analysis be suspicious. Don't just seek the peaks you expect, but be prepared to find peaks that you don't expect.	Take care to label each peak on the computer display or note it in your lab notebook when you have decided which element it comes from.

Remember, each X-ray is emitted with a very well-defined line energy ($\sim$ 1–5 eV wide) but the XEDS system degrades the line to a broad peak (FWHM $\sim$ 80–180 eV over the energy range detected by a typical Si(Li) detector). So we'll talk about peaks in the spectrum, which correspond to X-ray lines emitted from specific elements and which are identified as belonging to families of lines that are superimposed by the computer software on the displayed spectrum.

Third: go back to Chapter 4 and remind yourself about things like critical ionization energy, X-ray line energy, K, L, M families of X-ray lines, relative weights of lines, fluorescence yield, etc., because we're assuming you know all of this backwards.

Our peak analysis will always include the following steps

- Look first at the most intense peak since this should be easiest to identify as a K, L, or M_α peak; then work on down through the associated family lines. If you can't identify the most intense peak easily then you've got a problem; e.g., the calibration is off or the electronics or software are not functioning correctly, so it's time to ask for technical help and go for yet more coffee.
- The most intense peak is also the most likely to have associated artifacts, such as an escape peak (1.74 keV below that peak energy in a Si(Li) system) and a sum peak (at twice the peak energy), so you can quickly remove such small peaks from the unknown list (most software now automatically notes the energy where these artifact peaks should appear).
- Go to the next most intense peak not included in the above step and repeat the search. Then repeat this exercise until all peaks are identified.
- Always think about pathological overlaps; look for spurious peaks, system peaks, and artifact peaks.

Fourth: the bookkeeping; in choosing the possible K, L, or M lines that could be present at a specific energy, you can either use the computer-generated X-ray line markers on the display or consult an appropriate source such as the 'slide rules' offered by most commercial manufacturers, or find them on the Web at, e.g., URLs 1 and 2. We introduced DTSA in Chapter 33 and it is great for comparing simulated spectra with acquired spectra, as well as for checking the specific energies of X-ray lines, particularly in the more complex families of lines in spectra from heavier elements. There's much more about DTSA in the companion text.

Good bookkeeping is essential during the identification sequence we will now describe, particularly if your spectrum contains many peaks.

So now we've outlined the principles, let's get down to the specifics. Follow this 8-step process

1. If a K_α line matches the peak, look for the K_β line which has about $\sim$10% of the K_α intensity (this 10% is a line 'weight'; go back and check Table 4.1). With a modern XEDS, the K_β line *must* be present at X-ray energies above $\sim$1.74 keV (Si K_α), so long as it is not overlapped by a more intense peak from another element. Below this energy, your detector may not be able to resolve the two lines.
2. If a K_α and K_β pair fits the peaks and the K_α energy is > $\sim$ 8 keV (Ni K_α), look for the L lines at $\sim$ 0.9 keV if you are using a Be-window detector. For an UTW/windowless detector the L_α lines from Cl and above (> $\sim$ 0.2 keV) may be detectable but only if there's a lot of Cl, because (a) these relatively weak X-rays will be strongly absorbed and (b) their fluorescence yield is pretty abysmal. Ni L_α = 849 eV, Cl L_α = 200 eV.
3. If a K_α line does not fit, check for an L_α or M_α line fit since these are the most intense lines in the L and M families.
4. If an L_α line fits, there *must* be accompanying lines in the L family. The number of visible lines will vary depending on the intensity and energy of the L_α line, with more lines resolvable at higher line energies. The other lines in the family are all of lower intensity than the L_α line, and the following lines may be detectable (the number in parentheses is the weight relative to the L_α line); $L_{\beta 1}$ (0.7), $L_{\beta 2}$ (0.2), and $L_{\gamma 1}$ (0.08) lines at higher energies and possibly the Ll (0.04) line at lower energy. Other, even less intense, lines ($L_{\gamma 3}$ (0.03) and L_η (0.01)) may be visible if the L family is extraordinarily intense, but this is rare.
5. If the L lines fit, there *must* be a higher energy K_α/K_β pair, since the AEM beam energy is usually sufficient (> 200 keV) to generate the K lines from all the elements in the periodic table. Make sure you choose a broad enough energy range to display them on your computer.
6. The M lines are usually only visible for elements above La in the periodic table if a Be-window detector is used, and above about Nb, if a UTW detector is used. Again there has to be a lot of Nb to pick up the weak M line. La M_α = 833 eV, Nb M_α = 202 eV.

7. The M_α/M_β line overlap is difficult to resolve because all the M lines are < 4 keV. If an M_α/M_β line fits, look for three very small lines M_ζ (0.06), M_γ (0.05), and $M_{II}N_{IV}$ (0.01) lines, which will be more visible if more of the element is present.
8. If the M_α line fits there *must* be a higher energy L line family and possibly the very high energy K lines may be detectable; again, this depends on the detector (IG is better for high kV lines), computer display (out to ~ 80 keV), and the accelerating voltage (higher is better).

Figure 34.2 shows the families of lines expected in the display range from 0 to 20 keV, giving you some idea of the distribution of families of elemental lines that you

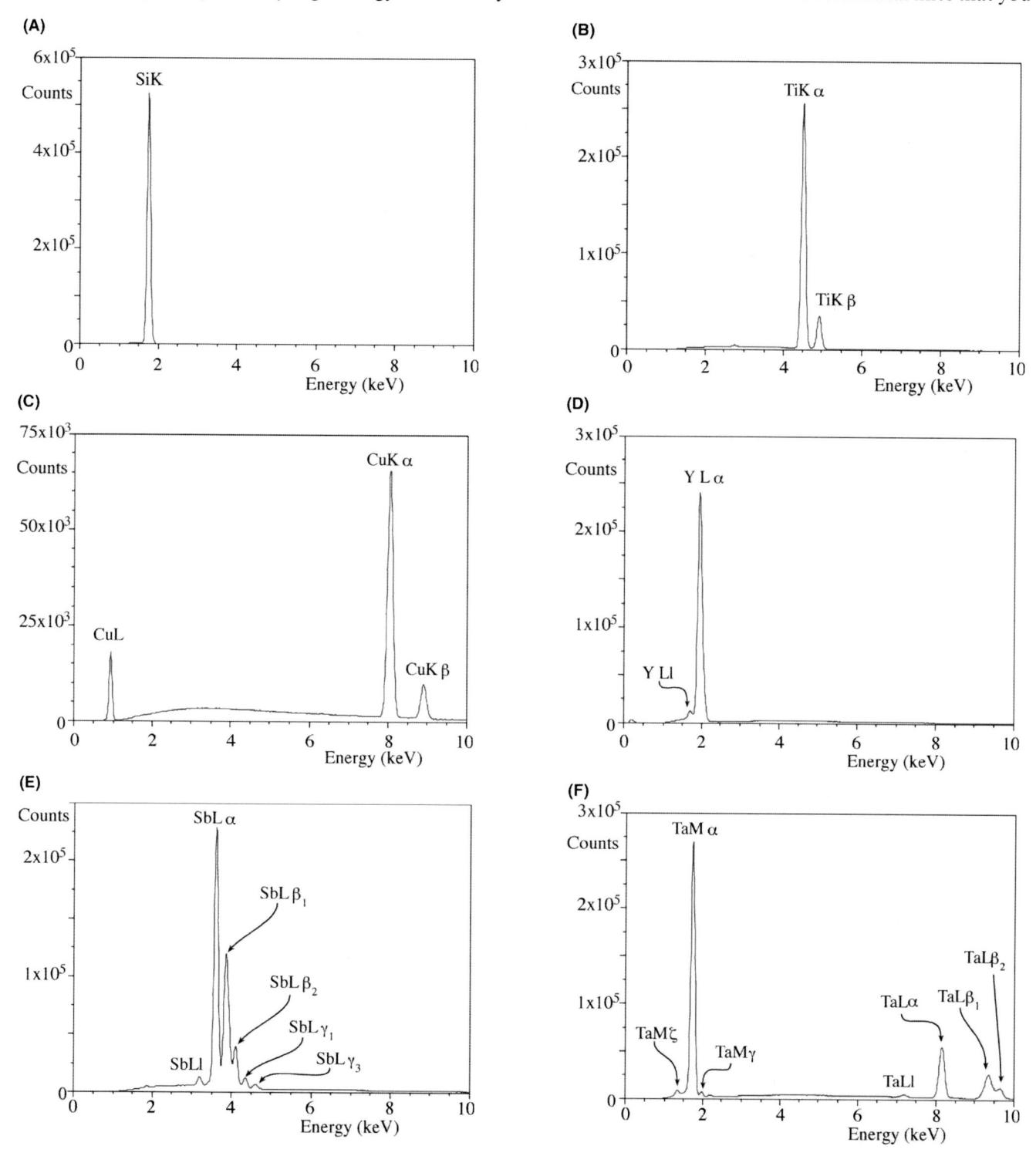

FIGURE 34.2. X-ray spectra from elements spanning much of the periodic table showing the families of characteristic lines for (A) Si, (B) Ti, (C) Cu, (D) La, (E) Sb, and (F) Ta. Starting with a single Si K_α line at low Z and low X-ray energy, the series progresses through the appearance of the families of L (Cu and La) and M (Sb and Ta) lines. Note the increasing separation of the peaks of a given family as both Z and keV increase.

34.3 PEAK IDENTIFICATION **629**

should find when you follow the procedure outlined above. For example, you can see for which elements you should expect to see only a single K line or resolve the K_α/K_β pair, and for which elements you should expect to see both K and L families or L and M families.

> **FAMILIES OF PEAKS**
> Looking for **families** of peaks. If a family member is missing your identification may be wrong.

Reasons for a missing family member

- It may be overlapped by another peak. This is the most likely cause and you can possibly resolve it by

  (a) re-gathering the spectrum with a longer time constant (analog system only)
  (b) using peak deconvolution software (see Section 34.4)
  (c) using a higher-resolution technique, such as EELS (see Chapter 38)

- Your computer display range may be too small so the peak is cut off (easy to solve, although if you followed our instructions this should never happen)
- The keV of the beam is too low to excite the line (should never be a problem in an AEM, only in an SEM)

Repeat the exercise: Go to the next most-intense peak that has not been identified by the eight steps in the first search. Continue this process *until all the major peaks are accounted for*. As you go, remember again to look for the escape peak(s) and sum peak associated with each *major* characteristic peak that you have conclusively identified. However, these artifacts and any CB peaks will be very small and before you worry about them you should make sure that the small peaks are statistically significant and we discuss how to do this for all minor peaks in Section 34.5 below. If you have a Si(Li) detector you'll always find a small Si-K internal-fluorescence peak, the Si escape peaks will lie at 1.74 keV below major peaks in the spectrum, and will not occur for elements below phosphorus. For an IG detector there will be Ge internal-fluorescence peaks (possible K and L) and there may be both Ge K and L line escape peaks at the appropriate energy below major peaks (9.89 keV for the Ge K_α escape and 1.19 keV for the Ge L_α escape). If you suspect a sum peak at twice the energy of any major peaks then re-acquire the spectrum at a much lower dead-time (< 20%) and see if the suspected sum peak disappears. If you suspect a CB peak then re-acquire the spectrum at a different accelerating voltage or specimen orientation and see if the small peak shifts.

Check for special cases: The relatively poor energy resolution of the XEDS detector means that there are several pairs of peaks that occur quite commonly in materials science specimens that cannot be resolved. These go by the delightful name of 'pathological overlaps' and include, inter alia

(a) the K_β and K_α lines of neighboring transition metals, particularly Ti/V, V/Cr, Mn/Fe, and Fe/Co
(b) the Ba L_α line (4.47 keV) and the Ti K_α line (4.51 keV)
(c) the Pb M_α (2.35 keV), Mo L_α (2.29 keV), and S K_α (2.31 keV) lines
(d) the Ti, V, and Cr L_α lines (0.45–0.57 keV) and the K lines of N (0.39 keV) and O (0.52 keV) detected in UTW/ATW or windowless XEDS systems.

> **PATHOLOGICAL OVERLAP**
> When it is impossible to separate two peaks even when you know they are both there.

These problems can often be solved by careful choice of the energy range on the computer display. For example, if you are only observing from 0 to 10 keV the S K/Mo L line overlap would be clarified by the presence or absence of the Mo K lines around 18 keV which, again, you should have seen in your first, broad energy-range, spectrum acquisition. If you suspect that any pathological peak overlaps are occurring in your spectrum, then re-gather under conditions that maximize the energy resolution of the detector system (i.e., longest (analog) time constant and low count rate (< 5 kcps)), and also maximize the display resolution to at most 5 eV per channel (you'll need more coffee).

34.4 PEAK DECONVOLUTION

If the overlap is still not resolvable, then you should run a peak deconvolution routine in your computer software (these are pretty standard and not much has changed since; see, e.g., Schamber 1981). Such routines are capable of detecting and resolving many of the classic materials science overlaps, such as the transition metal L lines and low-Z K lines and an example of such a deconvolution is shown in Figure 34.3.

In addition to deconvoluting any peak overlaps, it can be very useful to deconvolute the point-spread function of the XEDS detection system using a process called zero-peak deconvolution (which is analogous to the zero-loss peak deconvolution process used in EELS (Sections 37.5.A and 39.6)). Such a process is becoming increasingly popular in many imaging and spectroscopy techniques because of the development of various robust mathematical procedures (e.g., Janssen 1997)

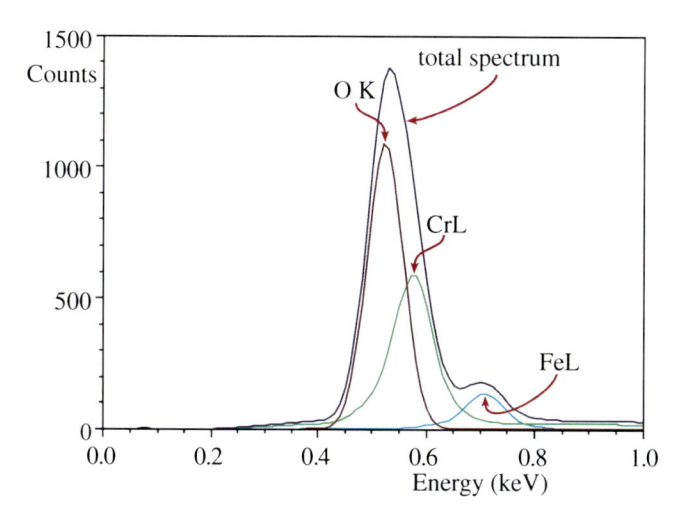

FIGURE 34.3. The total spectrum (dark blue) arises from the overlap of three Gaussian spectral peaks (the L_α lines of Fe (light blue) and Cr (green) and the O K_α line (brown)) from a mixed Fe-Cr oxide. Deconvolution of the individual contributions to the total spectrum is essential to determine the intensities in the three constituent peaks prior to any quantification attempt.

some of which are described in the companion text. In effect, this deconvolution removes the electronic-noise component of the characteristic peak width, giving a spectrum with nearer to noise-free resolution (Watanabe and Williams). This process requires that your XEDS system must display the noise peak at 0 eV in the spectrum (which not all manufacturers do) and this is called the 'strobe peak'. An iterative process can, in effect, remove the noise in your spectrum. As shown in Figure 34.4, this process obviously improves energy resolution but also improves the P/B ratio (and thus the minimum detection limit, as we'll see in Chapter 36) and can reveal peaks that would otherwise be masked by adjacent, more-intense, peaks.

Now, you always have to be careful with deconvolution, since any mathematical manipulation can introduce its own artifacts, while otherwise improving the spectrum quality. So it's best to practice deconvolution on both simple and complicated spectra that you know and understand well, until you feel confident that you understand the strengths and limitations of the procedure available in your particular software package.

Given all these steps and the multiple decisions that you have to make, it's clear that qualitative analysis can be an extraordinarily difficult procedure even for the

FIGURE 34.4. (A) The effects of deconvoluting the zero-energy strobe peak from an experimental XEDS spectrum of NIST SRM 2063. (B) The consequent reduction in peak FWHM (i.e., improvement in energy resolution) across the energy range 0–10 keV. (C) The improvement in P/B ratio of the major peaks in the SRM 2063 spectrum as a result of deconvolution. (D) Revealing the L_α peak from Ti in TiO_2, which is usually hidden by the O K_α.

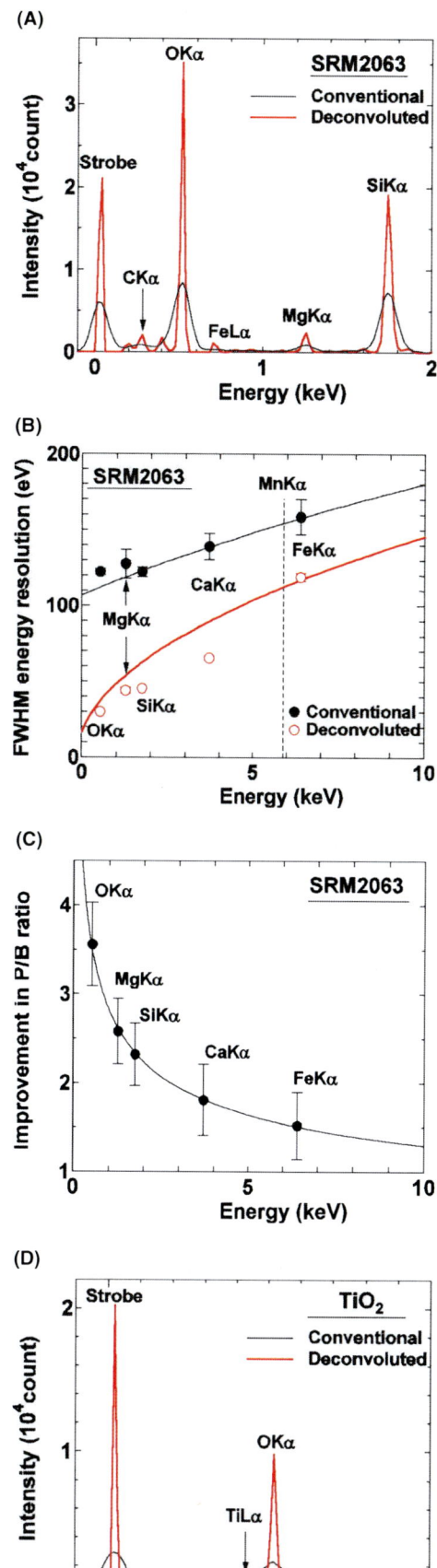

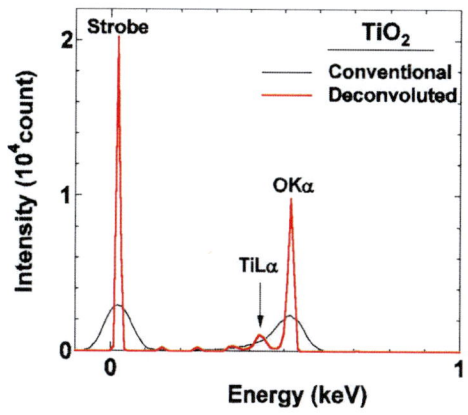

34.4 PEAK DECONVOLUTION

most experienced analyst. This complexity is one reason why even the best software gets it wrong sometimes (see Section 34.6). So, as we said at the start, be suspicious!

In summary, applying the 8-step process, with appropriate deconvolution where needed, should permit you to identify all the major peaks in your spectrum. There might still be minor peaks, which may or may not be statistically significant, and you have to decide whether you are going to identify or ignore these peaks. Now we'll tell you how to make this decision.

34.5 PEAK VISIBILITY

Small intensity fluctuations are often present in your spectrum that you cannot clearly identify as peaks. In this case, there is a simple statistical criterion (Liebhafsky et al.) that you can apply to ascertain if the peak is statistically significant or if it can be dismissed as random noise. You must count for a long enough time so that the bremsstrahlung intensity is smooth and any peaks are clearly visible, as summarized in Figure 34.5.

- Increase the display gain until the average background intensity is half the total full scale of the display, so the small peaks are more easily observed.
- Get the computer to draw a line under the peak to separate the peak and background counts.
- Integrate the peak (I_A) and background $((I_A^b))$ counts over the same number of channels; use FWHM if it can be discerned with any confidence; if not, then the whole peak integral will do.

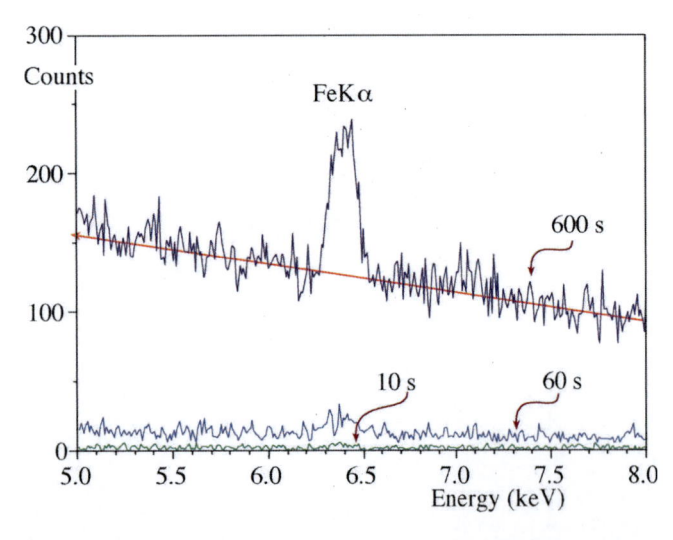

FIGURE 34.5. With increasing counting time, a clear characteristic Fe K_α peak develops above the background in these spectra from Si-0.2% Fe, thus demonstrating the need to acquire statistically significant counts before deciding if a peak is present or absent. As indicated (orange line), the background in the 600 s spectrum approximates to a straight line making the peak clearly visible. Note that Fe K_β is beginning to appear at 7.05 keV, although it is not yet statistically significant.

If $I_A > 3\sqrt{I_A^b}$ then the peak is statistically significant at the 99% confidence limit and must be identified. Remember, you'll still make an erroneous peak identification in ~ 1% of analyses using this criterion; just hope it isn't the one that derails your PhD or gets you fired!

If $I_A < 3\sqrt{I_A^b}$ then the peak is not significant and should be ignored.

If the insignificant peak is at an energy where you expect a peak to be present, but you think there is only a small amount of the suspected element in your specimen, then *count for a longer time* to see if the statistical criterion can be satisfied in a reasonable length of time. If this peak is a critical one, and it is often the minor or trace elements that are most important, then take *whatever time is necessary* to detect the peak. There is no reason not to gather the spectrum for many minutes or even an hour or more, so long as doing so does not change/damage or contaminate your specimen. Lunch breaks now become beneficial.

TIME

When you count for long times to search for characteristic peaks of low intensity, you will also begin to detect more easily the small peaks from the various spurious effects; e.g., CB peaks, Si or Ge internal-fluorescence peaks, and system peaks such as Fe and Cu. You also increase the possibility of contamination and beam damage to your specimen.

However, do *not* obtain more counts by raising the count rate above that which the processing electronics can handle, because you may introduce extra sum peaks and also degrade the energy resolution of the spectrum.

If you're worried about damaging the analysis area, as we stated at the beginning, it is best to spread the probe over as large an area as possible, either by defocusing the C2 lens in TEM mode or by using a scan raster in STEM mode.

Identifying the statistically significant peaks by the above method is one thing. Quantifying the amount of the element responsible for the peak is another matter and usually many more counts are required, as we'll see when we talk about detection limits in Chapter 36. However, once you're happy with the peak ID, you may be able to identify the phase/nanoparticle/precipitate that is being analyzed without any further work. For example, in the material that you are investigating, thermodynamics may tell you that there are only a few possible phases that can exist after the processing/thermal treatment given to it, and these phases may have very different chemistries. A glance at the relative peak intensities may be sufficient to conclude which phase you have just analyzed because, as we'll see below and in more detail in the next chapter, one of the marvelous advantages of thin-foil analysis in the AEM is that the peak intensities are often directly proportional to the elemental concentrations. As a result, quantification can be extremely simple.

To conclude this section, we'll look at two examples before going into the details of quantification in the next chapter.

The oxide-glass example: let's run a qualitative analysis on the spectrum in Figure 34.6. The spectrum is from a thin NIST oxide-glass film on a carbon support film on a Cu grid. X-rays were accumulated for 1000 s with a Be-window, IG detector at an accelerating voltage of 300 kV. Because of the Be window, we do not expect to see lines below ~0.8 keV and so the O K_α (0.52 keV) will not be detectable. The spectrum only contained peaks in the range from 0 to 10 keV and the first peak to be examined was the most intense high-energy peak, line #1, which was consistent with the Si K_α line at 1.74 keV. (The K_α/K_β pair cannot be resolved by the detector at this energy.) A similar treatment of the next most intense high-energy line (#2) at 3.69 keV produced a match with the Ca K_α and it is also possible to identify the associated K_β at 4.01 keV (the Ca L line is not detectable). The third most intense line #3 is at an energy of 6.4 keV. The K-line markers identified it, along with the smaller one to its right, as being the Fe K_α and K_β pair. No L line fit was reasonable (Dy L_α at 6.5 keV being the only alternative) and there are no M lines above about 4 keV. The Fe L line at 716 eV will not be detectable because of the Be window.

Next, the smaller peaks were tackled and the Cu K_α (and K_β) was identified at 8.04 keV, the Ar K_α at 2.96 keV (the K_β was too small to be visible), and the Mg K_α was the last to be identified at 1.25 keV.

Mg (as its oxide) is a common element in glasses, so this peak is expected. But Ar isn't such a common element in glasses. It probably arises from Ar implanted during ion-beam thinning and, as we discussed earlier and in detail back in Chapter 10, many specimen preparation processes can affect the chemistry of the specimen surface. You should also be aware that Ar K at 2.96 keV is often confused with the Al K sum peak (2×1.49 keV). But since there is no Al in the glass, and no major peak at half the Ar K energy, then Ar is the best answer.

> **SPECIMEN PREPARATION**
> Now you see why it is very useful to have some knowledge about your specimen and how it was prepared.

Likewise, Cu is not a common glass-forming element, but, since the specimen was on a Cu support grid, the Cu peaks are most probably due to post-specimen scatter of electrons or X-rays and so we cannot conclude that there is any Cu within the specimen. No escape or sum peaks were detectable.

> **ABSENCE OF THE Cu L_α Line**
> The absence of the line at 0.93 keV in Figure 34.6 is evidence that the thick Cu grid is responsible for a Cu line; the low-energy L X-rays will be absorbed in the grid itself before they can be detected.

The Fe-Cr-Ni example is shown in Figure 34.7 and this spectrum contains six Gaussian peaks, which can easily be identified following the procedure outlined above as the K_α and K_β pairs from Fe, Cr, and Ni. Even the average metallurgist will know that this specimen can only be some kind of stainless steel and this may be all the information that is required, making subsequent quantitative analysis redundant. But if you need more information, e.g., the specific grade of stainless steel, then you have to make measurements of the relative peak intensities, and this is the first step in the quantification procedure. In fact, we will see in Chapter 36 that the thin-foil quantification equation, to a first approximation, predicts that the amount of each element is directly proportional to the peak height. If you measure the relative heights of the K_α peaks in Figure 34.7 with a ruler, you can estimate the composition as ~ Fe-20% Cr-10% Ni which is within 10% of the classic 316 stainless composition of Fe-18% Cr-8 % Ni.

One real advantage of thin foil X-ray analysis is that you can get a good estimation of the composition of the analysis volume just by measuring the relative peak heights with a ruler.

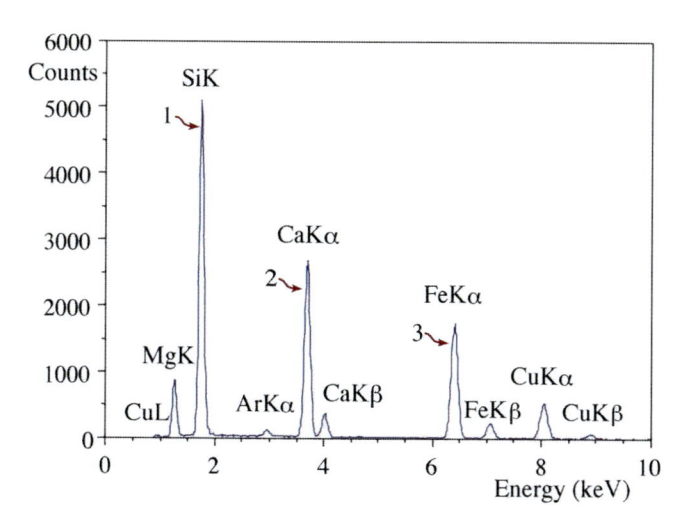

FIGURE 34.6. Energy-dispersive spectrum obtained at 300 kV from a thin oxide-glass film. The characteristic peaks were identified through the procedure outlined in the text.

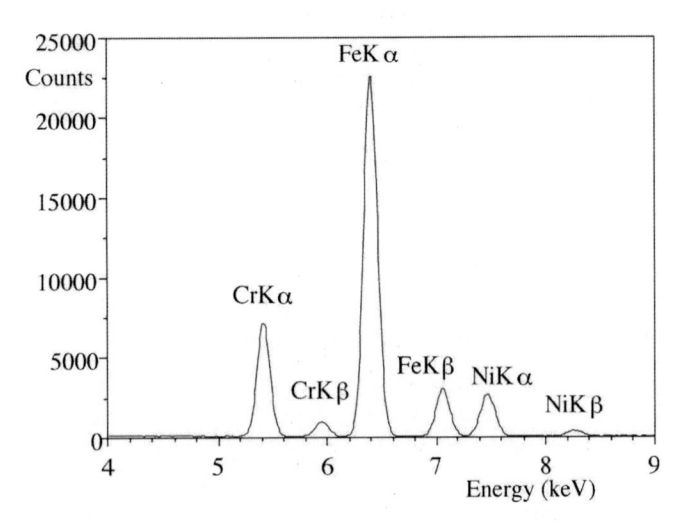

FIGURE 34.7. Spectrum from a stainless-steel foil in which the peaks are resolved and quite close in energy. To a first approximation, quantification is possible simply by measuring the relative heights of the K_α peaks.

Now there must be a good reason why, despite the fact that a 50-cent plastic ruler can give you a reasonable quantitative analysis of your spectrum in a few seconds, all AEMs have tens of thousands of dollars of computer hardware and software attached to the XEDS detector. This is because the stainless-steel spectrum is not really a challenge since all the peaks are close in energy, the peaks have lots of counts in them and there are no small peaks from trace elements. If the peaks were far apart in energy then relative X-ray absorption might occur, which can change the simple relationship between peak height and composition. Also if we really wanted to be sure that we had 304 rather than 316- stainless steel, then we would need a full quantification using the procedures described in the next chapter. If you were to do this, then you'd find that a full quantification would give a very similar result, but you could have much greater confidence in the true composition.

34.6 COMMON ERRORS

As we've seen, there's lots of room to misidentify peaks, particularly small ones that may be one of the many artifacts from the detector, the processing electronics, or the AEM-XEDS system. The first thing to realize is that if you (or even your advisor) can make mistakes, then the software can also do so. Remember, the software is only as good as the programmer. So while commercial automated peak-ID software is generally outstanding and getting better with every iteration, don't automatically believe it is always correct. By all means, as we said right at the start of Section 34.3, push the peak-ID button and get a quick analysis of your spectrum, but after you've done that, you might want to read the paper by Newbury which indicates some errors produced by a range of software systems. (See also the subsequent discussion noted in that reference.) The specimens used by Newbury to show the errors were often quite complex, with multiple peaks from rare, high-Z elements, but the article is instructive and illustrates the point that simply believing the software output is not always wise and suspicion is healthy.

As we've mentioned, it can be really useful to know something about your specimen chemistry *before* you put it into the AEM (and of course the TEM should be just about the last technique you use to study a completely unknown foil). It can also be very useful to remember how your specimen was thinned, since most thinning methods can change the surface chemistry of the foil. For example, electro-thinning methods can preferentially remove, or re-deposit one element from the foil or leave surface residue from the polishing medium. The thinner the foil is the more such changes in surface chemistry are exacerbated. Likewise, ion-beam thinning can result in the implantation of Ar, and FIB thinning can do likewise for Ga. So your foil may oxidize preferentially during thinning or pick up Cl from the perchloric-acid polishing solution, and so on. Ultramicrotomy is about the only method of thin-foil preparation that doesn't change the surface chemistry (although the freshly cut surface may be prone to rapid corrosion in the atmosphere and the defect structure is seriously changed from the bulk sample). So always be suspicious.

34.7 QUALITATIVE X-RAY IMAGING: PRINCIPLES AND PRACTICE

Individual-point analyses or multiple-point profiles across an interface are not the only ways to display X-ray data. As we saw in the previous chapter, we can produce X-ray images by a variety of methods and, if you are careful, such images can be used as compositional maps in which the intensity of the signal in the map is directly proportional to the generated X-ray intensity I_A. Under most circumstances, as we just described, we can take the next step and assume that, to some reasonable degree of accuracy, the X-ray intensity from element A in a thin foil is proportional to the concentration C_A. But there are some limitations to drawing a direct correlation, which we'll deal with in the next chapter on quantitative analysis. While there are obvious advantages to comparing maps of elemental distributions with other TEM images, this process is limited by the relatively poor statistics of X-ray acquisition, as should be eminently obvious to you by now. As you'll see in the next chapter, good quantification requires ~10,000 counts in the characteristic peak from element A, I_A. In early AEMs, such intensity would easily take you a minute or so to acquire, even if there was a large amount of A present in your specimen. At this acquisition rate, even a 56 × 56 pixel image would take 50 hours to gather, so imaging didn't see a lot of practitioners and we just had to make do with qualitative, noisy maps, as shown back in Figure 33.13. However, as we've told you already, a lot has changed in recent years to improve the X-ray acquisition rate. First, to generate more X-rays, we have intermediate-voltage FEGs and spherical-aberration correctors to permit larger final apertures while maintaining small probes sizes. Second, to permit greater throughput of X-rays, we have developed digital pulse-processing, larger X-ray collection angles, and SDD arrays. Consequently, particularly if you can get away with using a reasonably thick specimen, X-ray mapping is a viable option on most modern AEMs where digital-display technologies and new image-processing software also makes life much easier.

The older X-ray dot maps, such as in Figure 33.13, simply register a dot when an X-ray is registered in a particular energy window in the XEDS spectrum. It doesn't matter whether the X-ray is a characteristic X-ray, a bremsstrahlung X-ray, a spurious X-ray, or a system X-ray, it is still registered, so problems like thickness or atomic number effects can occur since thicker/higher-Z specimens will generate more bremsstrahlung (as well as more characteristic) X-rays. These problems are a lot greater in bulk-specimen, SEM X-ray imaging, which has grappled with these issues for many decades (summarized in 1990 by Newbury et al.). To account for all these effects, a full quantitative procedure has to be applied to the intensity in each pixel and we'll cover this in detail in the next chapter.

For qualitative mapping, gray-scale images are still acceptable, but for full quantification, there is no way around the use of full-color images because of the many more color signal levels that your eye can discern and the advantages of color overlays to compare maps of different elements, and an example is shown in Figure 34.8, where the final RGB color image shows the chemical inhomogeneity of the Au-Pd nanoparticles, thus giving insight into why they work as catalysts for the peroxide-synthesis process. (Similar qualitative maps were shown in Figures 33.14 and 33.15.)

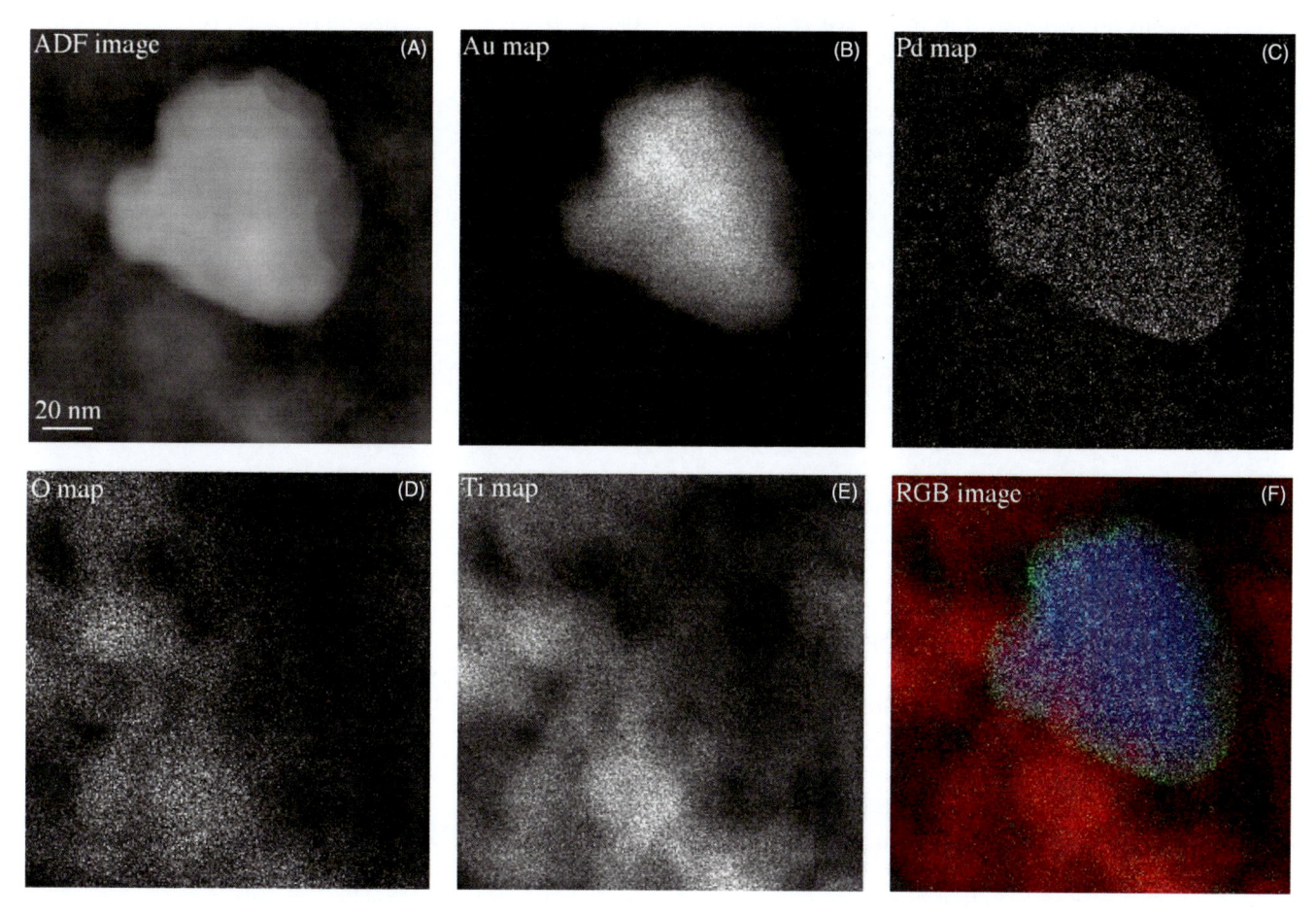

FIGURE 34.8. (A) STEM ADF image and qualitative X-ray maps showing the distribution of the (B) Au, (C) Pd, (D) O, and (E) Ti in a Au-Pd/TiO$_2$ catalyst nanoparticle. Such particles are used for peroxide synthesis. (F) The overlay of the color images from Ti (red), Pd (green), and Au (blue) reveals the core-shell nature of the particles with Pd on the outside and Au on the inside.

So how do you acquire a qualitative map like Figure 34.8?

- First you have to be in STEM mode, so your beam is scanning the specimen under digital control. Then select the largest probe size that will be still compatible with the desired X-ray image resolution. (Mapping with < 5 nm spatial resolution as in Figure 34.8 is challenging (see Chapter 36 on the factors controlling spatial resolution)). Only the best intermediate-voltage FEG instruments approach 1–2 nm, so a 5 nm probe would be a good start.
- Second, select a region of the X-ray spectrum that you want to map (e.g., set a window around a principal characteristic peak).
- Third, set a dwell time for the probe at each pixel that will permit sufficient X-ray counts to be acquired. Much more than a few seconds/pixel will make the acquisition time rather long, and < 0.1 s is only worthwhile if you have an intermediate-voltage FEG. (The best aberration-corrected FEG-AEMs can gather acceptable maps with dwell times of < 10 ms/pixel.) But, unless you have access to one of these instruments, start at about 1 s/pixel and gather a 64×64 pixel map and this will take you just over an hour. Do all the things we've told you to maximize the count rate: large beam diameter, large C2 aperture, and shortest (analog) pulse-processing time (this will minimize the chances of the processing electronics rejecting any of the few counts that are generated in such short dwell time).
- Last, start the scan and allow the map to commence building up on the computer display. If the intensity is acceptable and useful gray-scale information is obtained, it might be worthwhile to stop and re-gather the map for a longer period of time, e.g., for several hours, or even overnight, if it is a crucial map, However, for such long mapping times, you'll need to apply drift-correction software so that the map comes from the intended region of your specimen. Also it is obviously paramount that your specimen and AEM are clean, otherwise carbon contamination will build up to the point where the X-ray detection can be compromised, Also, as we've now told you innumerable times, specimen damage/contamination becomes an issue with longer scan times.

There are many other things you can do with the quantitative-analysis software and, given the time involved for even a simple qualitative map as just described, it's probably only worth doing fully quantitative maps. So we'll return to this technique in a lot more detail after you've learned all the steps necessary to translate the X-ray intensity into the elemental composition.

CHAPTER SUMMARY

One last time; doing the qualitative analysis first is not an option. The following steps are essential

- *Always be suspicious of any small peaks.*
- Get an intense spectrum across the energy range that contains all the characteristic peaks.
- Starting at the high-energy end of the spectrum, identify all the major peaks and any associated family lines and artifacts.
- If in doubt, collect for a longer time to decide if the intensity fluctuations are in fact peaks.
- Beware of pathological overlaps and be prepared to deconvolute any that occur.
- If you have the time, take a qualitative image using any crucial X-ray peaks.

GENERAL REFERENCES

Goldstein, JI, Newbury, DE, Joy, DC, Lyman, CE, Echlin, P, Lifshin, E, Sawyer, L and Michael JR 2003 *Scanning Electron Microscopy and X-ray Microanalysis* 3rd Ed. p355 Kluwer Academic Press New York.

SPECIAL REFERENCES

Jansson, PA, Ed. 1997 *Deconvolution of Image and Spectra* Academic Press San Diego.

Newbury, DE, Fiori, CE, Marinenko, RB, Myklebust, R., Swyt, CR and Bright, DS 1990 (A,B) *Compositional Mapping with the Electron Probe Microanalyzer* A: Anal. Chem. 62 1159–1166A, B: Anal. Chem. 62 1245–1254A. More on errors.

Liebhafsky, HA, Pfeiffer, HG, Winslow, EH and Zemany, PD 1972 *X-rays, Electrons and Analytical Chemistry* p349 John Wiley and Sons New York. Some statistics.

Newbury, DE 2005 *Misidentification of Major Constituents by Automatic Qualitative Energy Dispersive X-ray Microanalysis: A Problem that Threatens the Credibility of the Analytical Community* Microsc. Microanal. **11** 545–561. (See discussion by Burgess, S 2006 *idem* **12** 281 and Newbury, DE *idem* **1**, 282.) Discussion of errors produced using commercial software.

Schamber, FH 1981 *Curve Fitting Techniques and Their Application to the Analysis of Energy Dispersive Spectra* in *Energy Dispersive X-ray Spectrometry* 193–231 Eds. KFJ Heinrich, DE Newbury, RL Myklebust and CE Fiori NBS Special Publication 604 US Department of Commerce/NBS Washington DC.

Watanabe, M and Williams, DB 2003 *Improvements to the Energy Resolution of an X-ray Energy Dispersive Spectrum by Deconvolution Using the Zero Strobe Peak* Microscopy and Microanalysis Eds. D Piston, J Bruley, IM Anderson, P Kotula, G Solorzano, A Lockley and S McKernan 124–125.

URLs
1) http://microanalyst.mikroanalytik.de/index_e.phtml
2) http://www.cstl.nist.gov/div837/Division/outputs/DTSA/DTSA.htm

SELF-ASSESSMENT QUESTIONS

Q34.1 Why is it important to know the relative weights of characteristic X-ray family lines?

Q34.2 Why can't we predict the relative weights of the lines in different X-ray families (e.g., the K_α/L_α ratio) while we can predict the ratios within a family (e.g., K_α/K_β?)

Q34.3 Why should you identify the largest, high-energy X-ray peak first?

Q34.4 What factors determine whether or not an X-ray peak is visible?

Q34.5 What units give a good measure of the total efficiency of X-ray generation and detection?

Q34.6 Why should you do qualitative analysis anyhow?

Q34.7 What are the three kinds of resolution we worry about in XEDS analysis and how do they rank in importance?

Q34.8 Why is careful bookkeeping so important when doing qualitative analysis?

Q34.9 Why is it essential to calibrate your XEDS display and how often should you do it?

Q34.10 Why is Ray Dolby equally applauded by TEM and audio enthusiasts?

Q34.11 Why is it necessary to define a minimum criterion for peak visibility?

Q34.12 State the minimum visibility criterion.

Q34.13 Why would you choose a 99% confidence limit for peak visibility rather than a lesser value and what does this 99% confidence limit mean?

Q34.14 Why are you wasting your time if you gather spectra for very long times (such as overnight) in order to maximize the counts in the spectrum?

Q34.15 Why are characteristic-peak overlaps described as 'pathological'?

Q34.16 List a few key pathological overlaps that might affect you if you are a metallurgist.

Q34.17 Why do X-ray spectra get more complex as your specimens get higher in atomic number?

Q34.18 If you can't conclusively identify a small peak in the spectrum, what should you do?

Q34.19 If you can't conclusively identify a large peak in your spectrum, what should you do?

Q34.20 Can you think of an occasion when qualitative analysis might preclude the necessity for future quantitative analysis?

TEXT-SPECIFIC QUESTIONS

T34.1 Using Figure 34.2 as a basis, list the key characteristics of each of the principal families of lines that we see in XEDS spectra between 0 and 10 keV.

T34.2 Why don't we see N X-ray lines?

T34.3 There isn't an Fe K_β peak in Figure 34.5. Should you be concerned? If you are concerned, what should you do to lower your blood pressure?

T34.4 If you acquire spectra from a Ni jet-polished disc, why does the NiK_α/NiL_α ratio vary as you move the probe away from the thin edge surrounding the hole in the specimen?

T34.5 If you are analyzing an electropolished Al-Cu thin foil, why might the Cu signal increase substantially at the thinnest edges of the foil? (Hint: go back and look at Chapter 10.)

T34.6 Why is it difficult to analyze spectra from transition elements adjacent to one another in the periodic table, especially if the lower-Z transition metal is present in significantly greater amounts?

T34.7 When acquiring a qualitative spectrum from an unknown specimen, explain why you use as large a probe size as possible.

T34.8 When acquiring a qualitative spectrum from an unknown specimen, explain why you use as short a time constant as possible.

T34.9 When acquiring a qualitative spectrum from an unknown specimen, explain why you count for as long as possible.

T34.10 No. 13 of Murphy's Law of Microanalysis* states that 'The probability of detecting argon in aluminum decreases with time'. Justify the validity of this law. (*Copyright Kevex Inc.; reproduced with permission.)

T34.11 Look at the spectrum in Figure 34.6 and explain why

- you cannot confuse the Si K_α peak with the Si internal fluorescence artifact peak?
- the Cu K and L lines are from the Cu support grid and not from Cu in the specimen?
- the Cu L_α peak is present but not the Fe L_α peak?
- the Ar peak is not an artifact? Explain from whence it came.
- the Ca K_α and K_β peaks (at 3.69 and 4.01 keV, respectively) are not, in fact, the Sn L_a and L_β peaks (at 3.66 and 4.13 keV)?
- the Ca K_α and K_β peaks are not in fact the Te L_α and L_β peaks (at 3.77 and 4.03 keV)?
- there are no detectable escape and sum peaks?

T34.12 Look at the following spectra (which are really two different 'magnifications' of the vertical (counts) scale of the same spectrum to reveal both the small peaks and more intense peaks). This was taken from a thin-foil specimen in a 200-keV AEM with an ATW-XEDS Si(Li) detector. Carry out a qualitative analysis and identify unambiguously all the peaks labeled as #1–#12. When you have successfully finished the qualitative analysis of the above spectrum, use DTSA and try to reproduce this same spectrum.

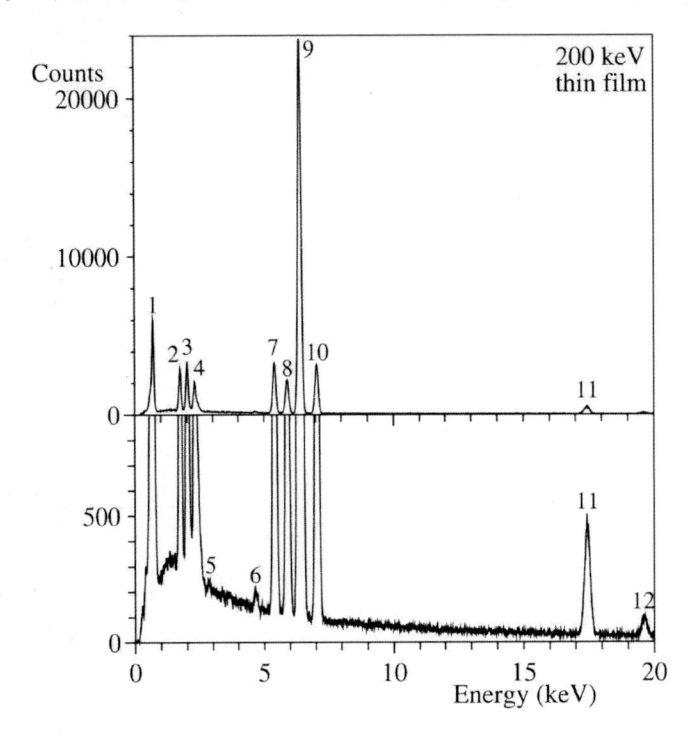

35
Quantitative X-ray Analysis

CHAPTER PREVIEW

Now you've got an idea of how to acquire XEDS spectra and images from thin foils. You understand what factors limit the useful information they may contain and what false and misleading effects may arise. Also you know how to be very sure that a certain characteristic peak is due to the presence of a certain element and the occasions when you may not be so confident. Having obtained a spectrum or image that is qualitatively interpretable, it turns out to be a remarkably simple procedure to convert that information into quantitative data about the distribution of elements in your specimen; this is what we describe in this chapter.

This chapter is a little longer than the average. You may find that you can skip parts of it as you work through it the first time. We have decided to keep the material together so as to be a more useful reference when you are actually doing your analyses on your microscope. This aspect of XEDS is poised to change significantly in the future and the prospects for much improved quantification are hinted at and covered in detail in the companion text.

35.1 HISTORICAL PERSPECTIVE

Quantitative X-ray analysis in the AEM is a straightforward technique. What is surprising is that, given its simplicity, relatively few users take the trouble to extract quantitative data from their spectra, or produce quantitative images, despite the fact that numerical data are the basis for all scientific investigations. Before we describe the steps for quantification, you should know a little about the historical development of quantitative X-ray analysis, because this will emphasize the advantages of thin-foil analysis over analysis of bulk specimens, which was in fact the driving force for the development of the first commercial analytical TEMs.

Historically, X-ray analysis in electron-beam instruments started with the study of bulk specimens in which the electron beam was totally absorbed, as opposed to 'thin' specimens through which the beam penetrates. The possibility of using X-rays generated by a focused electron beam to give elemental information about the specimen was first described by Hillier and Baker in 1944, and the necessary instrumentation was built several years later by Castaing in 1951. In his extraordinary Ph.D. dissertation, Castaing not only described the equipment but also outlined the essential steps to obtain quantitative data from bulk specimens. The procedures that Castaing proposed still form the basis of the quantification routines used today in the EPMA and may be summarized as follows. Castaing assumed that the concentration C_i of an element i in the specimen generates a certain intensity of characteristic X-rays. However, it is very difficult in practice to measure this generated intensity so Castaing suggested that a known standard of composition $C_{(i)}$ be chosen for element i. We then measure the intensity ratio $I_i/I_{(i)}$

I_i is the measured intensity emerging from (not generated within) the **specimen**.

$I_{(i)}$ is the measured intensity emerging from the **standard**.

Castaing then proposed that, to a reasonable approximation

$$C_i/C_{(i)} = [K]I_i/I_{(i)} \qquad (35.1)$$

where K is a sensitivity factor (not a constant) that takes into account the difference between the *generated* and *measured* X-ray intensities for both the standard and the unknown specimen. The contributions to K come from three effects

- Z The atomic number
- A The absorption of X-rays within the specimen
- F The fluorescence of X-rays within the specimen

The correction procedure in bulk analysis is often referred to as the ZAF correction. The necessary calculations, which have been refined over the years since

Castaing first outlined them, are exceedingly complex and best handled by a computer. If you're interested, there are several standard textbooks available which describe the ZAF and related procedures in detail, for example, those by Goldstein et al. and by Reed.

It was soon realized that, if an electron-transparent rather than a bulk specimen was used, the correction procedure could be greatly simplified. To a first approximation, in thin films, the A and F factors could be ignored and only the Z correction would be necessary. Furthermore, in thin specimens, the analysis volume would be substantially reduced, giving a much better spatial resolution. (We discuss this latter point in detail in the next chapter.)

These two obvious advantages of thin-foil analysis led to the development of the so-called Electron Microscope MicroAnalyzer (EMMA), pioneered by Duncumb in England in the 1960s. Unfortunately, the EMMA was far ahead of its time, mainly because the WDS was the only X-ray detector available. As we have seen back in Chapter 32, the classic WDS is handicapped by its poor collection efficiency, relatively cumbersome size, and slow, serial operation. These factors, particularly the poor efficiency, meant that a large probe size (~0.2 μm) had to be used to generate sufficient X-ray counts for quantification of the weak signal from thin foils and, therefore, the gain in spatial resolution over the EPMA was not so great. Also, the poor stability of the WDS meant that it was necessary to measure the beam current to make sure that the X-ray intensities from both standard and unknown could be sensibly compared. As a result of all these drawbacks, the EMMA never sold well and the manufacturer (AEI) soon went out of the EM business.

It is ironic that around this time the commercial developments that would transform TEMs into viable AEMs were all taking place. The XEDS detector was developed in the late 1960s, and commercial TEM/STEM systems appeared in the mid-1970s. However, before the demise of the EMMAs, they were to play a critical role in the development of the thin-foil analysis procedures that, surprisingly, we still use today. The EMMA at the University of Manchester, operated by Graham Cliff and Gordon Lorimer, was re-fitted with an XEDS system and they soon realized that the pseudo-parallel collection mode, the greater collection efficiency, and the improved stability of the XEDS removed many of the problems associated with WDS on the EMMA. Cliff and Lorimer (1975) showed that quantification was possible using a simplification of Castaing's original ratio equation, in which there was no need to incorporate intensity data from a standard, but simply ratio the intensities gathered from two elements simultaneously in the XEDS. This finding revolutionized thin-foil analysis and remains the basis for most quantifications today.

35.2 THE CLIFF-LORIMER RATIO TECHNIQUE

The basis for the Cliff-Lorimer technique is to rewrite equation 35.1 as a ratio of two elements A and B in a binary system. We have to measure the above-background characteristic intensities, I_A and I_B, simultaneously. This is trivial with an XEDS and, therefore, there is no need to measure the counts from a standard

> **THIN SPECIMENS**
> We assume that the specimen is thin enough so that we can ignore any absorption or fluorescence. This assumption is called the 'thin-foil criterion.'

The weight percents of each element C_A and C_B can then be related to I_A and I_B thus

$$\frac{C_A}{C_B} = k_{AB} \frac{I_A}{I_B} \quad (35.2)$$

This equation is the Cliff-Lorimer equation and the term k_{AB} is often termed the Cliff-Lorimer factor. As with K in equation 35.1, k_{AB} is a sensitivity factor, not a constant, so don't be fooled by the use of this letter. The k-factor varies according to your TEM/XEDS system and the kV you choose. Because we are ignoring the effects of absorption and fluorescence, k_{AB} is related only to the atomic-number correction factor (Z) in Castaing's original ratio equation. Now to obtain an absolute value for C_A and C_B, we need a second equation and, in a binary system, we simply assume that A and B constitute 100% of the specimen, so

$$C_A + C_B = 100\% \quad (35.3)$$

We can easily extend these equations to ternary and higher order systems by writing extra equations of the form

$$\frac{C_B}{C_C} = k_{BC} \frac{I_B}{I_C} \quad (35.4)$$

$$C_A + C_B + C_C = 100\% \quad (35.5)$$

You should also note that the k-factors for different pairs of elements AB, BC, etc., are related thus

$$k_{AB} = \frac{k_{AC}}{k_{BC}} \quad (35.6)$$

So long as you are consistent, you could define the composition in terms of atomic %, or weight fraction or any appropriate units. Of course the value of the k-factor would change accordingly.

Remember that Cliff and Lorimer developed the ratio approach to overcome the limitations of early

AEMs, particularly the low brightness of thermionic sources, the small collection angle of early detectors, electrical and mechanical instabilities (particularly the beam current), and the tendency of the early AEMs to contaminate the analysis area. Consequently, X-ray count rates were very low, thus limiting quantification. So it was very difficult to use the well-established, pure-element standards approach developed over the preceding 25 years for the EPMA because you'd have to keep changing specimens to measure the standards. In doing so, the probe current would change because, in a TEM, you have to switch off the beam to stop the vacuum changes burning out the gun (unlike in an EPMA where you can measure the current in situ and also isolate the gun automatically). Switching the gun on and off prevents any meaningful comparison of standards and unknowns. The ratio approach cancels out variations in the probe current incident on the analysis area; such variations arise from electron-gun/condenser-system instabilities, drift, and contamination buildup. Despite the fact that most of these problems have been minimized in modern AEMs, the ratio technique still remains the only quantitative thin-film analysis software available on commercial XEDS systems. Clearly, this is not an ideal situation and after describing the Cliff-Lorimer method, we'll discuss the ζ-factor, an alternative, improved approach, which combines the ease of application of the ratio method with the more rigorous aspects of pure-element (or other) thin-film standards (more about this in the companion text). The ζ-factor method requires in-situ measurement of the probe current, which, while standard on EPMAs for almost 50 years, has still not penetrated the design of commercial AEMs! So, despite its vintage, the Cliff-Lorimer equation remains the basis for quantitative analysis on all AEMs. Let's see how we use it in practice.

> **WT%**
> The convention is: define the units of composition as wt%.

35.3 PRACTICAL STEPS FOR QUANTIFICATION

First of all, you should try to use K_α lines, where possible, for the measured counts (I). (The K_β peak is combined with the K_α if the two K peaks cannot be resolved.) Use of L or M lines is more difficult because of the many overlapping lines in each family, but may be unavoidable if the K_α lines are too energetic and go right through your detector. (Think why you can't use L or M lines if the K lines are too *weak* for your detector.)

To gather characteristic X-ray intensities for quantification

- Keep your specimen as close to 0° tilt as possible to minimize spurious effects.
- If you have a wedge specimen, orient it so the thin portion of the wedge faces the detector, to minimize X-ray absorption (see Section 35.6).
- If the area of interest in your specimen is close to a strong two-beam dynamical diffraction condition, tilt it slightly to kinematical conditions.
- Accumulate enough counts in the characteristic peaks, I_A, I_B, etc. As we will see below, for acceptable errors, there should ideally be at least 10^4 counts above background in each peak.

While you can't always obtain 10^4 counts in a reasonable time before specimen drift, damage, or contamination limits your analysis, you should always choose the largest probe size which is consistent with maintaining the desired spatial resolution, so you get most current into your specimen. (Remember all the other ways to maximize the X-ray count rate that we discussed in Chapter 34.)

(The reason we worry a little about the diffraction conditions is that anomalous X-ray generation can occur across bend contours or whenever a diffracted beam is strongly excited. This point is not too critical because we quantify using a ratio technique. If the beam has a large convergence angle, which is usually the case, any diffraction effect is further reduced. However, we will see in Section 35.9 that, under certain conditions, there are some advantages to be gained from such crystallographic effects.)

Having accumulated a spectrum under these conditions, how do you quantify it? All you have to do is measure the peak intensities I_A, I_B, etc., and then determine a value for the k_{AB} factor. To determine the peak intensities, you first have to remove the background counts from the spectrum and then integrate the peak counts. Both of these steps are accomplished by various software routines in the XEDS computer system or in DTSA. There are advantages and disadvantages to each approach, so you should pick the one that is most suited to your problem.

35.3.A Background Subtraction

Remember, as we saw back in Section 4.2.B, we are not very precise in the terminology we use for the X-ray background intensity, so it can be confusing. 'Background' refers to the counts under the characteristic peaks in the spectrum displayed on your computer screen. These X-rays are generated by the 'bremsstrahlung' or 'braking-radiation' process as the beam electrons interact with the coulomb field of the nuclei in the specimen. The intensity distribution of the bremsstrahlung decreases continuously as the X-ray energy increases, reaching zero at the beam energy (go back and look a, Figure 4.6). Thus, the energy distribution can be described as a 'continuum,' although as we've seen, the phenomenon of coherent bremsstrahlung disturbs this continuum.

> **TERMINOLOGY**
> We tend to use these three terms 'background,' 'bremsstrahlung,' and 'continuum' interchangeably, although strictly speaking they have these specific meanings.

Remember also that the generated bremsstrahlung intensity is modified at energies below about 1.5 keV by absorption within the specimen and the detector, so we are usually dealing with a background in the spectrum that looks something like Figure 35.1. The best approach to background subtraction depends on two factors

- whether the region of interest in your spectrum is in the low-energy regime, where the intensity decreases rapidly with decreasing energy.
- if the characteristic peaks you want to measure are close together or isolated.

Window methods: In the simplest case of isolated characteristic peaks superimposed on a slowly varying background, you can easily remove the background counts by drawing a straight line below the peak, and defining the background intensity as that present below the line, as shown in Figure 35.2. So you get the computer first to define a 'window' in the spectrum spanning the width of the peak, and then draw the line between

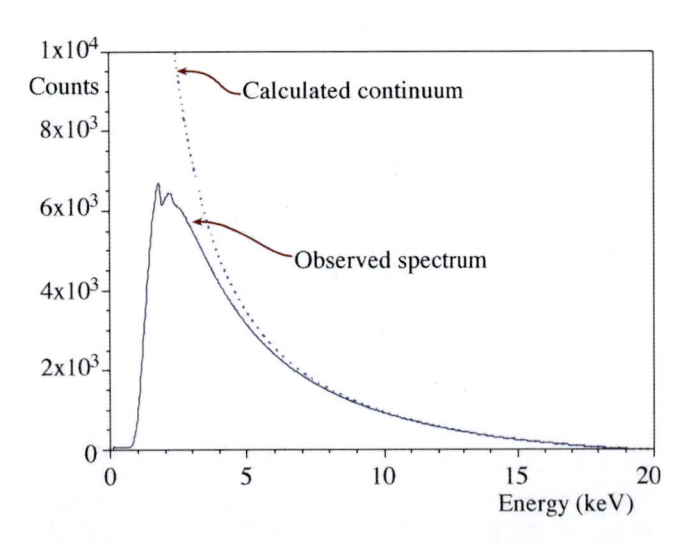

FIGURE 35.1. The theoretically calculated and experimentally observed bremsstrahlung intensity distribution as a function of energy. Both curves are similar until energies below ~2 keV when absorption within the specimen and the XEDS system reduces the detected counts. The best method of background removal depends on where in the spectrum your characteristic peaks are present.

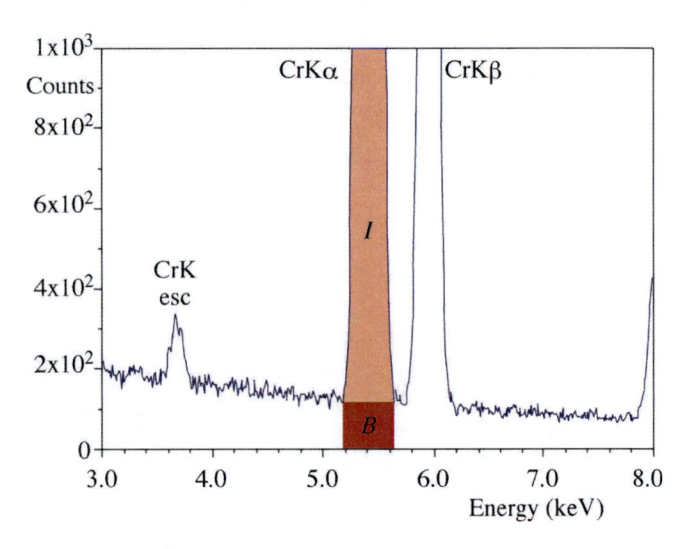

FIGURE 35.2. The simplest method of estimating the background contribution (B) to the counts in the characteristic peak (I); a straight line drawn beneath the Cr K_α peak provides a good estimate, if the counting statistics are good and the intensity approximates to a slowly varying function of energy. There should be no overlap with any other characteristic peak and the peak energy should be $> \sim 2$ keV.

the background intensities in the channels just outside the window. As with *all* spectral manipulations, this method gives better results with more counts in the spectrum. The background intensity variation is then less noisy, so it is easier to decide where the peak ends and the background begins and, furthermore, the background variation better approximates to a straight line.

Another, similarly primitive, approach involves averaging the bremsstrahlung counts above and below the characteristic peak by integrating the counts in two identical windows on either side of the peak, as shown in Figure 35.3. We then assume that the average of the two intensities equals the background counts under the

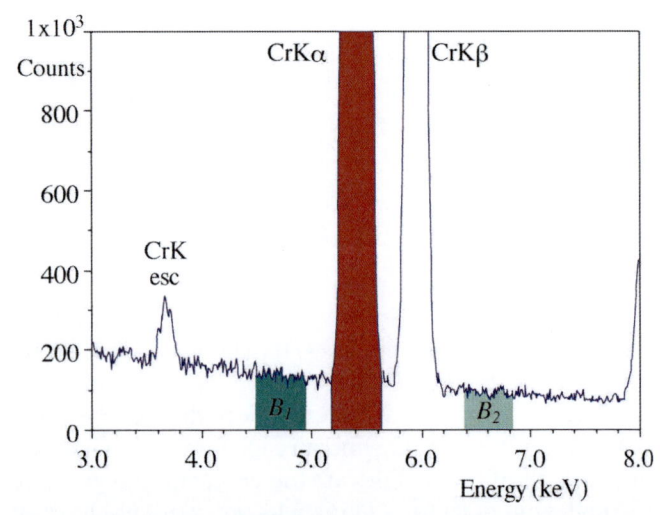

FIGURE 35.3. Background subtraction can be achieved by averaging the bremsstrahlung counts in two identical windows (B_1, B_2) on either side of the characteristic (Cr K_α and K_β) peaks. There should be no overlap with any other characteristic peaks and the peak energy should be >2 keV.

QUANTITATIVE X-RAY ANALYSIS

peak. This assumption is reasonable in the higher-energy regions of the spectrum and when the specimen is thin enough so that the bremsstrahlung is not absorbed in the specimen. This absorption happens because the bremsstrahlung X-rays with energies just above the peak energy preferentially fluoresce the characteristic-peak X-rays, resulting in a detectable reduction in the bremsstrahlung counts above the peak energy compared to below.

TOO THICK?
If you see this bremsstrahlung absorption effect in your spectrum, your specimen is too thick for Cliff-Lorimer quantification.

When you use the two-window approach, you must remember the window width you used, because *identical* windows must be used when subtracting the background both in the unknown spectrum and in the spectrum from the known specimen that you will use to determine the k-factor (see Section 35.4).

While the two techniques we just described have the advantage of simplicity, you can't always apply them to real specimens because the spectral peaks may overlap. Also, if the peaks lie in the low-energy region of the spectrum where the background is changing rapidly due to absorption, then neither of these two simple methods gives a good estimate of the background and more sophisticated mathematical approaches are required. We'll now discuss these methods.

Modeling the background: the bremsstrahlung distribution can be mathematically modeled, based on the expression developed by Kramers (1923). The number (N_E) of bremsstrahlung photons of energy E produced in a given time by a given electron beam is given by Kramers' law.

$$N_E = KZ \frac{(E_0 - E)}{E} \quad (35.7)$$

Here Z is the *average* atomic number of the specimen, E_0 is the beam energy in keV, and E is the X-ray energy in keV. The factor K in Kramers' law actually takes account of numerous parameters. These include

- Kramers' original constant.
- The collection efficiency of the *detector*.
- The processing efficiency of the *detector*.
- The absorption of X-rays within the *specimen*.

OPTIMUM WINDOW
The typical choice of window width is FWHM, but this throws away a substantial amount of the counts in the peak. FWTM gives better statistics, but incorporates more bremsstrahlung than characteristic counts; 1.2(FWHM) is the optimum window.

All these terms have to be factored into the computer calculation when you use this method of background modeling.

Be wary when using this approach because Kramers developed his law for bulk specimens. However, the expression is still used in commercial software, and seems to do a reasonable job.

Modeling the spectrum produces a smooth curve fit that describes the shape of the complete spectrum. This approach is particularly valuable if many characteristic peaks are present, since then it is difficult to make local measurements of the background counts by a window method. Figure 35.4 shows an example of a spectrum containing many adjacent peaks, with the background counts estimated underneath all the peaks.

Filtering out the background: another mathematical approach to removing the background uses digital filtering. This process makes no attempt to take into account the physics of X-ray production and detection as in Kramers' law. Rather it relies on the fact that the characteristic peaks show a rapid variation of counts as a function of energy (i.e., dI/dE is large), while the background exhibits a relatively small dI/dE. This approximation is valid even in the region of the spectrum below ~1.5 keV where absorption is strong. In the process of digital filtering, the spectrum intensity is filtered by convoluting it with another mathematical function. The most common function used is a 'top-hat' filter function, so called because of its shape. When the top-hat filter is convoluted with the shape of a typical X-ray spectrum, it acts to produce a second-difference spectrum, i.e. d^2I/dE^2 versus E. After the top-hat filter, the background with small dI/dE is transformed to a linear

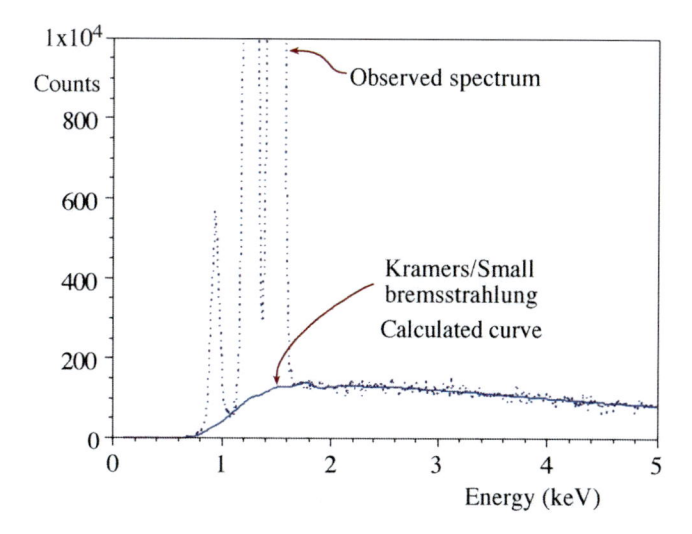

FIGURE 35.4. The bremsstrahlung intensity modeled using Kramers' law, modified by Small to include the effects of absorption of low-energy X-rays in the specimen and the detector. This method is useful when the spectrum contains many overlapping peaks, particularly in the low-energy range, such as the Cu L$_\alpha$ and the Mg and Al K$_\alpha$ lines shown in this spectrum (based on Chapter 34, can you determine which peak is which?).

35.3 PRACTICAL STEPS FOR QUANTIFICATION

function with a value of zero (thus it is 'removed'), while the peaks with large dI/dE, although distorted to show negative intensities in some regions, are essentially unchanged as far as the counting statistics are concerned. Figure 35.5A shows schematically the principle behind the filtering process and Figure 35.5B,C shows an example of a spectrum before and after digital filtering.

> **BACKGROUND REMOVAL**
> You must always take care to apply the same background-removal process to both the standard and the unknown.

In summary, you can remove the background by selecting appropriate windows to estimate the counts in the peak, or use one of two mathematical-modeling approaches. The window method is generally good enough if the peaks are isolated and on a linear portion of the background. The mathematical approaches are most useful for multi-element spectra and/or those containing peaks below ~1.5 keV. You should choose the method that gives you the most reproducible results (check this on a specimen for which you know the composition).

After removing the background, you have to integrate the peak intensities I_A, I_B, etc.

35.3.B Peak Integration

If you used a window method of background estimation, then the peak counts are obtained simply by subtracting the estimated background counts from the total counts in the chosen window. Therefore, if the computer drew a line under the peak as in Figure 35.2 then the peak intensity is that above the line.

- If you chose, e.g., an ideal window of 1.2 FWHM and averaged the background on either side of the peak then the average value must be subtracted from the total counts in the 1.2 FWHM window; always use the same window width for B and I.

FIGURE 35.5. (A) Digital filtering involves convolution of a top-hat filter function with the acquired spectrum. To obtain the filtered spectrum, each channel has the top-hat filter applied to it. The channels on either side of that being filtered (#8 in this case) are multiplied by the appropriate number in the top-hat function. So channels 1–5 and 11–15 are multiplied by –1 and channels 6–10 by +2. The sum of the multiplications is divided by the total number of channels (15) and allotted to channel #8 in the filtered spectrum at the bottom. The digital filtering process in (A) applied to a spectrum from biotite (B) results in the filtered spectrum (C) in which the background intensity is zero at all places, and the characteristic peaks remain effectively unchanged.

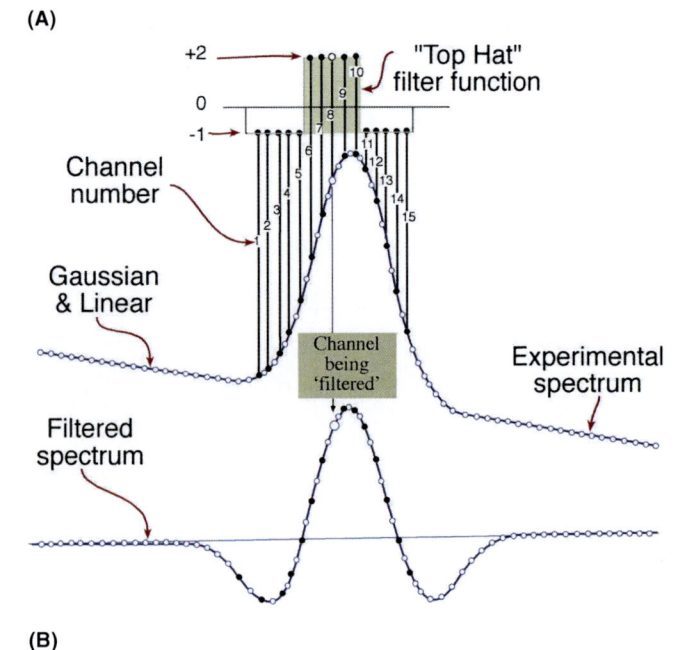

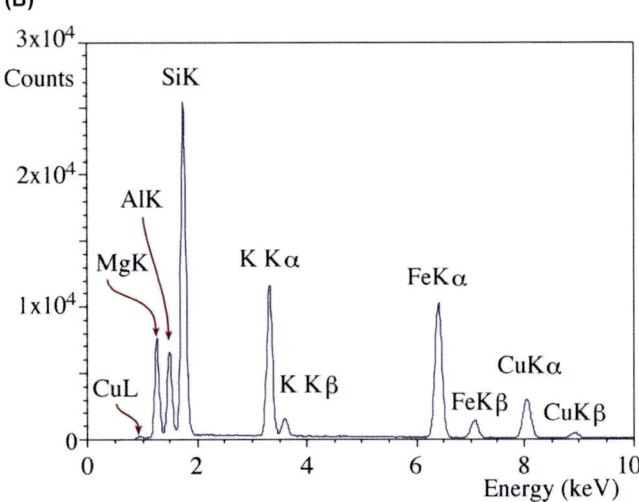

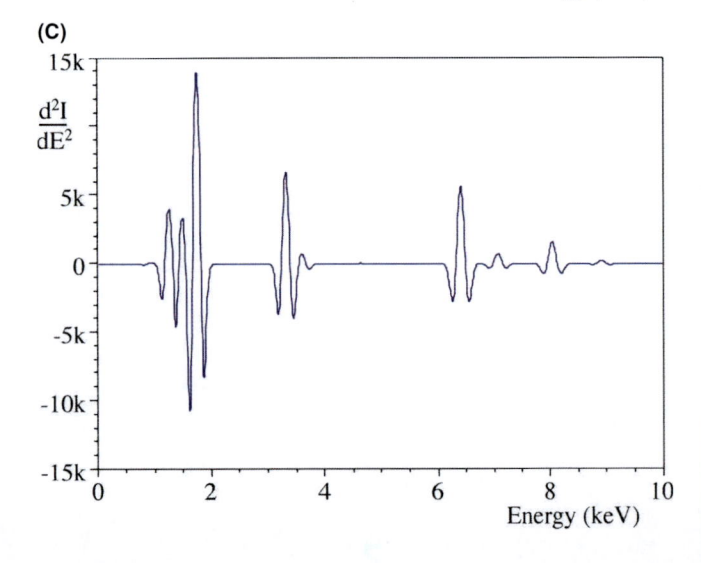

- If you used a Kramers' law fit, the usual method of peak integration is to get the computer to fit the peak with a slightly modified Gaussian, and integrate the total counts in the channels under the Gaussian.
- If a digital filter was used, you have to compare the peaks with those that were taken previously from standards, digitally filtered, and stored in a library in the computer. The library peaks are matched to the experimental peaks via a multiple least-squares fitting procedure and the counts determined through calculation of the fitting parameters.

Each of the two curve-matching processes is rapid. Each can be used to deconvolute overlapping peaks and each uses all the counts in the peak. The Kramers-law fit and the digital filter have much wider applicability than the simple window methods. However, these computer processes are not invariably the best, nor are they without error.

The Gaussian curve fitting must be flexible enough to take into account several variables

- The peak width can change as a function of energy or as a function of count rate.
- The peak distortion due to incomplete charge collection can vary.
- There may be an absorption edge under the characteristic peak if your specimen is too thick.

The creation of a library of spectra gathered under conditions that match those liable to be encountered during analysis (particularly similar count rates and dead times) is a tedious exercise. However, you do get a figure of merit for the 'goodness of fit' between the unknown spectrum and the standard. Usually, a χ^2 value is given which has no absolute significance, but is a most useful diagnostic tool. Typically, the χ^2 value should be close to unity for a good fit, although a higher value may merely indicate that some unidentified peaks were not accounted for during the matching process. What you have to watch out for is a sudden increase in χ^2 compared with previous values. This indicates that something has changed from your previous analyses. Perhaps your standard is not giving a good fit to the experimental spectrum and either a new library spectrum needs to be gathered or the experimental peak should be looked at carefully. For example, another small peak may be hidden under the major peak and would need to be deconvoluted from the major peak before integration proceeds. If you suspect a poor fit, you should make the computer display the 'residuals,' that is the counts remaining in the spectrum after the peak has been integrated and removed. As shown in Figure 35.6, you can easily see if a good fit was made (Figure 35.6A) or if the library peak and the experimental peak do not match well (Figure 35.6B).

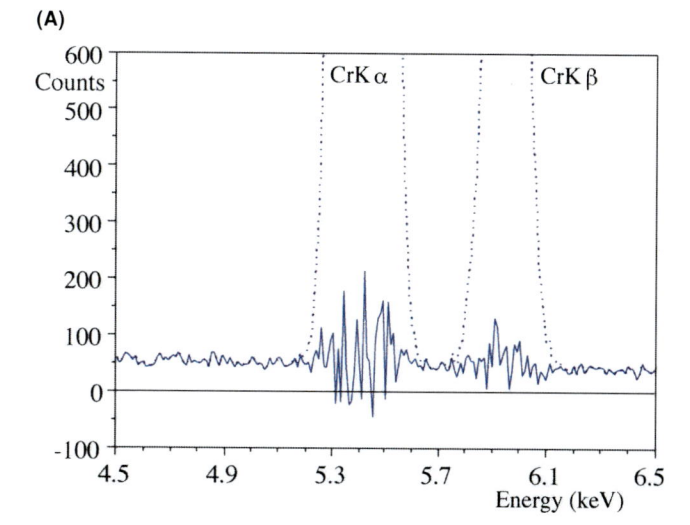

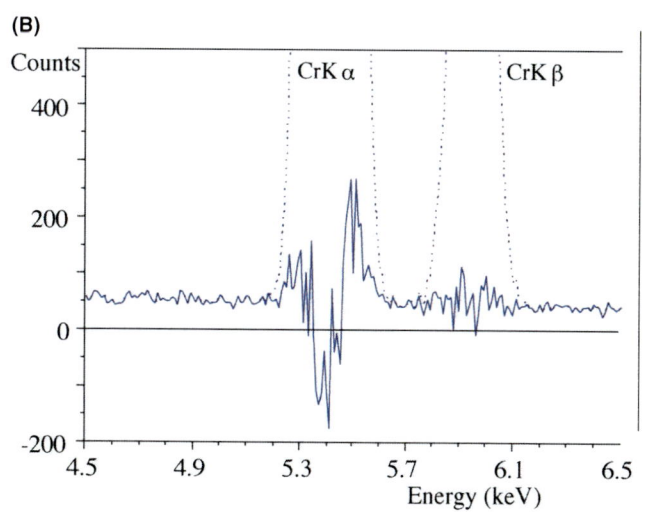

FIGURE 35.6. (A) A filtered Cr K-line family spectrum showing the residual background counts after the peaks have been removed for integration. The approximately linear residual intensity distribution indicates that the peaks matched well with the library standard stored in the computer. (B) A similar filtered spectrum showing the distorted residual spectrum characteristic of a poor fit with the library standard.

χ^2 OR CHI-SQUARED
Don't fear the math or the statistics. You'll rarely ever need to repeat it but you should know what your software is doing.

Any of the above methods is valid for obtaining values of the peak counts. They should all result in the same answer when used to quantify an unknown spectrum, so long as you apply the same method consistently to both the standard and the unknown.

Statham has reviewed the limitations of extracting peak intensities from X-ray spectra, with particular emphasis on low-keV lines. While aimed primarily at the EPMA community, almost all the issues in this paper are relevant to thin-film quantification.

Having obtained the peak counts, the next step is to insert the values into the Cliff-Lorimer equation and know the correct value of the k-factor. So we now need to discuss the various ways to obtain k_{AB}.

35.4 DETERMINING k-FACTORS

Remember that the k-factor is **not** a constant. It is a sensitivity factor that will vary not only with the X-ray detector, the microscope, and the analysis conditions, but also with your choice of background subtraction and peak-integration methods. So values of k-factors can be sensibly compared *only* when they were obtained under identical conditions. We will return to this point at the end of this section when we look at various sets of k-factors published in the literature. There are two ways you can determine k-factors

- Experimental determination using standards.
- Calculation from first principles.

The first method is slow and laborious but gives the most accurate values. The second method is quick and painless but the results are less reliable. You might wonder—can k-factors even depend on the previous user of the TEM?

> **k-FACTOR VERSUS ζ FACTOR**
> There are no generally accepted standards that meet all the above criteria for ideal k-factor determination, which is a major limitation to this approach and is overcome by the pure-element standards used in the ζ-factor method, which we describe later.

35.4.A Experimental Determination of k_{AB}

If you have a thin specimen of known composition, C_A, C_B etc., then all you have to do, in principle, is place that specimen in the microscope, generate a spectrum, obtain values of I_A, I_B, etc., and insert those values in the Cliff-Lorimer equation 35.2. Since you know C_A and C_B the only unknown is k_{AB}. However, there are several precautions that you must take before this procedure can be used

- The standard must be a well-characterized specimen, and it is usually best if it is single phase.
- The standard must be capable of being thinned to electron transparency. Ideally, when the specimen is thin there should be no significant absorption or fluorescence of the X-rays from the elements A, B, etc., that you wish to analyze.
- You must be sure that the thinning process did not induce any chemical changes (this is discussed in some detail in Chapter 10).
- It must be possible to select thin regions that are characteristic of the chemistry of the bulk specimen.
- You must be sure that the thin foil is stable under the electron beam at the voltage you intend to use for analysis.

This last point may often be the limiting factor in your choice of standards because, as we saw in Section 4.6, you have to take care to avoid not only direct knock-on damage, but also sputtering effects, which occur at voltages substantially below the threshold for direct atom displacement. Obviously, both these problems become greater as the beam voltage increases.

The National Institute of Standards and Technology (NIST) has issued a thin standard containing the elements Mg, Si, Ca, and Fe and O (SRM #2063 and subsequently #2063A). Unfortunately, X-rays from the lighter elements in this standard film are absorbed significantly in the film, and also in the detector and so a correction to the measured k-factor is necessary and NIST has not re-issued the standard.

It is best to use your own judgment in choosing standards, and also make use of the knowledge gained in previous k-factor studies.

Cliff and Lorimer's approach using mineral standards had three advantages.

- Crushing is an easy way to make thin flakes and does not affect the chemistry.
- The mineral stoichiometry is usually well known.
- The minerals chosen all contained Si, thus permitting the creation of a whole series of k_{ASi} factors.

The drawbacks are that the mineral specimens often contain more than one phase, or may be naturally non-stoichiometric. Clearly, some prior knowledge of the mineralogy of the specimen is essential in order to be able to select the right spectrum to use as a standard. Also Si K_α X-rays at ~1.74 keV are liable to be absorbed in the XEDS detector, so there may be a systematic difference in k-factors determined with different detectors. Finally, silicate minerals often exhibit radiolysis i.e., chemical changes due to beam-induced breaking of bonds.

> **GLASS STANDARDS**
> Can be made completely homogeneous. Reproducible (one batch). No channeling complications.

Several alternative approaches have been proposed that attempt to avoid the problems with k_{ASi}

- Wood et al. generated a series of k_{AFe} factors to overcome the Si absorption and the beam-sensitivity problems.
- Graham and Steeds used crystallized microdroplets which were routinely thin enough and retained their stoichiometry.
- Kelly used rapidly solidified droplets for the same reasons.
- Sheridan demonstrated the value of the NIST multi-element glasses.
- NIST created its own standard multi-element glass, as already noted.

Note also that there has been no new systematic determination of k-factors for many years, indicating the mature nature of the k-factor approach. But remember that you need factors for your TEM, your kV, etc.

> **DETERMINING k-FACTORS**
> A typical k-factor determination involves taking many spectra from different parts of the thin-foil standard. You must check both the homogeneity and the stability of the specimen. This is very time consuming, which is why so few analysts bother to do it properly.

In all cases, the bulk chemistry of the k-factor standard has to be determined by some technique with known accuracy, such as EPMA, atomic-absorption spectroscopy, or wet chemistry. Since all these techniques analyze relatively large volumes of material, it is best that the standard be single phase. However, because none of these techniques can determine if the specimen is homogeneous on a sub-micrometer scale, the only way to find out the level of homogeneity is to carry out many analyses within the AEM to confirm that any variation in your answer is within the expected X-ray statistical fluctuations.

Each spectrum should contain sufficient counts in the peaks of interest to ensure that the errors in the k-factor determination are at least less than ±5% relative and, if possible, less than ±3%. So, now we need to consider the errors associated with the X-ray spectra.

35.4.B Errors in Quantification: The Statistics

An unfortunate aspect of the simple Cliff-Lorimer ratio equation is that it has relatively large errors associated with it. The very nature of the thin foil minimizes the problems of absorption and fluorescence, but also generates relatively few X-ray photons per incident electron, compared with bulk specimens. This effect is compounded by the small collection angle of the XEDS detector and the end result is that poor counting statistics are the primary source of error in most AEM quantifications. The best way you can limit these errors is to use higher-brightness sources, large electron probes, and thicker specimens (unless absorption is a problem, or spatial resolution is paramount) and, of course, C_s-correction of the probe helps. In any case you should be prepared to count for a long time, assuming that specimen drift and/or contamination don't compromise your data.

The rest of this section is pure statistics. If you know it, then jump ahead.

> **GAUSSIAN STATISTICS**
> Experimental results show that the X-ray counts in the spectrum obey Gaussian statistics. Hence, we can apply simple statistics to deduce the accuracy of any quantification.

Given that our characteristic peak is Gaussian, then the standard deviation σ is obtained from

$$\sigma = N^{\frac{1}{2}} \quad (35.8)$$

where N is the number of counts in the peak above the background. For a single measurement, there is a 67% chance that the measured value of N will be within 1σ of the true value of N. This chance increases to 95% for 2σ and 99.7% for 3σ. If we use the most stringent condition, then the relative error in any single measurement is

$$\text{Relative Error} = \frac{3N^{\frac{1}{2}}}{N} 100\% \quad (35.9)$$

Clearly, the error decreases as N increases and hence, the emphasis throughout this chapter on the need to maximize the X-ray counts gathered in your spectra. Since the Cliff-Lorimer equation uses an intensity ratio, we can get a quick estimate of the error by summing the errors in I_A, I_B, and k_{AB} to give the total error in the composition ratio C_A/C_B.

Summing the errors in fact gives an overestimate of the error. Strictly speaking, we should add the standard deviations of the various terms in the Cliff-Lorimer equation in quadrature to give the standard deviation in the composition-ratio measurement σ_C using the expression

$$\left(\frac{\sigma_C}{C_A/C_B}\right)^2 = \left(\frac{\sigma_{k_{AB}}}{k_{AB}}\right)^2 + \left(\frac{\sigma_{I_A}}{I_A}\right)^2 + \left(\frac{\sigma_{I_B}}{I_B}\right)^2 \quad (35.10)$$

So we can determine the error for each datum point in this manner. If we are determining the composition of a single-phase region (for example, when determining a k-factor) then we can reduce the error by combining the results from n different measurements of the intensity ratio I_A/I_B. The total absolute error in I_A/I_B at a given confidence limit is obtained using the student's t distribution. For example, in this approach the error is given by

$$\text{Absolute Error} = \frac{(t_{95})^{n-1} S}{n^{\frac{1}{2}}} \quad (35.11)$$

where t_{95}^{n-1} is the student's t value at the 95% confidence limit for n measurements of k_{AB}. You can find lists of student's t values in any statistics text (see, e.g., Larsen and Marx) or on the Web (e.g., URL #1). Obviously, you could choose a lower or higher confidence level. S is the standard deviation for n measurements where

$$S = \left(\sum_{n=1}^{n} \frac{(N_i - N)^2}{n-1} \right)^{1/2} \quad (35.12)$$

So by increasing the number of measurements n, you can reduce the absolute error in k_{AB}. With enough measurements and a good homogeneous specimen, you can reduce the errors in the value of k_{AB} to $\pm 1\%$, as we will see in the example below. However, remember that this figure must be added to the errors in I_A and I_B. From equation 35.8 it is easy to determine that if we accumulate 10,000 counts in the peak for element A then the error at the 99% confidence limit is $[3(10,000)^{1/2}/10,000] \times 100\%$, which is ~3%. Using equation 35.10 and a similar value for I_B you get a total error in C_A/C_B of ~$\pm 4.5\%$. Now you see again why 'counts, counts, and more counts' is the mantra for thin-foil analysts.

If you take the time to accumulate 100,000 counts for I_A and I_B the total error is reduced to ~$\pm 1.7\%$, which represents about the best accuracy that can be expected for quantitative XEDS analysis in the AEM. This is an accuracy that almost no one ever takes the time to achieve.

It is appropriate here to go through an illustration of a k_{AB} determination using experimental data. Before deciding that a particular specimen is suitable, it should be checked for its level of homogeneity: there is a well-established criterion for this. If we take the average value N of many composition determinations, and all the data points fall within $\pm 3(N)^{1/2}$ of N then the specimen is homogeneous. In other words, this is our definition of 'homogeneous.' There are more rigorous definitions but the general level of accuracy in thin-foil analysis is such that there is no need to be so stringent.

An example: A homogenized thin foil of Cu-Mn solid solution was used to determine k_{CuMn}. The specimen was first analyzed by EPMA and found to be 96.64 wt% Cu and 3.36 wt% Mn. Since our accuracy is increased by collecting many spectra, a total of 30 were accumulated ($n = 30$ in equation 35.12). In a typical spectrum the Cu K_α peak contained 271,500 counts above background and the Mn K_α peak contained 10,800 counts. So if we insert these data into the Cliff-Lorimer equation we get

$$\frac{96.67}{3.36} = k_{CuMn} \frac{271,500}{10,800}$$

$$k_{CuMn} = 1.14.$$

To determine an error on this value of the k-factor, equation 35.11 must be used. The student's t analysis of the k-factors from the other 29 (yes 29!) spectra gives an error of ± 0.01 for a 95% confidence limit. This error of about $\pm 1\%$ relative is about the best that can be achieved using the experimental approach to k-factor determination, but remember that 30 individual spectra had to be accumulated from different regions of a well-characterized thin foil.

STUDENT'S t DISTRIBUTION
More statistics but with much more interesting origins.

So it takes a real effort to get quantitative data with errors $< \pm 5$–10% relative. Bear this in mind when you read any paper in which compositions are given to better than a rounded $\pm 5\%$ (such as the EPMA data in our own example).

THIN-FOIL COMPOSITIONS
Any thin-foil composition data given with decimal points must be really scrutinized to see if they were obtained via a procedure such as that outlined here. Otherwise they are just plain wrong.

Tables 35.1 and 35.2 summarize many of the available k-factor data in the published literature. You should go and read the original papers, particularly if you want to find out what standards and what conditions were used in their determination.

35.4.C Calculating k_{AB}

While it is clear that many of the values in these k-factor tables are very similar, the differences cannot be

TABLE 35.1 Experimentally Determined k_{ASi} and k_{AFe} Factors for K_α X-rays

Element (A)	k_{ASi} (1) 100 kV	k_{ASi} (2) 100 kV	k_{ASi} (3) 120 kV	k_{ASi} (4) 80 kV	k_{ASi} (5) 100 kV	k_{ASi} (5) 200 kV	k_{AFe} (6) 120 kV	k_{ASi} (7) 200 kV
Na	5.77	3.2	3.57 ± 0.21	2.8 ± 0.1	2.17	2.42		3.97 ± 2.32
Mg	2.07 ± 0.1	1.6	1.49 ± 0.007	1.7 ± 0.1	1.44	1.43	1.02 ± 0.03	1.81 ± 0.18
Al	1.42 ± 0.1	1.2	1.12 ± 0.03	1.15 ± 0.05			0.86 ± 0.04	1.25 ± 0.16
Si	1.0	1.0	1.0	1.0	1.0	1.0	0.76 ± 0.004	1.00
P			0.99 ± 0.016				0.77 ± 0.005	1.04 ± 0.12
S			1.08 ± 0.05		1.008	0.989	0.83 ± 0.03	1.06 ± 0.12
Cl					0.994	0.964		1.06 ± 0.30
K		1.03	1.12 ± 0.27	1.14 ± 0.1			0.86 ± 0.014	1.21 ± 0.20
Ca	1.0 ± 0.07	1.06	1.15 ± 0.02	1.13 ± 0.07			0.88 ± 0.005	1.05 ± 0.10
Ti	1.08 ± 0.07	1.12	1.12 ± 0.046				0.86 ± 0.02	1.14 ± 0.08
V	1.13 ± 0.07			1.3 ± 0.15				1.16 ± 0.16
Cr	1.17 ± 0.07	1.18	1.46 ± 0.03				0.90 ± 0.006	
Mn	1.22 ± 0.07	1.24	1.34 ± 0.04				1.04 ± 0.025	1.24 ± 0.18
Fe	1.27 ± 0.07	1.30	1.30 ± 0.03	1.48 ± 0.1			1.0	1.35 ± 0.16
Co							0.98 ± 0.06	1.41 ± 0.20
Ni	1.47 ± 0.07	1.48	1.67 ± 0.06				1.07 ± 0.006	
Cu	1.58 ± 0.07	1.60	1.59 ± 0.05		1.72	1.50	1.17 ± 0.03	1.51 ± 0.40
Zn	1.68 ± 0.07				1.74	1.55	1.19 ± 0.04	1.63 ± 0.28
Ge	1.92							1.91 ± 0.54
Zr								3.62 ± 0.56
Nb							2.14 ± 0.06	
Mo	4.3		4.95 ± 0.17				3.8 ± 0.09	
Ag	8.49		12.4 ± 0.63				9.52 ± 0.07	6.26 ± 1.50
Cd	10.6				9.47	6.2		
In								7.99 ± 1.80
Sn	10.6							8.98 ± 1.48
Ba					29.3	17.6		21.6 ± 2.6

TABLE 35.2 Experimentally Determined k_{ASi} and k_{AFe} Factors for L X-rays

Element (A)	k_{ASi} (8) 100 kV	k_{ASi} (5) 100 kV	k_{ASi} (5) 200 kV	k_{ASi} (9) 100 kV	k_{AFe} (6) 120 kV	k_{ASi} (7) 200 kV
Cu		8.76	12.2			
Zn		6.53	6.5			8.09 ± 0.80
Ge						4.22 ± 1.48
As						3.60 ± 0.72
Se						3.47 ± 1.11
Sr					1.21 ± 0.06	
Zr					1.35 ± 0.1	2.85 ± 0.40
Nb					0.9 ± 0.06	
Mo				2.0		
Ag	2.32 ± 0.2				1.18 ± 0.06	2.80 ± 1.19
In					2.21 ± 0.07	2.86 ± 0.71
Cd		2.92	2.75			
Sn	3.07 ± 0.2					
Ba		3.38	2.94			3.36 ± 0.58
Ce				1.4		
Sn	3.1 ± 0.2			1.3		
W	3.11 ± 0.2			1.8		3.97 ± 1.12
Au	4.19 ± 0.2	4.64	3.93		3.1 ± 0.09	4.93 ± 2.03
Pb	5.3 ± 0.2	4.85	4.24	2.8		5.14 ± 0.89

All L-line k-factors use the total L counts from the L_α and L_β lines.
Sources: (1) Cliff and Lorimer (1975), (2) Wood et al. (1981), (3) Lorimer et al. (1977), (4) McGill and Hubbard (1981), (5) Schreiber and Wims (1981), (6) Wood et al. (1984), (7) Sheridan (1989), (8) Goldstein et al. (1977), (9) Sprys and Short (1976).

accounted for by X-ray statistics alone. Some of the differences arise due to the choice of standard and the reproducibility of the standard. Other differences arise because the data were obtained under different conditions, such as different peak-integration routines. Therefore, the point made at the beginning of this section is worth repeating: the *k*-factors are not standards; they are sensitivity factors.

The only conditions under which you can expect the *k*-factors obtained on different AEMs to be identical are if you use the *same* standard at the *same* accelerating voltage, *same* detector configuration, *same* peak integration, and *same* background-subtraction routines. Even then there will be differences if one or more of the measured X-ray lines are not gathered by the detector with 100% efficiency; the X-ray may be either absorbed by the detector or it may be too energetic and pass straight through the detector.

You may not be able to obtain a suitable standard. For example, you might be working in a system in which no stoichiometric phases exist. You might not need a standard because accuracy might not be critical but you still need a quick analysis. Under such circumstances you can calculate an approximate *k*-factor. The programs necessary to calculate k_{AB} are stored in the XEDS computer and will give you a value of *k* in a fraction of a second. The calculated value should be accurate to within ±20% relative. This level of accuracy might be all you need to draw a sensible conclusion about the material you are examining (in which case, you can almost rely on a simple peak-height measurement). In general, if you can avoid the tedious experimental approach, then do so.

> **QUICK**
> Calculating *k*-factors is the recommended approach when a quick answer is required and the highest accuracy is not essential.

The expression for calculating the *k*-factor from first principles is derived in the paper by Williams and Goldstein. The derivation gives a good illustration of the relationship between bulk and thin-film analysis, and provides insight into the details of X-ray interactions with solids. However, at this stage you don't need to know the details of this derivation so we'll simply state the final expression

$$k_{AB} = \frac{1}{Z} = \frac{(Q\omega a)_B A_A}{(Q\omega a)_A A_B} \quad (35.13)$$

The subscripts A, B denote the elements A, B of atomic weight A_A and A_B. This expression derives from the physics of X-ray generation. An electron passing close to an atom in the specimen has first to ionize that atom and this is governed by Q_A, the ionization cross section (sometimes given by σ; go back and check equation 4.1). An ionized atom does not necessarily give off a characteristic X-ray when it returns to ground state and the fraction of ionizations that do generate an X-ray is governed by the fluorescence yield for the characteristic X-rays, ω_A (go back and check equation 4.6) The remaining term '*a*' is the relative-transition probability. This term takes account of the fact that if a K-shell electron is ionized and returns to ground state through X-ray emission, it can emit either a K_α or K_β X-ray. You may remember that we listed the relative weights of the various K, L, and M families of X-ray lines in Table 4.1.

> **THE *k*-FACTOR CALCULATION**
> Elements A and B; atomic weight A_A and A_B
> Ionization cross sections Q_A and Q_B
> Fluorescence yield ω_A and ω_B
> Relative transition probability, *a*
> Detector efficiency, ε_A and ε_B

As we mentioned at the start of the discussion on quantification, the Cliff-Lorimer *k*-factor for thin-foil analysis is related to the atomic-number correction factor (Z) for bulk specimen analysis. From equation 35.13, we can easily see what experimental factors determine the value of *k*

- The accelerating voltage is a variable since *Q* is strongly affected by the kV.
- The atomic number affects ω, *A*, and *a*.
- The choice of peak-integration method will also affect *a*.

Therefore, in order to calculate and compare different *k*-factors, it is imperative to define these conditions very clearly, as we have taken pains to emphasize.

Equation 35.13 assumes that equal fractions of the X-rays generated by elements A and B are collected and processed by the detector. This assumption will only be true if the same detector is used and the X-rays are neither strongly absorbed by, nor pass completely through, the detector. However, as we have already seen in Chapter 32, X-rays below ~1.5 keV are absorbed significantly by Be window and X-rays above ~20 keV pass through a 3 mm Si detector with ease. Under these circumstances, it is necessary to modify the *k*-factor expression, equation 35.13, in the following manner

$$k_{AB} = \frac{1}{Z} = \frac{(Q\omega a)_A}{(Q\omega a)_B} \frac{A_B}{A_A} \frac{\varepsilon_A}{\varepsilon_B} \quad (35.14)$$

The symbol ε represents simply a detector efficiency term plotted back in Figure 32.7 that we can write as follows.

$$\varepsilon_A = \exp\left(-\frac{\mu}{\rho}\bigg]_{Be}^{A} \rho_{Be} t_{Be}\right) \exp\left(-\frac{\mu}{\rho}\bigg]_{Au}^{A} \rho_{Au} t_{Au}\right)$$
$$\exp\left(-\frac{\mu}{\rho}\bigg]_{Si}^{A} \rho_{Si} t_{Si}\right) \left\{1 - \exp\left(-\frac{\mu}{\rho}\bigg]_{Si}^{A} \rho_{Si} t'_{Si}\right)\right\}$$
(35.15)

The first term accounts for absorption (via the mass absorption coefficients, μ/ρ) of X-rays from element A passing through the Be window. It should of course be modified for different windows and it disappears for windowless detectors. The second term covers absorption in the Au contact layer and the third term accounts for the Si dead layer. These two terms will have different values for different contact elements, IG dead layers, and SDDs which have very thin dead layers and side contact layers. The last term adjusts the k-factor for X-rays that *do not* deposit their energy in the active region of a detector which has density ρ and thickness t'. Typical values of t' (∼ 3 mm for Si(Li) and IG but ∼ 1 mm for SDDs) were discussed in Chapter 32. An IG detector will more efficiently stop high-energy X-rays, since it is designed to detect them preferentially, while an SDD will be less efficient since it is generally much thinner than a Si(Li) or IG. In fact, much of the effort over the last 20 years to improve detector technology that we discussed in detail in Chapter 32, minimizes the effects of equation 35.15 on the k-factor.

While equations 35.14 and 35.15 are simple for a computer to solve, the values that have to be inserted in the equations for the various terms are not always well known, or cannot be measured accurately. For example, we do not know the best value of Q for many elements in the range of voltages typically used in the AEM (100–400 kV). There are considerable differences of opinion in the literature concerning the best way to choose a value for Q. The two major approaches used are

- Assume various empirical parameterization processes (e.g., Powell).
- Interpolate values of Q to give the best fit to experimental k-factors (Williams et al.).

The other major variable in equation 35.15 is the Be-window thickness which is nominally 7.5 μm but in practice may be 3–4× thicker. Tables 35.3A and 35.3B list calculated k-factors obtained using various expressions for Q. As you can see, the value of k may easily vary by >±10%, particularly for the lighter and the heavier elements. This variation is due to the uncertainties in the detector-efficiency terms in equation 35.14. The values of k_{AB} for the L lines are even less accurate

TABLE 35.3A Calculated k_{AFe} Factors for K_α X-rays Using Different Theoretical Cross Sections

Element A	k_{MM}*	k_{GC}*	k_P*	k_{BP}*	k_{SW}*	k_Z*
Na	1.42	1.34	1.26	1.45	1.17	1.09
Mg	1.043	0.954	0.898	1.03	0.836	0.793
Al	0.893	0.882	0.777	0.877	0.723	0.696
Si	0.781	0.723	0.687	0.769	0.638	0.623
P	0.813	0.759	0.723	0.803	0.671	0.663
S	0.827	0.776	0.743	0.817	0.688	0.689
K	0.814	0.779	0.755	0.807	0.701	0.722
Ca	0.804	0.774	0.753	0.788	0.702	0.727
Ti	0.892	0.869	0.853	0.888	0.807	0.835
Cr	0.938	0.925	0.917	0.936	0.887	0.909
Mn	0.98	0.974	0.970	0.979	0.953	0.965
Fe	1.0	1.0	1.0	1.0	1.0	1.0
Co	1.063	1.069	1.074	1.066	1.096	1.079
Ni	1.071	1.085	1.096	1.074	1.143	1.23
Cu	1.185	1.209	1.227	1.19	1.31	1.24
Zn	1.245	1.278	1.305	1.255	1.44	1.32
Mo	3.13	3.52	3.88	3.27	3.84	3.97
Ag	4.58	5.41	6.23	4.91	5.93	6.28

TABLE 35.3B Calculated k_{AFe} Factors for L X-rays Using Different Theoretical Cross Sections

Element	k_{MM}*	k_P*	k_{BP}*	k_{SW}*	k_Z*
Sr*	1.73	1.33	1.32	1.64	1.39
Zr*	1.62	1.26	1.24	1.51	1.33
Nb*	1.54	1.21	1.18	1.43	1.28
Ag*	1.43	1.16	1.09	1.26	1.26
Sn	2.55	2.09	1.93	2.21	2.30
Ba	2.97	2.52	2.25	2.49	2.83
W	3.59	3.37	2.68	2.80	3.88
Au	3.94	3.84	2.94	3.05	4.43
Pb	4.34	4.31	3.05	3.34	4.97

All L-line k-factors use the total L counts from the L_α and L_β lines.
Cross sections used in the calculations are: MM (Mott-Massey); GC (Green-Cosslett); P (Powell); BP (Brown-Powell); SW (Schreiber-Wims); Z (Zaluzec).

than for the K lines mainly because the values of Q for the L lines are somewhat speculative. There are no data available for calculated k-factors for M lines. Under these circumstances, experimental determination is the only approach (or choose different materials to study). This point again emphasizes the advantages of K-line analysis where possible. If you are unfortunate enough to have heavy elements (say Z > 60) in your specimen, the L or M lines, which may be the strongest in a spectrum from a Si(Li) detector, will undoubtedly give rise to greater errors than the K lines, which may only be detectable with an IG system.

AFTER A SERVICE

If your detector is replaced or serviced, which is not an unusual occurrence on an AEM, then the new detector parameters must be inserted into the software.

The combination of uncertainties in Q and in the detector parameters is the reason why calculated k-factors are not very accurate, usually no better than ±10–20% relative. The computer system attached to the AEM should have predetermined values of all the terms in equations 35.14 and 35.15 stored in its memory. You don't usually have control over which particular parameters are being used. However, you should at least find out from your technical support the sources of the values of Q, ω, and a in your computer. You should then carry out a cross-check calculation with a known specimen to ensure that the calculated k-factor gives a reasonable answer.

> **BLACK BOXES**
> All software packages use preset values in their calculations and these may vary from package to package.

We cannot recommend a best set of values for Q, ω, and a, but the values of Q given by Powell, ω from Bambynek et al., and a from Schreiber and Wims have often been used. Also we can't give you specific detector parameters, so you should read the literature from your XEDS manufacturer. The values of μ/ρ that are still widely accepted are those determined by Heinrich although there is still considerable uncertainty in μ/ρ values for low-energy X-rays from the lighter elements. If you use the DTSA program from NIST (see Section 1.6), you may find that it predicts a worse value.

Figure 35.7A and B shows a comparison of the two methods of k-factor determination. The experimental data are shown as individual points with error bars and the solid lines represent the range of calculated k-factors, depending on the particular value of Q used in equation 35.14. The relatively large errors possible in the calculated k-factors are clearly seen and comparison of the K-line data in Figure 35.7A with the L-line data in Figure 35.7B again emphasizes the advantages of using K lines for the analysis where possible. Similar data for M lines are almost non-existent.

We can summarize the k-factor approach to analysis in the following way

- The Cliff-Lorimer equation has the virtue of simplicity; all you have to do is specify all the variables and treat the standard and unknown in an identical manner.
- You are better off calculating k_{AB} if you prefer speed to accuracy; experimental determination is best if you wish to have a known level of confidence in the numbers that you produce.

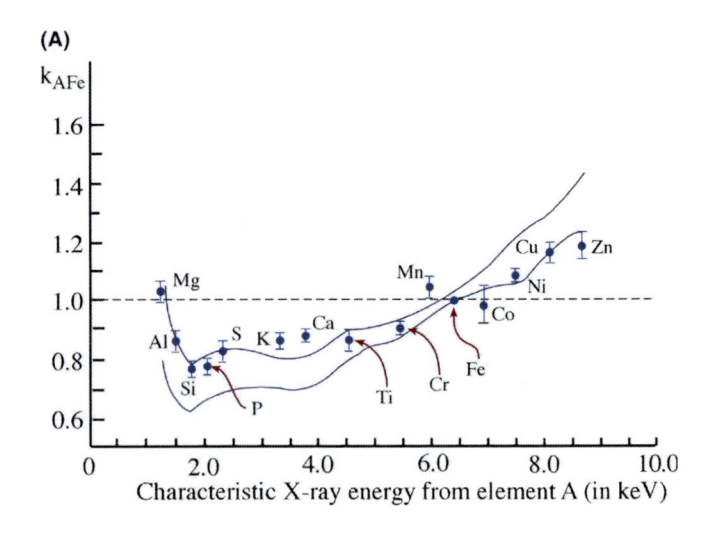

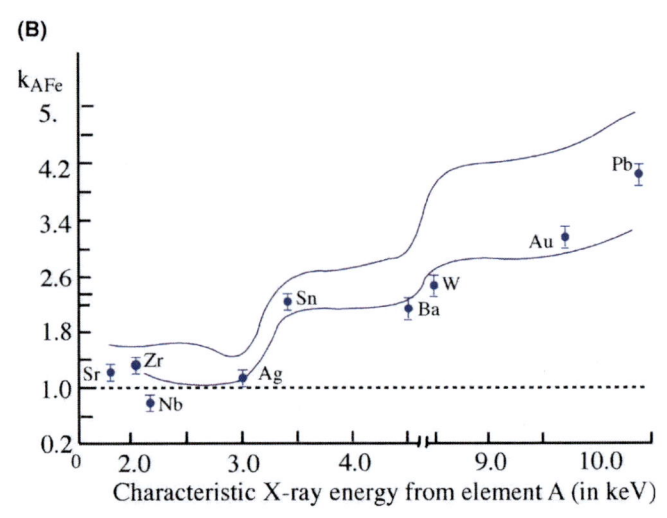

FIGURE 35.7. (A) Experimental k_{AFe} factors as a function of X-ray energy for the K_α X-rays from a range of elements A with respect to Fe. The solid lines represent the spread of calculated k-factors using different values for the ionization cross section. (B) Similar data to (A) for L_α lines from relatively high-Z elements. The errors in the calculated values of k are large, reflecting the uncertainties in L-line ionization cross sections.

35.5 THE ZETA-FACTOR METHOD

The Cliff-Lorimer ratio method is > 30 years old and, while it is simple in concept, the need for a k-factor combined with the difficulties of finding the right standard specimen, and calculating an accurate enough k-factor to solve a specific problem are significant limiting factors. These can be overcome if, instead of using a ratio method, we go back to the fundamentals of X-ray analysis as originally developed for the EPMA (based on equation 35.1) wherein pure-element standards are used. Pure-element standards have the distinct advantages of being easy to fabricate and they don't change composition during thinning or under beam damage. This approach, developed by Watanabe and Horita in Japan, is termed the zeta (ζ) factor method. A review of the development of the method and the pros and cons of ζ-factors versus *k*-factors was given by Watanabe and Williams in 2006.

The major experimental drawback to ζ-factors is that the method requires in-situ measurement of the probe current hitting your specimen. Unfortunately, despite the fact that beam-current measurement is standard on almost all SEMs and every EPMA, commercial AEMs don't come equipped with this capability, emphasizing the point made some while ago in Chapter 33 that AEMs are really modified TEMs and XEDS analysis is still a bit of an afterthought.

But, if you can measure the current in situ, e.g., with a specimen holder within which a Faraday cup is embedded, life gets much easier. In a thin-foil specimen, we can assume that the characteristic X-ray intensity is proportional to the mass thickness, ρt, if X-ray absorption and fluorescence are negligible. Therefore, we can define the ζ-factor for pure element A as

$$\rho t = \zeta_A \frac{I_A}{C_A} \qquad (35.16)$$

Where

$$\zeta \equiv \frac{A}{CN_0 Q \omega a i} \qquad (35.17)$$

where the only new terms are the beam current i, and Avogadro's number N_0. Under these conditions, the ζ-factor is dependent on the X-ray energy, the kV, and the beam current. The first two are constant for any experiment and the third you have to measure. The ζ-factor is independent of specimen thickness, composition, and density which, as we shall see later, makes any absorption correction trivial.

So we can write similar equations for all the other pure elements in our specimen

$$\rho t = \zeta_B \frac{I_B}{C_B} \qquad (35.18)$$

When we know the ζ-factors for A and B, C_A, C_B, and ρt can be expressed from equations 35.16 and 35.18, assuming $C_A + C_B = 1$ in a binary system:

$$C_A = \frac{I_A \zeta_A}{I_A \zeta_A + I_B \zeta_B},$$
$$C_B = \frac{I_B \zeta_B}{I_A \zeta_A + I_B \zeta_B} \quad \rho t = I_A \zeta_A + I_B \zeta_B \qquad (35.19)$$

Therefore, we can determine C_A, C_B, and ρt simultaneously just by measuring X-ray intensities. It is simple to rearrange equation 35.19 to compare with the Cliff-Lorimer ratio equation (equation 35.2) and equation 35.20 is just as simple to apply, but *k*-factors are no longer required, just ζ-factors.

$$\frac{C_A}{C_B} = \frac{I_A \zeta_A}{I_B \zeta_B} \qquad (35.20)$$

As we'll see below, absorption and fluorescence-correction terms can be combined directly with equation 35.19. To determine the ζ-factors, you just measure X-ray characteristic intensities (above the background)

TABLE 35.4 ζ-Factor Values Estimated from Experimental X-ray Spectra from the NIST SRM2063a Glass Thin Film in a 200-keV FEG-STEM JEM-2010F with an ATW Detector and in a 300-keV FEG-DSTEM VG HB603 with a Windowless Detector.

Element (Z)	ζ-factor (kg electron/m/photon) 200 keV ATW	300 keV Windowless	Element (Z)	ζ-factor (kg electron/m/photon) 200 keV ATW	300 keV Windowless
N	16,505.2±1,537	720.2±58.0	Mn	1,752.2±41.8	706.0±17.3
O	4,092.3±205.5	583.5±31.3	Fe	1,790.5±42.7	721.1±17.7
F	20,548.2±4,852.8	891.6±214.5	Co	1,919.0±45.8	772.0±18.9
Ne	5,345.6±702.1	635.8±85.5	Ni	1,950.0±46.5	783.2±19.2
Na	2,819.6±221.9	571.7±46.3	Cu	2,163.4±51.6	867.3±21.2
Mg	1,847.9±95.0	501.4±26.6	Zn	2,300.1±54.9	920.1±22.5
Al	1,510.2±56.2	483.6±18.6	Ga	2,541.8±60.7	1,014.4±24.8
Si	1,369.5±41.3	467.5±14.6	Ge	2,762.1±65.9	1,099.5±26.9
P	1,691.2±50.4	484.3±13.4	As	3,009.5±71.8	1,194.7±29.2
S	1,488.6±40.0	507.6±14.2	Se	3,328.6±79.4	1,317.7±32.2
Cl	1,465.5±37.3	528.0±13.9	Br	3,531.1±84.3	1,397.7±34.2
Ar	1,532.7±37.8	579.3±14.9	Kr	3,890.2±92.8	1,534.6±37.6
K	1,405.5±34.2	545.7±13.8	Rb	4,182.3±99.8	1,644.2±40.2
Ca	1,377.6±33.2	544.3±13.6	Sr	4,476.8±106.8	1,755.7±43.0
Sc	1,522.5±36.5	607.0±15.0	Y	4,786.6±114.2	1,871.6±45.8
Ti	1,553.4±37.2	622.8±15.4	Zr	5,185.4±123.7	2,019.8±49.4
V	1,632.3±39.0	656.5±16.1	Nb	5,578.9±133.1	2,164.8±53.0
Cr	1,654.6±39.5	666.5±16.4	Mo	6,088.3±145.3	2,353.4±57.6

from pure-element thin films with known composition and thickness, rather than trying to find a suitable set of standards, such as those listed in the references for Tables 35.1 and 35.2. Pure-element standards are more routinely available than the various multi-element, thin-film standards that have been used for k-factor determination over the last 30 years or so. You can determine an entire set of the ζ-factors for K-shell X-ray lines from a single spectrum from the National Institute of Standards and Technology (NIST) thin-film glass standard reference material (SRM) 2063. Table 35.4 lists such a set of ζ-factors obtained from the NIST standard SRM 2063a, which is a thinner version of SRM 2063. Its composition is known to a high degree of accuracy, as is its thickness and density but, unfortunately, it is now out of production.

There's much more about the ζ-factor in the companion text. Despite its obvious advantages, the ζ-factor approach is not yet commercially available; however, it can be downloaded from the book Web site (URL #2).

35.6 ABSORPTION CORRECTION

The point at which the simple Cliff-Lorimer approach breaks down is when the thin-foil criterion is invalid. Then the X-ray counts from your specimen are not a function of Z alone and absorption and (very occasionally) fluorescence invalidate the simple criterion. The effects of absorption are much more of a problem than fluorescence, so let's look at absorption first.

Preferential absorption of the X-rays from one of the elements in your specimen means that the detected X-ray counts will be less than the generated counts and so C_A is no longer simply proportional to I_A. So you have to modify the k-factor to take into account the reduction in I_A. This problem can arise (a) if your specimen is too thick, (b) if one or more of the characteristic X-rays has an energy less than ~1–2 keV (i.e., light-element analysis) or (c) when you have X-ray lines in your spectrum that differ in energy by > 5–10 keV (because the lower energy X-ray is much more likely to be absorbed than the higher energy one).

If we define k_{AB} as the true sensitivity factor when the specimen thickness t = 0, then the effective sensitivity factor for a specimen in which absorption occurs is given by k_{AB}^* where

$$k_{AB}^* = k_{AB}(\text{ACF}) \quad (35.21)$$

So we can write

$$\frac{C_A}{C_B} = k_{AB}(\text{ACF})\frac{I_A}{I_B} \quad (35.22)$$

The absorption-correction factor (ACF) is the A term in equation 35.13 and we can write it as

$$\text{ACF} = \frac{\int_0^t \left\{ \varphi_B(\rho t)\, e^{-\left(\frac{\mu}{\rho}\right)_{\text{Spec}}^B \rho t \csc \alpha} \right\} d(\rho t)}{\int_0^t \left\{ \varphi_A(\rho t)\, e^{-\left(\frac{\mu}{\rho}\right)_{\text{Spec}}^A \rho t \csc \alpha} \right\} d(\rho t)} \quad (35.23)$$

In this expression $\varphi(\rho t)$ is the depth distribution of X-ray production (which is the ratio of the X-ray emission from a layer of element A/B (of thickness $\Delta \rho t$ at depth t in the specimen with density ρ) to the X-ray emission from an identical, but isolated film). The term $\left.\frac{\mu}{\rho}\right]_{\text{Spec}}^A$ is the mass-absorption coefficient of X-rays from element A in the specimen and α is the detector take-off angle. Since the values of μ/ρ are often from old publications, the units are usually given in cm²/gm rather than kg/m², so you may have to use ρ in gm/cm³ and t in cm, rather than SI units (kg/m³ and m, respectively). Obviously, the value of the ACF is unity when no absorption occurs. Typically, if the ACF is >10% we define the absorption as significant, since 10% accuracy is routinely attainable in quantitative analysis using experimental k-factors. Let's now look at each of the terms and the problems associated with determining their value.

Again, we recommend that you use the values of μ/ρ given by Heinrich. The value of μ/ρ for a particular X-ray (e.g., from element A) within the specimen is the sum of the mass-absorption coefficients for each element times the weight fraction of that element, so

$$\left.\frac{\mu}{\rho}\right]_{\text{spec}}^A = \sum_i \left(\frac{C_i \mu}{\rho}\right)_i^A \quad (35.24)$$

where C_i is the fractional concentration of element i in the specimen such that

$$\sum_i C_i = 1 \quad (35.25)$$

The absorption of X-rays from element A by all elements i in the specimen is summed. The summation includes self-absorption by element A. Elements that may not be of interest in the experiment or that might not be detectable may still cause absorption.

The NiO-MgO example of this phenomenon occurs when Mg is being quantified in homogeneous NiO-MgO. The Mg K_α X-rays will be absorbed by oxygen, even if the O K_α X-ray is not of interest or cannot be detected because a Be-window detector is being used. This effect is shown in Figure 35.8, which shows an increase in the intensity ratio (Ni K_α/Mg K_α) as a function of thickness due to the increased absorption of the

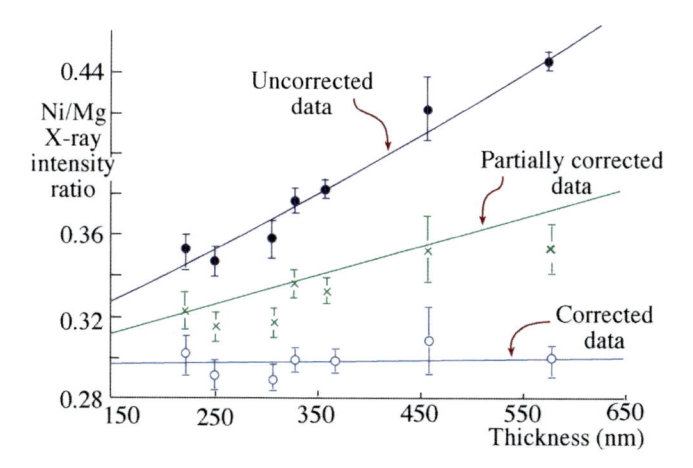

FIGURE 35.8. The upper curve shows the raw Ni K_α/Mg K_α intensity ratio as a function of thickness in a homogeneous specimen of NiO-MgO. The slope indicates strong absorption of Mg K_α X-rays. The middle curve shows the effect of correcting for absorption of the Mg K_α X-rays by Ni and the bottom line shows the effect of a further correction for absorption of the Mg K_α by O to give the expected horizontal line.

Mg K_α X-rays. (Absorption appears in an exponential term.) If we correct for the absorption by Ni, the slope of the line is reduced, but only when the effects of absorption by oxygen are taken into account does the slope become zero, as it should be for a homogeneous specimen.

In equation 35.22, we assume that the depth distribution of X-ray production $\varphi(\rho t)$ is a constant, and equal to unity. That is, a uniform distribution of X-rays is generated at all depths throughout the foil. This is a reasonable first approximation in thin foils, but in bulk specimens $\varphi(\rho t)$ is a strong function of t and so the measurement of $\varphi(\rho t)$ for bulk specimens is a well-established procedure. The few studies in thin specimens show an increase in $\varphi(\rho t)$ with specimen thickness, although the increase is no more than ~5% in foil thicknesses of < 300 nm. Therefore, the assumption appears reasonable since, if your specimen is thicker than 300 nm, you will have problems to worry about other than $\varphi(\rho t)$. The fact that we use a ratio of the two $\varphi(\rho t)$ terms in the absorption equation also helps to minimize the effects of this assumption.

We assume that $\varphi(\rho t)$ equals unity, then we can simply use equation 35.23 to give

$$\text{ACF} = \left(\frac{\left[\frac{\mu}{\rho}\right]^A_{\text{Spec}}}{\left[\frac{\mu}{\rho}\right]^B_{\text{Spec}}}\right) \left(\frac{1 - e^{-\left[\frac{\mu}{\rho}\right]^B_{\text{Spec}} \rho t \csc \alpha}}{1 - e^{-\left[\frac{\mu}{\rho}\right]^A_{\text{Spec}} \rho t \csc \alpha}}\right) \quad (35.26)$$

So we still need to know the values of ρ and t for our specimens.

The density of the specimen (ρ) can be estimated if you know the unit-cell dimensions, e.g., from CBED since

$$\rho = \frac{nA}{VN} \quad (35.27)$$

where n is the number of atoms of average atomic weight A in a unit cell of volume V, and N is Avogadro's number.

The absorption path length (t') is a major variable in the absorption correction. Fortunately, it is also the one over which you, the operator, have the most control. In the simplest case of a parallel-sided thin foil of thickness t at 0° tilt, the absorption path length, as shown in Figure 35.9, is given by

$$t' = t' \csc \alpha \quad (35.28)$$

where α is the detector take-off angle. To minimize this factor it is obvious that your specimen should be as thin as possible and the value of α as high as possible. There are many ways to determine the foil thickness, which we have discussed at various points in this text; they are summarized in Section 36.3. More recently, Banchet et al. (2003) have combined EELS measurements of relative specimen thickness with XEDS peak intensities to give a variation on the traditional, iterative absorption-correction process. No method is universally applicable, and few are either easy or accurate, so it's best to make thin specimens in the first place.

The value of α when your specimen is at the ideal 0° tilt is fixed by the design geometry of your AEM stage and the only way you can vary α is by tilting your specimen. As we have seen, there are good reasons not to tilt beyond about 10°, because of the increase in spurious X-rays, but if there is a severe absorption problem, then decreasing t' by tilting your specimen *toward* the detector is a sensible first step toward minimizing the problem. On some very old AEMs the detector may not be orthogonal to the axis of your specimen holder, in which case you've got a challenging exercise in solid geometry to determine α.

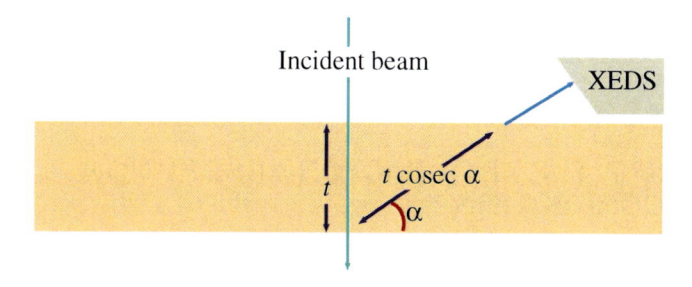

FIGURE 35.9. Relationship between the specimen thickness, t, and the absorption path length, $t \csc \alpha$, for a take-off angle α.

> **HOW THICK?**
> Remember: it's rare that you'll know your specimen thickness as well as you would like.

So far we've assumed that our specimens are parallel sided, but this is uncommon. Many thin-foil preparation methods result in wedge-shaped foils, and under these circumstances the detector must always be looking toward the thin edge of the specimen so that the X-ray path length is minimized, as we already described in Figure 32.15. The only way to ascertain if this is a problem is to measure the thickness at each analysis point. Because this is such a tedious exercise, you should get round it by using the ζ-factor method, as we explain in the next section.

Because the specimen density ρ (and therefore the values of μ/ρ) varies with your specimen composition the complete absorption-correction procedure is an iterative process. The first step is to use the Cliff-Lorimer equation without any absorption correction and thus produce values for C_A and C_B. From these values, the computer performs a first-iteration calculation of μ/ρ and ρ, and generates modified values of C_A and C_B, and iterates again. Usually the calculation converges after two or three iterations.

In summary, there is substantial room for error in determining the various terms to insert into the ACF. For example, the ACF for k_{NiAl} in Ni$_3$Al, which is a strongly absorbing system, varies from ~5.5 to ~12% when the specimen doubles in thickness from 40 to 80 nm. This change is still quite small and within the limits of all but the most accurate analyses. In FeNi, which is a weakly absorbing system, a similar change in thickness would change the ACF for k_{FeNi} from ~0.6 to ~1.3%, which is negligible. So while we've spent a fair bit of time introducing you to the absorption correction, the final message is clear.

> **CLIFF-LORIMER ACCURACY?**
> The Cliff-Lorimer approach only incurs large errors in strongly absorbing systems and/or very thick specimens.

35.7 THE ZETA-FACTOR ABSORPTION CORRECTION

So, if significant absorption is unavoidable, information about your specimen density and thickness is required at each analysis position, in order to apply the absorption correction. Obviously, this is the major limitation, since independent measurements are required for the specimen density and thickness and inaccuracies in such measurements may cause further errors in quantification. In fact, the ζ-factor method was originally proposed in order to overcome these limitations and difficulties associated with the absorption correction because, if we substitute equation 35.16 into equation 35.22, the ρt term can be eliminated

$$\frac{C_A}{C_B} = k_{AB} \left(\frac{I_A}{I_B} \left[\frac{(\mu/\rho)_{sp}^A}{(\mu/\rho)_{sp}^B} \right] \right.$$
$$\left. \left\{ \frac{1 - \exp\left[-(\mu/\rho)_{sp}^B \zeta_A (I_A/C_A) \csc(\alpha)\right]}{1 - \exp\left[-(\mu/\rho)_{sp}^A \zeta_A (I_A/C_A) \csc(\alpha)\right]} \right\} \right)$$
(35.29)

If you've determined ζ-factors, then the k-factor can be substituted in the above equation since

$$k_{AB} = \frac{\zeta_A}{\zeta_B} \quad (35.30)$$

So the ζ-factor overcomes the two major limitations of the Cliff-Lorimer method by avoiding the tedium of preparing multiple thin-foil standards and making multiple thickness and density measurements at each analysis point when significant X-ray absorption is occurring. That's why we introduce its depth in the companion text and suggest strongly that the serious X-ray analyst use this approach.

35.8 THE FLUORESCENCE CORRECTION

X-ray absorption and fluorescence are intimately related because a primary cause of X-ray absorption is the fluorescence of another X-ray (such as the fluorescence of SiK_α X-rays in the XEDS detector which gives rise to the escape peak). You might think, therefore, that fluorescence corrections should be as widespread as absorption corrections. However, this is not the case for the following reasons. Strong absorption effects occur when there is a small amount of one element whose X-rays are being absorbed by the presence of a relatively large amount of another element. The absorption of Al K_α X-rays by Ni in Ni$_3$Al is a classic example. In this case, Ni X-rays are indeed fluoresced as a result of the absorption of Al K_α X-rays. However, there is a relatively small increase in the total number of Ni X-rays because Ni is the dominant element; the relative decrease in the Al K$_\alpha$ intensity is large because Al is the minor constituent. In this particular example there is a further reason why fluorescence of Ni X-rays is ignored; it is the Ni L$_\alpha$ X-rays which are fluoresced by the absorption of Al K$_\alpha$ X-rays. The Ni L X-rays are not the ones that we use for analysis anyhow, since the higher energy Ni K X-rays are not absorbed or fluoresced.

In the rare case that fluorescence occurs to a degree that limits the accuracy of your analysis, read the detailed discussion given by Anderson et al. Practical examples of the fluorescence correction are hard to come by and a classic case is Cr in stainless steels where the minor Cr K$_\alpha$ line is fluoresced by the major Fe K$_\alpha$ line, giving rise to an apparent increase in Cr content as the foil gets thicker.

> **DON'T WORRY ABOUT FLUORESCENCE**
> Fluorescence is usually a minor effect and often occurs for X-rays that are not of interest. (So don't worry if you know you needn't!)

35.9 ALCHEMI

We told you early on in this chapter to acquire your X-ray spectra away from strong diffraction conditions. This is because of the Borrmann effect. Close to two-beam conditions, the Bloch waves interact strongly with the crystal planes and so X-ray emission is enhanced compared with kinematical conditions, thus negating the assumptions inherent in the Cliff-Lorimer equation, which assumes emission is constant with specimen tilt. However, we can make use of this phenomenon to locate which atoms lie on which crystal planes. The technique has the delightful (and wholly inappropriate) acronym ALCHEMI, which is a selective abbreviation of the expression 'atom location by channeling-enhanced microanalysis.'

ALCHEMI is a quantitative technique for identifying the crystallographic sites, distribution and types of substitutional impurities in crystals. The technique was first developed for the TEM by Spence (who, with archetypical antipodean humor, coined the acronym) and Taftø. Interestingly, channeling is also used for atom-site location in other analysis techniques (e.g., see Chu et al.).

The way to do ALCHEMI experimentally is to tilt your specimen to a strong two-beam condition and acquire a spectrum under strong channeling conditions, such that the Bloch wave is interacting strongly with a particular systematic row of atoms. You should choose the channeling orientation so that the specific crystal planes interacting strongly with the beam also contain the candidate impurity atom sites. So it helps a lot if you have some a priori ideas about where substitutional atoms are most likely to sit. This technique is therefore particularly well suited to layer structures. When the Bloch wave is maximized on a particular plane of atoms, the X-ray counts from the atoms in that plane will be highest. So start by finding the orientations 1 and 2 that give the most pronounced channeling effects for the atoms A and B, as shown schematically in Figure 35.10A. Usually a very small tilt is all that is necessary to get a different spectrum from the two planes.

If you are looking at two elements A and B and a substitutional element X then follow this procedure

- Measure X-ray intensities from each element in orientations 1 and 2.
- Then find a non-channeling orientation (3) where the electron intensity is uniform for both planes.

In this orientation we define the ratio k (NOT the Cliff-Lorimer factor) as

$$k = \frac{I_B}{I_A} \quad (35.36)$$

I_B is the number of X-ray counts from the element B in the non-channeling orientation. For the two channeling orientations 1 and 2, we define two parameters β and γ such that

$$\beta = \frac{I_B^{(1)}}{k I_A^{(1)}} \quad (35.37)$$

$$\gamma = \frac{I_B^{(2)}}{k I_A^{(2)}} \quad (35.38)$$

Now assuming we know from looking at the relative intensity changes in the spectra that the element X sits on specific sites, say it substitutes for atom B, then we define an intensity ratio term R such that

$$R = \frac{I_A^{(1)} I_X^{(2)}}{I_X^{(1)} I_A^{(2)}} \quad (35.39)$$

Then the fraction of atom X on B sites is given by

$$C_X = \frac{R-1}{R-1+\gamma-\beta R} \quad (35.40)$$

Similar expressions can be generated for X atoms on A sites, but in fact the fraction of X atoms on A sites must be $1-C_X$.

As you see, ALCHEMI can give a direct measure of the occupation of substitutional sites. However, the intensity differences in different orientations are often quite small and you need good X-ray statistics to draw sound conclusions (counts, counts, and more counts!). This makes ALCHEMI difficult to apply if high spatial resolution is also desired because, as we shall see in the next chapter, the conditions to give the best spatial resolution also give the worst counting statistics. Figure 35.10B shows the variation in X-ray emission across a bend contour highlighting the Borrmann effect which is the basis of ALCHEMI. A comprehensive review has been given by Jones, who extends the discussion of the technique and develops it in depth in the companion text.

35.10 QUANTITATIVE X-RAY MAPPING

As we have discussed throughout the previous chapters, gathering X-ray maps rather than individual or lines of spectra makes a great deal of sense in terms of getting unbiased elemental-distribution information about your specimen. The major difficulty with moving from qualitative to quantitative mapping is the need for sufficient counts for quantification. As noted in Section 35.4.B above, we recommend acquiring 10,000 counts in a characteristic peak in order for reasonable quantification (i.e., with errors $\sim \pm 10\%$ relative). A simple calculation will show how unrealistic this is if we are to acquire maps in a reasonable time. A minimum map to give a reasonable X-ray image is 128×128 pixels giving $> 16{,}000$ total pixels. Even if we acquire for only 1 s/pixel (if we are lucky we will acquire a few tens of total counts rather than a few thousands), then we will be mapping for 4.5 hours minimum and more likely days if we wish to see hundreds or thousands of counts/pixel. As we've stated too often, such long times introduce specimen drift, damage, contamination, and operator boredom and so conspire to make life very difficult. While overnight mapping with lower-resolution EPMA systems (where the drift and stability requirements are much less stringent and the X-ray count rate is far greater) is indeed a common occurrence, STEMs are not yet stable enough to do this while retaining nanometer-level resolution.

Nevertheless, there has been significant progress in quantitative mapping, particularly with the use of 2–300 kV FEG instruments, the development of

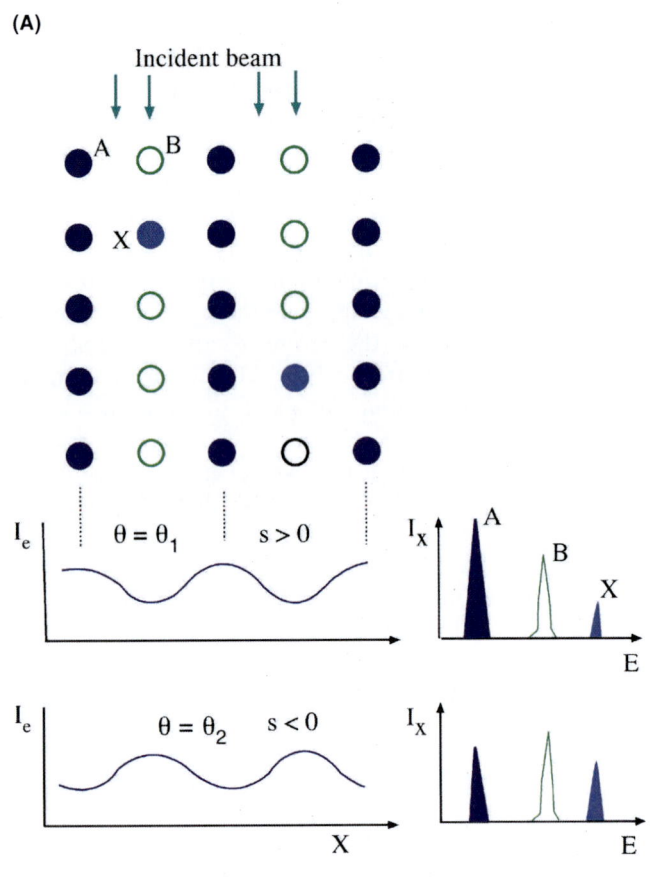

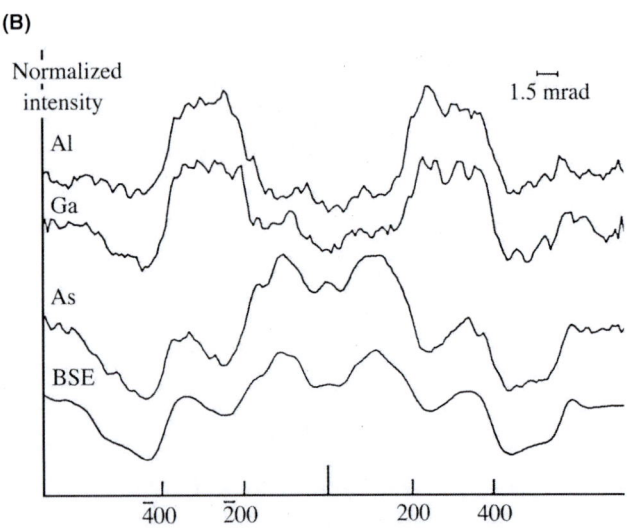

FIGURE 35.10. (A) ALCHEMI allows the determination of the site occupancy of atom X (light blue) in columns of atoms A (dark blue) and B (open circles). By tilting to $s > 0$ and then $s < 0$, the Bloch waves interact strongly with row A then row B giving different characteristic intensities, shown schematically in the spectra, from which the relative amounts of X in columns of A and B can be determined. (B) The Borrmann effect: the variation in the characteristic X-ray emission close to strong two-beam conditions as the beam is rocked across the 400 planes of GaAlAs. The X-rays from Al, which occupies Ga sites, follow the Ga emission variation while the As varies in an approximately complementary fashion. The BSE signal is inversely proportional to the amount of channeling so the As signal is strongest where the channeling is weakest.

higher solid angles of collection of X-rays, and general improvements in instrument design. So, as shown back in Figures 33.14 and 33.15, quantitative X-ray mapping is feasible. This process requires that the counts in the characteristic peak that is being mapped be integrated, subject to background subtraction, then processed via a Cliff-Lorimer or ζ-factor equation to turn the counts into composition. It is often best to exhibit the map as a color image because usually more than one element needs to be mapped or the composition changes need to be emphasized in which case overlays or side-by-side comparisons of the different maps permit easier understanding of the relative distributions of the various elements. Not surprisingly, quantitative mapping is improved by C_s correction since corrected probes have ~3–5× more current without loss of probe size. Thus the count rate is increased, or the acquisition time decreased, or both.

While it is certainly of interest to improve the stability of our TEM and XEDS systems and optimize the gathering of data over long periods of time, there are ways to minimize the limitations of low count rates and long acquisition times and this requires high-level computer control of the X-ray acquisition, implementation of spectrum imaging, or position-tagged spectrometry (check back in Sections 33.6.C and 33.6.D) and then manipulation of the resulting data cube using multivariate statistical analysis (MSA) to extract the maximum signal information and minimize the noise (which constitutes most of the signal in the channels in a spectrum acquired for a short period of time).

When this combination of SI and MSA is implemented, it is possible to gather spectra in as little as 100–500 ms/pixel giving mapping times of a few minutes to several tens of minutes for a 128 × 128 pixel image and it is equally feasible to contemplate larger maps. The SI/MSA combination is dealt with in great detail in the companion text. About the best that can be done with a modern C_s-corrected FEGSTEM, sophisticated data handling is shown in Figure 35.11 where segregation of trace elements to grain boundaries is mapped. A similar example was also shown back in Figure 33.15D which maps out differences in the composition of various small precipitates, a few nanometers in diameter. In Figure 35.11B, the data were acquired on a 300-keV STEM and the improvement in mapping quality obtained after MSA (Figure 35.11C) is clear. Addition of a C_s corrector to the STEM results in a further gain in spatial resolution, as shown in Figure 35.11D. This latter example of Zr segregating to a grain boundary in a Ni-base superalloy mapping reveals a typical enrichment of only ~1–2 atoms/nm² with a spatial resolution of ~0.5 nm!

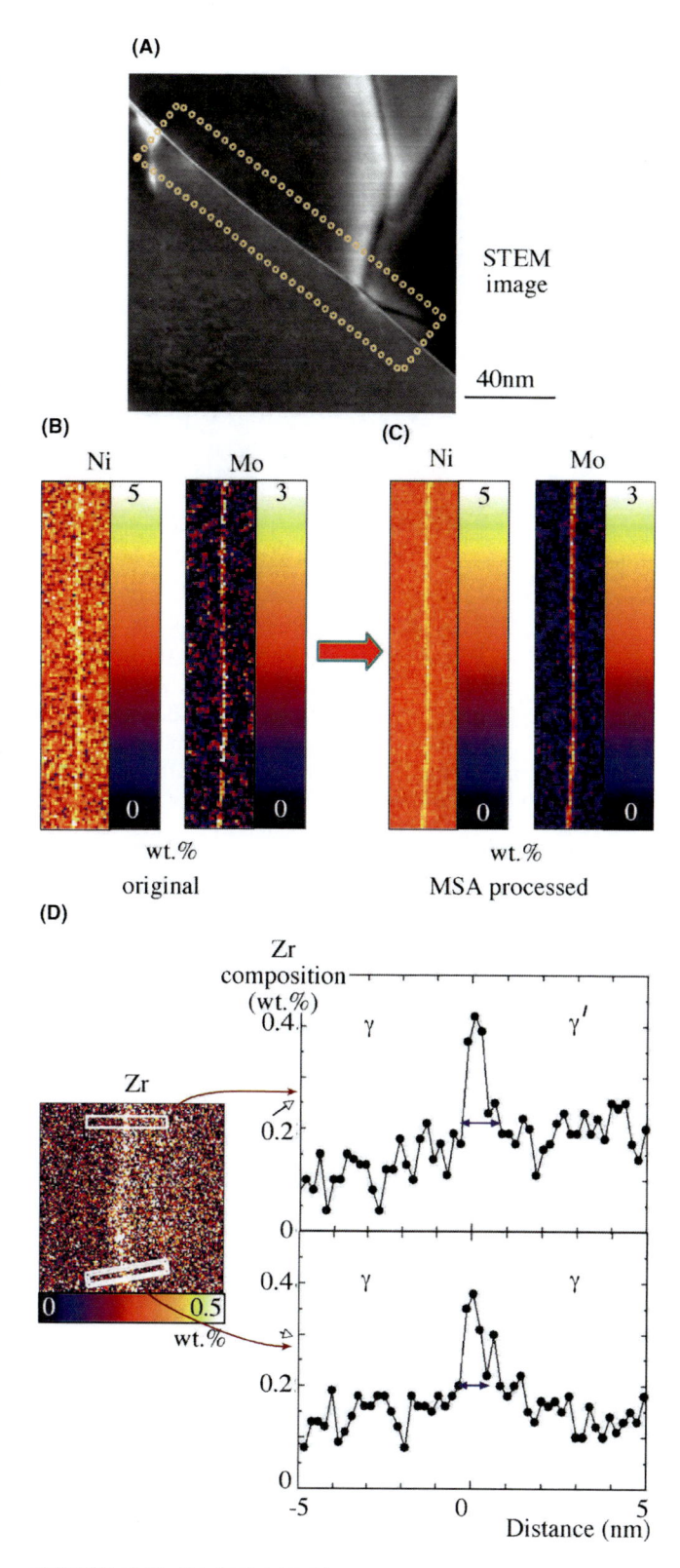

FIGURE 35.11. (A) STEM ADF image and (B) quantitative X-ray maps showing the segregation of trace amounts of Ni and Mo to grain boundaries in a low-alloy steel. (C) Applying MSA improves the quality of the maps. (D). Mapping the segregation of Zr to an interface in a Ni-base superalloy in a C_s-corrected STEM designed to give a 0.4-nm (FWTM) probe containing 0.5 nA. The Zr is present in the bulk alloy at ~0.04 wt% and without MSA processing could not be mapped. The composition profiles show that the Zr is localized to <1 nm at two different positions on the interface.

CHAPTER SUMMARY

The fact that much of this chapter is unchanged from the first edition sends a message that not much has changed in the last decade. This is unfortunate because, while quantitative analysis of spectra from thin foils can be straightforward, the standard Cliff-Lorimer approach has serious limitations. Most of the problems are overcome by the newer ζ-factor method, which is not yet commercially available but can be downloaded from the book web site (URL #2). Perhaps the greatest difficulty remains the need to know the specimen thickness in order to compensate for X-ray absorption and, again, the ζ-factor approach is invaluable in avoiding this. We can minimize absorption by making the thinnest possible specimens but then the number of X-ray counts may be so small that errors in the quantification are unacceptably large. The use of FEG sources, C_s correction, and improved TEM-EDS configurations with detector arrays to maximize the collection angle all help. With these latest advances, we can now perform quantitative X-ray mapping with a spatial resolution of a less than a nanometer and detection limits of a few atoms. There's much more about these exciting new aspects of quantitative analysis in the companion text.

GENERAL REFERENCES

Garratt-Reed, AJ and Bell, DC 2003 *Energy-Dispersive X-ray Analysis in the Electron Microscope* Bios (Royal Microsc. Soc.) Oxford UK.

Goodhew, PJ, Humphreys, FJ and Beanland, R 2001 *Electron Microscopy and Analysis* 3rd Ed. Taylor & Francis London. Instructive comparison of XEDS in SEM and TEM.

Goldstein, JI, Williams, DB and Cliff, G 1986 *Quantification of Energy Dispersive Spectra* in *Principles of Analytical Electron Microscopy* 155–217 Eds. DC Joy, AD Romig Jr. and JI Goldstein, Plenum Press New York. Introduction to many of the concepts in this chapter and the next one, including many worked examples.

Friel JJ and Lyman CE 2006 *X-ray Mapping in Electron-Beam Instruments* Microsc. Microanal. **12** 2–25. Detailed review of qualitative and quantitative mapping.

Jones, IP 1992 *Chemical Analysis Using Electron Beams* The Institute of Materials, London. The best source of examples of quantitative XEDS calculations.

Williams, DB and Goldstein, JI 1991 *Quantitative X-ray Microanalysis in the Analytical Electron Microscope* in *Electron Probe Quantitation* 371–398 Eds. KFJ Heinrich and DE Newbury Plenum Press New York. Derivations of the essential equations for thin-film quantification.

Zaluzec, NJ 1979 *Quantitative X-ray Microanalysis: Instrumental Considerations and Applications to Materials Science* in *Introduction to Analytical Electron Microscopy* Eds. JJ Hren JI Goldstein and DC Joy 121–167 Plenum Press NY. We've left this in so Nestor doesn't take us off his Web site.

CALCULATIONS

Anderson, IM, Bentley, J and Carter, CB 1995 *The Secondary Fluorescence Correction for X-Ray Microanalysis in the Analytical Electron Microscope* J. Microsc. **178** 226–239.

Bambynek, W, Crasemann, B, Fink, RW, Freund, HU, Mark, H, Swift, CD, Price, RE and Rao, PV *X-rayfluorescence Yields, Auger and Coster-Kronig Transition Probabilities* 1972 Rev. Mod. Phys. **44** 716–813.

Cliff, G and Lorimer, GW 1975 *The Quantitative Analysis of Thin Specimens* J. Microsc. **103** 203–207. The original paper based on crushed mineral standards (all borrowed from the desk drawer of Pam Champness, Lorimer's wife, who is a well-known mineralogist).

Heinrich, KFJ 1986 *Mass Absorption Coefficients for Electron Probe Microanalysis* in Proc. ICXOM-11 67–77 Eds. J Brown and R Packwood University of Western Ontario Canada.

Powell, CJ 1976 *Evaluation of Formulas for Inner-shell Ionization Cross Sections* in *Use of Monte Carlo Calculations in Electron Probe Analysis and Scanning Electron Microscopy* 97–104 NBS Special Publication 460 Eds. KFJ Heinrich, DE Newbury and H Yakowitz U.S. Department of Commerce/NBS Washington DC.

Watanabe, M and Williams, DB 2006 *The Quantitative Analysis of Thin Specimens: a Review of Progress from the Cliff-Lorimer to the New ζ-Factor Methods* J. Microsc. **221** 89–109.

Williams, DB. Newbury, DE Goldstein, JI and Fiori, CE 1984 *On the Use of Ionization Cross Sections in Analytical Electron Microscopy* J. Microsc. **136** 209–218.

Schreiber, TP and Wims, AM 1981 *Quantitative Analysis of Thin Specimens in the TEM Using a $\phi(\rho z)$ Model* Ultramicrosc. **6** 323–334.

Schreiber, TP and Wims, AM 1982 *Relative Intensity Factors for K, L and M Shell X-ray Lines* X-ray Spectrometry **11** 42–45.

Larsen, RJ and Marx, ML 2001 *An Introduction to Mathematical Statistics and its Applications* 3rd Ed. Prentice Hall Upper Saddle River NJ.

CHANNELING

Chu, W-K, Mayer, JM and Nicolet, M-A 1978 *Backscattering Spectrometry* Academic Press Orlando.

Goldstein, JI, Newbury, DE, Echlin, P, Joy, DC, Romig, AD Jr, Lyman, C., Fiori, CE and Lifshin, E 2003 *Scanning Electron Microscopy and X-ray Microanalysis* 3rd Ed. Springer New York.

SOME HISTORY

Castaing, R 1951 *Application des Sondes Électroniques a une Méthode d'Analyse Ponctuelle Chimique et Cristallographique* Thèses, Université de Paris ONERA Publication #55 Paris. A widely reproduced and historically significant publication.

Hillier, J and Baker, RF 1944 *Microanalysis by Means of Electrons* J. Appl. Phys. **15** 663–675.

Jones, IP 2002 *Determining the Locations of Chemical Species in Ordered Compounds; ALCHEMI* in *Advances in Imaging and Electron Physics* **125** 63–119.

Kramers, HA 1923 *On the Theory of X-ray Absorption and of the Continuous X-ray Spectrum* Phil. Mag. **46** 836–871.

Lorimer, GW, Al-Salman, SA and Cliff, G 1977 *The Quantitative Analysis of Thin Specimens: Effects of Absorption, Fluorescence and Beam Spreading* in *Developments in Electron Microscopy and Analysis* 369–371 Ed. DL Misell The Institute of Physics Bristol and London.

McGill, R. and Hubbard, FH 1981 *Quantitative Analysis with High Spatial Resolution* p30 Eds. GW Lorimer, MH Jacobs and P Doig The Metals Society London.

Reed, SJB 2005 *Electron Microprobe Analysis and Scanning Electron Microscopy in Geology* 2nd Ed. Cambridge University Press Cambridge UK.

Spence, JCH and Taftø, J 1983 *ALCHEMI - A New Technique for Locating Atoms in Small Crystals* J. Microsc. **130** 147–154.

Sprys, JW and Short, MA 1976 *Quantitative Elemental Analysis of 'Transparent' Particles in the TEM* Proc. 34th EMSA Meeting Ed. GW Bailey Claitors Baton Rouge LA 416–7

Statham PJ 2002 *Limitations to Accuracy in Extracting Characteristic Line Intensities From X-Ray Spectra* J. Research NIST **107** 531–546.

Wood, JE, Williams, DB and Goldstein, JI 1981 *Determination of Cliff-Lorimer k Factors for a Philips EM 400T* in *Quantitative Analysis with High Spatial Resolution* 24–30 Eds. GW Lorimer, MH Jacobs and P Doig The Metals Society London. k_{Fe} factors.

PROBLEMS WITH k_{ASi}

Goldstein, JI, Costley, JL, Lorimer, G, and Reed, SJB 1977 *Quantitative X-ray Microanalysis in the Electron Microscope SEM 1977* **1** 315–325 Ed. O Johari IITRI Chicago IL.

Graham, RJ and Steeds, JW 1984 *Determination of Cliff-Lorimer k Factors by Analysis of Crystallized Microdroplets* J. Microsc. **133** 275–280.

Sheridan, PJ 1989 *Determination of Experimental and Theoretical k_{ASi} Factors for a 200-kV Analytical Electron Microscope* J. Electr. Microsc. Tech. **11** 41–61.

Wood, JE, Williams, DB and Goldstein, JI 1984 *An Experimental and Theoretical Determination of k_{AFe} Factors for Quantitative X-ray Microanalysis in the Analytical Electron Microscope* J. Microsc. **133** 255–274.

URLS

1) mathworld.wolfram.com/Studentst-Distribution.html
2) http://www.TEMbook.com

SELF-ASSESSMENT QUESTIONS

Q35.1 Why do we have to correct the *k*-factor for absorption of X-rays?
Q35.2 Why is the fluorescence correction so small and generally ignored?
Q35.3 Why do we integrate only the K_α peak intensity rather than the $K_\alpha + K_\beta$ peaks if we can resolve them in the spectrum?
Q35.4 Why is the *k*-factor not a constant between different AEM-XEDS systems?
Q35.5 Why is the calculated *k*-factor generally inaccurate?
Q35.6 What's the largest contribution to the absorption correction and what does this tell you about ways to reduce your chance of having to do such?
Q35.7 What's the best way to minimize the errors in X-ray microanalysis?

Q35.8 What are the essential requirements for a good thin-foil standard and what does your list tell you about finding and selecting such standards?

Q35.9 What's the best way to measure the thickness of (a) your glass specimen, (b) your alloy foil, (c) your BN nanoparticle? (Hint: look ahead to Chapter 36.)

Q35.10 Why do quantitative analysis anyhow?

Q35.11 Why is it important to determine the errors in your quantitative analysis?

Q35.12 What is a typical ballpark quantification error in a simple binary (A–B) quantitative analysis?

Q35.13 What do you have to do to improve significantly on this error value?

Q35.14 When would you choose to use calculated k-factors rather than experimental ones?

Q35.15 When would you choose to determine your k-factors experimentally rather than calculate them?

Q35.16 List three ways to subtract the bremsstrahlung intensity from beneath the characteristic peaks in a spectrum.

Q35.17 Distinguish bremsstrahlung, continuum, and background X-rays.

Q35.18 What does ALCHEMI stand for and why is it anything but alchemy?

Q35.19 Why do we typically use wt% rather than at.% in X-ray microanalysis?

Q35.20 Why is it best to have as thin a window as possible on your XEDS detector but why is it generally impractical to have no window at all?

TEXT-SPECIFIC QUESTIONS

T35.1 If k-factors are not constants, what is the use of tables of k-factors, such as Tables 35.1 and 35.2?

T35.2 What are 'residuals' in the filtered spectra in Figure 35.6 and why are they useful?

T35.3 Distinguish top-hat apertures and top-hat filters and explain why both are useful in AEM-XEDS.

T35.4 Why would you use the two-window method of background subtraction as shown in Figure 35.5 rather than the one-window method in Figure 35.4?

T35.5 Why does the background intensity go to zero in Figure 35.4 and what experimental and instrumental factors affect the energy at which it goes to zero?

T35.6 Why do the values of the k-factors shown in Figure 35.7 decrease with decreasing atomic number and then increase again? (Hint: there's a good physics explanation for both of these trends.)

T35.7 Copy Figure 35.8 and extrapolate the three lines to lower thickness. At what value of thickness do they converge and why is this the case?

T35.8 Give three limitations to the k-factor approach that are overcome by the ζ-factor method?

T35.9 What is the single most cautionary lesson you can gain from Figure 35.10A?

T35.10 Why, in the AEM, can we measure elemental segregation phenomena generated in materials at lower temperatures compared with EPMA experiments? (Hint: think abut diffusion kinetics.)

T35.11 You have standard thin foils of Fe_2O_3, NiO, Ni_3Al, $CuSO_4$. Explain how you would determine k-factors for analysis of (a) FeS, (b) NiAl, (c) Al_2O_3, and (d) Al-Cu solid solution. List any specific concerns you may have with your determinations.

T35.12 Using Table 35.1, calculate reasonable first approximations for the k-factors for (a) Mn-Cr, (b) Mg-Al, and (c) Al-Cu. State any assumptions you make in your calculations.

T35.13 Using the DTSA software, practice generating X-ray spectra for different elements and compounds at different accelerating voltages (e.g., 100–300 kV) and different take-off angles (e.g., $\alpha = 20°$, 60°). Observe the differences in the background and characteristic spectra as a function of kV and α. Use the background-subtraction options to measure peak intensities and run practice quantifications.

T35.14 Using DTSA run quantification routines for a single binary specimen of your choice, but select a range of ionization cross sections with which to calculate a k-factor and compare the range of answers that you get for your k-factor and your quantification. What does this tell you about the limitations of calculated k-factors?

T35.15 Calculate the absorption correction factor (ACF) for Fe-10 wt% Al foils 10, 100, and 300 nm thick for different take-off angles of 20° and 70°. Then, evaluate the specimen thickness necessary to meet the thin-film criterion (> 10% absorption). Use 6.61 g/cm^3 for the specimen density and following data for the mass-absorption coefficients for Fe K_α, Fe L_α, and Al K_α lines (Heinrich 1986). If you wish, after completing the manual calculation compare your result with DTSA calculation of the same correction. (courtesy M. Watanabe)

Absorber	Fe K_α	Fe L_α	Al K_α
Fe	71.1	2,157	3,626
Al	96.5	2,936	397.5

Mass-absorption coefficient (cm^2/g)

36
Spatial Resolution and Minimum Detection

CHAPTER PREVIEW

Often when you do X-ray analysis of thin foils you are seeking information that is close to the limits of spatial resolution. Before you carry out any such analysis you need to understand the various controlling factors and in this chapter we explain these. Minimizing your specimen thickness is perhaps the most critical aspect of obtaining the best spatial resolution, so we summarize the various ways you can measure your foil thickness at the analysis point, but the quality of the TEM-XEDS system is also important.

A consequence of going to higher spatial resolution is that the X-ray signal comes from a much smaller volume of the specimen. A smaller signal means that you'll find it very difficult to detect the presence of trace constituents in thin foils. Consequently, the minimum mass fraction (MMF) in TEM is not as small as many other analytical instruments which have poorer spatial resolution. This trade-off is true for any analysis technique, and so it is only sensible to discuss the ideas of spatial resolution in conjunction with analytical detection limits. We'll make this connection in the latter part of the chapter. Despite the relatively poor MMF, it is possible to detect the presence of just a few atoms of one particular element if the analyzed volume is small enough, and so the TEM actually exhibits excellent minimum detectable mass (MDM). With the latest advances in XEDS and TEM technology, particularly C_s correction, X-ray analysis with atomic-column resolution and single-atom detection is now feasible in the same instrument.

36.1 WHY IS SPATIAL RESOLUTION IMPORTANT?

As we described in the introduction to Chapter 35, the historical driving force for the development of X-ray analysis in the TEM was the improvement in spatial resolution compared with the EPMA. This improvement arises for two reasons

- We use thin specimens, so less electron scattering occurs as the beam traverses the specimen.
- The higher electron energy (>100–400 keV in the TEM compared with 5–30 keV in the EPMA) further reduces scattering.

The latter effect occurs because the mean free path for both elastic and inelastic collisions increases with the electron energy. The net result is that *increasing* the accelerating voltage when using thin specimens *decreases* the total beam-specimen interaction volume, thus giving a more localized X-ray signal source and a higher spatial resolution, which is good (see Figure 36.1A). Conversely, with bulk samples, *increasing* the voltage *increases* the interaction volume and spatial resolution is at best ~0.5–1 μm which is not so good (see Figure 36.1B). There is increasing interest in reducing the spatial resolution of SEM-X-ray analysis using very low voltage electron beams and low-energy X-ray lines. While this is challenging, there has been considerable progress and a spatial resolution <100 nm with E_0 <5 keV is feasible. Aberration correctors and bolometer detectors will help even further but, for the best spatial resolution, there is still no alternative to thinning your specimen.

Much theoretical and experimental work was carried out in the early days of AEM to define and measure the spatial resolution of XEDS in the TEM, and we'll introduce some of these concepts. The ultimate aim, of course, is to push spatial resolution to the atomic scale and detection limits to the single-atom level. Both of these goals have been attained in EELS, as we'll see in the next several chapters, but, as ever, we are limited by the small number of X-ray counts generated in thin foils and the poor collection efficiency of the XEDS. C_s correctors and SDDs are helping here, as we shall see.

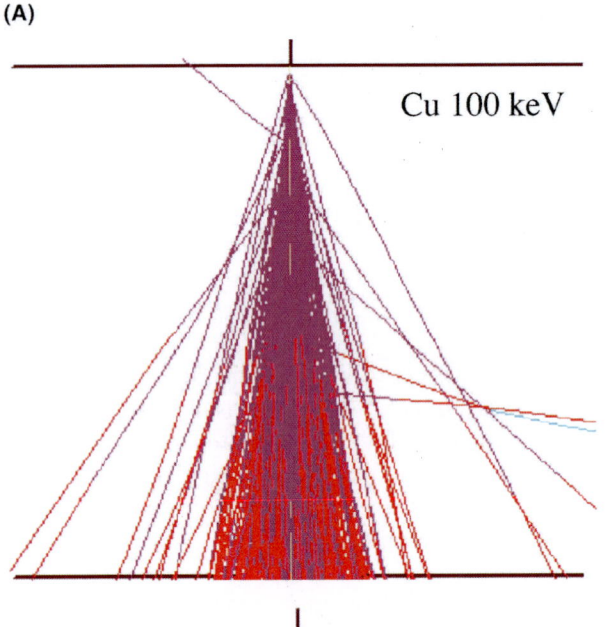

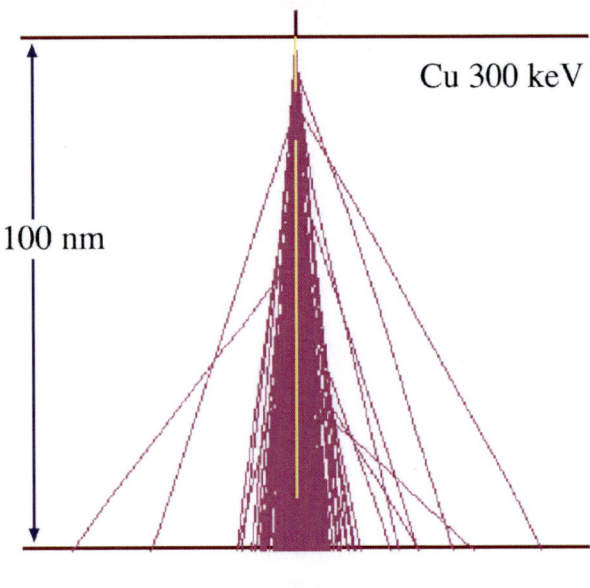

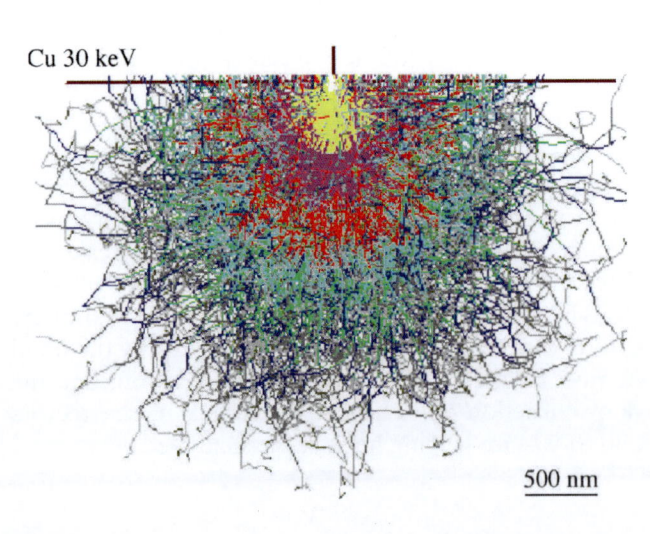

36.2 DEFINITION AND MEASUREMENT OF SPATIAL RESOLUTION

It has long been recognized that the analysis volume, and hence the spatial resolution, is governed by the beam-specimen interaction volume, since the XEDS can detect X-rays generated anywhere within that volume (you'll see later that this is different to the situation in EELS). The interaction volume is a function of the incident-beam diameter (d) and the beam spreading (b) caused mainly by elastic scattering within the specimen. Therefore, the measured spatial resolution (R) is a function of your specimen and this has made it difficult to define a generally accepted measure of R. Let's look first at d and b and how we define them.

> **SPATIAL RESOLUTION FOR XEDS**
> We can define this spatial resolution as the smallest distance (R) between two volumes in the specimen from which independent analyses can be obtained. The definition of R has evolved as AEMs have improved and smaller analysis volumes have become possible.

We've already discussed how to define and measure d in TEMs and STEMs way back in Chapter 5, so you need only remind yourself that the beam diameter d is defined as the FWTM of the Gaussian electron intensity. We can measure d directly from the TEM image or indirectly by traversing the beam across a sharp edge and looking at the intensity change on the STEM screen.

This definition takes account of only 90% of the electrons entering the specimen, so it is still an approximation. Remember that the electron-intensity distribution in the incident beam is Gaussian only if you are careful in your choice (small) and alignment of the C2 aperture and you restrict the beam to paraxial conditions (go back and read Section 6.5.A to decide if you should be at the Gaussian image plane or the disk of minimum confusion for best resolution). It is a little more difficult to define and measure b, so this needs more explanation.

FIGURE 36.1. (A) Monte Carlo simulations of 10^3 electron trajectories through a 100 nm Cu foil; (upper) 100 kV; (lower) 300 kV. Note the *improved* resolution at higher kV. (B) Conversely, in a bulk sample, the interaction volume at 30 kV is much larger than that at 10 kV, thus giving *poorer* X-ray spatial resolution at higher kV. The color in both sets of simulations reflects the change in energy of the electrons. Note the relatively constant energy in the thin foil compared with the rapid energy loss in the bulk sample.

36.2.A Beam Spreading

The amount that the beam spreads (b) on its way through the specimen has been the subject of much theoretical and experimental work. While results and theories differ in minor aspects, there is a general consensus that b is governed by the beam energy (E_0), foil thickness (t), and atomic number (Z). It turns out that the simplest theory for b gives a good approximation under most analysis conditions. This theory (sometimes called the 'single-scattering' model because it assumes that each electron only undergoes one elastic scattering event as it traverses the specimen) was first given in the seminal paper by Goldstein et al. and re-defined in SI units by Jones.

$$b = 8 \times 10^{-12} \frac{Z}{E_0} (N_v)^{1/2} t^{3/2} \quad (36.1)$$

where b and t are in m, E_0 is in keV, and N_v is the number of atoms/m³. In the original derivation, this latter term was given as $(\rho/A)^{1/2}$ and is confusing in that the density may vary considerably from point to point in a multi-phase alloy and is generally unknown anyhow. Furthermore, the atomic-weight dependence was the opposite to the atomic-number dependence which is counterintuitive. So using N_v is clearer and you can work the value out from the ratio of the number of atoms/unit cell to the volume of the unit cell, for which you need to know the lattice parameter. This definition again comprises 90% of the electrons emerging from the specimen, so it is consistent with our definition of d.

There is some question as to whether this single-scattering expression adequately describes the behavior of b for either very thin or very thick foils, but it has generally survived the test of time and its strength remains in its simplicity.

You should of course estimate/calculate b *prior* to spending an inordinate amount of time trying to do an experiment that is impossible for lack of sufficient resolution. Prior simulation of the expected resolution versus the necessary resolution to detect the phenomenon of interest can be very useful here, so let's discuss how best to do this, particularly when the specimen geometry is complex (e.g., multiple/overlapping phases) so equation 36.1 is difficult to apply.

FOR YOUR MAC/PC
We recommend that you keep this equation stored in the TEM computer (or your phone) so you can quickly estimate the expected beam spreading in your planned experiment.

When you can't apply equation 36.1, the best alternative is the Monte Carlo computer simulation, which we introduced in Section 2.5, as a way of modeling electron scattering. Such simulations are used in a wide variety of fields, including SEM and EPMA, as well as other nuclear-particle fields, as a quick search of the Web will reveal. A full description of Monte Carlo simulations is beyond the scope of this text but good reference books exist on the topic. In Joy's book, you'll find a code listing for a Monte Carlo simulation program which can be run on a PC. The public-domain Monte-Carlo programs that do the best job are WinCASINO and WinXRAY from Gauvin at McGill (see URLs #1 and #2); thin-foil versions of this software are under development. Until these are available, we recommend Joy's software. These simulations are now extremely rapid, and in a few minutes on a PC or Mac, they can provide all the information you need to estimate the beam spreading in more complex microstructures.

Basically, the Monte Carlo technique simulates, in a random manner (hence the name), a feasible set of electron paths through a defined specimen. After simulating ideally several thousand paths, an approximate value of b can be obtained by asking the computer to calculate the diameter of a disk at the exit surface of the specimen that contains 90% of the emerging electrons. This definition of b is consistent with that described at the start, and is the dimension of b given by equation 36.1. In fact, the schematic trajectories in Figure 36.1A and B are Monte Carlo simulations using Joy's software. Figure 36.2 shows Monte Carlo simulations of electron trajectories at three points across an interface between Cu and Au. Such a complex situation with elements of radically different Z cannot be easily handled by the single-scattering model estimates of b from equation 36.1.

While beam spreading is the main aspect of spatial-resolution theories, we mustn't forget that what we really want to know is the beam-specimen interaction volume, which corresponds to the size of the X-ray source. Monte Carlo simulations can help because in principle they can

- Incorporate the effects of different kVs and beam diameters
- Handle difficult specimen geometries, specimen tilt, thickness variations, and multi-phase specimens
- Automatically calculate the effect of the depth distribution of X-ray production, $\varphi(\rho t)$, on the X-ray source size
- Display the X-ray distribution generated anywhere in your specimen, as a function of all the variable parameters in equation 36.1, N_v, Z, and t. This tells you the relative contributions to your XEDS spectrum from different parts of the microstructure.

In addition to the theories of beam spreading that we've discussed, there are several more in the literature. A common feature of these theories is that they all predict a linear relationship between b and $t^{3/2}$ and an

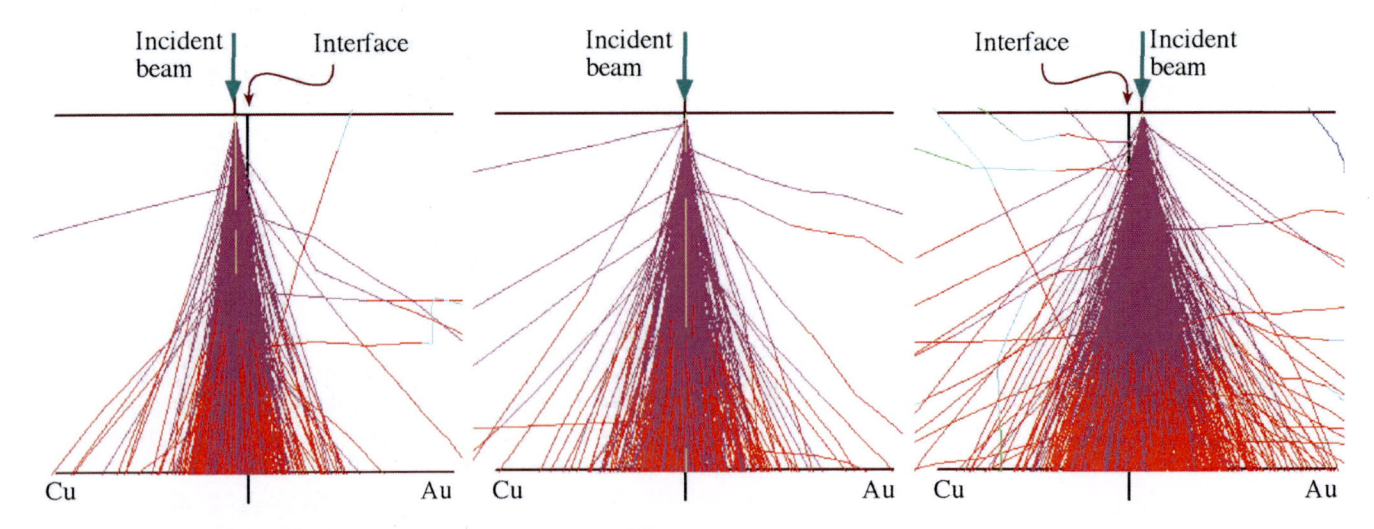

FIGURE 36.2. Monte Carlo simulation of electron trajectories across an interface between two metals of different Z, in which the scattering is very different. Note the rapid increase in the electron scattering in the higher Z region and therefore, X-rays would come from larger regions, thus lowering the local spatial resolution.

inverse relationship between b and E_0. If you're interested in the details of the various theories you'll find a discussion in the Goldstein et al. 1986 paper. However, we'll also see (look ahead to Figure 36.9 in Section 36.3.E) that there are ways to determine the spatial resolution on-line while you're doing your analyses and/or mapping and this is undoubtedly the best approach, since it combines the simple equations we'll now discuss with actual experiments rather than calculations, which make assumptions about your specimens.

36.2.B The Spatial-Resolution Equation

Now we've defined d and b, all we have to do is combine them to come up with a definition of R. If the intensity distribution of the incident beam is Gaussian, and if the beam emerging from the specimen retains a Gaussian form, it is reasonable to add b and d in quadrature (just as we did for image resolution back in Section 6.6.B) to give a value for R

$$R = \left(b^2 + d^2\right)^{\frac{1}{2}} \quad (36.2)$$

Gaussian beam-broadening models are also available, based on equation 36.1, which permit convolution of the Gaussian descriptions of d and b to come up with a definition of R. Based on the Gaussian model and experimental measurements, Michael et al. proposed that the definition of R be modified so as not to present the worst case (given by the exit-beam diameter) but to define R midway through the foil, as shown in Figure 36.3

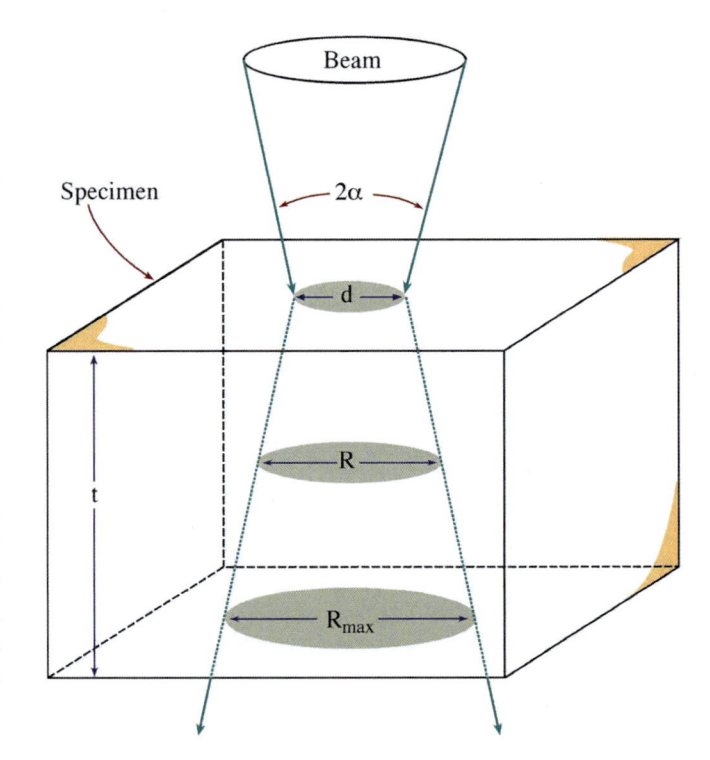

FIGURE 36.3. Schematic diagram of how the incident beam size and the beam spreading combine to degrade the exit-probe diameter to R_{max}, thus defining R.

$$R = \frac{d + R_{max}}{2} \quad (36.3)$$

where R_{max} is given by equation 36.2.

Like all definitions of spatial resolution, there is no fundamental justification for the choice of various factors, such as the FWTM diameter and the selection of the mid-plane of the foil at which to define R. Similarly,

666

this approach ignores any contribution of electron diffraction in crystalline specimens and beam tailing beyond the 90% limit. Nevertheless, the definition has been shown to be consistent with experimental results and sophisticated Monte Carlo simulations (Williams et al.). Finally, this definition retains the advantage of the original single-scattering model, i.e., it has a simple form and is easily amenable to calculation.

> **RESOLUTION IN XEDS**
> Equation 36.3 is the formal definition of the X-ray spatial resolution.

36.2.C Measurement of Spatial Resolution

Experimental measurements of the spatial resolution, such as composition profiles measured across atomically sharp interfaces, are very useful (go back and look at the profiles in Figure 1.4D). Several other kinds of specimens have been proposed but using interphase interfaces retains its validity since there are fewer unknowns than for the other specimens. If thermodynamic equilibrium exists on either side of the interface, the solute content of each phase is well defined. Also, interphase interfaces are common to many engineering materials, as is evident from many images of such defects throughout this text.

In order to compare experimental and calculated measurements of R, you have to understand how we relate the measured composition profile across the interface to the actual discrete profile shape, shown schematically in Figure 36.4. We do this by deconvolution of the beam shape from the measured profile. The finite beam size d and the effect of b degrade the sharp profile to a width L which is related to R by the following equation

$$R = 1.414L \qquad (36.4)$$

Assuming this relationship holds, we just measure the distance L between the 2% and 98% points on the profile, as shown in Figure 36.4. This spread contains 90% of the beam electrons, consistent with our assumption of a 90% (FWTM) incident-beam diameter. In practice, you will find it difficult to measure the 2% and 98% points because of the errors in the experimental data. So you should measure the distance from the 10% to the 90% points on your profile, corresponding to the beam spread containing 50% of the electrons (FWHM), then multiply this distance by 1.8 to give the FWTM.

Nevertheless, this definition is easy to remember, relatively easy to measure, consistent with the definitions of b and d, and, most importantly, gives a number that is close to the experimentally measured degradation of discrete composition changes introduced by the beam-specimen interaction.

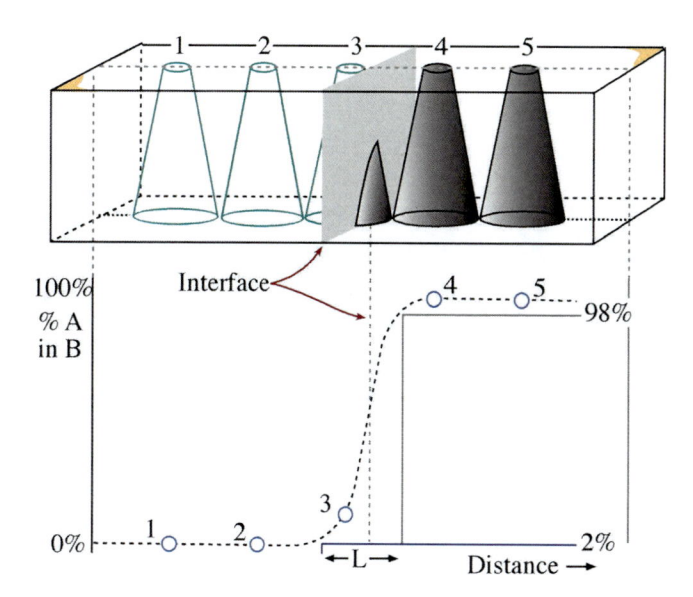

FIGURE 36.4. Schematic diagram showing a composition profile measured across an interface at which an atomically discrete composition change occurs (like the simulation in Figure 36.2). The measured spatial resolution can be defined in terms of the extent (L) of the measured profile between the 2% and 98% points.

It is obvious from equation 36.2, that if we want to improve spatial resolution, then both d and b must be minimized. Unfortunately, if we minimize d we reduce the input beam current: for thermionic sources, if d <10 nm, count rates will become unacceptably low. However, with a FEG, sufficient current (~1 nA) can be generated in a small enough (1 nm) beam to permit quantitative analyses with high spatial resolution.

> **DEFINING R, b, AND d**
> Note that this definition of R, like the definitions of b and d that we have used, is arbitrary.

So if you have a thermionic source TEM

- Your specimen has to be thick enough that sufficient counts are generated for quantification and b will be the main contributor to R.
- Alternatively, you may have to increase the beam size such that d dominates rather than b.

A large beam is needed in that example in order to generate sufficient beam current to get a reasonable X-ray count rate at 100 kV. This is why 200–300 kV FEG TEMs are the best high-resolution analysis instruments

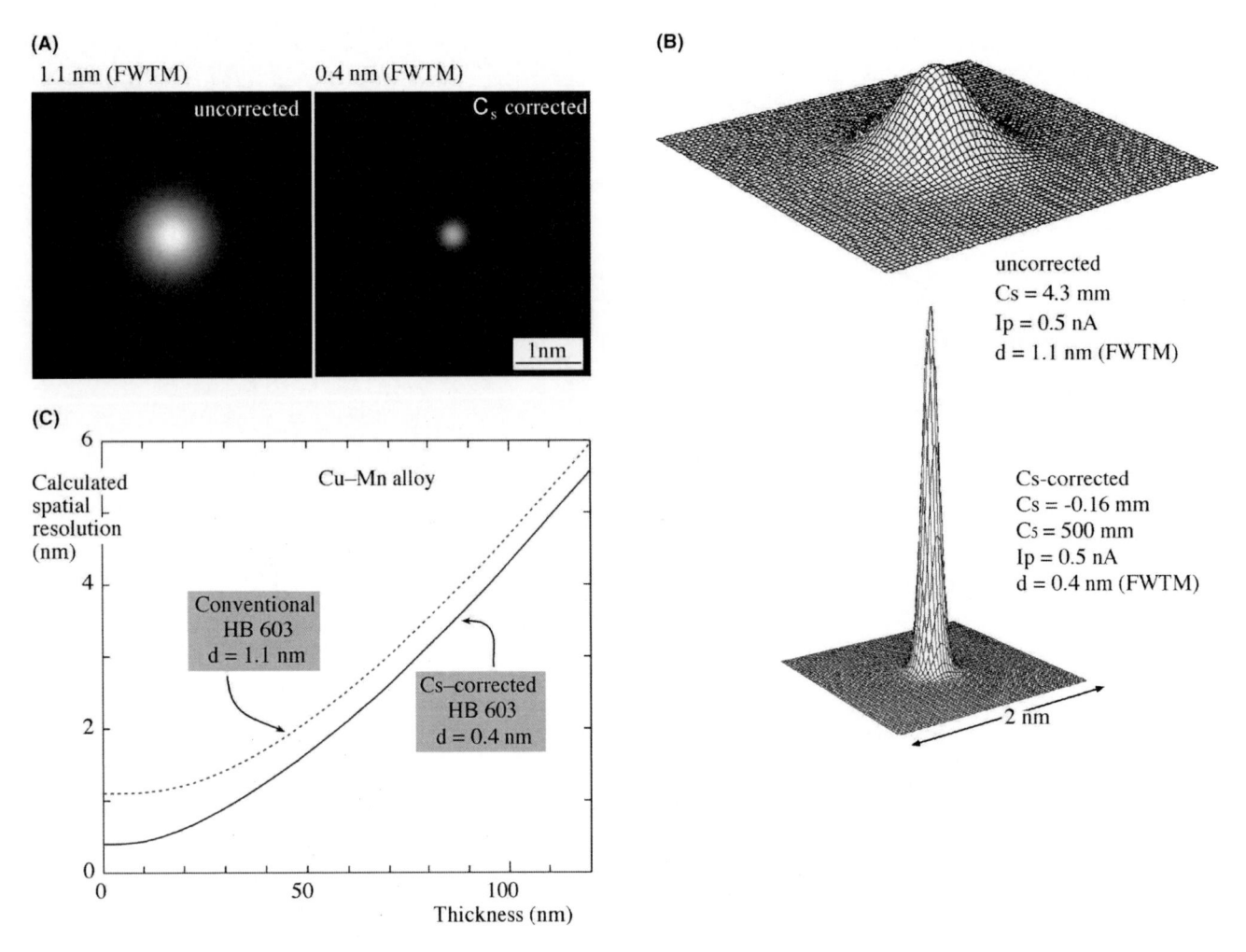

FIGURE 36.5. Simulated probe images for a VG HB-603 300-kV FEG STEM with and without C_s correction; (A) probe dimensions in plan view, (B) 3D intensity distributions. The simulations assume the same probe current (0.5 nA) and C_s correction results in a ~3× decrease in the FWTM probe size from ~1.1 to ~0.4 nm. (C) Calculation of the effect of C_s correction on the spatial resolution in a Cu-Mn alloy as a function of decreasing foil thickness. The resolution at zero foil thickness is improved also by ~3× from just >1 to ~0.4 nm. Compare Figure 36.5B with Figure 2.11 and wonder!

and C_s correction further improves the resolution (go back and compare the segregation-profile widths in Figure 35.11C and D) because you can keep the same probe current while reducing the probe size by a factor of ~3×. New X-ray detectors with larger collection angles would also help.

There are some practical factors that can also limit your experimental spatial resolution and the most important is specimen drift. If your specimen or probe drifts for mechanical or electrical reasons, then drift-correction software should be used. If you're planning to carry out analysis at the highest spatial resolution where you're obliged to count for long times to accumulate adequate X-ray intensity, then such software is indispensable.

In summary, the spatial resolution R is a function of both the beam size and the beam spreading. You can get a good estimate of R from equation 36.3. The theories all indicate a $t^{3/2}$ dependence for b, so thin specimens are essential for the best resolution. Intermediate-voltage FEG sources, especially when augmented with C_s correctors, give sufficient beam current in sub-nm probes to generate reasonable counts even from very thin specimens and invariably give the best spatial resolution which, as you can see from the calculations in Figure 36.5, can approach atomic dimensions.

36.3 THICKNESS MEASUREMENT

Given the $t^{3/2}$ dependence of the beam spreading, you can see the importance of knowing t when estimating the spatial resolution. You already know that t is also an essential parameter in correcting for the absorption of characteristic X-rays, as we saw in Section 35.6. Furthermore, you should remember that knowledge of t is important in high-resolution phase contrast imaging and CBED. You'll also see in Chapter 39 that minimizing t is critical to obtaining the best ionization-edge spectra in EELS. So, in almost all TEM techniques, your specimen has to be as thin as possible to get the best results (although some CBED studies and many in-situ experiments are notable exceptions to this generalization).

So let's take the opportunity here to summarize the methods available for measuring thickness. The methods are many and varied, and a full discussion of the most important techniques will be found in other parts of this book. The first point to consider is, what is t?

> **THE REAL THICKNESS**
> The thickness we are interested in is t; this is the thickness through which the beam penetrates. It is not necessarily the same as t_0.

This value of t depends both on the tilt of the specimen γ, and the true thickness at zero tilt, t_0. As shown in Figure 36.6, for a parallel-sided foil

$$t = \frac{t_0}{\cos \gamma} \tag{36.5}$$

If your specimen is wedge-shaped, then t and t_0 will vary in an arbitrary fashion depending on the foil shape.

36.3.A TEM Methods

In the TEM you can always make an estimate of your specimen thickness if it is wedge-shaped (and crystalline). By tilting to two-beam conditions for strong dynamical diffraction, the BF and DF images both show thickness fringes, as we saw in Section 24.2. These fringes occur at regions of constant thickness. The intensity in the BF image falls to zero at a thickness of $0.5\xi_g$ at $\mathbf{s} = 0$. Therefore, to determine t, all you have to do is look at the BF image and count the number (n) of dark fringes from the edge of the specimen to the analysis region. At that point $t = (n-0.5)\xi_g$ assuming that the thinnest part at the edge is $< 0.5\xi_g$ thick. (Be very careful with this assumption.) Remember that the value of ξ_g varies with diffracting conditions and so the $\mathbf{g}$ vector has to be specified. You can calculate ξ_g from the expression

$$\xi_g = \frac{\pi \Omega \cos \theta}{\lambda f(\theta)} \tag{36.6}$$

where Ω is the volume of the unit cell, λ is the electron wavelength, and $f(\theta)$ is the atomic scattering amplitude. Remember also that if you're not exactly at $\mathbf{s} = 0$ then the effective extinction distance ξ_{eff} must be used.

A related method relies on the presence of an inclined planar defect adjacent to the analysis region. The projected image of the defect, again under two-beam conditions, will exhibit fringes, which can be used to estimate the local thickness, or the projected width, w, of the defect image using the expression

$$t_0 = w \cot \delta \tag{36.7}$$

as shown in Figure 36.7 in which δ is the angle between the beam and the plane of the defect. Again, you have to compensate geometrically to measure t rather than t_0 if the foil isn't normal to the beam, and then

$$t = w(\cos \delta - \tan \gamma) \tag{36.8}$$

Of course, neither of these methods is applicable to non-crystalline materials, and it is not always possible to find a suitable inclined defect next to the analysis region in a crystal. Furthermore, two-beam conditions are not recommended for analysis because of the dangers of anomalous X-ray emission (see Section 35.9 on ALCHEMI for both an explanation of, and an exception to, this generalization). More insidious is the fact that oxidation, during or after specimen preparation,

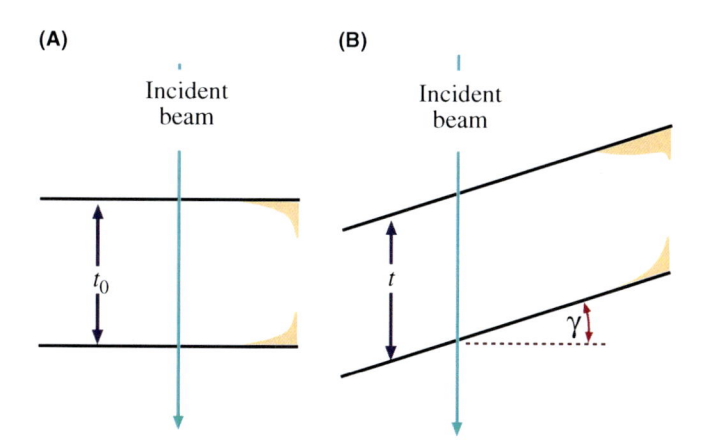

FIGURE 36.6. (A) The specimen thickness t_0 is equal to t, the distance traveled by the beam through a parallel-sided foil at zero tilt. (B) The beam travels a longer distance, t, in a specimen tilted through an angle γ and thus the beam will spread more in a tilted foil.

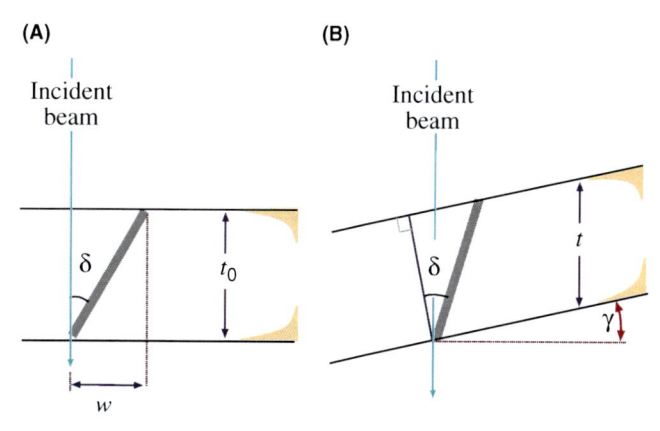

FIGURE 36.7. The parameters required to measure the specimen thickness t_0 from a planar defect (projected width, w), inclined to the incident beam by angle δ. Comparison of (A), a specimen normal to the beam, with (B) a specimen tilted through an angle γ gives some indication of the complexity of determining the appropriate thickness, t, to put into the beam-spreading equation.

means that your crystalline specimen may be coated with an amorphous layer which will not be measured by these diffraction-contrast techniques.

Another method related to the TEM-image contrast involves measurement of the relative transmission of electrons. The intensity on the TEM screen decreases with increasing thickness, all other things being equal. You can use a Faraday cup to calibrate the intensity falling on the screen and from this you can get a crude measure of relative thickness, which can be converted into an absolute measure of t if some absolute method is used for calibration. But you must be careful to make all the intensity measurements on your specimen under the same diffraction conditions and the same incident-beam current but with no objective aperture. The only advantage of this approach is that it is applicable to all materials, both amorphous and crystalline, but it is tedious and not very accurate.

36.3.B Contamination-Spot Separation Method

This method, common in old (S)TEMs, but also commonly used by makers of dirty specimens, relies on the propensity of old instruments or contaminated foils to generate carbon peaks on both top and bottom surfaces of the specimen, at the point of analysis. If you tilt your specimen by a large enough angle (γ), you can see discrete contamination spots (Figure 36.8). Their separation r, at a screen magnification M, is related to t_0 by the following expression

$$t_0 = \frac{r}{M \sin \gamma} \quad (36.9)$$

Matters get a little more complicated if the specimen itself is tilted by an angle ϵ when the contamination is deposited. Then, as in the case of tilted planar defect, you have to be careful to measure the thickness which will determine the beam spreading and, as we've taken pains to point out, this is not t_0 (the thickness at zero tilt) but t

$$t = \frac{r}{M} \frac{\cos \epsilon}{\sin \gamma} \quad (36.10)$$

Although this method is straightforward, it relies on highly undesirable contamination, which obscures the

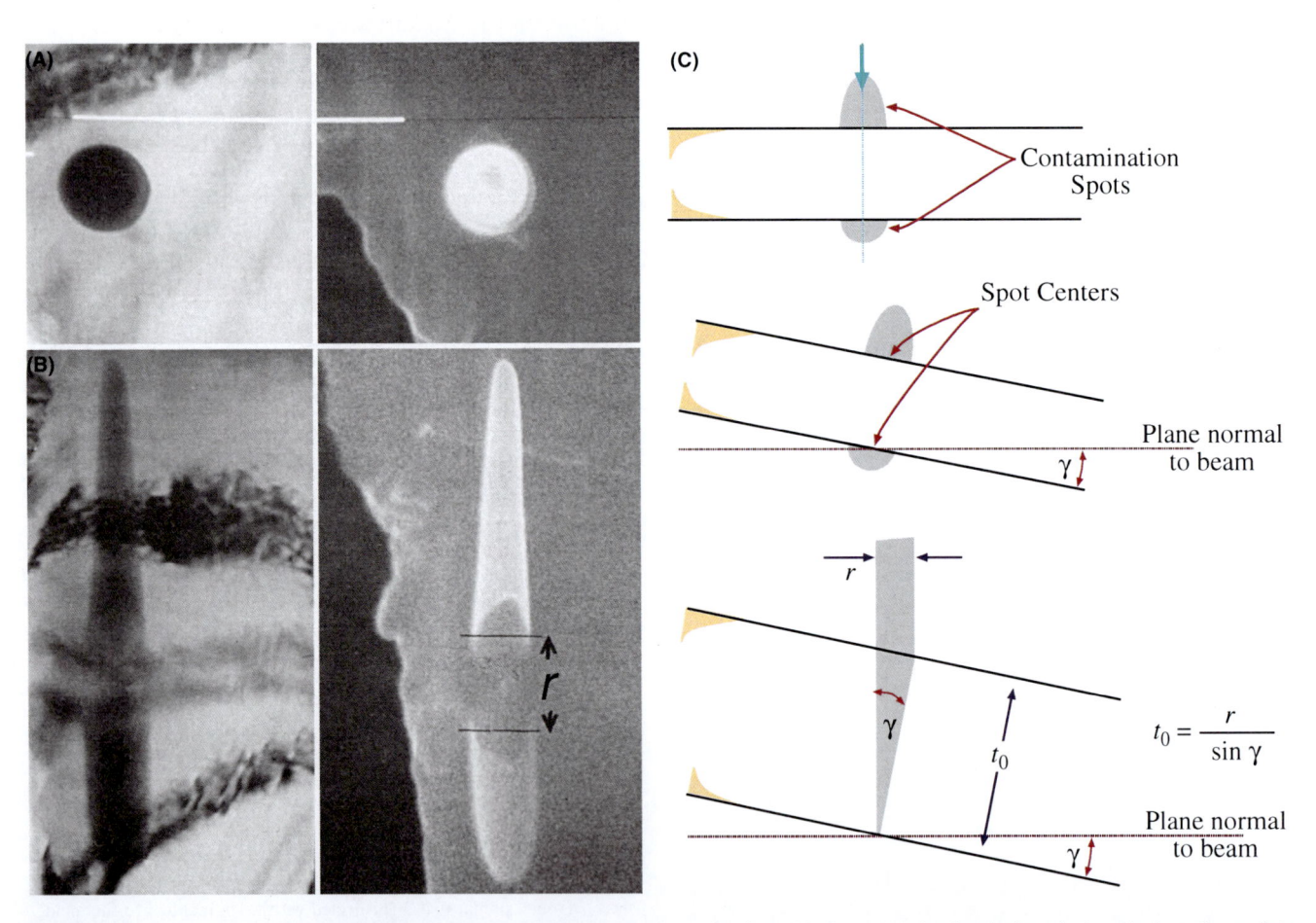

FIGURE 36.8. The contamination-spot separation method for thickness determination; (A) the contamination is deposited on both surfaces of the specimen at zero tilt and the separation (r) is only visible in (B) when the specimen is tilted sufficiently through an angle γ. STEM BF images are on the left and STEM SE images on the right; the SE mode gives the best contrast. (C) Geometrical diagrams of (A) and (B) showing how to determine t_0 from r, the projected separation of the contamination spots.

670 .... SPATIAL RESOLUTION AND MINIMUM DETECTION

very area you're looking at. Contamination degrades the spatial resolution and increases the X-ray absorption. In fact, we spend a lot of time and effort trying to minimize contamination, so it would be perverse to propose it as a useful way of determining t. The only redeeming feature is that this method measures t exactly at the analysis point and the shape of the spots can indicate if the beam or the specimen has drifted during analysis. If you find yourself even thinking about using this method, then your TEM should not be used for analysis or you should clean up your specimen-preparation act.

36.3.C Convergent-Beam Diffraction Method

The CBED pattern which is visible on the TEM screen when a convergent beam is focused on the specimen can also be used to determine the thickness of crystalline specimens. In Section 21.2, we described the procedure to extract the thickness from the K-M fringe pattern obtained under two-beam conditions. The CBED pattern must come from a region thicker than $1\xi_g$ or else fringes will not be visible. Also the region of the foil should be relatively flat and undistorted.

Remember that for a totally clean, crystalline specimen, CBED is the way to determine t at specific points in your specimen.

36.3.D Electron Energy-Loss Spectrometry Methods

Thickness information is present in the electron energy-loss spectrum since the intensity of inelastically scattered electrons increases with specimen thickness. In essence, you have to measure the intensity under the zero-loss peak (I_0) and ratio this to the total intensity in the spectrum (I_T). The relative intensities are governed by the mean free path (λ) for energy loss. A parameterization formula for λ (Malis et al. 1988) and other EELS methods, are discussed in detail in Section 39.5.

We can apply the EELS parameterization to any specimen, amorphous or crystalline. But the main advantage is that it is possible to measure the thickness sufficiently quickly that, unlike all the methods described so far, the EELS method can also produce maps of specimen thickness. Given that we have emphasized the value of compositional mapping over point analyses or line profiles, we have to conclude that EELS is best.

> **EELS FOR t**
>
> The EELS approach is highly recommended because it is applicable over a wide range of thicknesses and you can produce thickness maps of thin foils.

36.3.E X-ray Spectrometry Method

Since we're talking about X-ray spectrometry, it's good to know that there are X-ray methods which can also determine thickness. We can categorize these approaches, which have been developed to solve the absorption-correction problem (see Section 35.6), into two types: the first is the extrapolation method, which determines the absorption correction by extrapolation of X-ray intensity ratios to zero thickness; the second uses the difference in relative X-ray absorption between two emitted X-ray lines (K and L, or L and M) from the same element. Unfortunately, the extrapolation method is not easily applicable to thin foils where compositions vary locally, since you have to obtain a series of X-ray intensities from different thickness areas (by moving the incident beam or by tilting the specimen). In the intensity-ratio method, the essential requirement of two different X-ray lines from a single element limits the application to specimens which contain elements with $Z > 20$ (Ca).

All of these problems are solved using the ζ-factor approach, as described in detail in 2006 by Watanabe and Williams and in the companion text. The general expression (equation 35.18) for ρt from the ζ-factor analysis can be modified for N different elements in the specimen, thus

$$\rho t = \sum_j^N \frac{\zeta_j I_j A_j}{D_e}, \quad C_A = \frac{\zeta_A I_A A_A}{\sum_j^N \zeta_j I_j A_j}, \cdots$$

$$C_N = \frac{\zeta_N I_N A_N}{\sum_j^N \zeta_j I_j A_j} \qquad (36.11)$$

An iterative process is required to solve these equations for both composition and thickness determination. However, the iteration is straightforward and converges rapidly; ~10–15 iterations converge with < 0.001 wt% and 0.01 nm differences in composition and thickness, respectively, which are clearly far more than sufficient tolerance values for termination. Obviously, if X-ray absorption is negligible in a specific material, the initial mass-thickness and compositions are the final values and the iteration is no longer necessary. The essential point here is that in the ζ-factor method, the absorption-corrected compositions can be determined simultaneously with the specimen mass-thickness by *only using X-ray intensity data*.

This method is so quick and versatile that, like the EELS methods, it also permits direct mapping of the thickness at the same time as the composition is being mapped. Therefore, there is nothing to stop you mapping out the spatial resolution at the same time as doing your quantitative mapping, as shown in Figure 36.9. Both the ζ-factor and EELS methods can handle amorphous and crystalline specimens and Ohshima et al. compare the two methods.

In summary, there are many methods for determining t, but none is universally convenient, accurate, and

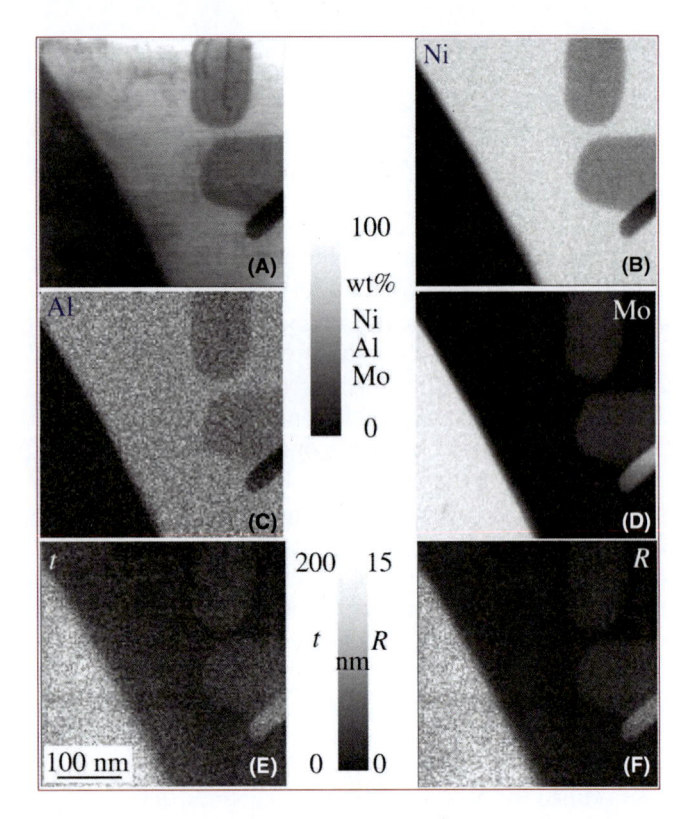

FIGURE 36.9. (A) STEM image and X-ray maps showing the quantitative distribution of (B) Ni, (C) Al, and (D) Mo in precipitates in a Ni-base superalloy. Using the ζ-factor the variation in thickness, t, across the foil can be mapped out (E). Knowing t, the spatial resolution, R, can also be mapped (F). Note the complex interaction of different atomic numbers and the variations in thickness in the resulting variations in R.

applicable. Beware; the various methods also measure different thicknesses, such as only the crystalline thickness, ignoring porosity and/or amorphous surface/oxide films, or the full thickness including porosity and surface films, or just the mass-thickness. Mitchell gives a good case study of how to handle such problems. The EELS and ζ-factor methods both have the possibility of widespread, real-time use and can produce thickness maps, so we recommend these. CBED is very useful for individual point analyses of crystals.

36.4 MINIMUM DETECTION

Minimum detection is a measure of the smallest amount of a particular element that can be detected with a defined statistical certainty. Minimum detection and spatial resolution are intimately related.

A TRUISM
It is a feature of any analysis technique that an improvement in spatial resolution is balanced by a worsening of the detection limit (all other factors being equal).

As the spatial resolution improves, the analyzed volume is smaller and, therefore, the signal intensity is reduced. This reduction in signal intensity means that the acquired spectrum will be noisier and small peaks from trace elements will be less detectable and more easily confused with artifact peaks. Accordingly, in the AEM, the price that is paid for improved spatial resolution is a relatively poor minimum detection. By way of comparison, Figure 36.10 compares the size of the analyzed volume in an EPMA, a TEM/STEM with a thermionic source, and a dedicated STEM with a FEG. The enormous reduction in the beam-specimen interaction volume explains the small signal levels that we obtain in the TEM. However, as we've noted on several occasions, C_s correction gives more current in a smaller probe so it offsets the traditional compromise. But you should now understand why we have spent so much time emphasizing the need to optimize your beam current through use of higher-brightness sources, optimizing the specimen-detector configuration, and so on.

MINIMUM DETECTION
One definition of minimum detection: the minimum mass fraction (MMF) that can be measured in the analysis volume. MMF represents the smallest concentration of an element (e.g., in wt% or ppm).

Alternatively, the minimum detectable mass (MDM) is sometimes used; the MDM describes the smallest amount of material (e.g., in mg or atoms) we can detect.

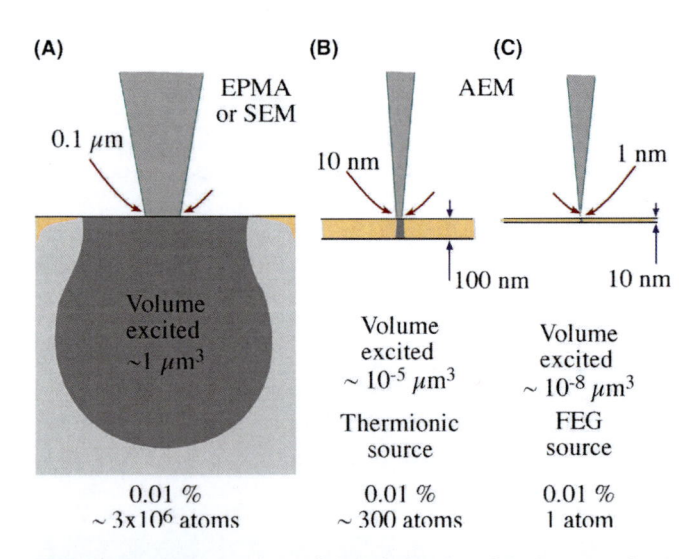

FIGURE 36.10. Comparison of the relative size of the beam-specimen interaction volumes in (A) a SEM/EPMA, (B) a thermionic source AEM, and (C) a FEG-AEM with bulk, thin, and ultra-thin specimens, respectively. The MMF (~0.01%) in each analyzed volume would correspond to ~10^7 atoms, ~300 atoms, and <1 atom, respectively.

We'll use the MMF since materials scientists are more used to thinking of composition in terms of wt% or at.%.

36.4.A Experimental Factors Affecting the MMF

We can relate the MMF to the practical aspects of analysis through the expression of Ziebold

$$\text{MMF} \propto \frac{1}{\sqrt{P(P/B)n\tau}} \quad (36.12)$$

Here P is the X-ray count rate in the characteristic peak (above background) of the element of interest, P/B is the peak-to-background count-rate ratio for that peak (defined here in terms of the same width for both P and B), and τ is the analysis time for each of n analyses.

To increase P you can increase the current in the beam by increasing the probe size and/or choosing a thicker analysis region. To increase P/B you can increase the operating voltage (E_0), which is easy, and decrease instrumental contributions to the background, which is not so easy (Lyman et al.). Improvements in TEM design, such as using a high-brightness, intermediate-voltage source, a C_s corrector if possible, and a larger collection angle for the XEDS will also increase P. To increase P/B, you need a stable instrument with a clean vacuum environment to minimize or eliminate specimen damage and contamination. Improved stage design, to minimize stray electrons and bremsstrahlung radiation, both of which contribute background to the detected spectrum, will also help to increase P/B, as we discussed back in Chapter 33.

Remember that the Fiori definition of P/B is not the one used in Ziebold's equation (36.12). If you actually want to calculate the MMF, go back and check the original references.

The other variables in equation 36.12, are the time of analysis (τ) and how many analyses (n) that you do, which are entirely within your control as operator. Usually both n and τ are a direct function of your patience and the recommended coffee break is usually the maximum time for any one analysis. With computer control of the analysis procedure, however, there should really be no limit to the time available for analysis. Particularly when detection of very small amounts of material is sought, τ should be increased to very long times. As computer control and stage stability improve, acquisitions of several hours or overnight are becoming feasible. Of course, the investment of so much time in a single analysis is dangerous unless you have judiciously selected the analysis region, and you are confident that the time invested will be rewarded with a significant result. Obviously, you should minimize factors that degrade the quality of your analysis with time, such as contamination, beam damage, and specimen drift. Therefore, you should only carry out long analyses if your TEM is clean (preferably UHV) and your specimen is also clean and stable under the beam. Any specimen drift must be corrected by computer control during the analysis, unless your specimen is uniformly thin and homogeneous in composition, in which case why bother analyzing it?

36.4.B Statistical Criterion for the MMF

We can also define the MMF by a purely statistical criterion. We discussed in Section 34.5 that we can be sure a peak is present if the peak intensity is greater than three times the standard deviation of the counts in the background under the peak. From this criterion we can come up with a definition of the detection limit which, when combined with the Cliff-Lorimer equation (assuming Gaussian statistics), gives the MMF (in wt%) of element B in element A as

$$C_B(\text{MMF}) = \frac{3(2 I_B^b)^{1/2} C_A}{k_{AB}(I_A - I_A^b)} \quad (36.13)$$

where I^b_A and I^b_B are background intensities for elements A and B; I_A is the raw integrated intensity of peak A (including background); C_A is the concentration of A (in wt%); and k_{AB}^{-1} is the reciprocal of the Cliff-Lorimer k-factor. However, if we express the Cliff-Lorimer equation as

$$\frac{C_A}{k_{AB}(I_A - I_A^b)} = \frac{C_B}{(I_B - I_B^b)} \quad (36.14)$$

and substitute it into equation 36.13, the MMF is

$$C_B(\text{MMF}) = \frac{3(2 I_B^b)^{1/2} C_B}{I_B - I_B^b} \quad (36.15)$$

Experimentally, low count-rates from thin specimens mean that typical values of MMF are in the range 0.1–1%, which is rather large compared with some other analytical techniques. The best compromise in terms of improving MMF while maintaining X-ray spatial resolution is to use high operating voltages (300–400 kV) and thin specimens to minimize beam broadening. The loss of X-ray intensity, P (or I), a consequence of using thin specimens, can be compensated in part by the higher voltages and/or by using an FEG where a small spot size of 1–2 nm can still be maintained. Obviously C_s-corrected TEMs will help because of their ability to put even more current into the same size probe. Figure 36.11 summarizes the classic

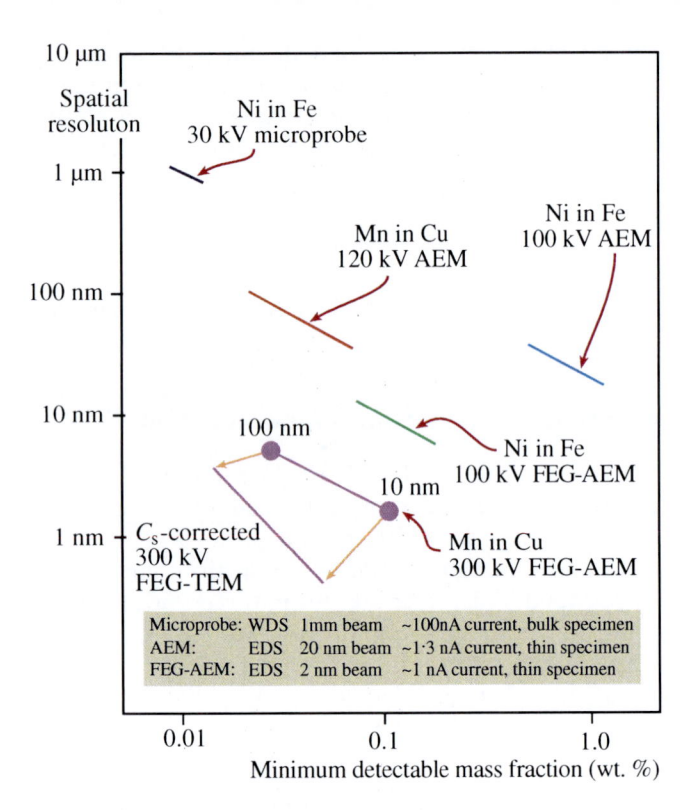

FIGURE 36.11. Calculation of the relationship between MMF and spatial resolution, R, for the EPMA and a range of AEMs. The inverse relationship between the MMF and R is clear, although it is also apparent that the high-brightness sources and high-kV electron beams in the AEM can compensate for the decreased interaction volume in a thin foil. C_s correction results in an enormous improvement in both resolution and sensitivity.

compromise between resolution and detection and how instrumentation improvements have continued to push the limits over the past few decades.

36.4.C Comparison with Other Definitions

The MMF definition is not the only way we can measure detection limits. Currie has noted at least eight definitions in the analytical-chemistry literature. Currie defined three specific limits.

- The decision limit: Do the results of your analysis indicate detection or not (L_C)?
- The detection limit: Can you rely on a specific analysis procedure to lead to detection (L_d)?
- The determination limit: Is a specific analysis procedure precise enough to yield a satisfactory quantification (L_q)?

For I_B counts from element B in a specific peak window and I_B^b in the background it can be shown that

$$L_C = 2.33\sqrt{I_B^b} \quad (36.16)$$

$$L_d = 2.71 + 4.65\sqrt{I_B^b} \quad (36.17)$$

$$L_q = 50\left\{1 + \left(1 + \frac{I_B^b}{12.5}\right)^{\frac{1}{2}}\right\} \quad (36.18)$$

If there are sufficient counts in the background

$$L_d = 4.65\sqrt{I_B^b} \quad \text{when } I_B^b > 69 \quad (36.19)$$

$$L_d = 14.1\sqrt{I_B^b} \quad \text{when } I_B^b > 2500 \quad (36.20)$$

Comparison of these definitions with the statistical criterion in the previous section shows that $C_{MMF} \approx L_d$. So, if you want to quantify an element, not just determine that it is present (L_d), then you need substantially more (~3×) of the element in your specimen. Rather than do the experiment yourself, it is possible to simulate spectra from small amounts of element B in A (or vice versa), using DTSA, as described in the companion text. We recommend that you simulate your analysis before embarking on a time-consuming experiment, which may be futile because the amount of the element you are seeking is below the MMF.

36.4.D Minimum-Detectable Mass

The MMF values of fraction of a percent may seem poor compared with other analytical techniques which report ppm or even ppb detection limits. However, it's a different matter if you calculate what the MMF translates to in terms of the minimum detectable mass (MDM).

Using data for the MMF of Cr in a 304L stainless steel measured in a VG HB-501 AEM with an FEG, Lyman and Michael obtained an MMF of 0.069 wt% Cr in a 164 nm foil with a spatial resolution of 44 nm and a 200 s counting time. The electron beam size was 2 nm (FWTM) with a beam current of 1.7 nA. In this analysis, an estimated 2×10^4 atoms were detected. The MDM was less than 10^{-19} g. If the counting time is increased by a factor of 10 and if the operating voltage is increased to 300 kV, the spatial resolution would improve to ~15 nm and the MMF would improve to ~0.01 wt%. Thus about 300 atoms could be detected. For a foil thickness of 16 nm (1/10th the above measured thickness), the MMF would degrade to ~0.03 wt%. However, the spatial resolution would improve to about 2 nm. For this case, about 20 atoms would be detected corresponding to less than 10^{-22} g, which is an amazing figure by any standard. Experimental verification of this was reported in 1999 by Watanabe and Williams: 2–5 atoms of Mn were detected in a 10-nm thick Cu-Mn alloy film. With the advent of C_s-correction and improved computer data analysis routines, single-atom

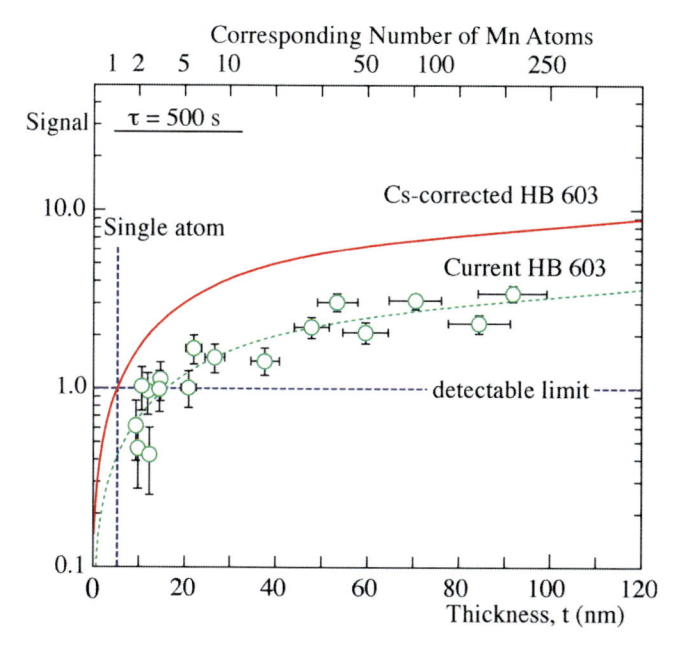

FIGURE 36.12. Calculation of the number of Mn atoms detectable in a Cu-0.1 wt% Mn foil as a function of foil thickness (dotted green line) based on experimental Mn K_α counts (green circles) in a 300-keV FEG STEM. When the Mn K_α signal is undetectable the ordinate axis value = 1 and this occurs when there are between 2 and 5 Mn atoms in the analysis volume (see top axis) which can be calculated knowing the foil thickness and bulk chemistry. A C_s corrector (red line) is calculated to improve the MDM from several atoms to ~1 atom right at the detection limit.

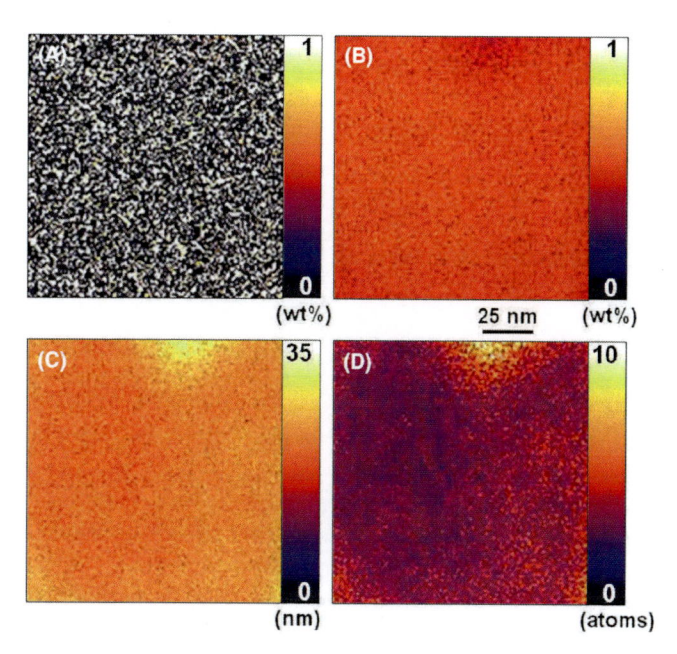

FIGURE 36.13. A series of quantitative maps obtained from a homogeneous Cu-0.5 wt% Mn foil in a C_s-corrected, 300-kV, UHV, FEG STEM. (A) Mn composition map from the original spectrum image, (B) Mn composition map from the spectrum image enhanced by MSA noise reduction, (C) thickness map, and (D) map of a number of Mn atoms. Note the look-up tables with each map. In (D) the dominating purple color corresponds to ~ 2–3 atoms. These maps were quantified by the ζ-factor method.

detection is now a distinct possibility and Figure 36.12 shows calculated improvements in detection limits for Mn atoms in solution in Cu in a C_s-corrected 300-kV FEG TEM. Figure 36.13 shows how mapping of a homogeneous solid solution can detect a few atoms at each pixel. This last figure summarizes just about everything we have discussed for AEM quantification and C_s-corrected mapping: (a) ζ-factor quantification (the image is still noisy), (b) MSA data manipulation (the image is much less noisy), (c) the importance of good thin foils (specimen thickness is very uniform and < 20 nm in most areas), and (d) MDM close to 1–2 atoms is attainable (even when mapping). Just to remind you how good this is, go back and take a look at Figure 36.10. Consider that in the EPMA with a ~1 µm³ excitation volume and a 0.01 wt% MMF, ~3 million atoms are detected in the analysis volume. So XEDS in the best C_s-corrected, intermediate-voltage, UHV, FEG TEM has an MDM detection limit that is *several million times better* than an EPMA.

MDM

In AEM it is useful to define the MDM as the minimum number of atoms detectable in the analyzed volume.

While the best XEDS-TEM combinations are approaching atomic-level detection and sub-nm spatial resolution, it is not yet possible to detect single atoms within individual atomic columns as is achievable in EELS in similar, C_s-corrected, intermediate-voltage FEG TEMs (see Chapter 39) so there is still room for improvement, e.g., through larger collection angles and more sophisticated data processing.

CHAPTER SUMMARY

Optimizing spatial resolution and minimum detection in the same experiment is always a compromise. You must decide which of the two criteria is more important for the result you're seeking

- To get the best spatial resolution, operate with the thinnest foils and the highest energy electron beam. Use an FEG if possible and a C_s corrector if you're lucky.

- To measure the specimen thickness use the ζ-factor method or the parameterized EELS approach. If neither is possible, use CBED for a crystalline foil. If you're reduced to contamination spots, find a better TEM or make cleaner specimens.
- To get the best MMF, use the brightest electron source, the largest possible beam, and thickest specimen, and count for as long as possible with the shortest time constant.
- If you want the best resolution *and* MMF, a C_s-corrected, intermediate-voltage, UHV, FEG TEM is essential, along with a clean specimen and computer-controlled drift correction; patience is also equally essential.

CLASSICS

Berriman, J, Bryan, R, Freeman, R and Leonard, K R 1984 *Methods for Specimen Thickness Determination in Electron Microscopy* Ultramicrosc. **13** 351–364. Old, but still useful, review of thickness measurements in the TEM.

Goldstein, JI, Williams, DB and Cliff, G 1986 *Quantification of Energy Dispersive Spectra* in *Principles of Analytical Electron Microscopy* 155–217 Eds. DC Joy, AD Romig Jr. and JI Goldstein, Plenum Press New York. Introduction to many of the concepts in this chapter and the preceding one.

Jones IP 1992 *Chemical Microanalysis Using Electron Beams* Institute of Materials London. Redefined Goldstein et al. equation in SI units on p173.

Joy, DC 1995 *Monte Carlo Modeling for Electron Microscopy and Microanalysis* Oxford University Press New York. The only textbook covering this essential topic

THICKNESS AND RESOLUTION

Malis, T, Cheng, SC and Egerton RF 1988 *The EELS Log-ratio Technique for Specimen-Thickness Measurement in the TEM* J. Electr. Microsc. Tech. **8** 193–200.

Michael, JR, Williams, DB, Klein, CF and Ayer, R 1990 *The Measurement and Calculation of the X-ray Spatial Resolution Obtained in the Analytical Electron Microscope* J. Microsc. **160** 41–53.

Mitchell, DRG 2006 *Determination of Mean Free Path for Energy Loss and Surface Oxide Film Thickness Using Convergent Beam Electron Diffraction and Thickness Mapping: a Case Study Using Si and P91 Steel* J. Microsc. **224** 187–196.

Williams, DB, Michael, JR, Goldstein, JI and Romig AD Jr. 1992 *Definition of the Spatial Resolution of X-ray Microanalysis in Thin Foils* Ultramicrosc. **47** 121–132.

P/B, ζ, AND MMF

Currie, LA 1968 *Limits for Qualitative Detection and Quantitative Determination. Application to Radiochemistry* Anal. Chem. **40** 586–593. Detection limits and different definitions.

Goldstein, JI, Costley, JL, Lorimer, G. and Reed, SJB 1977 *Quantitative X-ray Microanalysis in the Electron Microscope SEM 1977* **1** 315–325 Ed. O Johari IITRI Chicago IL. Seminal paper.

Lyman, CE, Goldstein, JI, Williams, DB, Ackland, DW, Von Harrach, S, Nicholls, AW and Statham, PJ 1994 *High Performance X-ray Detection in a New Analytical Electron Microscope* J. Microsc. **176** 85–98. Description of what is still the best X-ray analysis instrument in the world.

Lyman, CE and Michael, JR 1987 *A Sensitivity Test for Energy-dispersive X-ray Spectrometry in the Analytical Electron Microscope* in *Analytical Electron Microscopy-1987* 231–234, Ed. DC Joy, San Francisco Press San Francisco CA. Spatial resolution versus sensitivity limits.

Ohshima, K, Kaneko, K, Fujita, T and Horita, Z 2004 *Determination of Absolute Thickness and Mean Free Path of Thin Foil Specimen by ζ-Factor Method* J. Electron Microsc. **53** 137–142.

Watanabe, M and Williams, DB 1999 *Atomic-Level Detection by X-ray Microanalysis in the Analytical Electron Microscope* Ultramicrosc. **78** 89–101.

Watanabe, M and Williams, DB 2006 *The Quantitative Analysis of Thin Specimens: a Review of Progress from the Cliff-Lorimer to the New ζ-Factor Methods* J. Microsc. **221** 89–109. The ζ-factor approach for quantification, mapping, thickness determination, and everything else.

Ziebold, TO 1967 *Precision and Sensitivity in Electron Micro-probe Analysis* Anal. Chem. **39** 858–861. Just what it says!

URLs

1) http://montecarlomodeling.mcgill.ca/software/winxray/contacts.html

SELF-ASSESSMENT QUESTIONS

Q36.1 Why do we spend so much time discussing the spatial resolution of XEDS?
Q36.2 Define R, b, and d.
Q36.3 Why are there so many variable definitions of the spatial resolution in the literature?
Q36.4 What's the most important factor controlling the spatial resolution?
Q36.5 Why do you sometimes have little control over this specific factor?
Q36.6 Why is it challenging to measure the spatial resolution experimentally?
Q36.7 List the various methods of determining your specimen thickness and under each method list its most important advantage and its greatest disadvantage.
Q36.8 What's the most important factor in controlling the detection limits in any experiment?
Q36.9 What's the difference between the MMF and the MDM?
Q36.10 Why is a high peak (P) intensity more important than a high P/B ratio when trying to improve the detection limit?
Q36.11 Why do we choose 90% of the exit electron distribution to define the spatial resolution? Why not choose 100%? Why not choose 50% which is commonly used when calculating probe-limited image resolution?
Q36.12 Why is an interphase interface often chosen as the ideal feature across which to measure the experimental spatial resolution of analysis?
Q36.13 Can you suggest other specimens that might offer similar advantages?
Q36.14 Why is an FEG the best electron source to use if you want the highest spatial resolution?
Q36.15 Is an FEG necessarily the best source to use if you want to obtain the highest analytical sensitivity? If not, why not?
Q36.16 What improvements might be gained in both spatial resolution and analytical sensitivity from using an aberration-corrected electron probe?
Q36.17 Why isn't it a good idea to rely on the contamination-spot method to estimate your specimen thickness?
Q36.18 Define the decision limit, the detection limit, and the determination limit.
Q36.19 If a 300-keV FEG AEM can detect 0.01 wt% of an element in a foil ~5 nm thick, estimate how many atoms this represents in the analyzed volume. State any assumptions; be brief.

TEXT-SPECIFIC QUESTIONS

T36.1 Why does a higher voltage give higher spatial resolution in a thin specimen in the AEM but lower spatial resolution in a thick, EPMA specimen as shown in Figure 36.1?
T36.2 Figure 36.1A does not take into account electron-diffraction effects; why does this not seriously compromise our estimation of spatial resolution?
T36.3 The situation in Figure 36.3 assumes that all the incident beam is confined on the entrance surface of the specimen in a circular probe of diameter b. List several factors that can make this assumption unreasonable. (Hint: go back and look at Figure 33.5.)
T36.4 Look at Figure 36.4. Is there any advantage to be gained by moving the interaction cones closer together (i.e., taking more point analyses in the profile)? (Hint: go and look at equation 36.3.)
T36.5 Estimate, from Figure 36.4, the maximum angle at which the interface could be tilted before the spatial resolution profile is degraded beyond the usual experimental limits.
T36.6 Tilting the specimen (see Figure 36.6) degrades the spatial resolution. What other disadvantages occur when the specimen is tilted, and under what analytical conditions is there an advantage to tilting the specimen?
T36.7 To a first approximation, calculate what it would take to detect a single atom of element B in the analyzed volume of element A such as in Figure 36.10. State any assumptions.
T36.8 If you compare the left and right diagrams in Figure 36.10, it is clear that if a FEG AEM is to exhibit comparable analytical performance to an EPMA, it has to be millions of times more sensitive, since the analyzed volume is smaller by such a factor. Indicate what technical differences exist between the two techniques such that this extraordinary improvement in signal detection and generation actually occurs.
T36.9 Explain clearly why the trend in Figure 36.11 is common to all microanalysis techniques, i.e., improving spatial resolution invariably results in degrading the minimum detection limit.
T36.10 Which method(s) would you use to determine the thickness of (a) SiO_2 glass, (b) SiO_2 crystal, (c) Cu-4% Al? Justify your choice of method in each case.
T36.11 Use DTSA to determine the minimum amount of P impurity detectable in a spectrum from otherwise pure Fe. Your specimen is 100 nm thick, and you are operating at 200 kV with a take-off angle of 20°. Courtesy M. Watanabe
T36.12 Using the results of the previous question (MMF of P in Fe), calculate a number of detectable P atoms in Fe (MDM) for a LaB_6-AEM (incident beam size, $d = 10$ nm) and a FEG-AEM ($d = 2$ nm). For the density and the atomic weight to calculate the beam broadening, use the values for Fe, since the detection limits of P are low enough to ignore any effects of this element (if you have answered correctly). Courtesy M. Watanabe

CHAPTER SUMMARY

37

Electron Energy-Loss Spectrometers and Filters

CHAPTER PREVIEW

Electron energy-loss spectrometry (EELS) is the analysis of the energy distribution of electrons that have come through the specimen. These electrons may have lost no energy or may have suffered inelastic (usually electron-electron) collisions.

These energy-loss events tell us a tremendous amount about the chemistry and the electronic structure of the specimen atoms, which in turn reveals details of their bonding/valence state, the nearest-neighbor atomic structure, their dielectric response, the free-electron density, the band gap (if there is one), and the specimen thickness.

In order to examine the spectrum of electron energies, we invariably use a magnetic-prism spectrometer. As we saw with X-rays, it is now common to form images of the various EELS signals in addition to gathering spectra and to do this we use an energy filter, which is based on the same magnetic-prism concept. Energy-filtered TEM (EFTEM) is perhaps the most powerful AEM technique, as should become apparent. The magnetic prism/energy filter is a highly sensitive device with an energy resolution <1 eV, even when the electron-beam energy is as high as 300 keV.

In this chapter, we'll describe the operational principles of both the spectrometer and filter, how to focus and calibrate them, and how to determine the collection semi-angle (β). This angle affects the quality and interpretation of much of your experimental data. In subsequent chapters, we'll go on to look at the spectra and images in more detail and the information they contain.

The EELS technique is an excellent complement to, and is now more widely used than the somewhat simpler XEDS, since it offers substantially more information than mere elemental identification and is well suited to the detection of light elements, which are difficult to analyze with XEDS.

37.1 WHY DO EELS?

So why should we do EELS when XEDS can identify and quantify the presence of all elements above Li in the periodic table with a spatial resolution approaching a few atoms and analytical sensitivity close to the single-atom level? Well, in fact EELS does even more than XEDS in that it can detect and quantify *all* the elements in the periodic table and is especially good for analyzing the light elements. Furthermore, EELS offers even better spatial resolution and analytical sensitivity (both at the single-atom level) in addition to providing much more than just elemental identification as indicated in the preview. So the next question is, why bother with XEDS at all? The answer to this is that EELS, as you will see, can be a challenging experimental technique; it requires very thin specimens to get the best information and understanding and processing the spectra and images requires somewhat more of a physics background than XEDS.

The main point that you need to understand is that XEDS and EELS are highly complementary techniques; most AEMs come equipped with both kinds of spectrometer.

37.1.A Pros and Cons of Inelastic Scattering

When a high-energy electron traverses a thin specimen it can either emerge unscathed or it loses energy by a variety of processes that we first discussed way back in Chapter 4. EELS separates these inelastically scattered electrons into a spectrum, which we can interpret and quantify, form images or DPs from electrons of specific energy, and also combine the spectra and images via spectrum imaging, as we've already seen for XEDS (see Section 33.6.C). Throughout the book, we've already seen some contrasting aspects of inelastic scattering

- Kikuchi lines and HOLZ lines occur in DPs; the electrons in these lines are diffracted very close to the Bragg angle, and give us much more accurate crystallographic information than the SAD/CBED (spot/disk) pattern. In thick specimens, many of the electrons in these lines are inelastically scattered, so thicker specimens can be useful.
- Conversely, the background intensity that surrounds the direct beam in DPs, obscuring faint spots in SADPs and fine detail in CBDPs is due to inelastic scatter. So if you can remove (i.e., filter out) these electrons, you can considerably enhance the quality of your DPs.
- Chromatic aberration is due to energy-loss electrons following different paths through the objective lens and limits the image resolution in thick specimens. You can avoid this by using very thin specimens, but if you're stuck with a thick specimen then you can restore the quality of your images by filtering out inelastic electrons thus removing specimen-induced chromatic-aberration effects.
- Specimen damage, which is usually undesirable, is often caused by inelastic interactions and there isn't much you can do about this, as we discussed back in Chapter 4.
- In Chapter 4, we also discussed many of the other ways that electrons could lose energy going through the specimen, producing X-rays and other phenomena such as plasmons and phonons. These and other interactions produce energy-loss electrons, which contain useful information about the electronic structure and elemental makeup of your specimen.

In fact, if your AEM has an energy-loss spectrometer or filter, then inelastic scattering, in general, is something you should want to happen in your specimens because, if it didn't happen, TEM would be a much less useful technique and this book would be several chapters shorter (so we at least are happy).

The technique of EELS predates X-ray spectrometry. In fact, the experimental pioneers of EELS, Hillier and Baker (1944), were the same two scientists who first proposed and patented the idea of X-ray spectrometry in an electron-beam instrument, similar to the EPMA. If you want to read a brief history of the technique, see the classic book by Egerton. We'll refer to the second edition of Egerton's text on many occasions. In contrast to X-ray analysis, EELS was relatively slow to develop, but is now perhaps the dominant spectrometry technique on probe-forming TEMs for the reasons we gave at the start of the chapter.

Once you've finished this set of chapters, you should find Egerton's book to be highly informative and likewise, the multi-author text on energy-filtered imaging, edited by Reimer. Every 4 years since 1990, there has been an international, focused workshop on EELS and related techniques in the TEM (URL #1) and the proceedings have been published in special issues of various EM journals, as noted in the reference list.

37.1.B The Energy-Loss Spectrum

Let's start as we did for XEDS by looking at a typical spectrum such as Figure 37.1. We won't go into detail here but will save that for the subsequent chapters. For the time being, it's worth pointing out that we split up the spectrum into the low-loss and high-loss regions, with ~ 50 eV being the somewhat arbitrary break point. It isn't particularly apparent from the figure, but we'll see later that the low-loss region contains electronic information from the more weakly bound conduction and valence-band electrons, while the high-loss region contains primarily elemental information from the more tightly bound, core-shell electrons and also details about bonding and atomic distribution. For the time being, you should note that

- the zero-loss peak is very intense, which can be both an advantage and a hindrance
- the intensity range is enormous; this graph uses a logarithmic scale as the only way to display the whole spectrum
- the low-loss regime containing the plasmon peak (see Chapter 38) is relatively intense

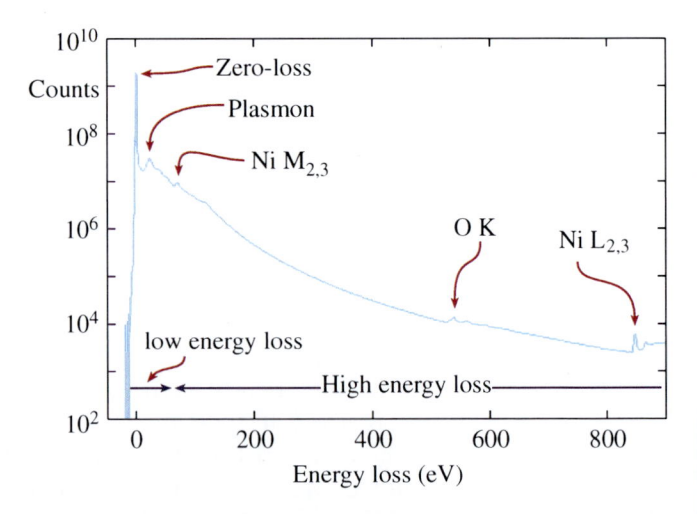

FIGURE 37.1. An EELS spectrum displayed in logarithmic intensity mode. The zero-loss peak is an order of magnitude more intense than the low energy-loss portion (characterized by the plasmon peak), which is many orders of magnitude more intense than the small ionization edges identified in the high energy-loss range. Note the relatively high (and rapidly changing) background.

- the element-characteristic features called ionization edges (see Chapter 39), are relatively low in intensity compared to the background
- the overall signal intensity drops rapidly with increasing energy loss, reaching negligible levels above ~2 keV, which really defines the energy limits of the technique (and this is about the energy when XEDS really comes into its own, emphasizing again their complementarity).

37.2 EELS INSTRUMENTATION

Throughout this and the subsequent chapters, we'll distinguish between spectrometers and energy filters (so called because they filter out electrons of specific energy). The former primarily produce spectra and are great for the dedicated spectroscopist. The latter can produce spectra but are designed to create images and so are more useful to microscopists who find spectra boring and who want filtered images and DPs to compare with standard ones produced by their TEM. There is only one kind of spectrometer commercially available, manufactured by Gatan, Inc. termed a parallel-collection EELS or PEELS. The PEELS is a magnetic-prism (sometimes called a magnetic-sector) system and is mounted on a TEM or STEM after the viewing screen or post-specimen detectors. There are two kinds of filters which result in radically different instruments which, nevertheless, perform similar functions. We'll describe these two types in some detail throughout the chapter.

The post-column Gatan Image Filter (GIF) is a development of their magnetic-prism PEELS. The in-column filter, a magnetic variant of the original Castaing-Henry magnetic prism/electrostatic mirror, is exemplified by the Omega (Ω) filter, pioneered by Zeiss and also now used by JEOL. The in-column filter, as the name implies, is integrated into the TEM and sits between the specimen and the viewing screen/detector (rather than being an optional addition, like the PEELS/GIF). Magnetic spectrometers, along with electrostatic or combined electrostatic/magnetic systems, have been the subject of serious reviews by Metherell and Egerton. If you're an instrument enthusiast, you should try to read these articles but if you just want a more concise summary, then try Egerton's chapter in Ahn's edited book.

ENERGY FILTERS

Two types of commercial filters are presently manufactured: the post-column filter and the in-column filter.

Surface scientists use electron spectrometers to measure exceedingly small (meV) energy losses in low-energy electron beams reflected from the surfaces of samples in UHV instrumentation, such as XPS and Auger systems. We will deal only with transmission EELS studies of high-voltage TEM beams. Recent advances in monochromators for electron guns, combined with spherical and chromatic aberration correction, mean that energy resolution <100 meV is now possible and in the near future this will open a whole new field of TEM-based EELS studies, complementing the established, surface-science techniques but with much better lateral resolution.

Whether you use a Gatan post-column PEELS, a GIF, or an Ω filter, the electron passes through one or more magnetic spectrometers, so we'll start by discussing the principles of this basic tool before moving on to the more complex filtering systems.

37.3 THE MAGNETIC PRISM: A SPECTROMETER AND A LENS

The magnetic prism spectrometer is preferred to an electrostatic or combined magnetic/electrostatic spectrometer, for several reasons

- It is compact and easily interfaced to the TEM. (Remember the WDS problem.)
- It offers sufficient energy resolution to distinguish spectra from all the elements in the periodic table and so is ideal for analysis.
- Electrons in the energy range 100–400 keV, typical of AEMs, can be dispersed sufficiently to detect the spectrum electronically, without limiting the energy resolution.

The basic PEELS-TEM interface and ray paths are shown in Figure 37.2 and a picture of a Gatan spectrometer (actually a combination imaging spectrometer), which would be installed beneath the camera system of a TEM, is shown in Figure 37.3. Because these spectrometers are so widespread, many of the numerical values in this chapter are taken from the Gatan literature (see URL #2). For the details of operation you should, of course, read the instruction manual and Brydson gives a concise summary of all the experimental steps.

From Figure 37.2B, you can see that electrons are selected by a variable entrance aperture (diameters: 1, 2, 3, or 5 mm in the Gatan system). (Obviously, you have to make sure the screen is raised and any on-axis detectors or cameras are removed in order to detect the spectrum.) The electrons travel down a 'drift tube' (Figure 37.2A) through the spectrometer and are deflected through ≥90° by the magnetic field. Electrons that have lost

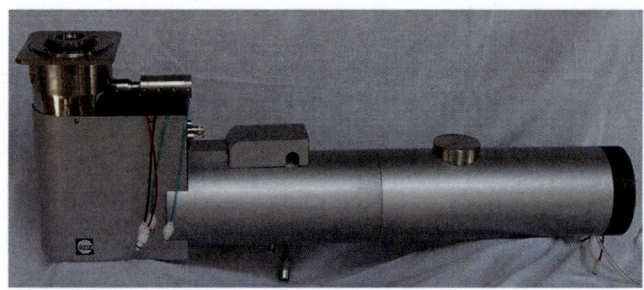

FIGURE 37.3. A Gatan Tridiem PEELS which interfaces below the viewing screen of an AEM.

> **NOTATION**
> Although we've consistently used the letter E for energy, energy-loss should, therefore, be denoted by ΔE since it's a change in energy. However, it is a convention in the EELS literature to use E interchangeably for both an energy loss (e.g., the plasmon loss E_p) and a specific energy (e.g., the critical ionization energy E_C). So we will use $\mathcal{E}$ (note the different font) but remember, it really means a **change** in E.

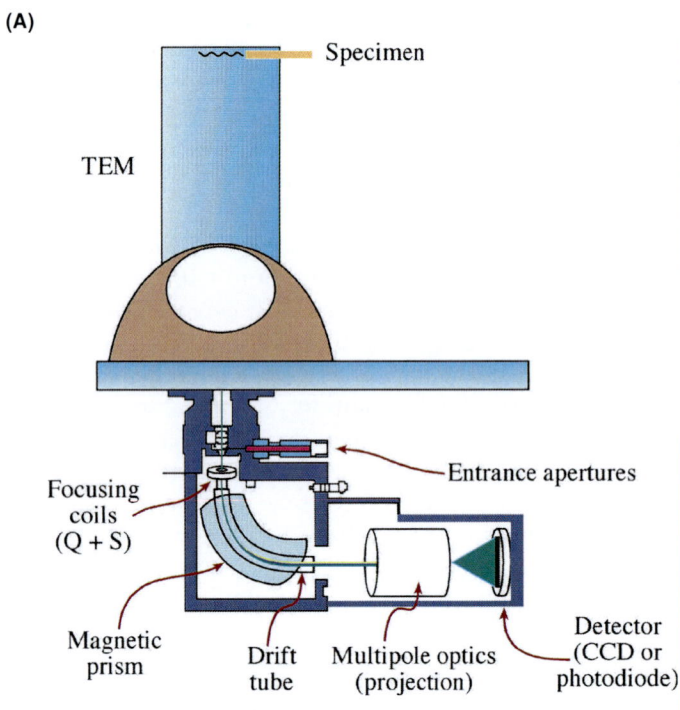

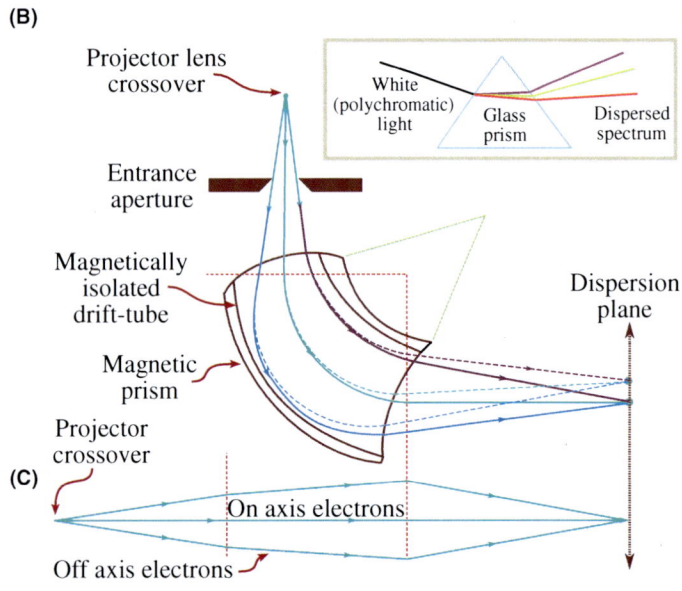

FIGURE 37.2. (A) Schematic diagram showing how a PEELS is interfaced below the viewing screen of a TEM and the position of the various components. (B) Ray paths through a magnetic prism spectrometer showing the different dispersion and focusing of the no-loss and energy-loss electrons in the image (dispersion) plane of the spectrometer. The inset shows the analogy with the dispersion of white light by a glass prism. (C) The lens focusing action in the plane normal to the spectrometer.

Now if you look at Figure 37.2B, you'll see that electrons suffering the same energy loss but traveling in both on-axis and off-axis directions are also brought back to a focus in the dispersion (or image) plane of the spectrometer. So the prism also acts as a magnetic lens. This focusing action is not seen in the otherwise-analogous glass prism (if you recall your high-school physics lab, you had to use a post-prism convex lens to focus the spectrum and separate the individual colors (i.e., frequencies/energies)). We'll give many examples of EEL spectra in the subsequent chapters.

37.3.A Focusing the Spectrometer

Because the spectrometer is also a lens, you have to know how to focus it, and how to minimize the aberrations and astigmatism that are inherent in any magnetic lens. The latest spectrometers are fully corrected for third-order aberrations and the alignment, compensation for stray-AC fields, and focusing are all software controlled. So read the manual, because we will not reproduce it here, merely describe the principles.

The spectrometer has to focus the electrons because off-axis electrons experience a different magnetic field to on-axis electrons. The spectrometer is an axially *asymmetric* lens unlike the other TEM lenses. The path length of off-axis electrons through the magnet also varies, and the magnet has to be carefully constructed to ensure correct compensation for different electron paths so that focusing occurs. This correction is achieved by machining the entrance and exit faces of the spectrometer so they are not normal to the axial rays, as shown in

energy are deflected further than those suffering zero loss. A spectrum is thus formed in the dispersion plane, consisting of a distribution of electron intensity (I) versus energy loss ($\mathcal{E}$). You can see that this process is closely analogous to the dispersion of white light by a glass prism shown in the inset.

Figure 37.2B. These non-normal faces also act to ensure that electrons traveling out of the plane of the paper in Figure 37.2B are also focused in the dispersion plane, as shown in Figure 37.2C (because it focuses in two planes we call it 'double focusing'). The faces of the prism are curved to minimize aberrations and, like their counterparts within the TEM column, the spectrometer lenses continue to improve as higher-order aberrations are minimized. The many quadrupoles, sextupoles, and other focusing electronics and lenses are not shown, but the length of the spectrometer in Figure 37.3 (and looking ahead to Figure 37.15) gives you some idea of its complexity.

As with any lens, the spectrometer takes electrons emanating from a point in an object plane and brings them back to a point in the image (dispersion) plane. Because the spectrometer is an asymmetric lens, we have to fix both the object distance and image distance if we want to keep the spectrum in focus. The object plane of the spectrometer or filter depends on the detail of the machine you are using

- In DSTEMs with no post-specimen lenses the object plane for a post-column spectrometer is the plane of the specimen.
- In a TEM/STEM, or a DSTEM with post-specimen lenses, the object plane for a post-column spectrometer is the back-focal plane of the projector lens, which can contain either an image or a DP.
- In a TEM/STEM, or a DSTEM with post-specimen lenses, the object plane for an in-column filter is the back-focal plane of the first projector (or intermediate) lens, which can contain either an image or a DP.

In a TEM, the projector-lens setting is usually fixed, so the object plane is fixed and the manufacturer usually sets this plane to coincide with the differential pumping aperture separating the column from the viewing chamber. In some DSTEMs there are no post-specimen lenses so the object plane of the spectrometer is the plane of the specimen. In this case, it is essential that you keep your specimen height constant.

In practice, the back-focal plane of the projector may move slightly as you change operating modes (for example, from TEM to STEM) and so you have to be able to adjust the spectrometer. You do this by looking at the electrons that come through the specimen without losing any energy. These electrons have a Gaussian-shaped intensity distribution called the zero-loss peak (ZLP), which we'll talk about more in the next chapter. You can see the ZLP on the computer display of the EELS system and the software focuses it by adjusting a pair of pre-spectrometer quadrupoles until it has a minimum width and maximum height.

The consistency of the information passed through the spectrometer is described by the transmissivity, which is the imaged area (object radius) as a function of the solid scattering angle for a given energy resolution. A perfect spectrometer would uniformly transmit electrons of a specific energy loss over the whole spectrum or, more importantly, over the whole image. If the ZLP is scanned across the spectrometer slit, then a uniform intensity should fall on the detector. Deviations from uniformity reflect the aberrations in the spectrometer; see the paper by Uhlemann and Rose for the details.

37.3.B Spectrometer Dispersion

We define the dispersion as the distance in the spectrum (dx) between electrons differing in energy by dE. It is a function of the strength of the magnetic field (which is governed by the strength (i.e., size) of the spectrometer magnet) and the energy of the incident beam E_0. For the Gatan magnet, the radius of curvature (R) of electrons traveling on axis is about 200 mm, and for 100-keV electrons dx/dE is ~ 2 μm/eV. For PEELS this dispersion value is inadequate and typically electrons with an energy range of about 15 eV would fall on each 25-μm wide diode. Therefore, the dispersion plane has to be magnified $\sim 15\times$ before the spectrum can be detected with resolution closer to 1 eV. This magnification requires post-spectrometer lenses and four quadrupoles are used. The dispersion should be linear across the PDA (photo-diode array); you can check this by measuring the separation of a known pair of spectral features (e.g., zero loss and C K edge) as you displace the spectrum across different parts of the PDA.

37.3.C Spectrometer Resolution

We define the energy resolution of the spectrometer as the FWHM of the focused ZLP and, while it might seem trivial, you should remember to focus the spectrometer every time you acquire a spectrum or filtered image. The type of electron source determines the resolution. As we saw back in Chapter 5 (Table 5.1), at ~ 100 keV a W source has the worst energy resolution (~ 3 eV), a LaB$_6$ is slightly better at ~ 1.5 eV, a Schottky field emitter can give ~ 0.7 eV, and a cold FEG gives the best value of ~ 0.3 eV. These values will all get slightly worse at higher keV. Because of the high emission current from thermionic sources, the energy resolution is limited by electrostatic interactions between electrons at the filament crossover. This electron-electron interaction is called the Boersch effect. We can partially overcome this limit by undersaturating the filament and using only the electrons in the halo. Then a LaB$_6$ source can attain a resolution of ~ 1 eV but at the expense of a considerable loss of current, which we can compensate for by increasing the beam size and/or the C2 aperture. There are other ways you can improve the energy resolution, for example, dropping the kV and the probe current. Figure 37.4A shows data from a cold FEG operated at 200 keV delivering a FWHM of 370 meV (0.37 eV) which is about the best that can be obtained under standard operating conditions.

If you operate at a higher voltage you should also expect a degradation of energy resolution as the kV increases, approximately tripling from 100 to 400 kV.

Because the magnetic prism is so sensitive, external magnetic fields in the microscope room may limit the resolution. If you have an older PEELS, you may see a disturbance to your spectrum if you sit in a metal chair and move around or if you open metal doors into the TEM room, so fine, hand-carved chairs are *de rigeur* for old EELS operators.

> **ENERGY RESOLUTION**
> For comparison: for XEDS it's >100 eV; for EELS it's <1 eV.

The best energy resolution requires a small projector crossover and a small (1 or 2 mm) entrance aperture giving a collection semi-angle of ~ 10 mrads (see Section 37.4.B below). A larger entrance aperture degrades the resolution because the off-axis beams suffer aberrations. The resolution may change as you deflect the ZLP onto different regions of the PDA, although this should not happen if your spectrometer is properly aligned.

37.3.D Calibrating the Spectrometer

We can calibrate the spectrometer (in terms of eV/channel, just like an XEDS) by placing an accurately known voltage on the drift tube, or changing the accelerating voltage slightly, both of which displace the spectrum by a known, fixed amount. Figure 37.4B shows images of the ZLP displaced by a known amount, thus defining both the resolution and the dispersion of the spectrometer at the same time. Usually, calibration is automatically handled by the software but, if you have a really early system, you can (as in XEDS) look for features in a spectrum from a known specimen that occur at specific energies, such as the ZLP at 0 eV and the Ni L_3 ionization edge at 855 eV. Modern electronics are reasonably stable and the calibration doesn't shift substantially but, unless you have the PEELS, which automatically compensates for energy and current drift, you have to check these regularly throughout an operating session since, if shifts of even a few eV occur, they are of the same order as the energy resolution of the spectrometer.

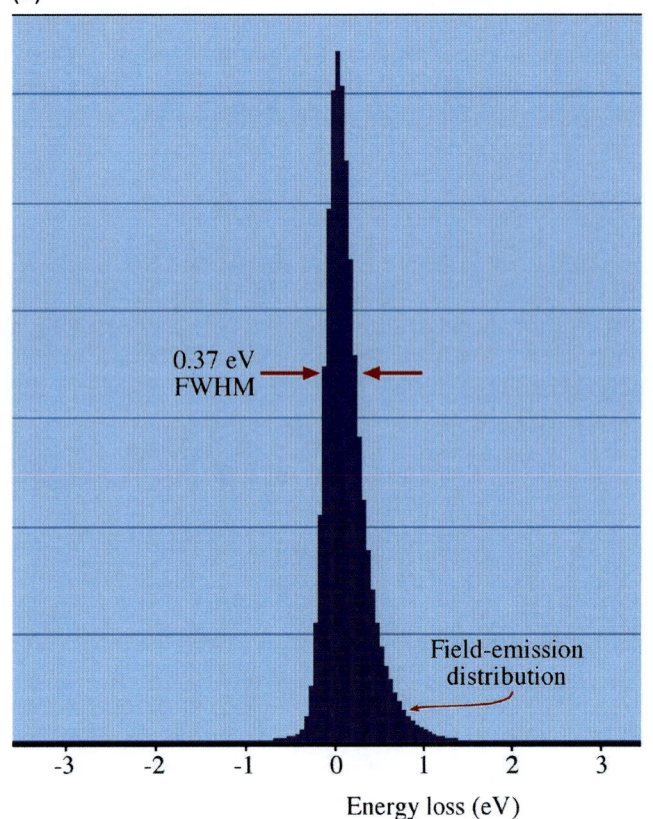

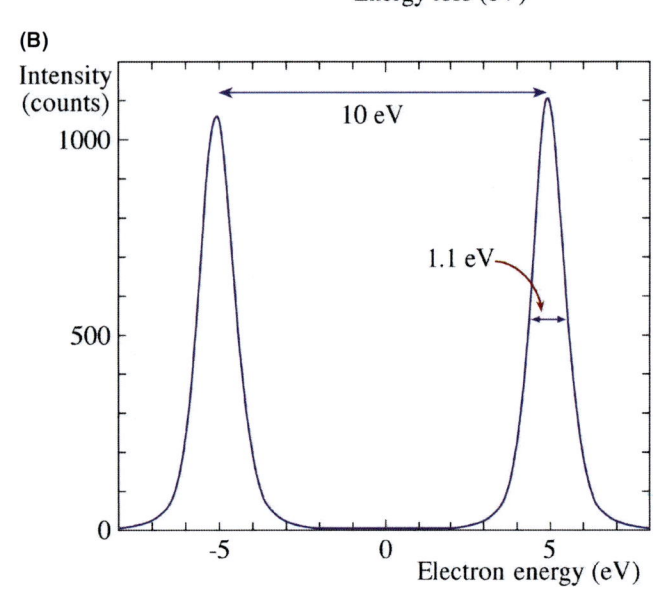

FIGURE 37.4. (A) The energy resolution (0.37 eV) of a cold FEG at 200 keV with 150 pA of current determined from the FWHM of the ZLP. The peak is not symmetrical because of electrons tunneling out from the tip with a slight (< 1 eV) loss of energy. (B) The intensity profile across the ZLP exposed on a CCD camera, then displaced by 10 eV and re-exposed. The resolution, defined by the FWHM, in this case is 1.1 eV and if the number of channels between the peak centroids is counted (e.g., 100 channels) then the dispersion (0.1 eV/channel) is easy to calculate.

The energy resolution decreases slightly as the energy loss increases, but it should be no worse than ~1.5× the ZLP width up to 1000 eV energy loss.

37.4 ACQUIRING A SPECTRUM

To gather a spectrum, such as shown in Figure 37.1, we need a recording device in the dispersion plane of the spectrometer. Historically this was photographic film and early commercial PEELS had a semiconductor PDA but now a CCD is used, in common with both

GIFs and Ω filters (we talked about CCDs in Chapter 7). CCDs show lower gain variation, ~30× better sensitivity, higher dynamic range, and improved energy resolution compared with PDAs. It's worth a brief aside to mention serial EELS or SEELS which was the first commercial method of recording spectra around 1980. While it offered some advantages, SEELS was slow and tedious since each energy loss channel was recorded sequentially and it was rapidly superseded by PEELS which gathers the whole energy range simultaneously. The Gatan PEELS uses a YAG scintillator coupled via fiber optics to a PDA or CCD in the dispersion plane of the spectrometer, as shown in Figure 37.5. Since many PDA-PEELS systems are still in use, it is worth describing its operation and limitations, The PDA consists of 1024 electrically isolated and thermoelectrically cooled Si diodes, each ~25 μm across. The integration time to gather a spectrum can vary from a few msec to several hundred seconds depending on the intensity of the signal. Because each diode saturates at ~16,000 counts you have to select an integration time that avoids saturation during any single acquisition, and you can then sum as many individual integration times as you need to give a spectrum of the desired quality. The CCD detector is 100 × 1340 array of 20 μm pixels and does not saturate rapidly like the PDA so acquisition is much more straightforward. After integration, the whole spectrum is read out via an amplifier through an A/D converter and into the computer display system. The Gatan software offers a variety of standard acquisition conditions suited to different types of spectra.

37.4.A Image and Diffraction Modes

When using any spectrometer or filter in a TEM/STEM, we can operate in one of two modes, and the terminology for this is confusing. If we operate the TEM such that an image is present on the viewing screen then the back-focal plane of the projector lens contains a DP, which the post-column spectrometer uses as its object. From the spectroscopist's viewpoint, therefore, this is termed 'diffraction mode' or 'diffraction coupling,' but from the microscopist's viewpoint, it is more natural to call this 'image mode' since we are looking at an image on the screen. Conversely, if you adjust the microscope so a DP is projected onto the screen (which includes STEM mode in a TEM/STEM), then the (post-column) spectrometer/GIF object plane contains an image, and the terminology is reversed. Likewise, as we'll see for an in-column filter, the back-focal plane of the intermediate lens on which the filter is focused can contain either an image (diffraction mode; there's a DP on the TEM screen) or a DP (image mode; there's an image on the TEM screen).

- The spectroscopist uses the term *image coupling* and the microscopist says *diffraction mode*.
- In this text, *image mode* means an image is present on the TEM screen; naturally, we use the microscopists' terminology. You should *not* use this mode for any spectroscopy, only for EFTEM imaging.
- You should use *diffraction (or STEM) mode* for all spectroscopy and imaging, except for EFTEM imaging.
- Both sets of terms appear in the earlier literature, often without precise definition, so it can be rather confusing, but the microscopists have generally won this conflict.

37.4.B Spectrometer-Collection Angle

The collection angle (as before, we really mean semi-angle) of the spectrometer (β) is a most important variable in several aspects of EELS, so you should know β for all your usual operating modes. If you do gather spectra with different β it is difficult to make sensible comparisons without considerable post-acquisition processing. Poor control of the collection angle is the most common error in quantification, although it is less important the higher the value of β that you use. The detailed intensity variations in the spectrum depend on the range of electron-scattering angles gathered by the spectrometer. Under certain circumstances, the effective value of β can be modified if the beam-convergence

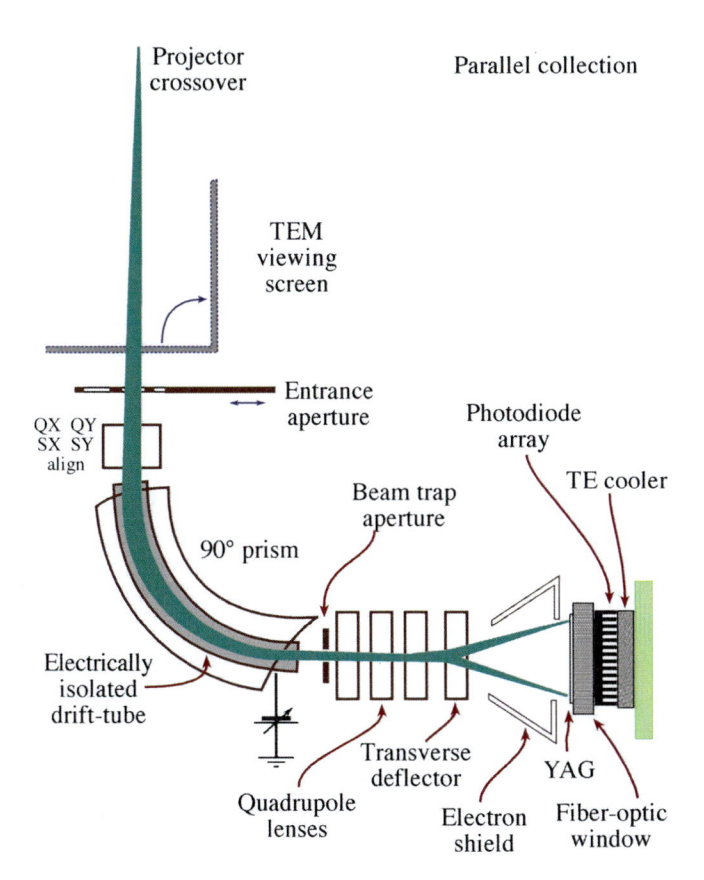

FIGURE 37.5. Schematic diagram of parallel collection of the energy-loss spectrum onto a YAG scintillator fiber-optically coupled to a semiconductor PDA.

angle α is > β, but we'll discuss that when we talk about quantification of ionization edges in Chapter 39. Likewise, we'll see that there is a characteristic or most-probable scattering angle for specific loss processes and typically, β should be 2–3× that angle. So, if in doubt, make β larger. The value of β is affected by your choice of operating mode, and so we will describe how to measure β under different conditions that you may encounter and, while the Gatan software can calculate the value for you, it's good to know the principles behind the black box. Let's start by considering the most simple definition for β, as illustrated in Figure 37.6.

SEMI-ANGLE β
β is the semi-angle subtended at the specimen by the entrance aperture to the spectrometer or filter.

Dedicated STEMs. In a basic DSTEM the situation is straightforward if there are no post-specimen lenses because, as shown in Figure 37.6, β can be calculated from simple geometry. Depending on the diameter (d) of the spectrometer entrance aperture and the distance from the specimen to the aperture (h), β (in radians) is given by

$$\beta \approx \frac{d}{2h} \quad (37.1)$$

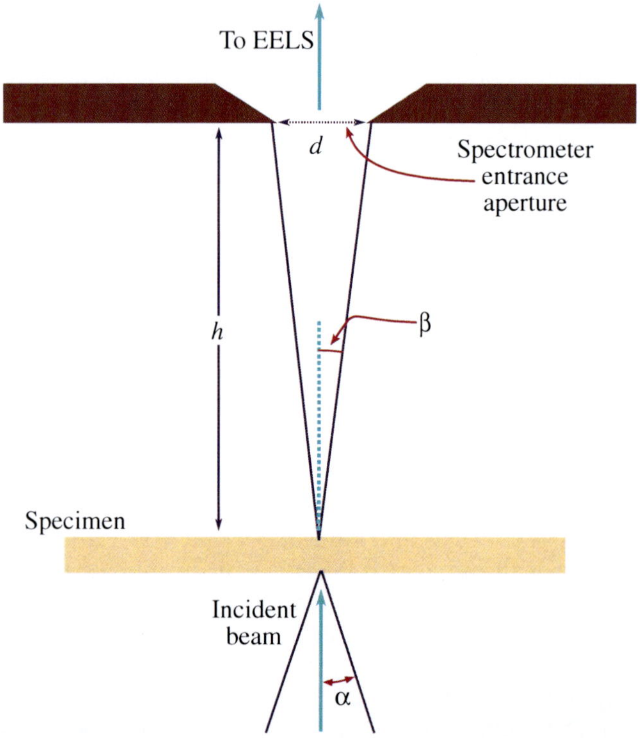

FIGURE 37.6. Schematic diagram showing the definition of β in a DSTEM in which no lenses exist between the specimen and the spectrometer entrance aperture.

This value is approximate and assumes β is small. Since h is not a variable, the range of β is controlled by the number and size of the spectrometer apertures. Therefore, if h is ~100 mm, then for a 1-mm diameter aperture, β is 5 mrads. If there are post-specimen lenses and apertures, the situation is similar to that in a TEM/STEM, as we discuss below.

TEM-image mode. Remember that, in image mode, a magnified image of the specimen is present on the viewing screen or detector. In contrast to what we just described for a dedicated STEM, the angular distribution of electrons entering the spectrometer aperture below the center of the TEM screen is *independent* of the entrance-aperture size. This is because we control the angular distribution of electrons contributing to any TEM image by the size of the objective aperture in the back-focal plane of the objective lens. If we don't use an objective aperture then the collection angle is very large (>~100 mrads) and need not be calculated accurately because we'll see that small differences in a large β do not affect the spectrum or subsequent quantification.

If, for some reason, you do wish to calculate β in image mode with no aperture inserted, you need to know the magnification of the DP in the back-focal plane of the projector lens (which is the front-focal plane of the spectrometer). As you will recall, this magnification is controlled by the camera length L of the DP, and this is given by

$$L \approx \frac{D}{M} \quad (37.2)$$

where D is the distance from the projector crossover to the recording plane and M is the magnification of the image in that plane. So if D is about 500 mm and the screen magnification is 10,000× then L is 0.05 mm. Thus we can show

$$\beta \approx \frac{r_0}{L} \quad (37.3)$$

where r_0 is the maximum radius of the DP in the focal plane of the spectrometer. Typically, r_0 is ~5 μm, and so β is 0.1 rads or 100 mrad which, as we just said, is so large that we rarely need to know it accurately. In fact, in TEM-image mode without an objective aperture, if you just assume β = 100 mrads, any calculation or quantification you do will be effectively independent of β.

If you insert an objective aperture and you know its size and the focal length of the objective lens then β can easily be calculated geometrically. To a first approximation, in a similar manner to equation 37.1, β is the objective-aperture diameter divided by twice the focal length of the objective lens, as shown in Figure 37.7. For example, with a focal length of 3 mm and a 30 μm diameter aperture, β is ~ 5 mrads, which is good if you need high-energy resolution.

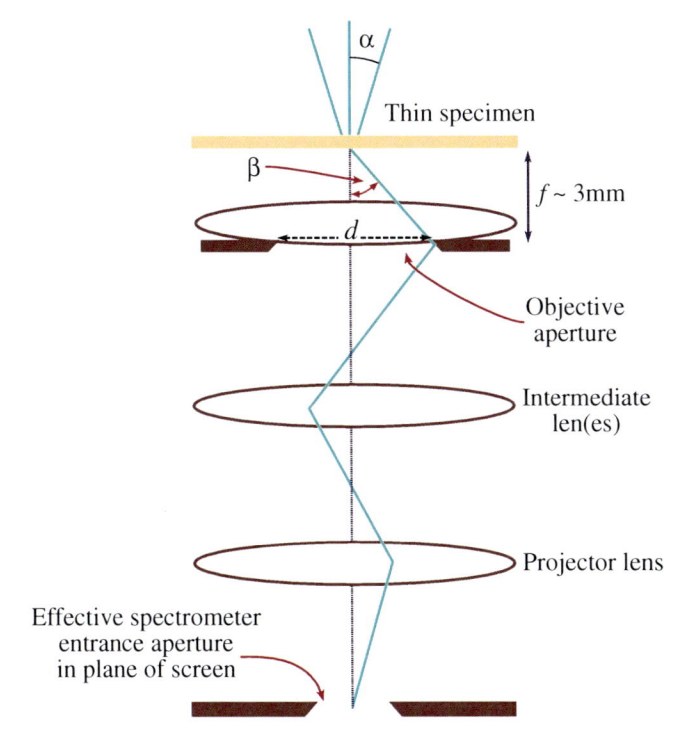

FIGURE 37.7. The value of β in TEM image mode is governed by the dimensions of the objective aperture in the BFP of the objective lens.

If you insert an objective aperture, a normal BF image can be seen on the TEM screen and the information in the spectrum is related to the area of the image that sits directly above the spectrometer-entrance aperture. However, as we'll see in Section 37.4.C, there is some considerable error (~ 100 nm) due to chromatic aberration. We will return to this point in Section 39.10 when we discuss the spatial resolution. Remember also that with the objective aperture inserted, you cannot do XEDS, so simultaneous EELS and XEDS is not possible in this mode.

TEM/STEM diffraction mode. In diffraction (also STEM) mode, the situation is a little more complicated. We focus the spectrometer on an image of the specimen; so we see a DP on the screen and also in the plane of the spectrometer-entrance aperture. Under these circumstances, you control β by your choice of the spectrometer-entrance aperture.

If a small objective aperture is inserted, it is possible that it may limit β; the effective value of β at the back focal plane of the projector lens is β/M where M is the magnification of the image in the back focal plane of the projector lens.

You have to calibrate β from the DP of a known crystalline specimen, as shown in Figure 37.8. Knowing the size of the spectrometer-entrance aperture, the value of β can be calibrated because the distance (b) that separates the 000 spot and a known hkl maximum is twice the Bragg angle, $2\theta_B$. If the effective aperture diameter in the recording plane (equivalent to the

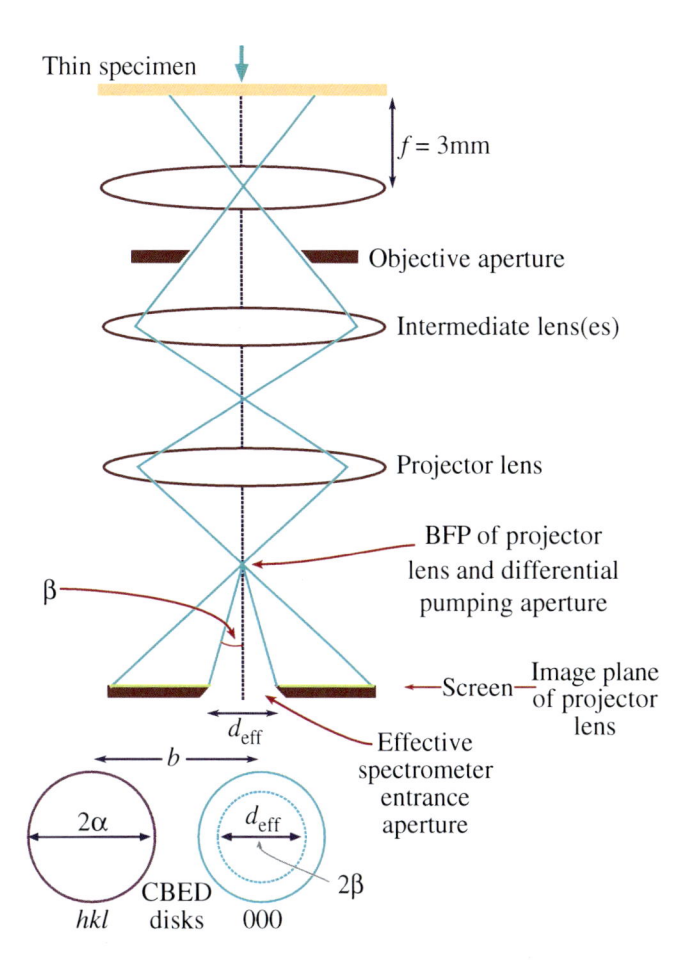

FIGURE 37.8. The value of β in TEM/STEM diffraction mode is determined by the dimensions of the spectrometer entrance aperture, projected into the plane of the DP. The dimensions can be calibrated by reference to a known DP.

STEM detector collection angle back in Section 22.6) is d_{eff} and $b = 2\theta$ then

$$\beta = \frac{d_{\text{eff}}}{2}\frac{2\theta_B}{b} \qquad (37.4)$$

The effective entrance aperture diameter d_{eff} at the recording plane is related to the actual diameter d by

$$d_{\text{eff}} = \frac{dD}{D_A} \qquad (37.5)$$

where D is the distance from the projector crossover to the recording plane and D_A is the distance between the crossover and the entrance aperture. Alternatively, β can be determined directly if the camera length on the recording plane (L) is known since

$$\beta = \frac{D}{D_A}\frac{d}{L} \qquad (37.6)$$

D_A is typically 610 mm for Gatan PEELS systems but D varies depending on the TEM; you have control over d and L. For example, if D is 500 mm and L is 800 mm then for a 5 mm diameter aperture, β is ~ 5 mrads.

37.4 ACQUIRING A SPECTRUM

If we choose a camera length such that the image of the specimen in the back focal plane of the spectrometer is at a magnification of 1×, then, in effect, we have moved the specimen to the focal plane of the spectrometer. This special value of L is equal to D, which you should know for your own microscope. Then β is simply the entrance aperture diameter divided by D_A (610 mm).

In summary, the collection angle is a crucial factor in EELS and the spectrometer/filter software should calculate the values of β in the various operating conditions that you'll encounter.

- Generally, large collection angles will give high intensity but poor energy resolution.
- If you collect your spectrum in image mode without an objective aperture then you won't compromise your energy resolution but spatial resolution is poor, as we'll see below.
- If you're in diffraction mode and you control β with the entrance aperture, then a large aperture (high intensity, high β) will lower the resolution and vice versa.
- Generally, smaller collection angles also give a higher signal-to-noise ratio in the spectrum.

37.4.C Spatial Selection

Depending on whether you're operating in image or diffraction mode, you obtain your spectrum from different regions of the specimen. In TEM-image mode, we position the area to be analyzed on the optic axis, above the entrance aperture. The area selected is a function of the aperture size demagnified back to the plane of the specimen. For example, if the image magnification is 100,000× at the recording plane and the *effective* entrance aperture size at the recording plane is 1 mm, then the area contributing to the spectrum is 10 nm. So, you might think that you can do high spatial resolution analysis without a probe-forming STEM. However, if you're analyzing electrons that have suffered a significant energy loss, they may have come from areas of the specimen well away from the area you selected, because of chromatic aberration. This displacement d is given by

$$d = \theta \, \Delta f \quad (37.7)$$

where θ is the angle of scatter, typically < 10 mrads, and Δf is the defocus error due to chromatic aberration given by

$$\Delta f = C_c \frac{\mathcal{E}}{E_0} \quad (37.8)$$

where C_c is the chromatic-aberration coefficient. So if we take a typical energy loss $\mathcal{E}$ of 284 eV (the energy required to eject a carbon K-shell electron) and we have a beam energy E_0 of 100 keV, then the defocus due to chromatic aberration (with C_c = 3 mm) will be close to 10 μm which gives an actual displacement, d, of 10^{-4} mm or 100 nm. This figure is very large compared to the value of 10 nm, which we calculated without considering chromatic-aberration effects.

In TEM diffraction mode, you select the area of the specimen contributing to the DP in the usual way. You can use either the SAD aperture, which has a lower limit of about 5 μm, or you can form a fine beam as in STEM so that a CBED pattern appears on the screen. In the latter case, the area you select is a function of the beam size and the beam spreading, but is generally < ~50 nm wide. Therefore, this method is best for high spatial resolution EELS, just as for XEDS analysis. But rather than just selecting a single point, it is much better to keep the beam scanning and gather a spectrum at every point (i.e., spectrum imaging; see Section 37.8). This conclusion is true for both in-column and post-column filtering.

> **THE PRICE**
> While TEM-image mode might be good for gathering spectra with a large β and high-energy resolution, the price you pay is much poorer spatial resolution, so we don't recommend it.

37.5 PROBLEMS WITH PEELS

There are several standard tests you need to perform to determine that all is well with the PEELS PDA and electronics, just as we described for XEDS back in Chapters 32 and 33.

37.5.A Point-Spread Function

In a PEELS, you can reduce the magnification of your spectrum so that the ZLP occupies a single PDA channel or pixel on the CCD. Any intensity registered outside that channel is an artifact of the system and is called the point-spread function (PSF). This function acts to degrade the inherent resolution of the magnetic spectrometer. The ZLP may spread on its way through the YAG scintillator and the fiber optics before hitting the PDA or CCD. Figure 37.9A shows the PSF of a PEELS and clearly there is intensity well outside a single channel, although a CCD offers considerably better PSF performance than the PDA. The PSF broadens features in your spectrum such as ionization edges, but you need to remove its effect by deconvolution (see Section 39.6) thus restoring the resolution of the spectrum to that inherent in the beam, in the same way we described for X-ray spectra in Section 34.4, except that the commercial software is available and it makes sense to deconvolute the PSF from any spectrum that you gather. The concept is essentially the same as the point-spread function we discussed for HRTEM. Figure 37.9B and C demonstrates that such energy deconvolution can produce a significant sharpening of the spectrum and, just as in XEDS, there is no reason not to run a ZLP deconvolution routinely to get the best resolution in your spectrum, so long as you are confident (e.g., by testing the software on spectra from known specimens (see Section 39.6)) that such processing does not introduce its own artifacts; which is a good segue into the next section.

37.5.B PEELS Artifacts

Almost all the artifacts in a PEELS system are a consequence of the PDA, which is why CCD detectors have been introduced. If you have a PDA system, then the individual diodes will differ slightly in their response to the incident electron beam and therefore, there will be a channel-to-channel gain variation in intensity. If you spread the beam uniformly over the array using at least the 3 mm aperture and looking at the diode readouts, you can see any variation, as shown in Figure 37.10. You have to divide your experimental spectrum by this response spectrum to remove any gain variation. Alternatively, and this is recommended, you can gather two spectra with slight energy shifts (~1–2 eV) or spatial shifts between them and superimpose them electronically. Using a 2-D CCD array, as in the GIF, also removes this problem.

If you gather many spectra and superimpose them to avoid saturation of the PDA, you'll generate readout noise. Random readout noise or shot noise from the electronics chain is minimized by taking fewer readouts, and by thermoelectric cooling of the PDA. Individual diodes may fail and give high leakage currents which appear as spikes in the spectrum. The fixed-pattern noise is a function of the three-phase readout circuitry. All these effects will appear when there is no current falling on the diodes and together they constitute the dark current (see Figure 37.11). The dark current is

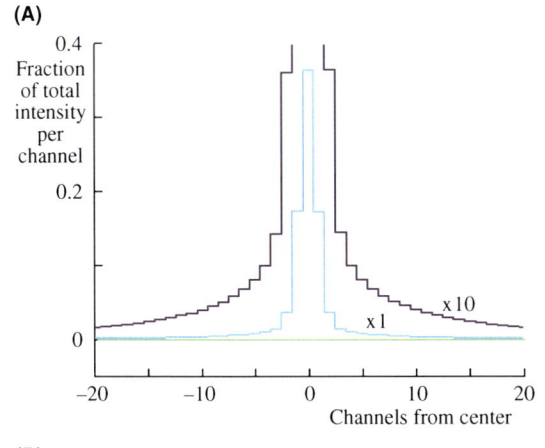

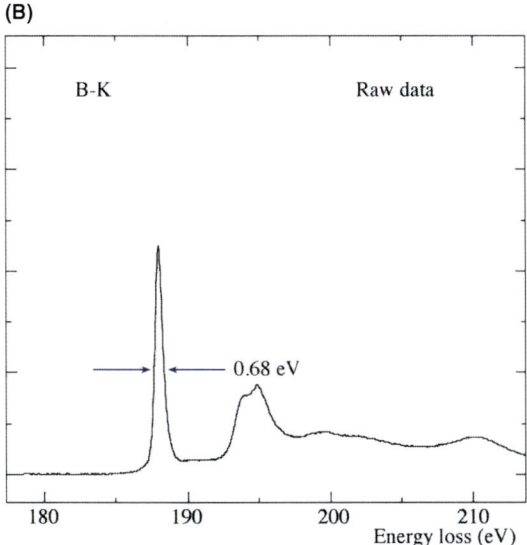

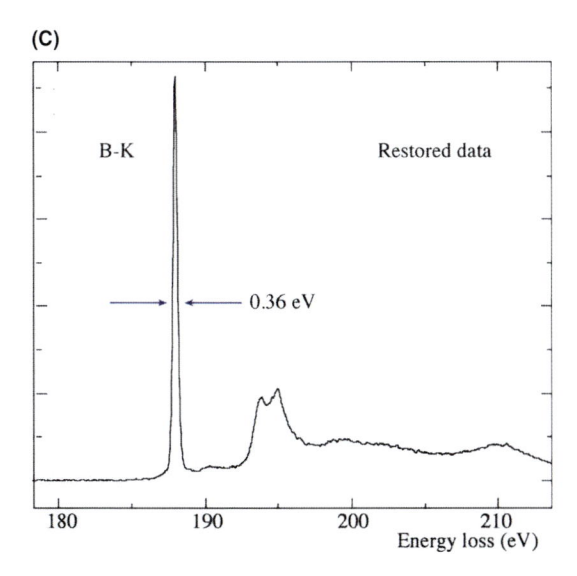

FIGURE 37.9. (A) The point-spread function showing the degradation of the intense well-defined ZLP through spreading of the signal as it is transferred from the scintillator via the fiber optic coupling to the PDA or CCD. The peak should occupy a single channel but is spread across several channels. (B, C) The improvement in energy resolution in the boron K edge from a BN nanotube as a result of ZLP deconvolution. The raw data (B) indicate a resolution of 0.68 eV for the B-K edge and this is improved (C) to 0.36 eV after deconvoluting the point-spread function.

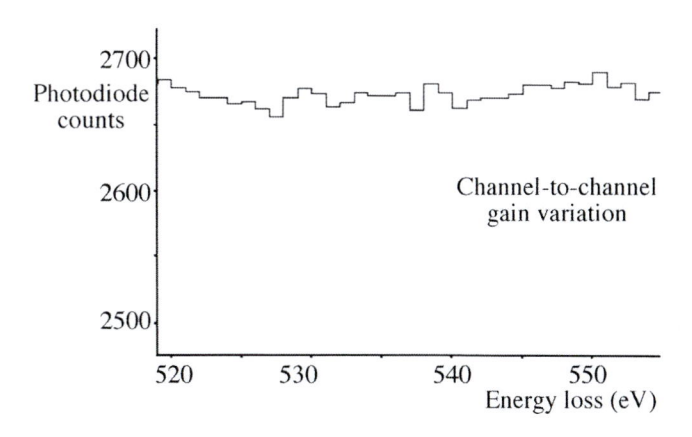

FIGURE 37.10. The variation in the response of individual diodes in the PEELS detection system to a constant incident electron intensity. The channel-to-channel gain variation is clear and each detector array has its own characteristic response function.

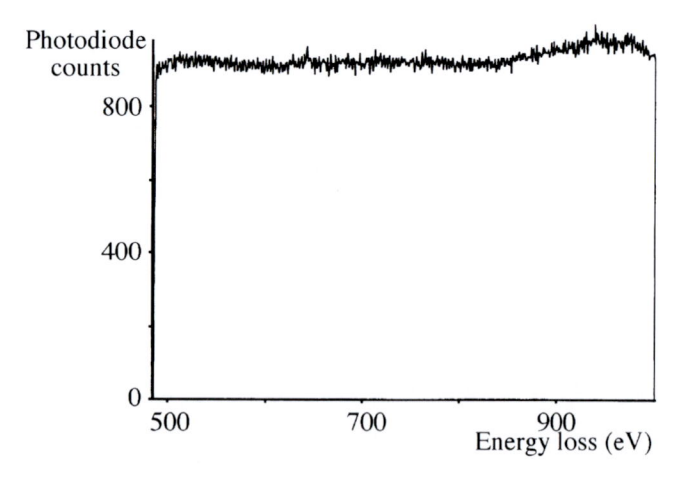

FIGURE 37.11. The intensity of the dark current which flows from the PDA when no electron beam is present.

small unless you have a bad diode and it is only a problem when there are very few counts in your spectrum or you have added together multiple (e.g., 10 or more) spectra. A CCD detector has a higher dark current than a cooled PDA. Figure 37.12 shows some of these effects and how to remove them.

When the diodes are cooled, only ~95% of the signal is read out in the first integration, ~4.5% on the second, ~0.25% on the third, and so on. This incomplete readout can introduce an artifact if you saturated the diodes with an intense signal like the ZLP. This residual peak then shows up as a ghost peak in the next readout and decays slowly over several readouts. So, if a ghost peak appears, just run several readouts and it will disappear; this way you'll never confuse a ghost with a genuine ionization edge or other spectral feature. You can also get a ghost peak on a CCD detector if you overexpose the scintillator.

Increasing the keV means more electrons are generated in the scintillator and the sensitivity should be linearly related to the electron energy. If you're doing quantitative analysis, check that the YAG responds linearly to different intensities by comparing the zero-loss intensity measured in a single 1s readout with that recorded say in 40 readouts each of 0.025 s. In each case, subtract the dark current. Obviously the ratio of these two intensities should be unity, for all levels of signal falling on the YAG. If you're more intent on low-loss spectrometry or fine-structure studies, this non-linearity is not important.

USING A CCD
If you use a PEELS, GIF, or another filter with a CCD detector, all these artifacts are absent, except the ghost peak and dark current.

We can summarize the PDA artifacts and how we eliminate them in Table 37.1.

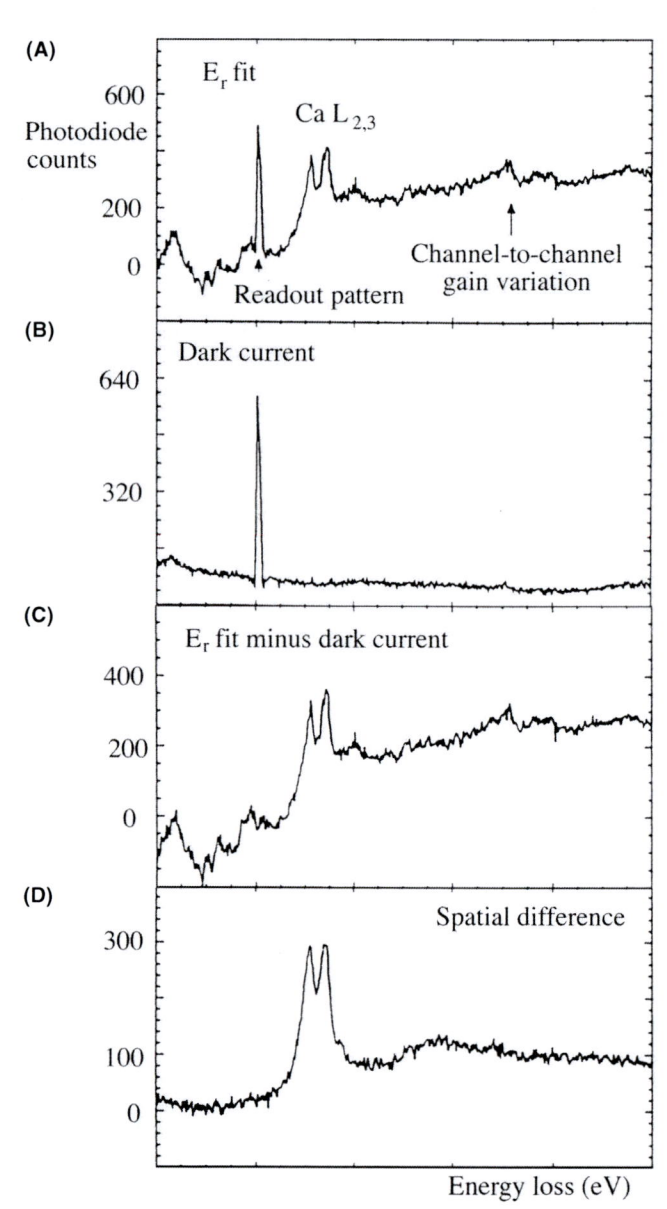

FIGURE 37.12. (A) A Ca $L_{2,3}$ edge spectrum showing both channel to channel gain variation and a faulty diode with a high leakage current which appears as a spike in the spectrum. The spike is referred to as the readout pattern and is present in every recorded spectrum. Subtracting the dark current (shown in (B)) removes the spike (C) and a difference spectrum (D) removes the gain variation, leaving the desired edge spectrum.

37.6 IMAGING FILTERS

Energy-filtered TEM (EFTEM) is sometimes termed energy-filtered imaging (EFI) or electron-spectroscopic imaging (ESI) and is perhaps the most powerful AEM technique, as will become apparent. To perform EFTEM, you basically select (or filter out) electrons of a specific energy coming through the spectrometer and form either an image or a DP. Doing EFTEM is perfectly acceptable in TEM image mode (unlike spectroscopy) so long as you keep the energy slit small to minimize chromatic aberration and, to select the

TABLE 37.1 PEELS Artifacts and How to Eliminate Them

Artifact	Source	Elimination
High leakage current (spike)	Bad diodes	Subtract dark current
Channel to channel gain variation	Different diode responses	Gather spectra on different portions of array and superimpose
Internal scanning noise	Electronics readout	Adjust the electronics and subtract the dark count
Ghost peak	Saturation of diode	Run several readouts
Non-linear response	YAG scintillator is damaged	Zero-loss intensity from different readout numbers with same total integrated time should be identical; if not, replace scintillator

electrons with a specific energy, you change the gun voltage slightly such that those electrons come on-axis and you don't need to keep re-focusing the objective lens.

At the start of the chapter, we noted that there are two types of energy filters: the in-column (Ω) filter and the post-column GIF and both produce EFTEM images. In-column filters are placed in the heart of the TEM imaging system, between the intermediate and projector lenses such that the recording CCD detector only receives electrons that have come through the filter. So all images/DPs consist of electrons of a specific selected energy. (You can, of course, turn off the filter and use the microscope like a normal TEM, but since there are so many advantages to filtered imaging, your reasons for doing this would be questioned.) You can also record a spectrum as we'll show, but the in-column filters are primarily imaging tools and it is both feasible and desirable to operate all the time in filtered mode.

The post-column GIF is added below the TEM viewing screen, just like a PEELS and therefore you can choose to use it or not. The GIF can be seen as either a more flexible instrument or one that limits you to having to decide whether or not to filter your images. Such differences, while perhaps pedantic, have been known to lead to fights in bars at M&M conferences. The best solution, of course, is to have one of each type of filter in your laboratory, but this takes some of the fun out of going to bars at M&M. Let's examine each type of filter in more detail.

37.6.A The Omega Filter

Zeiss first used a mirror-prism system originally devised by Castaing and Henry in 1962. The drawback to the mirror-prism is the need to split the high-voltage supply and raise the mirror to the same voltage as the gun. So Zeiss now uses a magnetic Ω filter, as does JEOL, currently the only other TEM manufacturer with an in-column filter. The filter is placed in the TEM column between the intermediate and projector lenses and consists of a set of magnetic prisms arranged in an Ω shape which disperses the electrons off axis, as shown in Figure 37.13A but, in the end, brings them back onto the optic axis before entering the final projector lens. Although it is not shown, each of the four prisms is machined with curved faces (like Figure 37.2B) to reduce aberrations. There isn't much to see externally on the TEM except an asymmetry on the side of the (somewhat taller) column, which you can also see if you go back and look at the picture of the Zeiss UHRTEM in Figure 1.9.

The multiple steps necessary for ETEM imaging via an in-column filter are shown in Figure 37.13B. As shown in this figure, we usually project an image into the prism which is focused on a DP in the back-focal plane of the intermediate lens (i.e., image mode in our terminology). Therefore, the entrance aperture to the spectrometer selects an area of the specimen and β is governed by the objective aperture. Electrons following a particular path through the spectrometer (the red ones in Figure 37.13B) can be selected by the post-spectrometer slit. Thus, only electrons of a given energy range, determined by the slit width, are used to form the image projected onto the TEM CCD. EFTEM has several advantages over conventional TEM images; there's much more about EFTEM in the companion text.

If you change the focus of the projector lens onto the dispersion plane of the filter (where the energy-selecting slit is located in Figure 37.13B) and then remove this slit, you'll see a spectrum on the viewing screen/CCD. The spectrum appears as a line of varying intensity (see Figure 37.14A), which you can imagine as looking down from above on a traditional spectrum display such as Figure 37.1. Since the spectrum is recorded digitally on the CCD, it is simple to select a line through this spectrum and have the computer display a traditional spectrum of counts versus energy loss, as shown in Figure 37.14B.

As with PEELS, you can also change the TEM optics and project a DP into the prism, thus producing an energy-filtered DP on the CCD, as we've already described back in Chapter 20. If you then use the slit to select a portion of the DP, you get an energy-loss spectrum showing not only the intensity distribution as a function of energy but also the angular distribution of the electrons. Such angular (or momentum)-resolved

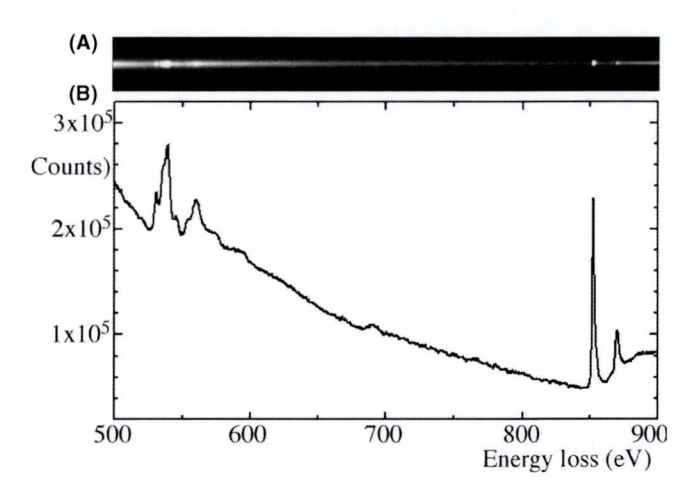

FIGURE 37.14. (A) Spectrum from an Ω filter. The axis of the line is energy loss and the intensity varies along the line. (B) Conventional spectrum of intensity versus energy-loss obtained from (A).

EELS is a whole separate field of study which we'll discuss in Section 40.8.

37.6.B The GIF

The GIF is a Gatan PEELS with an energy-selecting slit after the magnet and a 2D slow-scan CCD detector, rather than a 1D PDA, as the detector. Figure 37.15A shows a schematic diagram of a GIF interfaced to a TEM and Figure 37.15B shows an exploded view of a GIF (which shows the internal parts of the spectrometer shown back in Figure 37.3). Compared to a standard spectrometer, there are many more quadrupoles and sextupoles in the optics of the GIF. The dispersion of the spectrometer onto the slit has to be magnified and the quadrupoles after the slit have two functions. Either we project an image of the spectrum in the plane of the selecting slit onto the CCD or we compensate for the energy dispersion of the magnet and project a magnified image of the specimen onto the CCD. In the first mode, the system is operating like a standard PEELS; in the second, it produces images (or DPs) containing electrons of a specific energy selected by the slit. While the spectrometer is double focusing as we saw, the aberration correction is good only for a single plane so astigmatism is introduced and this has to be corrected by a further combination of sextupoles and octupoles. Obviously, such a large number of variable sextupoles and quadrupoles could be a nightmare to operate without appropriate computer control and this is built into the system software. One potential operational difficulty is that the magnification of the GIF system is such that the actual TEM screen magnification needs to be rather small in order to observe a filtered image with reasonable magnification. More recent AEMs satisfactorily compensate for this magnification differential, but others do not and

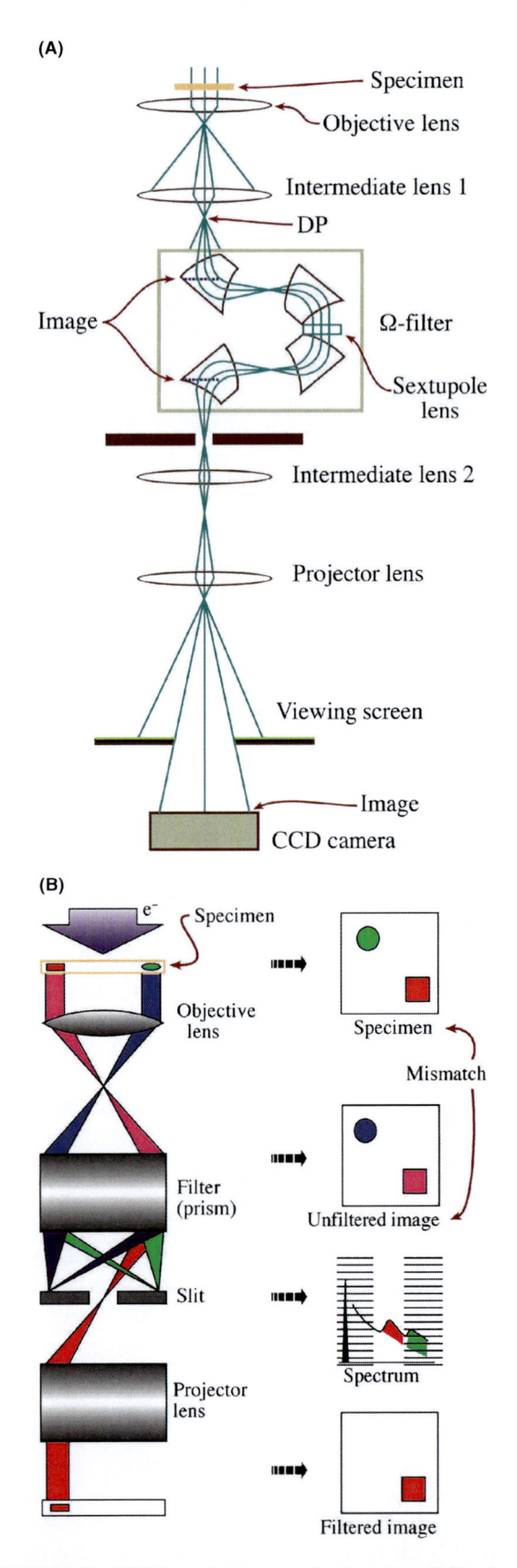

FIGURE 37.13. (A) Schematic diagram of an in-column Ω filter inserted in the imaging lens system of a TEM. (B) Schematic ray diagram of the steps needed to create an EFTEM image.

(A)

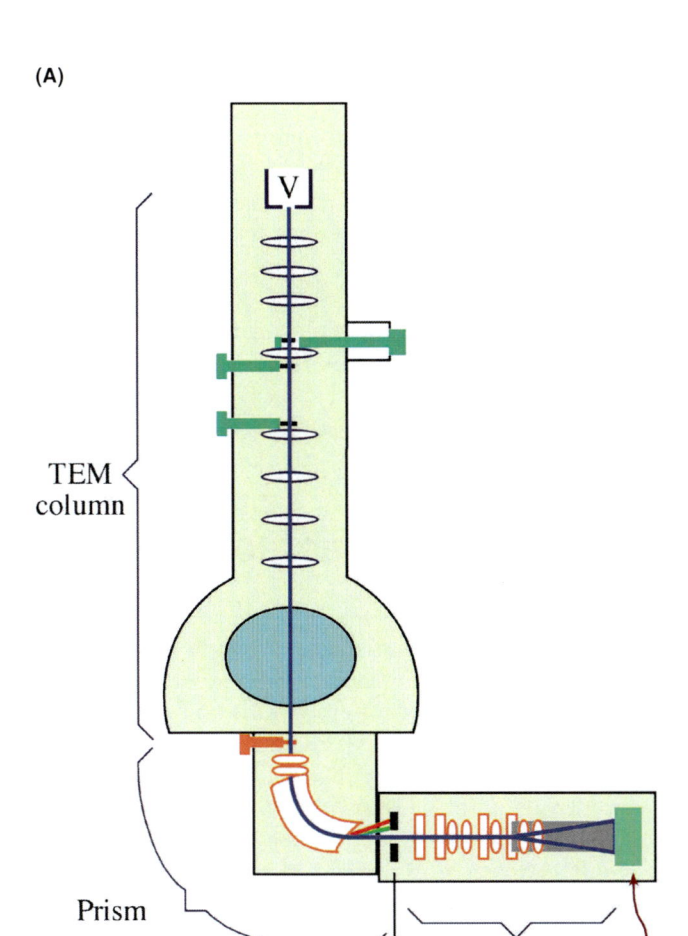

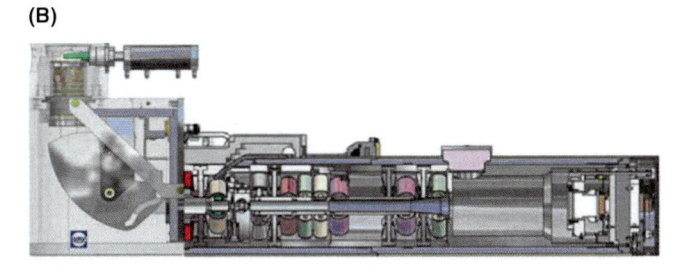

FIGURE 37.15. (A) Schematic diagram of how a post-column imaging filter is attached to the TEM column below the viewing chamber in the same position as a PEELS. (B) Cross section showing the complex inner workings of the Gatan (Tridiem) imaging filter (GIF).

then you have to move between TEM and GIF images and it is not so easy to operate in filtered mode all the time, which would be preferred.

37.7 MONOCHROMATORS

Energy resolution is obviously a key factor in EELS, much more so in fact than in XEDS where we still live with the miserable resolution of the solid-state detector. Resolution of a few eV is more than sufficient for ionization-loss spectrometry, but as we use EELS more to study such aspects as vibrational modes of atoms, inter/intra band excitations, fine structure, and electronic effects, sub-eV resolution is increasingly of interest. For example, Batson at IBM has pioneered high-resolution (HREELS) in a decades-long research project probing ever deeper into the electronic structure of Si and the Si/SiO$_2$ interface and other materials essential for modern electronics.

If your spectrometer is not a limiting factor, then the gun dictates the ultimate energy resolution. Thermionic sources with ~3 eV (W) and 1.5 eV (LaB$_6$) resolution, respectively, might be tolerable for ionization-loss spectrometry and basic low-energy plasmon studies. However, EELS really requires that you have a FEG. Either a cold FEG or a Schottky will give you sub-eV resolution, but this is still not sufficient in some cases. So commercial monochromators have become available which offer resolutions ~100 meV although at very low energy losses (<50 eV) it is even possible to get resolutions of ~25 meV. Basically, a monochromator is an EELS system fitted on the FEG source and the selecting slit refines the already narrow energy spread still further giving remarkably fine detail in the spectra. The monochromator is usually a Wien filter which has perpendicular electrostatic and magnetic fields which permit the chosen electrons to travel in a straight line down the TEM column. (Again, if you want to learn more, go back to Metherell's early review which has it all.) Figure 37.16A shows the ZLP at 200 keV obtained with and without a monochromator showing the reduction in FWHM. From Figure 37.16A, you can see that the FWHM of a commercial monochromated ZLP is much better than the 600 meV you can get with a Schottky gun and a little better than the ~ 300 meV delivered by a good cold FEG. The intensity of the beam tail is relatively low in comparison with that of a standard cold FEG. This low-intensity tail results in real improvements for low-loss spectrometry, as we'll see in the next chapter.

Figure 37.16B compares the quality of different spectra of the Co L$_{2,3}$ edges obtained on a variety of instruments and calculated (see Chapter 40 and the companion text for more details on the calculation). Clearly the degree of fine detail in the spectra improves as the energy range of the electron sources gets smaller. The monochromated FEG is similar in quality to the spectrum from the synchrotron source, if a little noisier. The degree of agreement between these latter two spectra and the calculated spectrum is encouraging.

The drawback to monochromation, of course, is that, in filtering out the tails of the Gaussian energy distribution to reduce the FWHM, we reduce the number of electrons significantly. So having spent serious dollars to get the brightest possible electron source for our AEM, we immediately throw away a large amount of it in pursuit of the best energy resolution. This compromise is a real limiting factor and you should rarely contemplate acquiring an AEM with a monochromator.

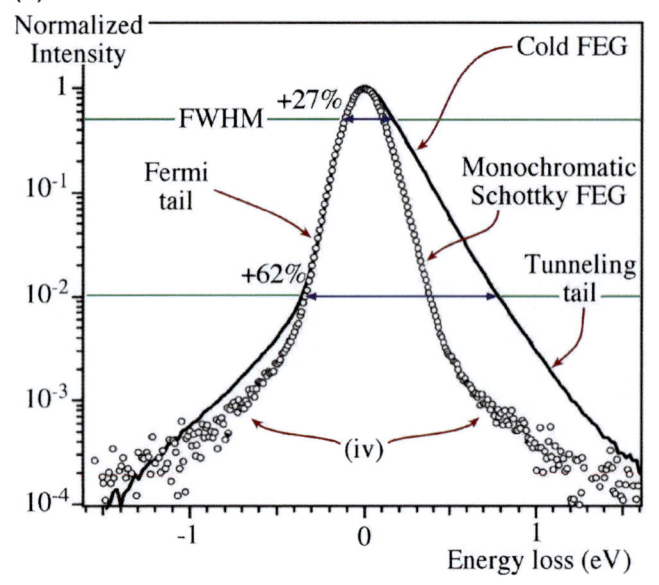

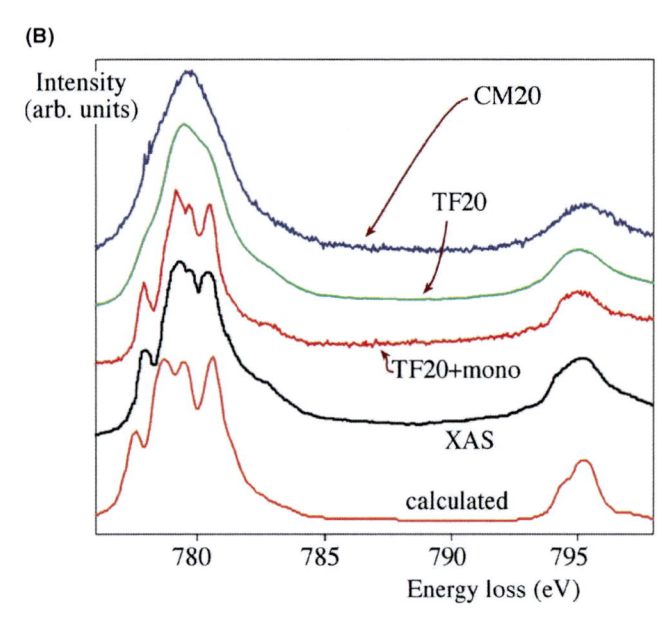

FIGURE 37.16. (A) Typical cold FEG ZLP with and without monochromation. Note that the ordinate is logarithmic so the FWHM is close to the top of the peak. (B) Comparison of Co $L_{2,3}$ spectra from: a Philips CM20 thermionic source, an FEI Tecnai TF20 with a cold FEG, a TF20 with a monochromator, a synchrotron (X-ray absorption spectrum), and a calculated spectrum using crystal-field theory. The improvement in resolution with monochromation is apparent.

> **WAIT BEFORE YOU MONOCHROMATE**
> If HREELS is the *raison d'etre* of the instrument remember that just about all other TEM/AEM operations will be degraded if you switch on the monochromator.

The one thing that you get in EELS is lots of electrons, particularly in the low-loss spectrum. In this case, monochromation is truly an advantage. Of course, it would also help if we could develop even brighter sources than cold FEGs.

There are also software alternatives to physical monochromation of the source and by various deconvolution routines it is possible to obtain spectra with a resolution of about 200 meV (e.g., the papers by Kimoto et al. and Gloter et al.). This approach is a very acceptable alternative to using a monochromator, unless you need the absolute best resolution for exploring the finest details of the electronic structure of nanomaterials (e.g., papers by Kimoto et al. in 2005, and Spence in 2006) when you need to lower both the keV and the beam current to produce sub-100 meV resolution.

37.8 USING YOUR SPECTROMETER AND FILTER

So now you've got a thin specimen in your microscope (you'll see later that for most EELS experiments, *very* thin is much better than merely thin) and you want to start acquiring spectra and filtered images. There are several ways to do this and basically they parallel what we've already described for XEDS

1. *Point analyses*: stop the STEM probe from scanning and position it on a selected point in the image over the entrance aperture and record a spectrum (see Figure 37.17A). This was the standard modus operandi for decades, but is (a) biased (because you pick what *you* think ought to be analyzed), (b) statistically poor (only one pixel selected out of a million or so in a typical STEM image), and (c) prone to damaging/contaminating the chosen area of interest by leaving the intense probe on the same point for long periods of time. So don't do point analyses, unless you're just looking quickly around the specimen to see what you can find. If you do, don't necessarily believe that what you find is significant!

2. *Line analyses*: take a series of spectra along a line that traverses some feature of interest (such as a planar interface or defect such as a grain boundary). This process is still biased by your choice of line but at least has the advantage of focusing on something that should provide useful information about the specimen. You can either plot the information from the spectra (e.g., the composition or the dielectric constant or whatever data you extract) or you can display the data as a spectrum-line, as shown in Figure 37.17B which shows changes in chemistry across a nanotube.

3. *TEM filtered images*: using a GIF or Ω filter, gather a filtered image or DP by using the slit to select electrons of a specific energy thus allowing only those electrons to fall on the viewing screen or CCD. If you use this method to filter out all the energy-loss electrons and produce a ZLP image, you can

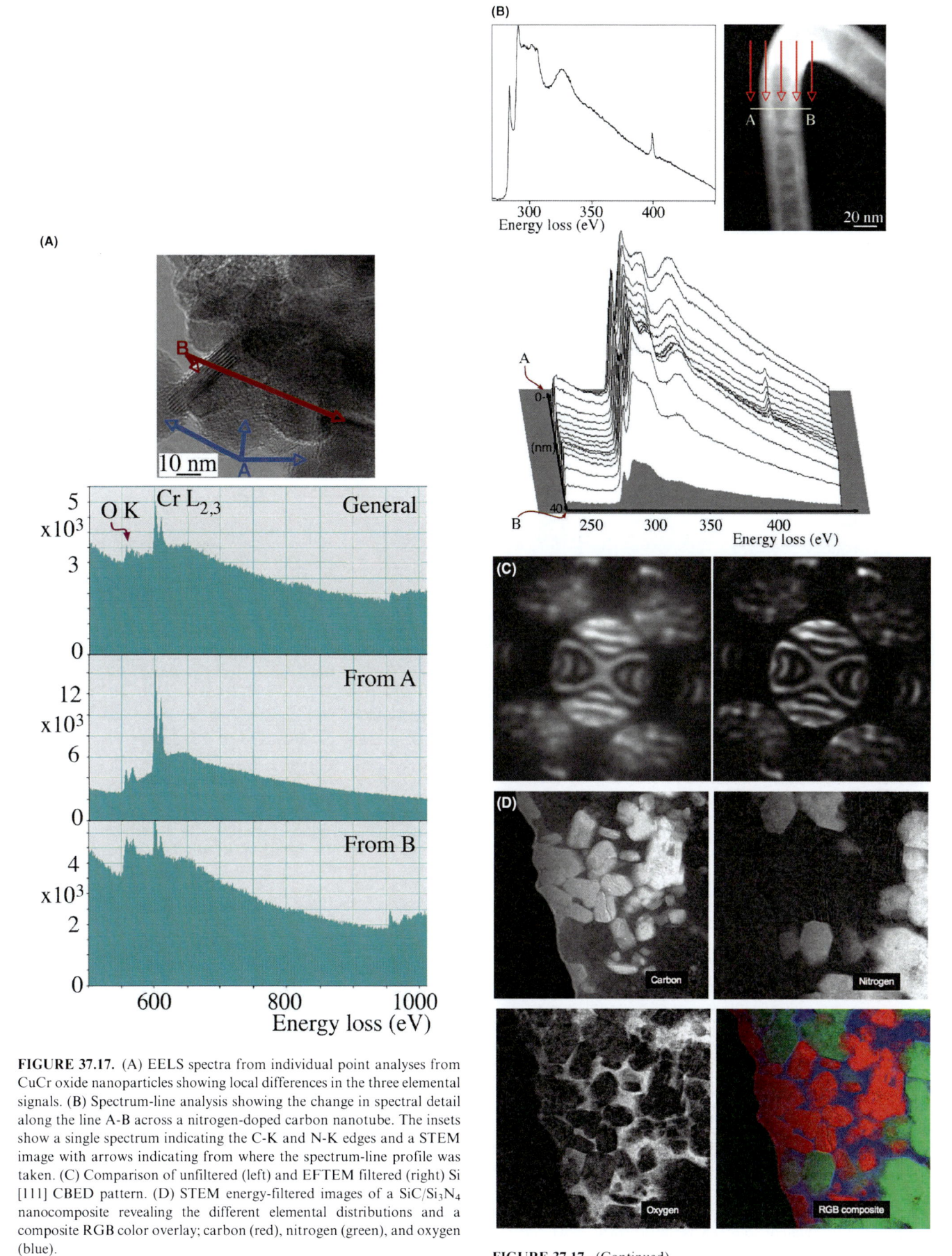

FIGURE 37.17. (A) EELS spectra from individual point analyses from CuCr oxide nanoparticles showing local differences in the three elemental signals. (B) Spectrum-line analysis showing the change in spectral detail along the line A-B across a nitrogen-doped carbon nanotube. The insets show a single spectrum indicating the C-K and N-K edges and a STEM image with arrows indicating from where the spectrum-line profile was taken. (C) Comparison of unfiltered (left) and EFTEM filtered (right) Si [111] CBED pattern. (D) STEM energy-filtered images of a SiC/Si$_3$N$_4$ nanocomposite revealing the different elemental distributions and a composite RGB color overlay; carbon (red), nitrogen (green), and oxygen (blue).

FIGURE 37.17. (Continued).

substantially enhance the quality of CBED patterns, as shown in Figure 37.17C (and also back in Figure 20.10). Also, as we'll see at various times, EFTEM improves mass-thickness contrast, phase contrast, and diffraction contrast. So why wouldn't you use this technique if you have a filter on your AEM?

4. *STEM filtered images*: scan the beam over an area of interest using the slit to select electrons of a specific energy thus allowing only those electrons to fall on the viewing screen or CCD. If you select electrons of a specific energy then you can produce composition maps (Figure 37.17D) similar to many of the XEDS composition maps we have already shown.

5. *Spectrum-images*: store a full spectrum at each pixel (obviously, you need to be operating in STEM to do this). It's much easier to do this in STEM because if you try and do this in TEM mode, you'll have to record hundreds of images at different selected energies and comparing complete images is challenging and time consuming, so not everyone does this; however, this situation is changing. In a similar manner to what we showed you for XEDS, after you've acquired a full spectrum image, you can go into the data cube and select whatever information you want. For example, you can view images at specific energies or spectra from specific points or lines in your specimen, thus removing the bias that affects your choice of spot or line analyses. EELS spectrum imaging was more easily implemented than the XEDS version because of the ease of acquiring large numbers of counts and the broad array of possible images was demonstrated by Hunt and Williams.

We'll use examples of these various methods throughout the next three chapters. Generally speaking, as we've emphasized for X-rays, it makes sense for you to form images of specific features in the spectrum rather than gather spectra from specific points, so that they can at least be compared with your TEM images. Since counts are not a problem in EELS (unlike for XEDS) imaging is relatively straightforward and fast. If you really want to optimize your information then acquire spectrum images.

CHAPTER SUMMARY

We use a magnetic-prism spectrometer for PEELS and a combination of one or more prisms plus imaging lenses for GIF/EFTEM. The prism is a simple device and very sensitive but you have to understand how it functions in combination with different TEM modes, You can operate with your TEM in imaging or diffraction (STEM) mode, or use a DSTEM. You have to know how the software focuses and calibrates the system and how it determines the collection angle, β. Once you can understand this you're in a position to acquire and analyze EEL spectra. So in the next chapter we'll tell you what these spectra look like and what information they contain. If you have an Ω filter or GIF you can routinely form images or DPs with electrons of specific E and it makes sense, if you can, to view all your images and DPs in filtered mode for reasons we've already mentioned in past chapters and that we'll reiterate in the subsequent chapters. Energy filtering is a rapidly evolving field and systems with reduced aberrations and better energy resolution such as the Mandoline filter on Zeiss's Sub-eV-Sub-Ångstrom Microscope (SESAMe) and Nion's UltraSTEM will offer far more exciting breakthroughs than we can describe here and Egerton gave a brief synopsis of a range of hardware and software advances in 2003.

THE INSTRUMENT

Brydson, R 2001 *Electron Energy-Loss Spectroscopy* 2001 Bios (Royal Microsc. Soc.) Oxford UK. Good introductory text, similar in content and level to much of this chapter and the subsequent ones.

Castaing, R. and Henry, L 1962 *Filtrage Magnétique des Vitesses en Microscopie Electronique* C. R. Acad. Sci. Paris **B255** 76–78. The original design for the mirror prism.

Egerton, RF 2003 *New Techniques in Electron Energy-Loss Spectroscopy* Micron **34** 127–139 A concise review of recent advances in instrumentation.

Egerton, RF, Yang, YY and Cheng, SY 1993 *Characterization and Use of the Gatan 666 Parallel-Recording Electron Energy-Loss Spectrometer* Ultramicrosc. **48** 239–250.

Hillier, J. and Baker, RF 1944 *Microanalysis by Means of Electrons* J. Appl. Phys. **15** 663–675. History – it's amazing to read that AEM was essentially all conceived more than 60 years ago.

Metherell, AJF 1971 *Energy Analysing and Energy Selecting Electron Microscopes* Adv. Opt. Elect. Microsc. **4** 263–361 Eds. R Barer and VE Cosslett Academic Press New York. Still the best review of spectrometers.

Uhlemann, S and Rose, H 1996 *Acceptance of Imaging Energy Filters* Ultramicrosc. **63** 161–167.

BACKGROUND DATA

Ahn, CC Ed. 2004 *Transmission Electron Energy Loss Spectrometry in Materials Science* 2nd Ed. Wiley-VCH Weinheim Germany. Updated version of the Disko et al. text (below).

Disko, MM, Ahn, CC and Fultz, B Eds. 1992 *Transmission Electron Energy Loss Spectrometry in Materials Science and the EELS Atlas* TMS Warrendale PA. Multi-author, practical text.

Egerton, RF 1986 *Electron Energy-Loss Spectroscopy in the Electron* Plenum Press New York. The bible of EELS; required reading for all serious spectroscopists.

Egerton, RF 1996 *Electron Energy-Loss Spectroscopy in the Electron Microscope* 2nd Ed. Plenum Press New York. Second edition of the EELS bible.

APPLICATIONS

Batson PE 2004 *Electron Energy Loss Studies of Semiconductors* in Transmission Electron Energy Loss Spectrometry in Materials Science 2nd Ed. 353–384 Ed. CC Ahn Wiley-VCH Weinheim Germany.

Gloter, A, Douiri, A, Tencé, M and Colliex, C 2003 *Improving Energy Resolution of EELS Spectra: an Alternative to the Monochromator Solution* Ultramicrosc. **96** 385–400.

Hunt, JA and Williams, DB *Electron Energy-Loss Spectrum Imaging* Ultramicrosc. **38** 47–73.

Kimoto, K, Kothleitner, G, Grogger, W, Masui, Y and Hofer, F 2005 *Advantages of a Monochromator for Bandgap Measurements Using Electron Energy-loss Spectroscopy* Micron **36** 185–189.

Spence, JCH 2006 *Absorption Spectroscopy with Sub-Angstrom Beams: ELS in STEM* Rep. Prog. Phys. **69** 725–758.

THE EELS WORKSHOP REPORTS

The quadrennial EELS workshops (1990, 1994, 1998, 2002, and 2006) pioneered and often organized and edited by Krivanek are a rich source of papers by the leading researchers in the field describing cutting-edge aspects of EELS and related techniques. Often the proceedings are published separately and are divided between methodology/instrumentation and practice. The respective journals (either complete volumes or part thereof) and editors are listed below.

Krivanek, OL Ed. 1991 Microsc. Microanal. Microstruct. **2** (# 2–3).

Krivanek, OL Ed. 1995a Microsc. Microanal. Microstruct. **6** 1.

Krivanek, OL Ed. 1995b Ultramicrosc. **59** (# 1–4).

Krivanek, OL Ed. 1999 Ultramicrosc. **78** (# 1–4).

Krivanek, OL Ed. 1999 Micron **30** (#2) 101.

Krivanek, OL Ed. 2003 Ultramicrosc. **96** (#2–4) 229.

Krivanek, OL Ed. 2003 J. Microsc. **210** 1.

Browning, ND and Midgley, P Eds. 2006 Ultramicrosc. **106** (#11–12).

Mayer, J Ed. 2006 Micron **37** 375.

URLs

1) http://www.energyloss.com/index.html
2) http://www.gatan.com/ Gatan's web site

SELF-ASSESSMENT QUESTIONS

Q37.1 What's the difference between a GIF and a PEELS system?
Q37.2 Why is it useful for the spectrometer to act as a lens also? Can we achieve the same combination when dispersing visible light?
Q37.3 What is the dispersion plane of the spectrometer and what is a typical value of the dispersion?
Q37.4 How would you measure the energy resolution of the spectrometer?
Q37.5 What's a typical value of the energy resolution and what governs the minimum possible value?
Q37.6 What factors can degrade (i.e., increase the value of) the energy resolution?
Q37.7 Why is the spectrometer collection angle (β) so important?
Q37.8 What controls β in image mode?
Q37.9 What controls β in diffraction mode?
Q37.10 Why might you have to integrate several spectra in your PEELS rather than just acquire a single spectrum?
Q37.11 What controls the spatial resolution in image mode?
Q37.12 What controls the spatial resolution in diffraction mode?
Q37.13 Why is diffraction mode in TEM equivalent to operating a dedicated STEM?
Q37.14 What is the point-spread function and why should you be concerned about it?
Q37.15 How do you correct for this artifact?
Q37.16 Why do you need to calibrate the spectrometer?
Q37.17 How do you calibrate the spectrometer (if the software can't do it for you)?
Q37.18 Why is it necessary to cool the diode array in a PEELS?

Q37.19 What's the difference between a GIF and an Ω filter?
Q37.20 What are the pros and cons of these two types of filter?
Q37.21 Why would you want to form an image from specific electrons in an EELS spectrum rather than just look at the spectrum?
Q37.22 Why is TEM image mode generally a bad choice for gathering spectra?

TEXT-SPECIFIC QUESTIONS

T37.1 Using Figure 37.1 and Figure 32.2A, contrast the principal characteristics of XEDS and EELS spectra, indicating the relevance of your observations to interpreting/quantifying the spectra.
T37.2 Examine Figure 37.2B. Why is the back focal plane of the projector lens used as the object plane of the spectrometer/lens?
T37.3 What is the object of the spectrometer/lens when an image is on the TEM screen?
T37.4 Electrons that have lost energy can be both an advantage and a disadvantage. Explain how such electrons are used to advantage in diffraction and in imaging, and likewise how they can degrade DPs and images (find examples (especially figures where possible) throughout the book to support your arguments).
T37.5 Using Figure 37.9 estimate what fraction of the single channel intensity is lost through the spreading of the point-spread function. What instrument modification might help reduce the point-spread function by keeping the probe more localized?
T37.6 Contrast the specific roles of the objective aperture and the selected area aperture in TEM and EELS.
T37.7 Describe the steps you would take to gather a spectrum with $\beta = 20$ mrad in (a) DSTEM, (b) TEM image mode, or (c) TEM diffraction mode.
T37.8 Can you think of circumstances in which you might get better energy resolution in a non-monochromated spectrum than in a monochromated one?
T37.9 Examine Figure 37.1 and estimate the absolute intensities of the zero-loss peak, the plasmon peak and the Ni L ionization edge above background. What does this tell you about the difficulties that we will encounter with processing EELS spectra? How does the P/B ratio compare with an XEDS spectrum?
T37.10 Although SEELS systems are no longer sold, can you think of an advantage to serial collection over parallel collection?
T37.11 Why is the camera length an important variable if you are operating in diffraction mode (Figure 37.8)? Choose two reasonable values of the camera length and calculate what effect this has on the operation of the spectrometer.
T37.12 Explain under what circumstances the spectrometer entrance aperture, rather than the objective aperture, might control the value of β in TEM mode (Figure 37.7).
T37.13 If the specimen is really thin so that the electrons suffer no significant energy losses while traversing the thin foil, can we totally ignore chromatic-aberration effects?
T37.14 Explain why the specimen you are looking at can affect the spatial resolution of the EELS analysis (in totally different ways to which the specimen controls the spatial resolution for XEDS).
T37.15 If your lab can't afford a monochromator, list all the other ways that you can improve the resolution of your spectra. Give the pros and cons of each method and list them according to their relative costs.
T37.16 What is a typical count rate in an EELS spectrum versus an XEDS spectrum? (If you can't find these data in the book, try a quick experiment on the AEM.)
T37.17 Would you expect a channel-to-channel gain variation in a scintillator-photomultiplier detector on an old SEELS and in a CCD camera in a GIF?
T37.18 How can you minimize the dark current?
T37.19 List the other common artifact in EELS spectra and explain how you would (a) recognize and (b) correct for each one.
T37.20 Why is it not a problem if one of the detectors in the PEELS diode array dies?
T37.21 When might it be better to perform a spectrum-profile analysis rather than an EFTEM analysis?
T37.22 Can you think of any circumstance when point analysis might be better than a line-profile or EFTEM analysis?

38
Low-Loss and No-Loss Spectra and Images

CHAPTER PREVIEW

The term 'energy loss' implies that we are interested only in inelastic interactions, but the spectrum will also contain electrons which have not lost any discernible energy, so we need to consider elastic scattering as well. In this chapter, we'll focus on the low-energy portion of the EEL spectrum which comprises

- The zero-loss peak, which primarily contains elastic, forward-scattered electrons, but also includes electrons that have suffered small energy losses. Forming images and DPs with the zero-loss electrons offers tremendous advantages over unfiltered images, particularly from thicker specimens.
- The low-loss region up to an (arbitrary) energy loss of ∼ 50 eV contains electrons which have interacted with the weakly bound, outer-shell electrons of the atoms. Thus, this part of the spectrum reflects the dielectric response of the specimen to high-energy electrons. We can also form images from these low-loss electrons that reveal information about the electronic structure and other characteristics of the specimen.

The energy-loss spectrum is more useful than an X-ray spectrum which only contains elemental information. However, this kind of spectrum is also far more complex. To understand its content you need a greater understanding of the physics of beam-specimen interactions, so we'll give you some hints about where to get the necessary education.

38.1 A FEW BASIC CONCEPTS

Back in Chapters 2–4, we described the difference between elastic and inelastic beam-specimen interactions and introduced the ideas of scattering cross sections and the associated mean free path. Remember, the cross section is a measure of the probability of a specific scattering event occurring and the mean free path is the average distance between particular interactions. It might be a good idea to re-read about those ideas before starting on this chapter. Briefly, you should recall that elastic scattering is an electron-nucleus interaction; the word elastic implies that there is no energy loss, although a change in direction, and hence in momentum, usually occurs. Elastic scattering occurs mainly as Bragg diffraction in crystalline specimens. Inelastic scattering is primarily an electron-electron interaction and entails both a loss of energy and a change of momentum.

We have to be concerned with both the amount of energy lost and the direction of the electrons after they've come through the specimen.

This latter point is one reason why the collection angle of the spectrometer, β, is so important.

Also you must remember to distinguish between the definitions of scattering that will keep appearing.

- *Single scattering* occurs when each electron undergoes at most one scattering event as it traverses the specimen.
- *Plural (>1 and < 20) scattering* implies that the electron has undergone a combination of interactions and lost energy from some or all of them.
- *Multiple (> 20) scattering* only occurs in very thick specimens or with very low energy electrons, so is irrelevant to TEM.

We'll see that the energy-loss spectrum is most understandable and more easily modeled when we can approximate everything to single scattering. This ideal is approached when we have a combination of very thin specimens and high accelerating voltages. In practice, most specimens are thicker than ideal and so we usually acquire plural-scattering spectra and we may have to remove the plural-scattering effects via deconvolution routines which are available in commercial and free EELS software packages (see URLs #1–3). We'll address the general topic of spectral simulation and manipulation later and in more depth in the companion text. We've already told you about deconvoluting out the PSF in Chapter 37, we'll see in Section 38.2.B, that you can do similar things if you want to remove the ZLP, and removing plural-scattering effects in high-loss spectra will be discussed in Section 39.6.

> **DECONVOLUTION WARNING**
> Whenever we mention deconvolution (and we'll do so quite often), remember that there is the danger of introducing artifacts into the part of the spectrum that you have just tried to simplify.

Typical energy losses: The principal inelastic interactions in order of increasing energy are phonon excitations, interband and intraband transitions, plasmon excitations, and inner-shell ionizations. We've already introduced these processes back in Chapter 4. The energy loss $\mathcal{E}$ of the principal scattering processes that we study in EELS are single-electron scattering (inter/intraband transitions), 2–20 eV, plasmon interactions 5–30 eV, and inner-shell ionizations, 50–2000 eV. Phonon excitations cause losses of ~ 0.02 eV so, even with the best energy resolution in monochromated AEMs, it's not possible to separate these from the ZLP, although the phonon-scattering angle can be quite large and (particularly for heavier elements) these electrons can be seen as the background intensity between the principal spots in an SAD pattern. But that's about all we'll say about phonon scattering, except that you should be able to work out why cooling your specimen will help to reduce their presence. We'll deal with the first three (low-loss) processes in this chapter and the ionization (high-loss) process in the next chapter. There are also ionization events with energy losses >2 keV but it is difficult to detect these because the signal is relatively weak and, at such energies, the X-ray signal is strong, so we tend not to do much EELS > 2 keV.

> **REMEMBER θ**
> The symbol θ in all cases refers to the scattering *semi-angle* even if we just say angle.

> **THE MOST IMPORTANT ANGLE IS θ_E**
> The so-called characteristic or most-probable scattering angle (for an energy loss $\mathcal{E}$)—it depends on the beam energy.

Typical scattering angles: It's a little difficult to be specific about the values of scattering angle, θ, because the angle varies with beam energy and the energy-loss process. We always assume that the scattering is symmetrical around the direct beam and there are two principal scattering angles that you should know. You can find derivations of the equations in Egerton's text.

$$\theta_E \approx \frac{\mathcal{E}}{2E_0} \quad (38.1)$$

If the beam energy is in eV the angle is in radians. This equation is an approximation (good to about ±10%) and it ignores relativistic effects and doesn't work for phonons, so you should only use it for rough calculations and be particularly suspicious above ~100 keV. We can be more precise and define θ_E as

$$\theta_E \approx \frac{\mathcal{E}}{(\gamma m_0 v^2)} \quad (38.2)$$

Here we have the usual definitions: m_0 is the rest mass of the electron, v is the electron velocity and γ is given by the usual relativistic equation (where c is the velocity of light)

$$\gamma = \left(1 - \frac{v^2}{c^2}\right)^{-\frac{1}{2}} \quad (38.3)$$

The other useful angle θ_c is the cut-off angle above which the scattered intensity is zero

$$\theta_c = (2\theta_E)^{\frac{1}{2}} \quad (38.4)$$

Be careful to calculate θ_c in radians, not milliradians. Knowing the characteristic scattering angle is obviously important if you want to gather an intense spectrum highlighting a particular energy loss. For example, a plasmon interaction with 100-keV electrons, causing a typical energy loss of 20 eV will have a characteristic scattering angle of ~0.1 mrads. Using a smaller β will cut off intensity in the spectrum which is why we told you in the last chapter to ensure that $\beta > 2$–$3\theta_E$. Knowing the cut-off angle (which is typically an order of magnitude greater than the characteristic angle at 100 keV) will give you a maximum useful value of β. If you use too large a value of β then there's the chance that you'll get unwanted electrons in the spectrum (e.g., diffracted beams) but you've got to try really hard to encounter this problem.

38.2 THE ZERO-LOSS PEAK (ZLP)

38.2.A Why the ZLP Really Isn't

If your specimen is thin enough for EELS, the predominant feature in the spectrum will be the ZLP, as shown in Figure 38.1. As the name implies, the ZLP consists primarily of electrons that have retained the incident-beam energy E_0. Such electrons may be forward scattered in a relatively narrow cone within a few mrads of the optic axis and constitute the 000 spot in the DP, i.e., the direct beam. If you were to tilt the incident beam so a diffracted beam entered the spectrometer then it, too, would give a ZLP.

> **MAGNITUDE OF ANGLES**
> The scattering angles for diffraction ($2\theta_B$) are relatively large (~20 mrad) compared to the smaller scattering angles in EELS. So the diffracted beams will only enter the spectrometer if you select them.

Actually, we can also measure the intensity and energy of electrons as a function of their angular distribution, and we'll discuss angular- or momentum-resolved EELS later, in Section 40.8.

Now the term ZLP is really a misnomer for two reasons. First, as we've seen, the spectrometer has a finite energy resolution (at best ~ 0.3 eV without monochromation) so the ZLP will also contain electrons that have energy losses below the resolution limit, which are mainly those that excited phonons. This is not a great loss since phonon-scattered electrons don't carry useful information; they only cause the specimen to heat up. However, it does explain why we shouldn't really call this the ZLP. Second, we can't produce a monochromatic (single-color, i.e., single-wavelength/energy) beam of electrons; the beam has a finite energy range about the nominal value E_0 (at best 10–100 meV even with a monochromator). Despite this imprecision, we will continue to use the term zero loss.

From a spectroscopist's point of view, the ZLP is more of a problem than a useful feature in the spectrum because, as we mentioned in the previous chapter, it is so intense that it can saturate the PDA or the CCD detector, creating a ghost peak. So if you don't need to collect the ZLP in the spectrum then deflect it off the detector (or use the attenuator in the Gatan system). Conversely, from the microscopist's standpoint, as we'll see, selecting the ZLP to form an image or DP from which most of the energy-loss electrons have been excluded is a *very* useful technique. Conversely (again) filtering out the ZLP and forming images with selected energy-loss electrons is also extraordinarily useful.

38.2.B Removing the Tail of the ZLP

The intense ZLP has a tail, either side of it (go back and look at Figure 37.4), ultimately limited by the energy resolution. On the low (negative) energy side of the peak, the point-spread function accounts for the tailing, but on the high (positive) energy side there are contributions from the low-loss (e.g., phonon) electrons we just discussed. It is sometimes necessary to remove this tail before studying the (very) low-loss spectrum, e.g., for dielectric-constant determination (see Section 38.3). There are various ways to remove this tail in Gatan's commercial software (e.g., comparison with reference spectra, subtraction versus deconvolution) and the software continues to improve. You must make sure you are displaying the spectrum with a high dispersion and deconvolute the point-spread function before doing anything else.

The best way to remove the tail, if you *really* need to study the spectrum close to the ZLP, is to use a monochromator and cut off the tail of the peak at the source, so that any intensity outside the ZLP is a true low-loss part of the spectrum. In this particular case, since the low-loss spectrum is relatively intense, the principal argument against monochromation (i.e., you throw away a lot of expensive electrons) is seriously weakened. We've already mentioned this in Section 37.7 and Figure 37.16 compares the energy spread in the ZLP before and after monochromation.

If you don't have a monochromator, removal of the ZLP peak is challenging and prone to artifacts. The

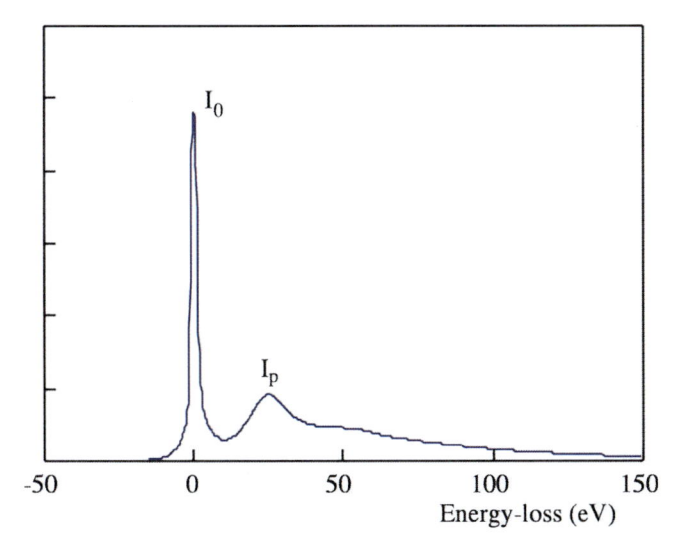

FIGURE 38.1. The low-loss spectrum showing an intense ZLP. The next most intense peak is a plasmon peak and the rest of the spectrum out to the high-loss (> 50 eV) region is relatively low intensity.

main problem is that the shape of the ZLP measured with the beam on the specimen is not always the same as the ZLP measured in the hole because of phonons and elastic-scattering effects.

38.2.C Zero-Loss Images and Diffraction Patterns

If we filter out all the electrons which have lost energy greater than the resolution of the spectrometer (typically $> \sim 1$ eV) then basically we have an elastic image or DP. In doing this we immediately remove chromatic-aberration effects from the image, since it is the imprecise focusing of energy-loss electrons that degrades the resolution of TEM images from thick specimens. You should go back and take a look at Section 6.5.B and equation 6.16 and you'll see that, if many electrons suffer a typical (e.g., plasmon) $\mathcal{E}$ of ~ 15 eV, your image resolution degrades from a few Å to several nanometers. For such $\mathcal{E}$ to occur, the specimen thickness has to be a goodly fraction of the plasmon mean free path (see Table 38.1), but that is not unusual. In addition to degrading the resolution by adding a diffuse component to otherwise focused TEM images, the energy-loss electrons also account for the diffuse intensity between spots in DPs. So filtering out these electrons should both increase the image contrast and improve the quality of the DPs.

The positive effect of energy filtering on resolution and *all* forms of TEM-image contrast has been known for several decades (see Egerton's early papers). This technique is particularly useful for enhancing the quality of images from thick biological (or polymeric) specimens in which the inelastic scattering is stronger than the elastic scattering (see Figure 38.2). However, diffraction contrast can also be enhanced by filtering of images from thick specimens (see Figure 38.3) but also in thin specimens because inelastically scattered electrons broaden the excitation error, **s**, thus reducing diffraction contrast. Sometimes enhancement of mass-thickness contrast images can be achieved by 'tuning' the spectrometer to select a specific range of energies (Figure 38.4). Contrast tuning (see Egerton's text) involves selecting an energy-loss window and sliding it around the spectrum while watching the image to find the best contrast. Tuning is useful in both the low and high-loss regions of

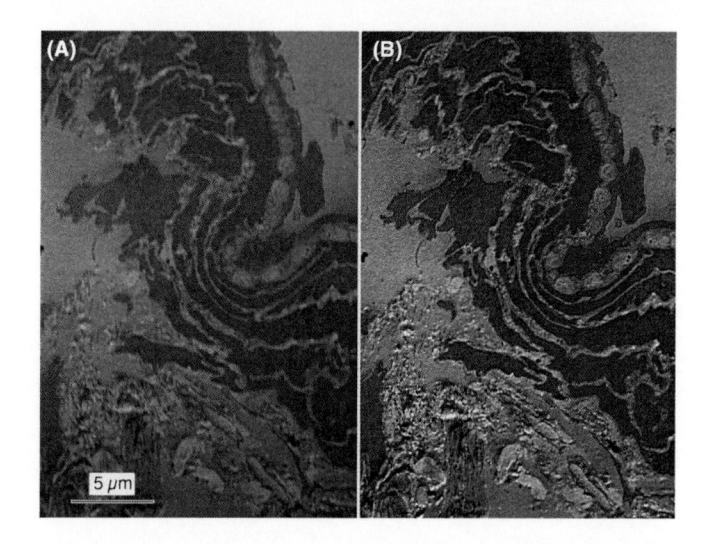

FIGURE 38.2. Comparison of (A) unfiltered and (B) filtered image of a thick biological section showing the enhanced contrast and resolution when the energy-loss electrons are removed.

TABLE 38.1 Plasmon-Loss Data for 100-keV Electrons for Several Elements (from Egerton, 1996)

Material	$\mathcal{E}_P$ (calc)(eV)	$\mathcal{E}_P$ (expt)(eV)	θ_E(mrad)	θ_C(mrad)	λ_P(calc) (nm)
Li	8.0	7.1	0.039	5.3	233
Be	18.4	18.7	0.102	7.1	102
Al	15.8	15.0	0.082	7.7	119
Si	16.6	16.5	0.090	6.5	115
K	4.3	3.7	0.020	4.7	402

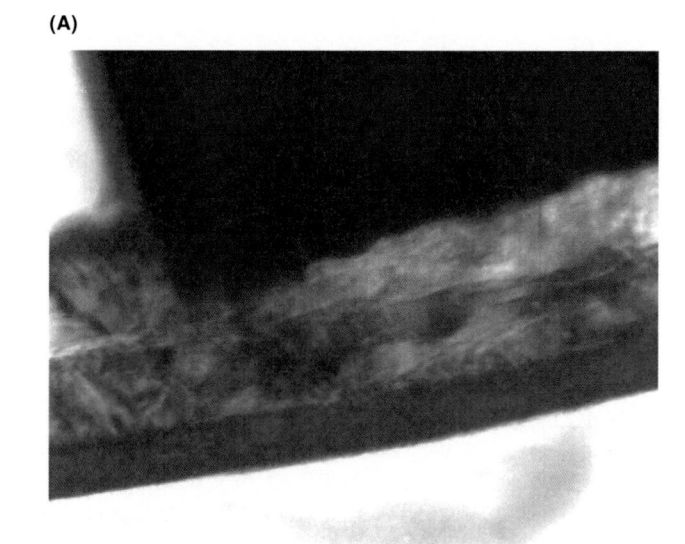

FIGURE 38.3. Comparison of (A) unfiltered and (B) filtered image of a thick crystalline specimen showing enhanced diffraction contrast when the energy-loss electrons are removed.

Unfiltered   Filtered (60±5eV)

(A)   (C)

(B)   (D)

Unfiltered/scaled   Filtered (0±5eV)

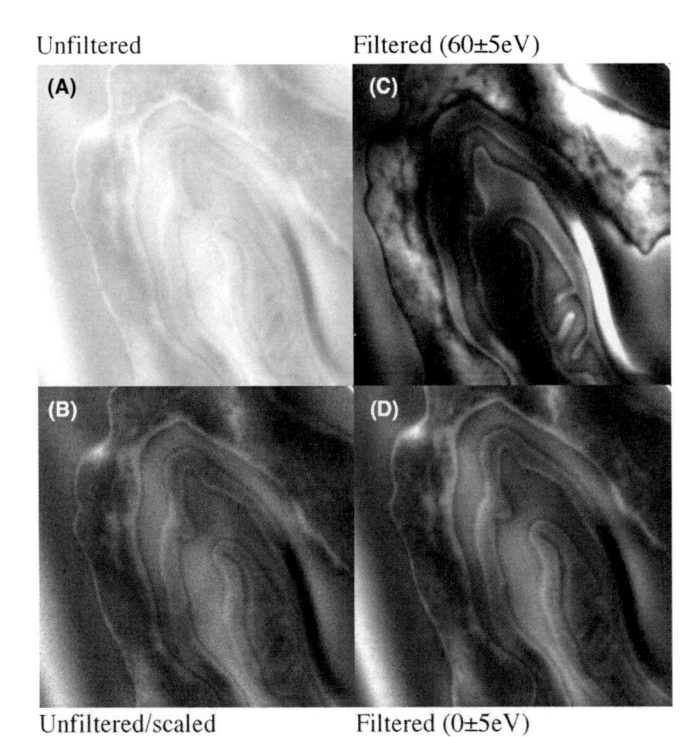

FIGURE 38.4. Contrast tuning of an image from a thick biological specimen to determine the region of the low-loss spectrum that gives the optimum contrast. (A) Unstained mouse epidermis (thickness = 0.1 μm, 100 keV). (B) Unfiltered image, digitally scaled to show the best contrast. (C) Filtered, contrast-tuned at 60±5 eV; the image has much better contrast than (B). (D) Filtered, contrast-tuned at 0±5 eV; the image has improved resolution and better contrast (B), but not as strong contrast as (C). Full width = 1 μm.

the spectrum and is typically done anywhere from 0 to 200 eV. As shown in Figure 38.4, there is significant possibility for contrast improvement in the low-loss region. In the high-loss region, selecting an energy window before an edge tends to reduce contrast from the edge electrons while windows after an edge tend to enhance contrast from the edge (see also jump-ratio imaging in Section 39.9).

High-resolution phase-contrast images (from thin specimens) should be more easily compared with theory if they are filtered to remove the diffuse background because we won't need so many 'fudge factors.' (See Chapter 28.) For a given thickness of specimen, the only alternative to reducing chromatic aberration is to go to higher voltages and this is perhaps an even more expensive option than buying a filter!

As we discussed back in Sections 37.8 and 20.5, if the energy-loss electrons are removed from SADPs and CBDPs it makes them much clearer (e.g., the paper by Midgley et al. and Reimer's chapter in Ahn's text), as shown back in Figures 20.10 and 37.17C. Energy filtering can also reveal extra diffraction information, such as the radial-distribution function for amorphous materials, which we'll tell you about later in Section 40.7.

If you've got a thick (several tens of nm) specimen (which is sometimes all you can manage to create), filtering may give many improvements

- Filtering can improve the image resolution.
- Filtering can enhance the image contrast no matter what mechanism is operating.
- Filtering can improve the contrast in diffraction patterns.
- Filtering can reveal finer detail in images and DPs.

So it would be really great if we could leave the filter switched on all the time, but this isn't always practically feasible.

Perhaps what is more surprising is why, given the tremendous advantage of filtering, simple zero-loss filters haven't been commercially available for decades? One experimental problem is that the energy-loss spectrometer is susceptible to external fields and the ZLP shifts over time, making continuous EFTEM imaging difficult: you have to continually re-align and re-focus the ZLP. Also, the best results, particularly for quantitative imaging (see the next chapter), require that your specimen has a similar thickness over the entire area of the image and that strong diffraction effects are minimized. Another possible reason is that filtering works best for thick specimens, which don't permit the TEM to perform at its (spherical-aberration) resolution limit of a fraction of a nanometer. It is perhaps not advantageous to remind users that most of their specimens are such that their TEM can't deliver anything like its best resolution, and that buying a 'better' TEM will have no beneficial effect on image quality or resolution for the great majority of (thick) specimens. But of course, this is only speculation...!

38.3 THE LOW-LOSS SPECTRUM

Look again at Figure 38.1, which is a typical low-loss spectrum and several points are immediately obvious

- After the ZLP, the plasmon peak is the next major peak.
- Apart from the plasmon, the spectrum is relatively featureless (the intensity changes are small).
- Despite the lack of features, there's still a lot of counts (check the units of the ordinate), so extracting useful data is still feasible and imaging should be relatively straightforward.

The cut-off energy for the low-loss spectrum, as we've already noted, is ~ 50 eV and the reason for this is that the other principal features of energy-loss spectra, namely, the ionization edges, don't appear (at least for solids) until $E > 50$ eV. In the low-loss spectrum, we are detecting beam electrons that have interacted with conduction and/or valence bands (hence another common term for the low-loss spectrum is 'valence spectrum'). These weakly bound electrons control many electronic properties of the specimen. In general, the low-loss spectrum is not as well understood as the high-loss spectrum and there has not been the same effort put into modeling low-loss spectra, as we'll describe for the higher-loss spectra in Chapters 39 and 40. However, things are beginning to change, as you'll see later in this chapter.

While we have just shown you the tremendous advantages of filtering out the plasmon peak in order to enhance contrast and resolution in TEM images and DPs, there is also much to be gained by imaging the plasmon peak and filtering out the ZLP and high-loss electrons. This approach was pioneered by Batson. Because of the strength of the signal, plasmon imaging is becoming a much more popular technique, particularly for mapping out low-loss properties of nanomaterials (e.g., see the papers by Eggeman et al. and by Ding and Wang).

38.3.A Chemical Fingerprinting

So what can we do with the low-loss spectrum? Because there are a lot of counts, we can use the shape of the spectrum to help identify specific phases or features in the TEM image with some degree of statistical certainty. The low-loss spectrum all the way up to ~50 eV, including any plasmon peaks (see Section 38.3.C), should be used for fingerprinting.

> **FINGERPRINTING**
> The low-loss spectrum can only be used for phase identification through a 'fingerprinting' process. You store the spectra of known specimens in a library in the computer.

So you overlay your unknown spectrum on one or more stored library-standard spectra. Figure 38.5 shows how low-loss spectra vary (A) for aluminum and various compounds and (B) for the main constituents of biological specimens. A collection of low-loss spectra from many elements and common compounds (mainly oxides) has been compiled in various databases, such as the EELS Atlas or on the Web at URL #4. Such sources can help considerably with fingerprinting unknown features in your image. As with any fingerprinting technique, including the forensic variety, you must be careful to decide when a 'match' is satisfactory. There is no 'black and white' here, only shades of gray, so don't convict unless the statistics are with you and there is strong supporting evidence from other techniques.

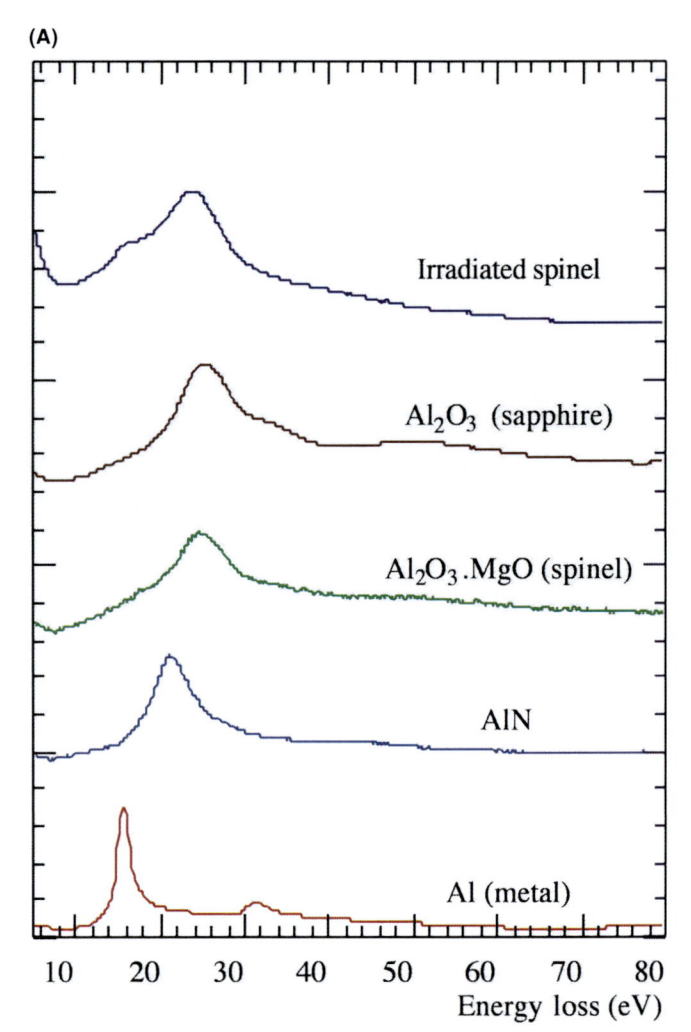

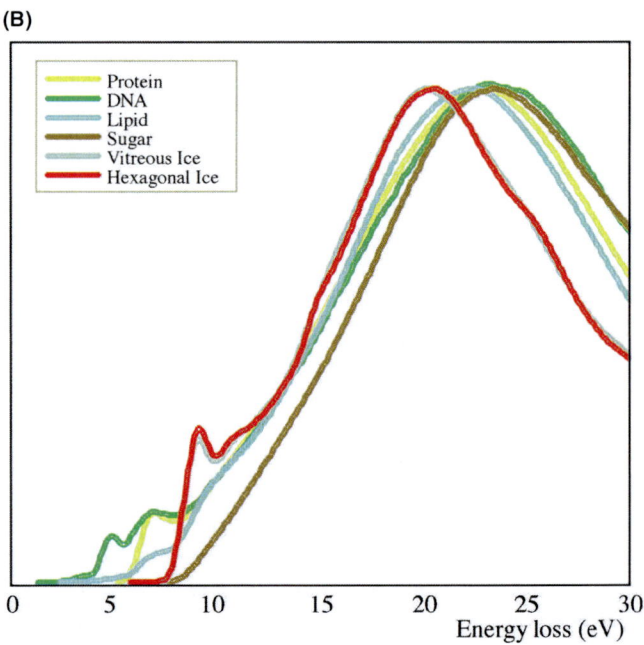

FIGURE 38.5. (A) The low-loss spectrum from specimens of Al and various compounds, showing differences in the intensity variations that arise from differences between the Al-Al, Al-O, and Al-N bonding. (B) Low-loss spectra from the principal components of cellular tissue.

38.3.B Dielectric-Constant Determination

We can view the energy-loss process as the dielectric response of the specimen to the passage of a fast electron. As a result, the very low energy spectrum (up to $E \sim 20$ eV) contains information about the dielectric constant or permittivity (ε). Localized dielectric-constant measurements are of great interest as the semiconductor industry explores high-dielectric materials, such as HfO_2, for the next generation of nanometer-scale gate oxides.

Assuming a free-electron model, the single-scattering spectrum intensity $I(E)$ is related to the imaginary (Im) part of the dielectric constant ε by the expression (modified from Egerton)

$$I(E) = I_0 \frac{t}{k} \text{Im}\left(-\frac{1}{\varepsilon}\right) \ln\left[1 + \left(\frac{\beta}{\theta_E}\right)^2\right] \quad (38.5)$$

where I_0 is the intensity in the ZLP, t is the specimen thickness, k is a constant incorporating the electron momentum and the Bohr radius, β is the collection angle (again note its importance), and θ_E is the characteristic-scattering angle. You can use a Kramers-Kronig analysis to analyze the energy spectrum in order to extract the real part of the dielectric constant from the imaginary part in equation 38.5 and details are given in Egerton's text. As usual, the advantage to doing this kind of measurement in the TEM is the high spatial resolution and the advantages of this are exemplified by the use of low-loss spectroscopy to determine optical gaps on BN nanotubes (Arenal et al.). Since you need a single-scattering spectrum, removing the plural scattering intensity by Fourier-logarithmic deconvolution is the first step (see Section 39.6) when determining the dielectric constant. The Gatan software package has the appropriate programs and public-domain software is also available, e.g., URLs #1, 5, and 6.

> **KRAMERS-KRONIG ANALYSIS**
> This analysis gives the energy-dependence of the dielectric constant and other information, which we usually obtain by optical spectroscopy.

The alternatives to EELS for this kind of work are various kinds of optical and other electromagnetic-radiation techniques. The part of the low-loss spectrum from ~ 1.5 to 3 eV is of great interest and corresponds to optical analysis of the dielectric response from the infra-red (~ 800 nm) through the ultra-violet (~ 400 nm) wavelength range. (This correspondence between EELS and optical spectroscopy only holds for small angles of scattering so the value of β you choose must be small ($< \sim 10$ mrads) thus lowering the intensity in the spectrum.) Higher energies correspond to various electronic transitions. Thus, in a single EELS experiment you can, in theory, substitute for a whole battery of optical-spectroscopy instrumentation (although optical spectroscopy techniques do offer even better energy resolution than EELS). Remember that EELS always offers better spatial resolution.

There is a tremendous similarity between the TEM-EELS approach and the valence (surface) EELS approach, including the need for Kramers-Kronig and deconvolution software. TEM-based EELS is in the extremely low-energy (i.e., low-frequency) range around 1 eV and below and corresponds to far infra-red spectroscopy which is well into the energy range of studies of bond vibrations. Higher up the energy-loss range corresponds to the visible and ultra-violet ranges.

If you don't have access to a monochromator, you can use software to remove the contribution of the tail of the ZLP but, as we mentioned above, be careful, because this process may introduce its own artifacts. An example of the correspondence between EELS and optical valence spectra is shown in Figure 38.6. In Figure 38.6A, the importance of initial deconvolution is demonstrated and the deconvoluted valence spectra are compared with ultra-violet spectra in Figure 38.6B. It is straightforward to assign the various peaks in the low-loss spectra to specific interband transitions and also to compare the data with band-structure calculations (e.g., van Bentham et al.). Thus, the electronic and optical properties can be obtained and you can, of course, select any of the features in the low-loss spectrum and form images with those electrons. So dielectric-constant imaging is feasible, as is imaging the various other low-loss signals which we'll now discuss.

> **FOR THE BEST LOW-LOSS SPECTROSCOPY**
> You need an FEG, a high-resolution, high-dispersion spectrometer and if you're really going to do it properly, a monochromator, so the tail of the ZLP does not mask the low-energy intensity.

38.3.C Plasmons

Plasmons are longitudinal wave-like oscillations that occur when a beam electron interacts with the weakly bound electrons in the conductance or valence band. You can think of plasmons as being like the ripples that spread out from where a pebble is dropped into a pond. But, unlike in a pond, the oscillations are rapidly damped, typically having a lifetime of about 10^{-15} s and so are quite localized to <10 nm. The plasmon peak is the second most dominant feature of the energy-loss spectrum after the ZLP. The small peak beside the ZLP in Figure 38.1 is a plasmon peak.

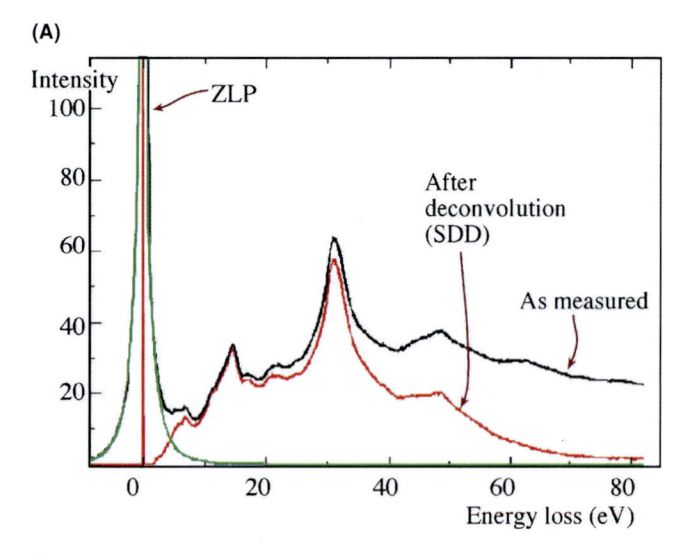

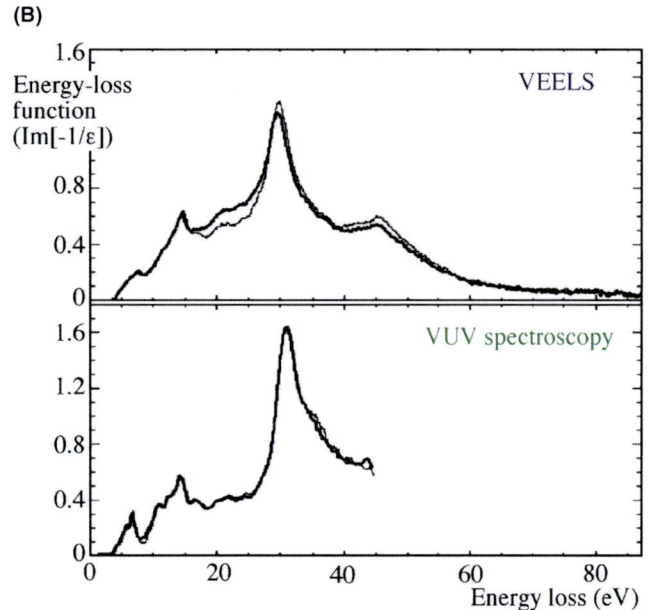

FIGURE 38.6. (A) Low-loss (valence) spectrum from SrTiO₃ before (black) and after (red) Fourier-log deconvolution. The extracted ZLP is shown in green. (B) Comparison of the change in the imaginary part of the complex dielectric function obtained from pairs of valence EELS (VEELS) spectra and valence ultra-violet (VUV) spectra from two different regions of SrTiO₃. The spectra show similar features but the VUV spectrum cannot be measured beyond ~ 45 eV.

If we assume the electrons are free (i.e., not bound to any specific atom or ion), then the energy E_P lost by the beam electron when it generates a plasmon of frequency ω_p is given by a simple expression

$$E_P = \frac{h}{2\pi}\omega_p = \frac{h}{2\pi}\left(\frac{ne^2}{\varepsilon_0 m}\right)^{\frac{1}{2}} \qquad (38.6)$$

where h is Planck's constant, e and m are the electron charge and mass, ε_0 is the permittivity of free space (remember the dielectric constant is the relative permittivity of a polarizable medium), and n is the free-electron density. Typical values of E_P are in the range 5–25 eV

and a summary of plasmon-loss characteristics is given in Table 38.1

- Plasmon losses dominate in materials with free-electron structures, such as Li, Na, Mg, and Al.
- Plasmon-like peaks occur to a greater or lesser extent in the low-energy spectra from all materials, including insulators, such as polymers and biological tissue.

So the 'free-electron' assumption is clearly not rigorous and we don't know everything about how this feature arises.

From equation 38.6, you can see that E_P is affected by n, the free-electron density. Interestingly, n may change with the chemistry of the specimen. Thus, measurement of the plasmon-energy loss can give indirect analytical information (see the next section).

The characteristic plasmon-scattering angle θ_E is very small, being typically < 0.1 mrads (as listed in Table 38.1), which means that the plasmon-loss electrons are strongly forward scattered. Their cut-off angle θ_c is much greater than θ_E so if you use a collection angle β of only 10 mrads, you will easily gather almost all the plasmon-loss electrons (again note the importance of knowing β in your system). Conversely, this means that even a small objective aperture will not stop plasmon-loss electrons entering the TEM imaging system. Plasmon-loss electrons also carry contrast information and, because they are the most intense energy-loss signal, they are the primary contribution to chromatic aberration in TEM images, which is why it is often a good move to filter them out. As we've already seen, Figure 38.2 shows the improvement in image contrast and resolution when the low-loss (primarily plasmon) spectrum is filtered out of the image of a specimen showing predominantly mass-thickness contrast. Likewise, Figure 38.3 shows a thick foil exhibiting primarily diffraction contrast. A similar improvement in resolution occurs when the many plasmon peaks are filtered out.

A typical value of the plasmon mean free path λ_P at AEM voltages is about 100 nm and so it is reasonable to expect at least one strong plasmon peak in all but the thinnest specimens. Likewise, the number of individual losses should increase with the thickness of your specimen and we can use the plasmon-peak intensity to estimate the specimen thickness. If indeed your specimen is so thin that only single scattering occurred, and the only significant scattering was a single plasmon event, then you should be very pleased because it's a great specimen for ionization-loss EELS (see the next chapter). Conversely, if your spectrum shows several plasmon peaks then it is too thick for ionization-loss studies. Under single-scattering circumstances we can assume

$$t = \lambda_p \frac{I_p}{I_0} \qquad (38.7)$$

where λ_P is the plasmon mean free path, I_P (see Figure 38.1) is the intensity in the first (and only) plasmon peak, and I_0 is the intensity in the zero-loss peak.

> **BALLPARK CALCULATION**
> A typical ballpark figure: if the intensity in the first plasmon peak is greater than one tenth the zero-loss intensity then your specimen is too thick for EELS quantification.

The method has advantages over other thickness measurement techniques in that you can apply it to any specimen, amorphous or crystalline, over a wide range of thicknesses. We'll tell you more about EELS thickness measurements and their role in ionization-loss spectrometry in Section 39.5.

If plural scattering is significant, then the spectrum becomes more difficult to interpret and other problems arise; e.g., your ionization-loss quantification results (next chapter) become unreliable.

Of course, one way around this problem is to use very thin foils, but often you can't produce thin-enough specimens. Murphy's law says that the area you're interested in will usually be too thick. Then you have to deconvolute the spectra, again using the Fourier-log approach (Section 39.6) to make the single-scattering assumption valid. As we've already noted, deconvolution brings its own problems.

Figure 38.7 shows the plasmon-loss spectra from (A) thin and (B) thick foils of pure Al and (B) also indicates how the Gatan software uses this information to come up with a measure of the local foil thickness. Since Al is a good approximation to a free-electron metal, the plasmon-loss process is the dominant energy-loss event. Plural-plasmon scattering in thicker foils is of concern because it limits the interpretation of high energy-loss spectra containing chemical information from ionization losses in which we are really interested (see Section 39.4).

The plasmon losses which we've just described all arise from interactions with the electrons in the interior of your specimen, but the incident electrons can also set up plasmon oscillations on the specimen surface. We can envisage these surface plasmons as transverse charge waves. Surface plasmons have about half the energy of volume plasmons (because the surface atoms are not so strongly bound). Generally, however, the surface plasmon peak is much less intense than the volume plasmon peak(s), even in the thinnest specimens, but you can still use them for imaging, as shown by Batson. With monochromators and aberration correctors, studies of surface plasmons, along with other low-loss features, will assume more importance in the TEM.

38.3.D Plasmon-Loss Analysis

As we just mentioned, the plasmon peaks contain chemical information, because the composition of the specimen may affect the free-electron density, n, which in turn changes the plasmon-loss peak position. Historically, this technique was the first aspect of EELS to produce quantitative analysis data, and it was used in a limited number of systems, mainly aluminum and magnesium alloys in which the plasmon-loss spectrum is dominant and consists of sharp Gaussian peaks (Williams and Edington). The lack of a more recent review gives some indication of the limitations of this approach (see below)!

The principle of plasmon-loss analysis is based on empirical observation of the shift in the plasmon-peak position ($\mathcal{E}_P$) with composition (C), giving an expression of the form

$$\mathcal{E}_P(C) = \mathcal{E}_P(0) \pm C\left(\frac{d\mathcal{E}_P}{dC}\right) \qquad (38.8)$$

where $\mathcal{E}_P(0)$ is the plasmon loss for the pure component. By creating a series of binary alloys of known composition we can develop a working curve which we can then use to calibrate measurements of $\mathcal{E}_P$ in unknown alloys.

Since plasmon-loss analysis demands the measurement of peak *shifts* rather than peak positions, you need an energy spectrum of the highest resolution and sufficient dispersion to measure the peak centroid accurately. The early plasmon-loss studies did not have

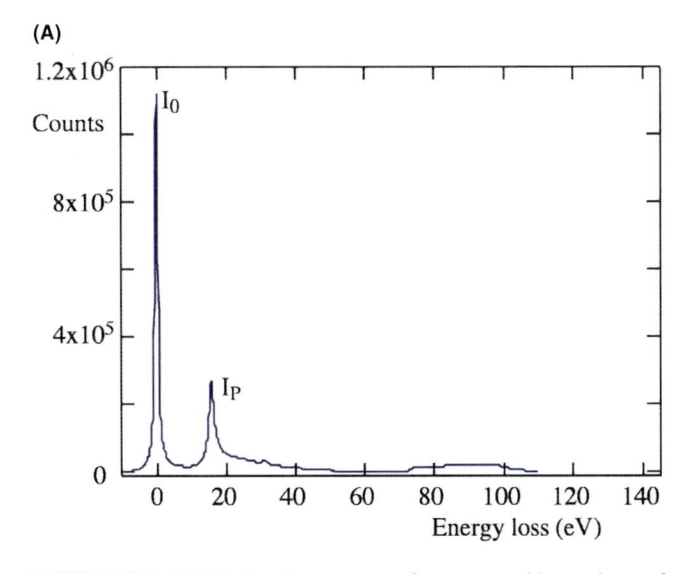

FIGURE 38.7. (A) The low-loss spectrum from a very thin specimen of pure Al showing the intense ZLP (I_0) and a small plasmon peak (I_P) at about 15 eV. (B) The low-loss spectrum from a thick specimen of pure Al showing several plasmon peaks, the first of which is almost as intense as the ZLP. The inset shows the calculation of the thickness from the Gatan software.

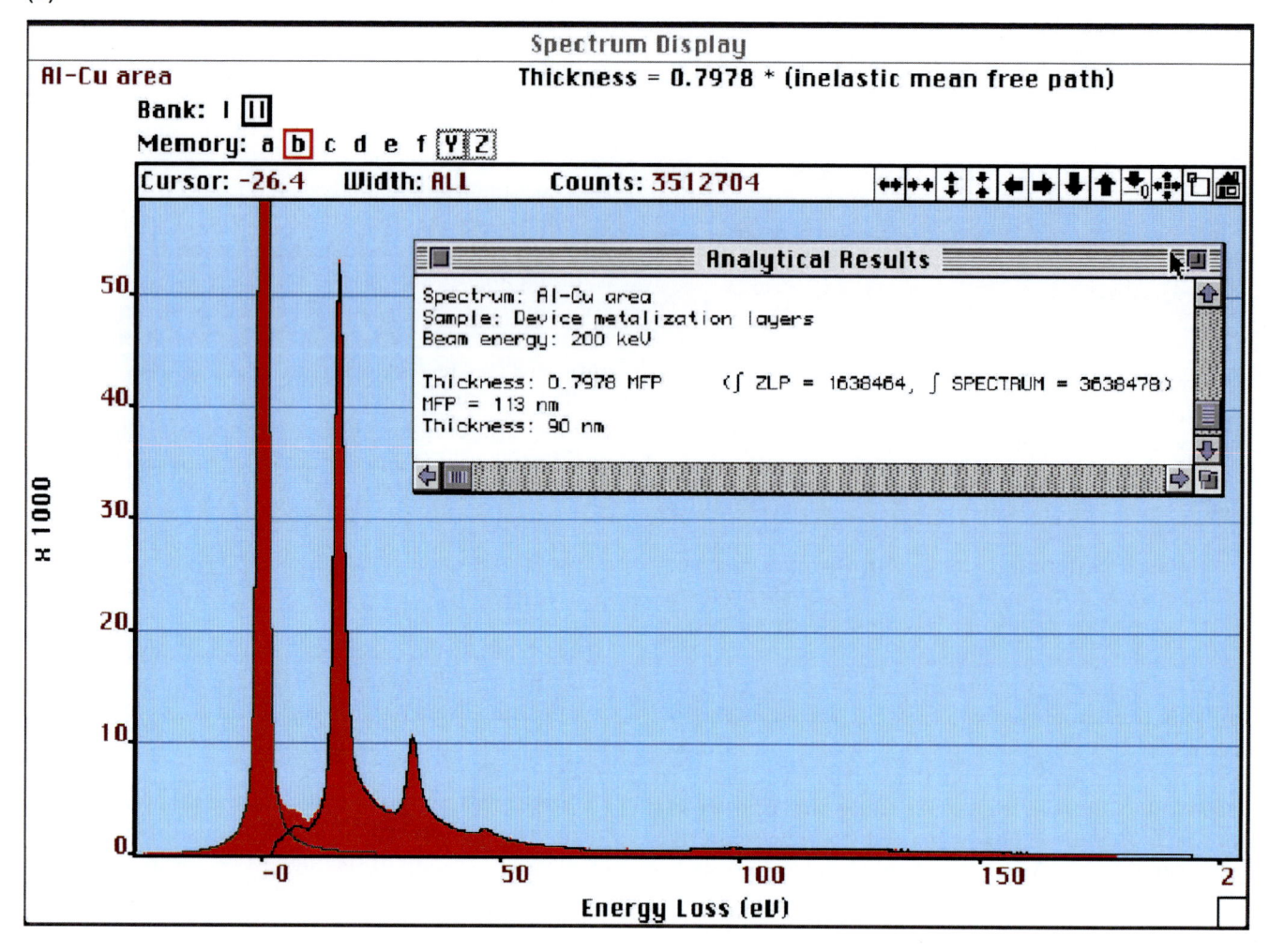

FIGURE 38.7. (Continued).

access to FEGs and so the resolution of the thermionic source was a limiting factor. Figure 38.8 illustrates some early plasmon-loss concentration data and the visible peak shifts that occur and also shows how we can use the plasmon peak shifts in Al-Li alloys and convert them into Li concentration data and also create Li compositions maps, which, given its low Z, is rather difficult to do with other analytical techniques.

Plasmon-loss spectrometry has reasonable spatial resolution and is relatively insensitive to specimen thickness and surface deposits. The spatial resolution is controlled by the localization of the plasmon oscillation which is only a few nm, since the plasmon disturbance is rapidly damped by the free electrons. The specimen thickness only affects the number and intensity of the plasmon peaks, not their position, as you can see in Figure 38.7. In fact you get the best results from plasmon-loss spectrometry when your specimen is about 1–2 mean free paths (λ_P) thick so that several, intense, Gaussian peaks are observable. There are, unfortunately, strong practical disadvantages, which account for the almost complete absence of plasmon-loss data since the advent of ionization-loss techniques in the mid-1970s

- We are limited to specimens showing well-defined peaks, and only binary specimens can be sensibly analyzed.
- The alloying element must produce a detectable change in E_P and this is not always the case. For example, the addition of 30 at.% Zn to Al scarcely changes E_P.

It is possible that application of modern detection and data processing techniques may improve the quality and ease of analyzing plasmon-loss spectra. While plasmon peak-shift analysis is limited, we can at least use the low-loss plasmon spectra for chemical fingerprinting, as we've already described, and we'll discuss the prospects for more quantitative interpretation of low-loss spectra via modeling in Section 38.4.

With increasing interest in the mechanical properties of nanoscale materials, the fact that strong scaling

38.3.E Single-Electron Excitations

A high-energy beam electron may transfer sufficient energy to a single electron in the valence band to change its orbital state, perhaps moving it to an unoccupied state in the conduction band. We call these events single-electron interactions and they result in inter/intraband transitions for the valence electrons, with energy losses of up to ~25 eV. An example of an interband transition is given in the spectra from different polymers which can be distinguished solely by their electronic differences (Hunt et al.), as shown in Figure 38.9A. Interactions with molecular orbitals, such as the π orbitals produce characteristic peaks in this low-energy region of the spectrum, sometimes causing shifts in the plasmon peak (either up or down depending on the relative energy of the interband transition and the plasmon loss), and that is why it is possible to use the intensity variation in this part of the spectrum to fingerprint a particular phase. A more challenging example is given in Figure 38.9B, which shows what can be done with a combination of cryo- and low-dose STEM to image a polymer nano-emulsion in an aqueous medium. The low-loss spectra reveal the electronic differences between the phases (including amorphous ice!) and the filtered images show the lobed shape of the emulsified particles. There is probably no other technique that could image such beam-sensitive material at such high resolution (Kim et al.).

If a beam electron gives a weakly bound, valence-band electron sufficient energy to escape the attractive field of the nucleus, then a secondary electron (SE) is created, of the sort used to give topographic images in the SEM and STEM. Typically, a SE requires < 20 eV to escape the surface and therefore the electrons causing SE emission appear in the same low-energy region of the spectrum as the inter- and intraband transitions.

38.3.F The Band Gap

In the region of the spectrum immediately after the ZLP, and before the rise in intensity preceding the plasmon peak, you can see a region of low intensity. If there are no interband transitions occurring, the intensity in this portion of the spectrum approaches the dark-current (noise) level of the detector. This low intensity implies that there is a forbidden-transition region, which is simply the band gap, between the valence and conduction bands in semiconductors and insulators. To determine the band gap, you need to strip off the tail of the ZLP (with all the consequent difficulties) and measure the energy range of the gap out to the rise in the initial low-loss spectrum. Figure 38.10A illustrates the variable band gap in spectra from specimens of Si, and its oxide and nitride. Mapping this change in the energy range in which no transitions occur gives band-gap images (Figure 38.10B) and several examples of this are given by Kimoto et al. As sub-nanometer-scale semiconductor technology advances,

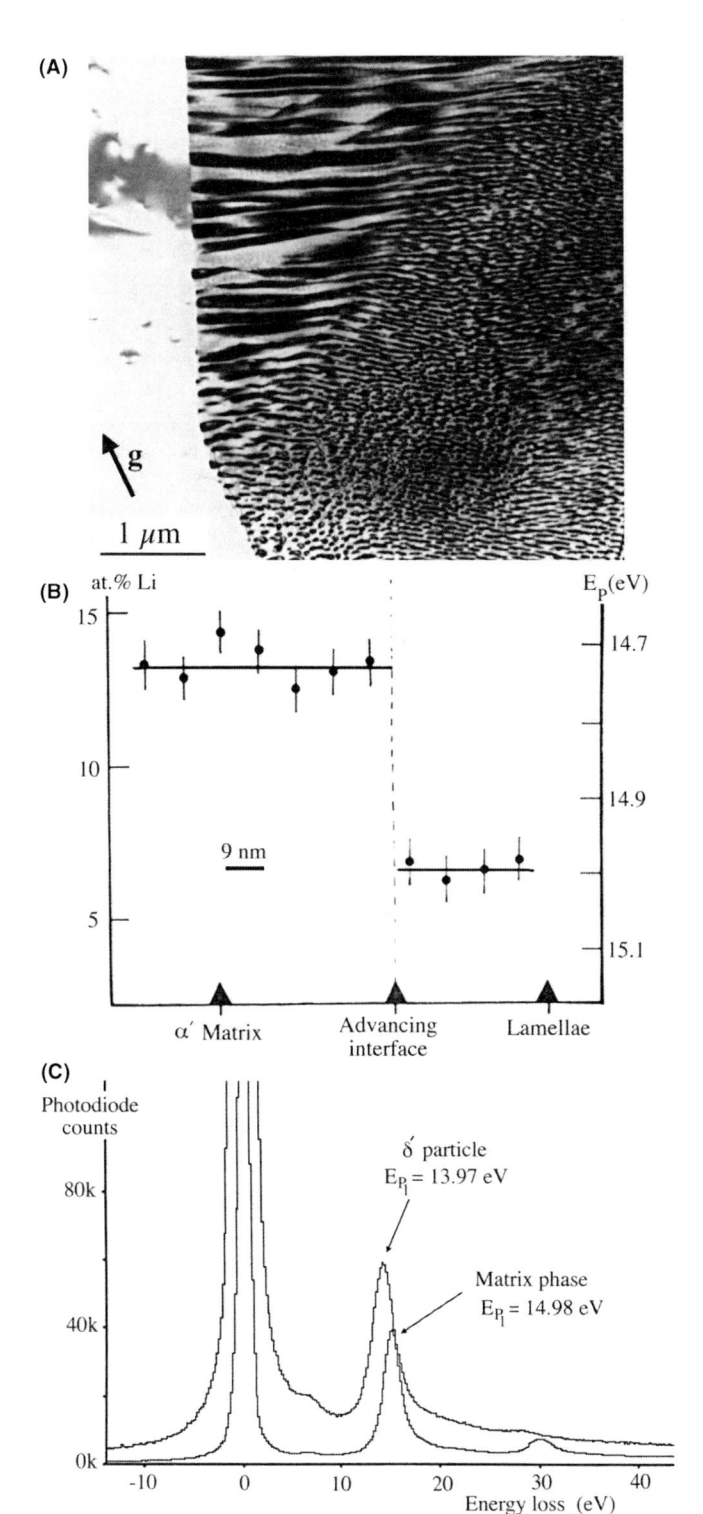

FIGURE 38.8. (A) A discontinuous precipitation reaction front in an Al-11 at.% Li specimen. (B) Experimental plasmon-loss measurements of the Li composition variation across the interface. (C) The shift in the plasmon peak for the matrix (5 at.% Li) and the precipitate (25 at.% Li) is clear in the two spectra.

correlations exist between the plasmon energy and elastic properties, hardness, valence-electron density, and cohesive energy is leading to a resurgence of interest in this part of the spectrum (e.g., Oleshko and Howe).

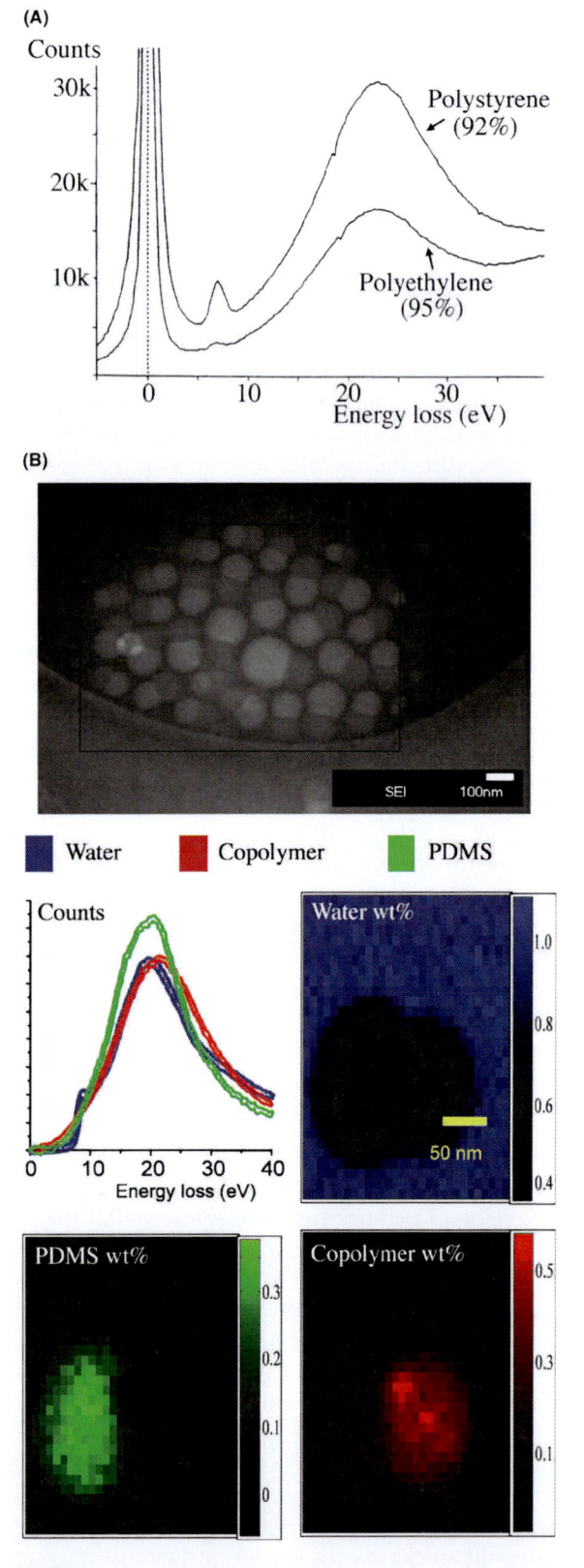

FIGURE 38.9. (A) The interband transition characteristic of polystyrene, clearly visible on the rise of the plasmon peak, compared with the absence of such a transition in polyethylene. (B) (top) A low-dose cryo-HAADF image of a two-phase polymer nano-emulsion in water; (bottom) low-loss spectra from amorphous ice (blue) polydimethyl siloxane (PDMS) (green) and a multi-phase co-polymer (red), together with the corresponding composition maps based on the differences in their low-loss spectra.

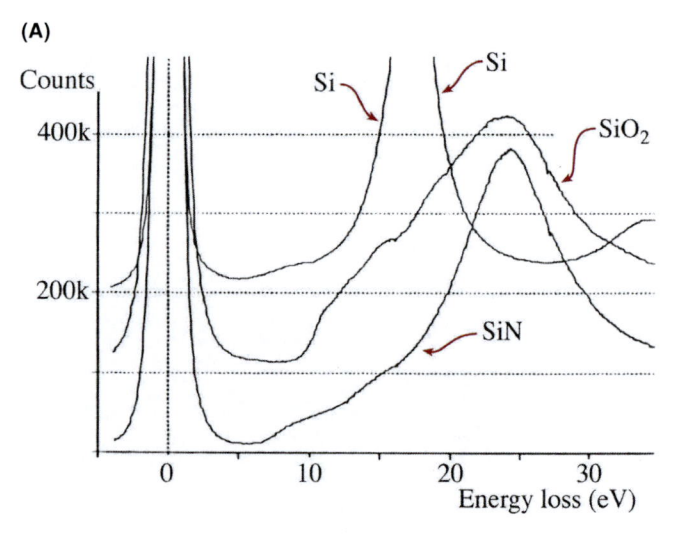

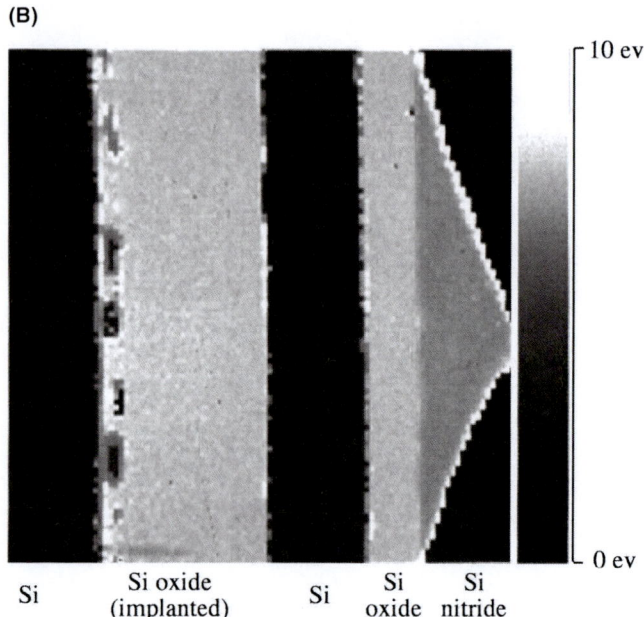

FIGURE 38.10. (A) Band-gap differences evident in the low-loss spectra of a Si semiconductor, SiO_2, and Si nitride (almost Si_3N_4) ceramic insulators and (B) the corresponding band-gap image (with scale on right; recorded at 90 K using 1024 channels and 150 ms dwell time). Note the Si islands in the oxide layer which were not visible in the TEM image.

the need for sub-nanometer resolution imaging of the band gap will increase and low-loss EEL images are the only way to visualize this electronic property.

38.4 MODELING THE LOW-LOSS SPECTRUM

As you now know, the low-loss spectrum has the advantage of significant intensity (so counting statistics are not a problem) and it contains useful data about your specimen, such as composition, bonding, the dielectric constant, the band gap, the free-electron density, and optical properties. With all this information you might have thought that we understood the spectrum very well and were able to model it with some degree of precision and could use the modeling to predict spectra from different materials. Perversely, we are better able to do this for the much lower intensity, high-loss spectrum, as we'll describe in Chapters 39 and 40. However, significant progress is being made in calculating the plasmon-loss energies and interband transitions. As we've seen, the plasmon peak is basically an oscillation of the free electrons, so equation 38.6 has been used over several decades to calculate the plasmon-loss energy, but this approach can't handle the effects of other low-loss features like interband transitions. French has developed software for low-loss modeling, called Electronic Structure Tools (see URL #1), which consists of a number of programs for the quantitative analysis of optical, VUV, and EELS spectra. Keast has shown good agreement between experimental and calculated low-loss spectra for a range of metals and ceramics, as shown in Figure 38.11, using abinitio methods, which we'll describe in some more detail in Chapter 40, and this topic is dealt with extensively in the companion text. Modeling of such spectra requires careful experimental control and for the data of Figure 38.11 the convergence angle of the (100-kV) beam was 8.3 mrad, the Gatan spectrometer collection angle was 5.8 mrad, 100 spectra (0.05 s per acquisition) were aligned, corrected for dark current and gain variations, and summed. The density-functional theory (see Section 40.5.A) calculations (using the random-phase approximation and neglecting local-field effects) were performed using the WIEN2k code. Exchange and correlation effects were treated using the generalized gradient approximation. The final spectrum was averaged over the different orientation components. So you get the idea that this is not straightforward!

Software for all aspects of low-loss analysis can be found at URLs #1 and 6.

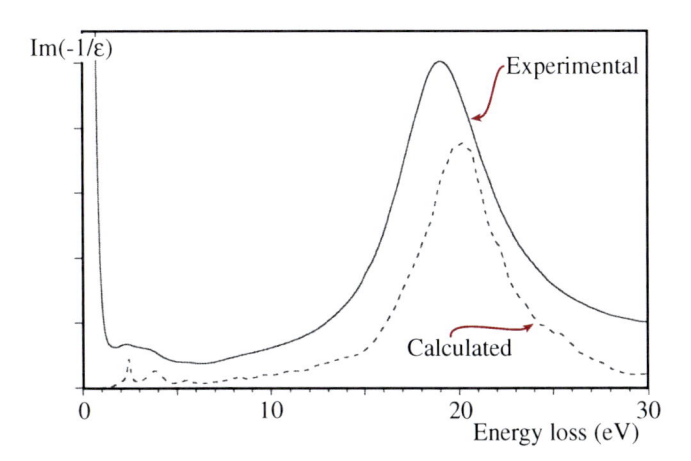

FIGURE 38.11. A comparison between the calculated (dashed) and experimental (full) low-loss spectrum from commercial MgB_2 particles.

CHAPTER SUMMARY

The low-loss (valence/plasmon) portion of the spectrum from 0 to 50 eV contains a wealth of useful information about the specimen.

- The ZLP is the most intense signal. If you filter out all the low-loss electrons apart from the ZLP, you get images and DPs which generally show higher resolution and better contrast than unfiltered ones because they are free of chromatic aberration and diffuse-scattering effects.
- The low-loss spectrum reflects beam interactions with loosely bound conduction and valence-band electrons.
- From different portions of the low-loss spectrum, you can measure the local dielectric constant of your specimen, the free-electron density, the thickness, the band gap, and observe inter/intraband transitions. You can also form images using energy-loss electrons which map out all these phenomena, generally with sub-nanometer resolution.
- The low-loss spectrum can be used to fingerprint (identify) specific elements, compounds, and biological tissue by comparison with the characteristics of standard spectra in databases.
- In some binary alloy systems of light elements, you can determine composition by measuring shifts in the plasmon-peak centroid. Plasmon imaging also has the potential for mapping nanoscale mechanical properties.
- We are getting much better at simulating the low-loss spectrum and understanding the various beam-specimen interactions that contribute to this high-intensity portion of the spectrum.

THE EELS ATLAS

Ahn, CC Ed. 2004 *Transmission Electron Energy-Loss Spectrometry in Materials Science and the EELS Atlas* 2nd Ed. Wiley-VCH Weinheim Germany. Buy this.

Ahn, CC and Krivanek, OL 1983 *EELS Atlas* Gatan Inc., 5933 Coronado Lane Pleasanton CA 94588. Buy this too (if you can find it).

SOME CALCULATIONS AND SPECIAL CONCEPTS

Egerton, RF 1976 *Inelastic Scattering and Energy Filtering in the Transmission Electron Microscope* Phil. Mag. **34** 49–65. One of the earliest indications of the power of EEL techniques.

Egerton, RF 1996 *Electron Energy Loss Spectroscopy in the Electron Microscope* 2nd Ed. Plenum Press New York. Includes the idea of high-contrast tuning.

Eggeman, AS, Dobson, PJ and Petford-Long AK 2007 *Optical Spectroscopy and Energy-Filtered Transmission Electron Microscopy of Surface Plasmons in Core-Shell Nanoparticles* J. Appl. Phys. **101** 024307–10.

Keast, VJ 2005 *Ab Initio Calculations of Plasmons and Interband Transitions in the Low-Loss Electron Energy-Loss Spectrum* J. Electron Spectrosc. Relat. Phenom. **143** 97–104.

Schattschneider, P and Jouffrey, B 1995 *Plasmons and Related Excitations* in Reimer, L Ed. *Energy-Filtering Transmission Electron Microscopy* 151–224 Springer New York. A thorough introduction to plasmons and related excitations.

APPLICATIONS

Arenal, R, Stéphan, O, Kociak, M. Taverna, D. Loiseau, A and Colliex, C 2005 *Electron Energy Loss Spectroscopy Measurement of the Optical Gaps on Individual Boron Nitride Single-Walled and Multi-walled Nanotubes* Phys. Rev. Lett. **95** 127601–127604.

Batson, PE 1982 *Surface Plasmon Coupling in Clusters of Small Spheres* Phys. Rev. Lett. 49 936–940.

Ding, Y and Wang, ZL 2005 *Electron Energy-Loss Spectroscopy Study of ZnO Nanobelts* J. Electr. Microsc. **54** 287–291.

Hunt, JA, Disko, MM, Behal, SK and Leapman, RD 1995 *Electron Energy-Loss Chemical Imaging of Polymer Phases* Ultramicrosc. **58** 55–64.

Kim, G, Sousa, A, Meyers, D, Shope, M and Libera, M 2006 *Diffuse Polymer Interfaces in Lobed Nanoemulsions Preserved in Aqueous Media* J. Am. Chem. Soc. **128** 6570–6571.

Kimoto, K, Kothleitner, G, Grogger, W, Matsui, Y and Hofer F 2005 *Advantages of a Monochromator for Bandgap Measurements Using Eelectron-Loss Spectroscopy* Micron **36** 185–189.

Midgley, PA, Saunders, M, Vincent, R and Steeds, JW 1995 *Energy-Filtered Convergent-Beam Diffraction: Examples and Future Prospects* Ultramicrosc. **59** 1–13.

Oleshko, VP and Howe, JM 2007 *In Situ Determination and Imaging of Physical Properties of Metastable and Equilibrium Precipitates Using Valence Electron Energy-Loss Spectroscopy and Energy-Filtering Transmission Electron Microscopy* J. Appl. Phys. **101** 054308–11.

Reimer, L 2004 *Electron Spectroscopic Imaging* in Transmission Electron Energy Loss Spectrometry in Materials Science and the EELS Atlas 2nd Ed. 347–400 Ed. CC Ahn Wiley-VCH Weinheim Germany.

Van Bentham, K, Elsasser, C and French, RH 2001*Bulk Electronic Structure of SrTiO3: Experiment and Theory* J. Appl. Phys. **90** 6156–6159.

Williams, DB and Edington, JW 1976 *High Resolution Microanalysis in Materials Science Using Electron Energy Loss Measurements* J. Microsc. **108** 113–145. Historical but not superceded!

URLs

1) http://www.lrsm.upenn.edu/~frenchrh/index.htm
2) http://www.hremresearch.com/Eng/download/documents/EELScatE2.html
3) http://www.gatan.com/answers2/index.php
4) http://www.cemes.fr/%7Eeelsdb/
5) http://www.cemes.fr/epsilon/home/main.php
6) http://www.deconvolution.com/

SELF-ASSESSMENT QUESTIONS

Q38.1 Distinguish the low-loss and high-loss regions of the spectrum.
Q38.2 What is usually the second most intense peak in any spectrum? What might be the second most intense peak in a spectrum from a very thick specimen?
Q38.3 List the characteristic scattering angles of the principal energy-loss processes and give ballpark values. How do these compare with other important scattering angles in TEM such as typical Bragg angles?
Q38.4 What's a typical value for a plasmon-energy loss?
Q38.5 What are inter- and intraband transitions and why do they result in relatively low energy losses?
Q38.6 Why is it important for the ZLP to be the most intense peak in the spectrum by a factor of 10 or more?
Q38.7 What's another expression for the 'permittivity of free space'?
Q38.8 What is meant by the 'free-electron density' and what role does it play in low-energy losses?
Q38.9 Why is the plasmon peak the most prominent energy-loss peak in the spectrum from a thin specimen?
Q38.10 What is the difference between the characteristic and the cut-off angle? Which is more important in EELS and why?

Q38.11 What electrons are in the ZLP?
Q38.12 Under what conditions would you wish to remove the tail of the ZLP?
Q38.13 Describe one other way to measure the dielectric constant apart from low-loss EELS. What are the relative advantages and disadvantages of the two approaches?
Q38.14 What's the best way to remove the tail of the ZLP?
Q38.15 What is fingerprinting and why should you be cautious about using it?
Q38.16 Why would you ever bother to form an image from which the energy-loss electron have NOT been removed?
Q38.17 Why would you ever bother to form a CBED pattern from which the energy-loss electrons have NOT been removed?
Q38.18 What is a Kramers-Kronig transformation? What information does it extract from the low-loss spectrum?
Q38.19 Why hasn't there been more use of plasmon-shift measurements for composition determination?
Q38.20 Explain why you might want to model the intensity in the low-loss spectrum.
Q38.21 Distinguish single, plural, and multiple scattering. Which is best for EELS and why?

TEXT-SPECIFIC QUESTIONS

T38.1 Distinguish the characteristic scattering angle, the cut-off angle, and the spectrometer collection angle. Explain why large differences in the characteristic scattering angle affect the information in the spectrum.
T38.2 Why does filtering out the energy-loss electrons improve the quality of images of specimens showing mass-thickness contrast?
T38.3 Why does filtering out the energy-loss electrons improve the quality of images of specimens showing diffraction contrast?
T38.4 Why does filtering out the energy-loss electrons improve the quality of diffraction patterns?
T38.5 What is contrast tuning and under what circumstance might you use it?
T38.6 Why do you think there's a residual interband transition peak in the polyethylene spectrum in Figure 38.9?
T38.7 Can you think of any other way to image the distribution of Li shown in Figure 38.8 (Hint: read Chapter 39 first)?
T38.8 Why were we able to use plasmon-peak shift measurements as an analysis technique over 30 years ago and why does nobody use it any more?
T38.9 Why does EELS low-loss determination of the dielectric constant compare with UV spectroscopy in terms of the valence states that can be determined? (Hint: work out the wavelength of electrons with a typical low energy loss.)
T38.10 Given that we typically think of the band gap as a non-spatially localized phenomenon due to overlap of the energy states above the atomic potential wells, explain how we can talk about band-gap imaging and the high spatial resolution of images such as Figure 38.10.
T38.11 Why would we want to calculate the intensity distribution in low-loss spectra?
T38.12 Given that the low-loss spectrum is so much more intense than the high-loss spectrum, why has there been relatively little theoretical and experimental work on this part of the spectrum compared with the high-loss regime?
T38.13 Estimate the relative intensities in the zero-loss and the low-loss regions of Figure 38.1 and then explain why we can approximate the total spectrum intensity to the sum of these two components.
T38.14 Study Figure 38.7, then draw diagrams showing how the spectral peaks continue to change in relative intensity with increasing thickness beyond that in Figure 38.7B.
T38.15 Why would you expect to see differences in the low-loss spectra from different compounds such as shown in Figure 38.5?
T38.16 Why do plasmon-like peaks occur in spectra from biological materials in which there are no free electrons?

CHAPTER SUMMARY

39
High Energy-Loss Spectra and Images

CHAPTER PREVIEW

The high energy-loss spectrum ($E > 50$ eV) consists primarily of ionization or core-loss edges on a rapidly decreasing plural-scattering background. Elemental-composition data and elemental maps can be extracted from these ionization edges. In this chapter, we'll examine how to get this information, quantify it, and image it. A good use for such data is light-element analysis wherein EELS complements XEDS. First, we'll remind you of the experimental variables over which you have control, because these are rather critical. Then we'll discuss how to obtain a spectrum and what it should look like if you're going to quantify it. Next, we'll discuss the various quantification routines which, in principle, are just as straightforward as those for XEDS but in practice require rather more sophisticated software, and we often need to use deconvolution routines. Elemental imaging is a powerful aspect of high-energy-loss EELS, particularly, because both the spatial resolution and minimum detection limits are somewhat better than XEDS and atomic-column spectroscopy and single-atom detection are more easily achievable in EELS.

39.1 THE HIGH-LOSS SPECTRUM

The high-loss portion of the spectrum above ~50 eV contains information from inelastic interactions with the inner or core shells. These interactions provide direct elemental identification in a manner similar to XEDS and other information, such as bonding and atomic position. We'll emphasize quantitative elemental analysis and mapping in this chapter and discuss the other features of the high-loss spectrum in the next chapter.

39.1.A Inner-Shell Ionization

When a beam electron transfers sufficient energy to a core-shell electron (i.e., one in the inner, more tightly bound K, L, M, etc., shells) to move it outside the attractive field of the nucleus, the atom is said to be ionized (go back and look at Figure 4.2). As you know from the earlier chapters on X-ray analysis, the decay of the excited atom back to its ground state may produce a characteristic X-ray or an Auger electron. So high-loss EELS and XEDS detect different aspects of the same phenomenon. Ionization is the primary event and X-ray emission is one of two secondary processes. We are interested in ionization losses precisely because the process is characteristic of the atom involved and so the signal is a direct source of chemical information, just like the characteristic X-ray. We call the ionization-loss signal in the EELS spectrum an 'edge,' rather than a peak, for reasons we'll describe shortly and we use the edges as the basis for elemental mapping.

> **EELS COMPLEMENTS XEDS**
> Detection of the high-energy electron that ionized the atom is independent of whether the atom emits an X-ray or an Auger electron. So EELS is **not** affected by the fluorescence-yield limitation that restricts light-element X-ray analysis. These differences explain, in part, the complementary nature of XEDS and ionization-loss EELS but also the much higher efficiency of EELS.

Ionization is a relatively high-energy process. For example, the lightest solid element, Li, requires an input of ~55 eV to eject a K-shell electron, and so the loss electrons are usually found in the high-loss region of the spectrum, above $E = 50$ eV. K-shell electrons require much more energy for ejection as Z increases because they are more strongly bound to the nucleus. The binding energy for electrons in the uranium K shell is about 99 keV. So we tend to use L and M edges when dealing with high-Z atoms (just like in XEDS) because the intensity of the K edges decreases substantially above ~2 keV. It's worth a short mention here about the nomenclature used for EELS edges. As for X-rays, where we have K, L, M, etc., peaks in the spectrum, we get ionization edges from K, L, M, etc., shell electrons. However, the much better energy resolution of the magnetic-prism spectrometer means that it is much easier to detect small differences in spectra that arise from the presence of different energy states in the shell. For example

- The K shell electron is in the 1s state and gives rise to a single K edge.

- In the L shell, the electrons are in either 2s or 2p orbitals, and if a 2s electron is ejected, we get an L_1 edge and a 2p electron causes either an L_2 or L_3 edge.

Depending on the ionization energy, the L_2 and L_3 edges may not be resolvable (they aren't in Al but they are in Ti), and so we call this edge the $L_{2,3}$. The full range of possible edges is shown schematically in Figure 39.1, and you can see that other dual edges exist, such as the $M_{4,5}$.

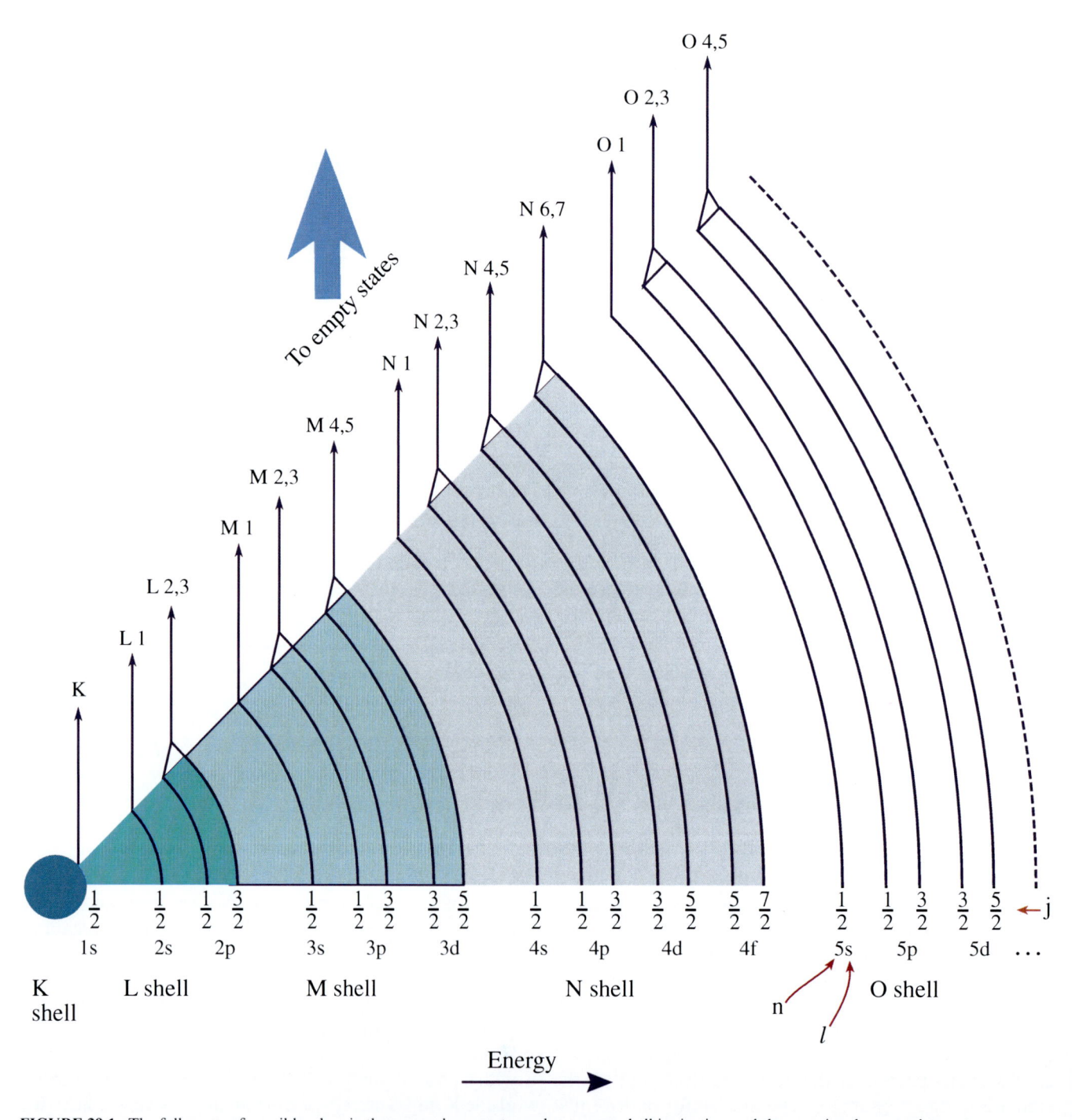

FIGURE 39.1. The full range of possible edges in the energy-loss spectrum due to core-shell ionization and the associated nomenclature.

Compared with plasmon excitation, which requires much less energy, the ionization cross sections are relatively small and the mean free paths relatively large. As a result, the ionization-edge intensity in the spectrum is much smaller than the plasmon peak, and becomes even smaller as the energy loss increases (look back to Figure 37.1). This is another reason for staying with the lower energy-loss (L and M) core edges. While the possibility of plural ionization events being triggered by the same electron is small in a typical thin foil, we'll see that the combination of an ionization loss and a plasmon loss is by no means uncommon. This phenomenon distorts the EEL spectrum and any filtered images.

If you go back and look at Figure 4.2, you can see that a specific minimum-energy transfer from the beam electron to the inner-shell electron is required to overcome the binding energy of the electron to the nucleus and ionize the atom. This minimum energy constitutes the ionization threshold, or the critical ionization energy, E_C.

We define E_C as E_K for a particular K-shell electron, E_L for an L shell, etc. Of course, it is also possible to ionize an atom by the transfer of $E > E_C$. However, the probability of ionization occurring becomes less with increasing energy above E_C because the ionization cross section decreases with increasing energy transfer. As a result, the ionization-loss electrons have an energy distribution that ideally shows a sharp rise to a maximum at E_C, followed by a slowly decreasing intensity above E_C back toward the background. This triangular shape is called an edge because you'll notice that, as shown in Figure 39.2A, it has a similar intensity profile to the absorption edges in X-ray spectroscopy. Often the term 'hydrogenic' is used for such a sharp edge-onset because this is what would arise from the ionization of the ideal single isolated hydrogen atom.

TRIANGULAR SHAPE
This idealized triangular or sawtooth shape is only found in spectra from isolated hydrogen atoms and is therefore called a hydrogenic ionization edge. Real ionization edges have shapes that approximate, more or less, to the hydrogenic edge.

In reality, because we aren't dealing with isolated atoms, but atoms integrated into a crystal lattice or an amorphous structure, the spectra are more complex. Ionization edges are superimposed on a rapidly decreasing background intensity from electrons that have undergone random, plural inelastic scattering events (Figure 39.2B). The edge may also show fine structure oscillations within ~50 eV of E_C (Figure 39.2C) which are due to bonding effects (termed energy-loss near-edge structure, ELNES). More than 50 eV after the edge, small intensity oscillations may be detectable (Figure 39.2D) due to diffraction effects from the atoms surrounding the ionized atom, and these are called extended energy-loss fine structure (EXELFS), which is analogous to extended X-ray absorption fine structure (EXAFS) in X-ray spectra, particularly, those generated from intense synchrotron sources.

ELNES AND EXELFS
Fine structure around the ionization edge onset is known as ELNES. Small intensity oscillations $>\sim50$ eV after the edge due to diffraction effects are called EXELFS.

Finally, as we noted earlier, the ionization-loss electrons may also undergo low-loss interactions. For example, they may create plasmons, in which case the edge contains extra plural-scattering intensity ~15–25 eV above E_C, as shown schematically in Figure 39.2E. So experimental ionization edges are far more complicated than the Gaussian peaks in an XEDS spectrum, but they also contain far more information about the specimen than a characteristic peak. From an XEDS spectrum, you only get *elemental* identification rather than *chemical* information, such as bonding which is contained both in the ELNES and the low-loss structure (although as we showed in Figure 32.9C, if the X-ray spectrometer has sufficiently high-energy resolution, it can detect such differences but the price to pay in the AEM is an unacceptably low count rate). Figure 39.3 shows a spectrum from BN on a C film. The various ionization edges show some of the features drawn schematically in Figure 39.2, in particular, strong ELNES on the B-K edge; we'll discuss these fine-structure effects more in Section 40.1 and how to form fine-structure images from them in Section 40.5.C.

39.1.B Ionization-Edge Characteristics

The angular distribution of ionization-loss electrons varies as $(\theta^2 + \theta_E^2)^{-1}$ and will be a maximum when $\theta = 0$, in the forward-scattered direction. The distribution decreases to a half width at the characteristic scattering angle θ_E given by equation 38.1. This behavior is essentially the same as for plasmon scattering but because we have relatively large values of E_C compared to E_P, we get larger characteristic scattering angles for ionization-loss electrons (e.g., for the typical, maximum core-loss energy that we would use for analysis $E = 2000$ eV so $\theta_E \sim 10$ mrad when $E_0 = 100$ keV).

The angular distribution varies depending on E, and because of the extended energy range of ionization-loss

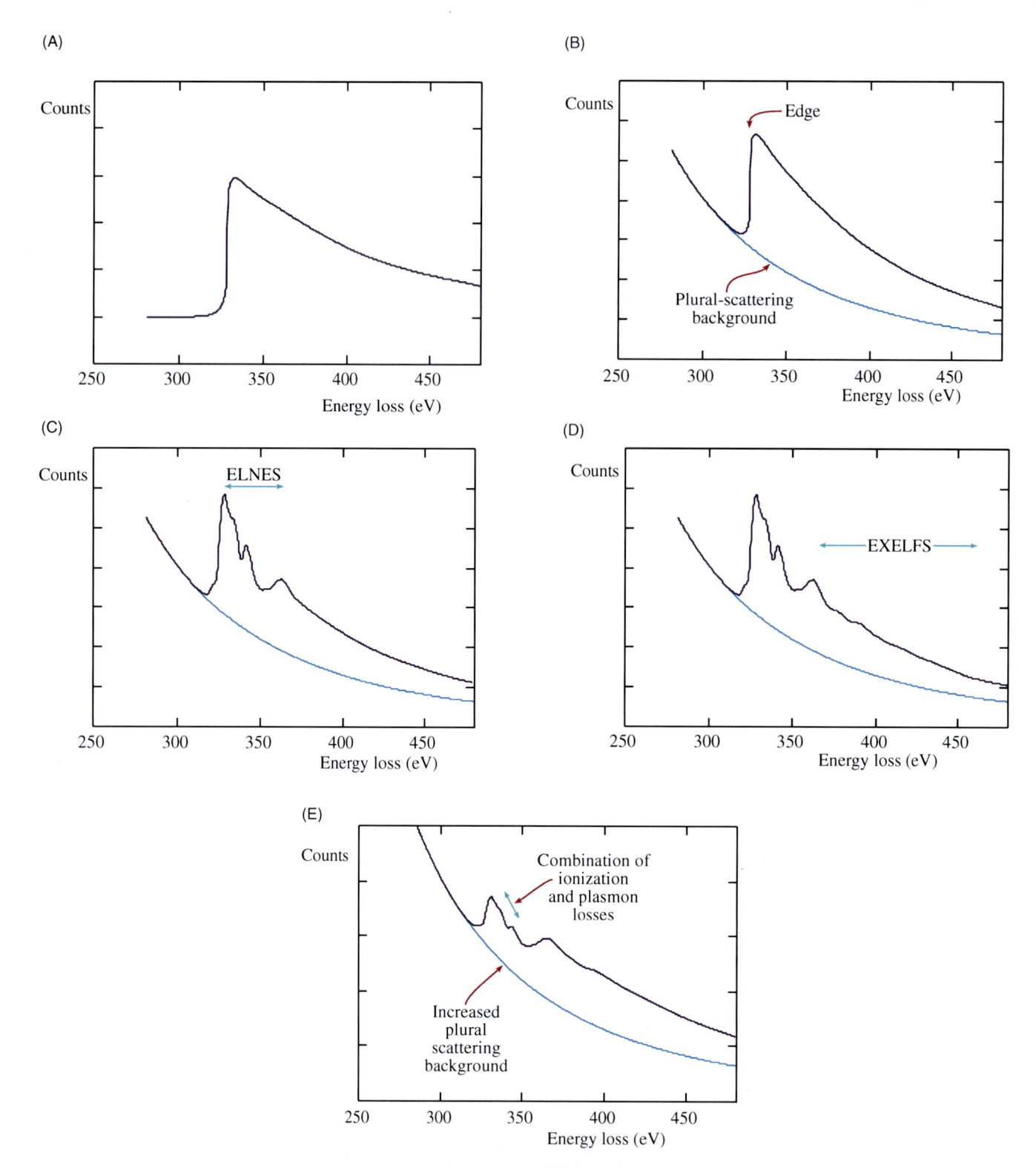

FIGURE 39.2. The characteristic features of an inner-shell ionization edge. (A) The idealized sawtooth (hydrogenic) edge. (B) The hydrogenic edge superimposed on the background arising from plural inelastic scattering. (C) The presence of ELNES. (D) The EXELFS. (E) In a thick specimen, plural scattering, such as the combination of ionization and plasmon losses, adds another peak to the post-edge structure and raises the background level.

electrons above E_C, this can be quite complicated. For $E \sim E_C$ the scattering intensity drops rapidly to zero over about 10 mrads (at θ_c), but as E increases above E_C the angular-intensity distribution drops around $\theta = 0°$, but increases at larger scattering angles, giving rise to the so-called Bethe ridge. However, this effect is not really important for the kind of analytical studies that we are emphasizing in this chapter.

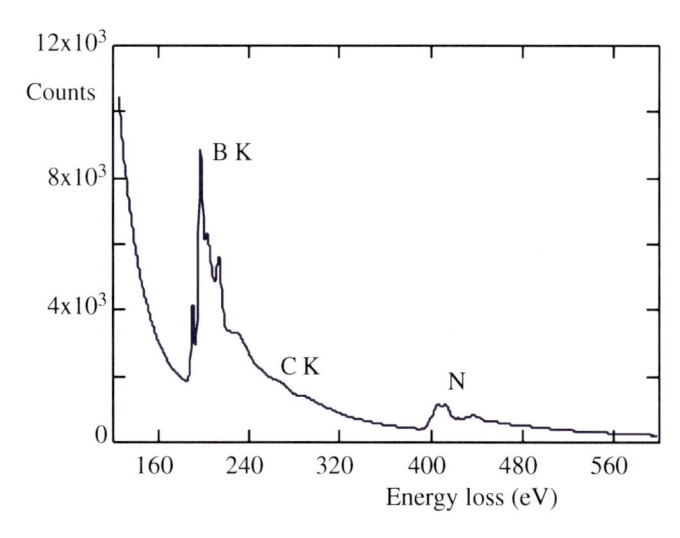

FIGURE 39.3. High energy-loss spectrum from a thin flake of BN sitting on the edge of a hole in an amorphous-carbon support film. The B K and N K edges are clearly visible superimposed on a rapidly decreasing background. A very small C K edge is also detectable at ~280 eV.

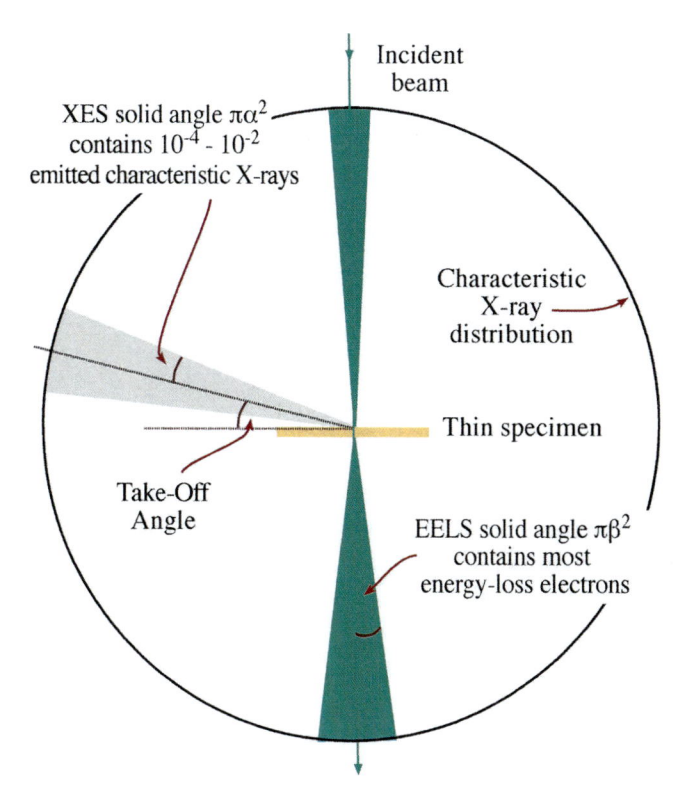

FIGURE 39.4. Comparison of the relative efficiencies of collection of EELS and XES. The forward-scattered energy-loss electrons are very efficiently collected with even a small EELS collection angle. In contrast, only a small fraction of the uniformly emitted (4π sr) characteristic X-rays is detected by the XEDS.

So, the distribution of characteristic scattering angles for the core-loss electrons that we use for analysis span the range from ~0.2 to 10 mrads and the scattering cut-off angles range from ~25 to 200 mrads (equation 38.4). In other words, like the plasmon-loss electrons, the ionization-loss electrons are very strongly forward-scattered. Consequently, efficient collection of most inelastically scattered electrons is straightforward, since a spectrometer entrance aperture angle (β) of 10 mrads will collect the great majority of such electrons. As a result, collection efficiencies in the range 50–100% are not unreasonable, which contrasts with XEDS where the isotropic generation of characteristic X-rays results in very inefficient collection. Figure 39.4 compares the collection of X-rays and energy-loss electrons and Figure 39.5 shows the variation in collection efficiency for ionization-loss electrons as a function of both β and energy.

While the K edges in Figure 39.3 show sharp onsets, like an ideal hydrogenic edge, not all edges are similar in shape. Some edges have much broader onsets, spread over several eV or even tens of eV. The edge shape in general depends on the electronic structure of the atom but, unfortunately, we can't give a simple relationship between edge types and specific shapes. The situation is further complicated because the edge shapes change depending on whether or not certain energy states are filled or unfilled. For example, if you go back and look at Figure 37.1, the Ni L edge shows two sharp peaks, which are the L_3 and L_2 edges. (We'll discuss these details much more in Section 40.1.) These sharp lines arise because the ejected L shell electrons don't entirely escape from the atom and have a very high probability of ending up in unfilled d-band states. In contrast, in Cu where the d band is full, the $L_{2,3}$ edge does not show these intense lines. Similar sharp lines appear in the $M_{4,5}$ edges in the rare earths. As if this were not enough, the

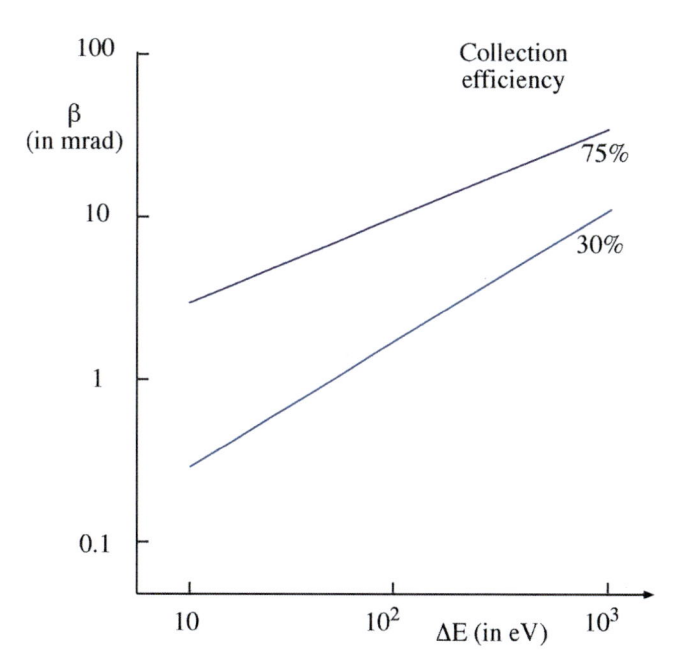

FIGURE 39.5. Variation in the collection efficiency of ionization-loss electrons as a function of energy loss and spectrometer collection angle. A 10 mrad collection angle will gather over 75% of all the incident-beam electrons that ionized C atoms and lost ~285 eV.

39.1 THE HIGH-LOSS SPECTRUM

details of the fine structure and edge shapes are also affected by bonding. For example, the Si edge in a spectrum from SiO₂ is different from the Si edge from pure Si. To sort all this out it's best if you consult the 2004 EELS Atlas (by Ahn and related references in the previous chapter) which contains representative edges from all the elements and many oxides.

Now that we've covered both the low and high energy-loss processes, we can summarize the characteristics of the energy-loss spectrum by examining a complete spectrum from NiO containing both low and high-loss electrons, as shown in Figure 39.6. In this figure, we also compare the spectrum to the energy-level diagram for NiO. You can see that

- The ZLP is above the potential wells since these electrons don't interact with the atom.
- The plasmon peak comes from interactions with the valence/conduction band electrons just below the Fermi level (E_F).

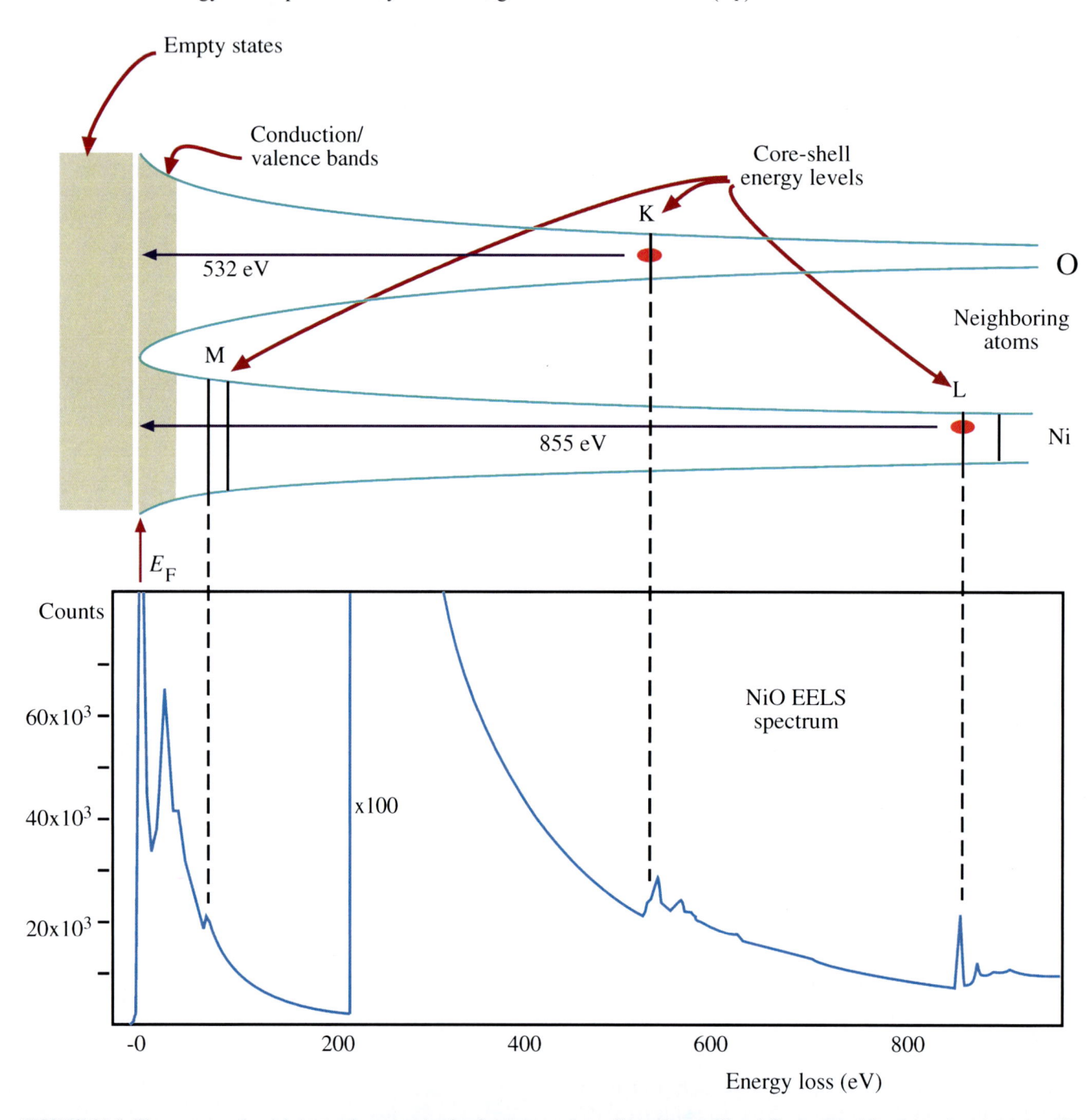

FIGURE 39.6. The correspondence between the energy levels of electrons surrounding adjacent Ni and O atoms and the energy-loss spectrum. The deeper the electrons sit in the potential well the more the energy needed to eject them. The ZLP is above the Fermi energy E_F, the plasmon peak is shown at the energy level of the conduction/valence bands where plasmon oscillations occur in the loosely bound electrons. The critical ionization energy required to eject electrons in specific shells is shown (Ni L: 855 eV and O K: 532 eV).

720 .... HIGH ENERGY-LOSS SPECTRA AND IMAGES

- The relative energy levels of the ionized shell (K, L, or M) control the position of the ionization edge in the spectrum. The closer to the nucleus, the deeper the potential well and the more the energy required to eject the electron.
- There will be a different density of states in the valence (3d) band of the Ni atom compared to the s band of the O atoms at the top of the potential wells.
- The core electrons could also be given enough energy to travel into the empty states, well above E_F and, in this case, we see ELNES after the ionization edge. We'll discuss more details of such fine structure in the spectrum in Chapter 40.

Despite the very high collection efficiency of the spectrometer, the ionization edges still show relatively low intensity, particularly as $\mathcal{E}$ increases. The edges have an extended energy range well above E_C and ride on a rapidly varying, relatively high background. All these factors, as we shall see, combine to make quantitative analysis using EELS a little more challenging than XEDS. However, for the lighter elements the X-ray fluorescence yield drops to such low values, and absorption becomes so strong, even in thin specimens, that EELS is the preferred technique. Experimentally, the choice between the two is not always simple, but below oxygen in the periodic table, EELS shows better performance than XEDS and for elements below boron, there is no sensible alternative to EELS for nanometer-scale analysis.

39.2 ACQUIRING A HIGH-LOSS SPECTRUM

From what we've described about the various EEL spectrometers and filters and the complexity of the spectra, it should be clear that there are many variables to control when acquiring a spectrum (see Brydson's monograph for a detailed description). Computer control via the Gatan software now makes this process very straightforward. We'll start by summarizing the major parameters relevant to acquiring high-loss spectra and images and indicate reasonable values for each parameter.

- *Beam Energy E_0*: It's best to use the highest E_0, unless doing so causes displacement damage or significant surface sputtering. A higher E_0 reduces the scattering cross section and so you get reduced edge intensity. Conversely, as E_0 increases, the plural-scattering background intensity falls faster than the edge intensity and so the signal to background increases. The increase in signal to background varies with the particular edge but it is never a strong variation; so while we recommend using the highest kV, it's not a good reason (on its own) to justify purchasing a 300-keV TEM.
- *Convergence angle α*: You know how to control α with the C2 aperture and/or the C2 lens, but α is only important in quantification if it is larger than β. So if you operate in TEM image or diffraction mode with a broad, parallel beam, rather than STEM mode, you can ignore α; otherwise, use the correction factor we give later in Section 39.7.
- *Beam size and current*: You control these factors by your choice of electron source, C1 lens, and C2 aperture. As usual, the beam size is important in determining spatial resolution in STEM mode, and the beam current controls the signal intensity.
- *Specimen thickness*: The specimen must be thin because this minimizes plural-scattering contributions to the spectrum and quantification is more straightforward.

SPECIMEN THICKNESS
Making your specimen as thin as possible is the most important part of EELS.

If your specimen is too thick then you'll have to use deconvolution procedures to remove the effects of plural scattering. So we'll tell you more about how to determine your specimen thickness from the spectrum and how to decide if you need to deconvolute the spectrum.

- *Collection angle β*: You know from Section 37.4 how to measure β in all operating modes. If you need lots of intensity and are happy with poor spatial resolution, use TEM image mode with no objective aperture ($\beta > \sim 100$ mrads). A small spectrometer entrance aperture ensures good energy resolution at the same time. If you want a small β to prevent contributions to the spectrum from high-angle scattering, use diffraction mode (TEM or STEM) and keep the small entrance aperture for good energy resolution. In the STEM case you also get good spatial resolution.

ENTRANCE APERTURE
Remember that a 5 mm diameter entrance aperture gives $\beta \sim 5$ mrad at a camera length of ~ 800 mm.

Generally, for analysis $\beta \sim 1$–10 mrads is fine, so long as it's less than the Bragg angle for your specimen orientation; but for EELS imaging, which we discuss in Section 39.9, 100 mrads may be necessary to get the necessary signal intensity.

- *Energy resolution*: ΔE is limited by your electron source unless you have a monochromator. Elemental analysis and imaging (the topic of this chapter) do not require the best ΔE, so ~5 eV would suffice. You really need the best ΔE for low-loss and fine-structure studies, which are probably the most useful and widespread aspect of EELS (see the surrounding chapters). Use an FEG source and a PEELS/imaging filter if you want to do this, especially if you're lucky enough to have access to a monochromator.
- *Energy-loss range and spectrum dispersion*: The full spectrum extends out to the beam energy E_0, but the useful portion only extends to ~2 keV. Above this E, the intensity is very low, and XEDS is both easier and more accurate. Since you rarely need to collect a spectrum above ~2 keV a minimum of 2048 channels in the computer display, giving 1 eV/channel is a good starting dispersion. You can easily select a higher display resolution if you want to look at a more limited region of the spectrum or if you want to see detail with ΔE <1 eV. Typically, you're only examining a limited portion of the spectrum anyhow and you set this by putting the necessary voltage on the drift tube or changing the high voltage.
- *Dwell time*: If you have a PEELS with a PDA, set the integration time so that at the maximum intensity you don't saturate the diodes: i.e., stay below 16,000 counts per acquisition in the most intense channel and sum as many spectra as you need to give sufficient counts for analysis.
- *Number of acquisitions*: Again, if you have PEELS/PDA, multiple acquisitions may be necessary to get sufficient counts in the edge, but remember that multiple acquisitions may give rise to minor artifacts, as we discussed in Section 37.5.

Before you analyze a particular spectrum, you should check four things

- Focus and align the ZLP and check the spectrometer resolution.
- Look at the low-loss (plasmon) portion of the spectrum; this gives you an idea of your specimen thickness.
- Look for the expected ionization edges. If you can't see any edges, your specimen is probably too thick or you need to raise the display gain.
- It's probably worth deconvoluting out the PSF prior to any quantification.

The first of these tasks is not critical, as we noted earlier. Regarding the second task, we noted back in Chapter 38 that, to a first approximation, if the plasmon-peak intensity is less than about one tenth the ZLP, then the specimen is thin enough for analysis. Otherwise, you'll probably have to deconvolute plural-scattering effects from your experimental spectrum. For the third task, you should ideally see discrete edges on a smoothly varying background, but you need to see at least a change in slope in the background intensity at the expected E_C. If the background intensity is too noisy it will make quantification more difficult, so acquire sufficient counts to generate a smoothly varying background.

THE JUMP RATIO
An important parameter in determining the quality of your spectrum is the signal-to-background ratio which in EELS we call the jump ratio.

The jump ratio is the ratio of the maximum edge intensity (I_{max}) to the minimum intensity (I_{min}) in the channel preceding the edge onset, as shown in Figure 39.7 (which is a well-defined edge from a suitably thin film of amorphous carbon). If the jump ratio is above ~5, for the carbon K edge at 284 eV from a 50 nm carbon film at 100 kV, then your system is operating satisfactorily. You should keep a standard thin, amorphous-carbon film available as a standard reference specimen and occasionally check that the jump ratio remains the same. We'll see that jump-ratio imaging is one method of acquiring filtered images from ionization edges. The jump ratio increases as E_0 increases. If you can't get such a jump ratio from a standard, thin, carbon film, then probably you need to realign the spectrometer. The actual ionization-loss edges from your real specimen, that you may wish to quantify or use to form images, will probably be nothing like this ideal edge, but the EELS software programs are more than capable of

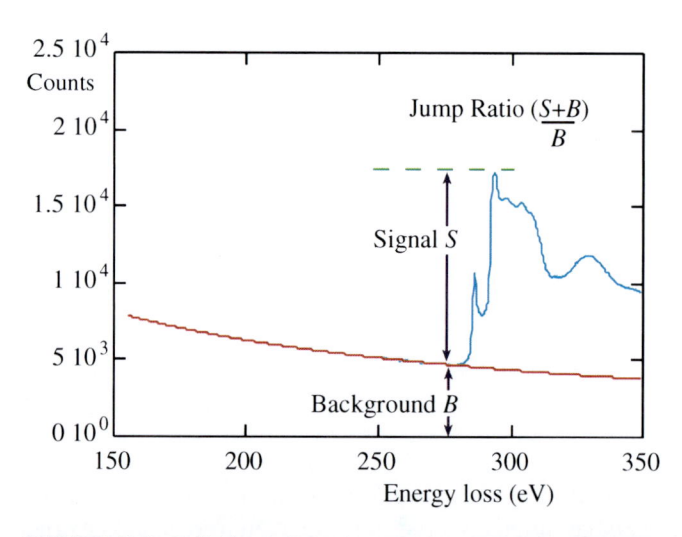

FIGURE 39.7. Definition of the jump ratio of an ionization edge which should be about 5–10 for the carbon K edge if the EELS is well aligned and the specimen really thin. This spectrum shows an adequate jump ratio.

handling much smaller edges riding on much higher backgrounds.

39.3 QUALITATIVE ANALYSIS

As with XEDS, you should always carry out a qualitative analysis first to ensure that you have identified all the features in your spectrum. Then you can decide which edges to use for subsequent quantitative analysis and imaging.

Qualitative analysis using ionization edges is very straightforward. Unlike XEDS, there are actually very few artifacts that can be mistaken for an edge. The most prominent artifact that may lead to misidentification is the ghost peak from diode saturation (see Section 37.5) which is easily removed. So long as you calibrate the spectrum to within a few eV you can unambiguously identify the edge energy.

> **IONIZATION EDGE**
> We identify the ionization edge as the energy loss at which there is a discrete increase in the slope of the spectrum; this value is the edge onset, i.e., E_C, the critical ionization energy.

You have to be careful here: sometimes you'll see the edge energy defined somewhat arbitrarily halfway up the edge, e.g., at the π^* peak on the front of a C-K edge. There is no strict convention, and very often L and M edges do not have sharp onsets anyhow. Examination of a portion of a spectrum, such as that shown back in Figure 39.3, is usually sufficient to let you draw a definite conclusion about the identity of the specimen, which in this case is BN on a C support film. In addition, it is wise to compare your spectrum with reference spectra from the EELS Atlas that we've mentioned several times before or through an on-line database, such as URL #1.

Remember that there are families of edges (K, $L_{2,3}$, $M_{4,5}$, etc.) just as there are families of peaks in X-ray spectra but, as with X-ray spectra, you might not be able to resolve all the edges in a single family. Given that above ~2 keV the edges are usually too small to be detected, it is in fact very rare that you would expect to see more than one family of lines from a given element (the Si L edge at ~100 eV and the Si K edge at 1.7 keV should both be visible in the same spectrum). As a rule of thumb, quantification is equally easy with K and L edges, but the accuracy of K-edge quantification is slightly better. Up to $Z = 13$ (Al) we usually use K edges because any L edges occur at very low energy and are masked by the plasmon peak. Above $Z = 13$ you can use either K or L edges. Sometimes there is the question of which edge is most visible. The K edge onset is generally a bit sharper than the L edge which consists of both the L_2 and L_3 edges and so may be somewhat broader, but this is not always the case.

L edges for $Z = 19–28$ (e.g., the Ca-L edge in Figure 37.12 and the Cr-L edge in Figure 39.13) and $Z = 37–45$ are characterized by intense near-edge structure called white lines. M edges for $Z = 55–69$ have similar intense lines.

These white lines are so named because they appeared as lines of varying intensity in photographically recorded, energy-loss spectra; they also appear that way if you look at the spectra from in-column filters (see Figure 37.14A). More details will be given in Section 40.1. If you have to use the M, N, or O edges without any white lines, you should know that they are very broad, with an ill-defined threshold, and quantification is best achieved with standards, as we'll see shortly.

The energy-loss spectrum clearly does not lend itself to a quick 'semi-quantitative' analysis; so we can't follow our XEDS approach. For example, the spectrum in Figure 39.3 comes from equal numbers of B and N atoms, but the intensities in the B and N edges are markedly different. This difference arises because of the variation in ionization cross section with E, the strongly varying nature of the plural-scattering background, and the edge shape, which causes the C and N K edges to ride on the tails of the preceding edge(s).

The Ti-nitride and Ti-carbide example: Sometimes qualitative analysis is often all that you need to do. Figure 39.8 shows images and spectra from two small precipitates in an alloy steel. The spectra show a Ti L_{23} edge in both cases and C and N K edges in Figure 39.8A and B, respectively. It does not take much effort to deduce that the first particle is TiC because it is the only known carbide of Ti, but the nitride could be either TiN or Ti_3N. To determine which of the two it is, you have to carry out full quantification, which we'll discuss shortly. You should note that such clear discrimination between TiC and TiN in Figure 39.8B would be difficult using windowless XEDS because the energy resolution is close to the separation of the Ti L ($E = 452$ eV) and the N K ($E = 392$ eV) X-ray peaks. In addition, the DPs from both phases are almost identical, so this problem is a perfect one for EELS.

39.4 QUANTITATIVE ANALYSIS

To quantify the spectrum or to form a quantitative image, you have to integrate the intensity (I) in the ionization edge(s) by removing the plural-scattering background. Then you have to determine the number of atoms (N) responsible for I. N is related to I by a sensitivity factor termed the partial ionization cross section (σ). We'll see that σ plays a similar role to the k_{AB} factor in X-ray analysis. If you go back and look at Figure 39.2,

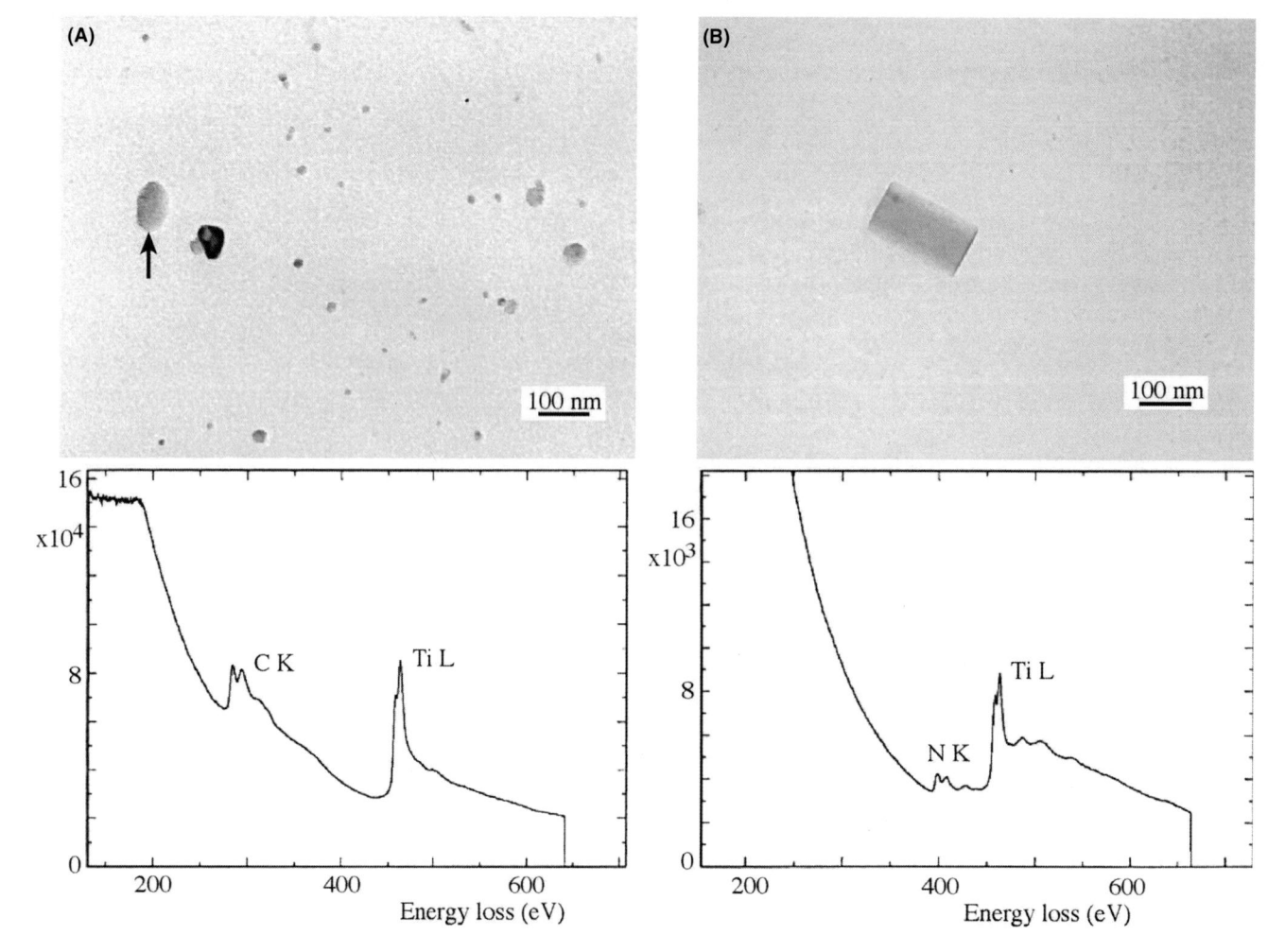

FIGURE 39.8. Images of small precipitates on an extraction replica from a stainless steel specimen, and the corresponding ionization edges showing qualitatively the presence of Ti, C, and N. Thus the precipitates can be identified as (A) TiC and (B) TiN, respectively.

you'll see how an ionization edge is built up from several contributions. The process of quantification in essence involves stripping away (or ignoring) the various contributions until you're left with Figure 39.2A, which contains the single-scattering or hydrogenic-edge intensity.

39.4.A Derivation of the Equations for Quantification

The equations we use for quantitative analysis and imaging have been derived, refined, and applied by Egerton and co-workers. The following derivation is a summary of the full treatment by Egerton given in his textbook.

We'll assume that we are quantifying a K edge, although the basic approach can be used for all edges. The K-edge intensity above background, I_K, is related to the probability of ionization, P_K, and the total transmitted intensity, I_T, by

$$I_K = P_K I_T \tag{39.1}$$

This equation assumes that the intensities are measured over the complete angular range (0–4π sr), which of course, is not the case, but we'll correct this later. In a good thin specimen we can approximate I_T to the incident intensity, neglecting backscatter and absorption effects. Now, this is the *important* point: if we assume also that the electrons contributing to the edge have only undergone a single ionization event, then we can easily obtain an expression for P_k.

$$P_K = N \sigma_K \exp\left(-\frac{t}{\lambda_K}\right) \tag{39.2}$$

where N is the number of atoms *per unit area* of the specimen (of thickness t) that contribute to the K edge. The assumption of a single ionization (i.e., scattering) event is reasonable, given the large mean free path (λ_k) for ionization losses; and it explains why you have to make thin specimens. Assuming single scattering also means that the exponential term is very close to unity and so

$$I_K \approx N\sigma_K I_T \tag{39.3}$$

and therefore

$$N = \frac{I_K}{\sigma_K I_T} \tag{39.4}$$

Thus, we can measure the absolute number of atoms per unit area of the specimen simply by measuring the intensity above background in the K edge and dividing it by the total intensity in the spectrum multiplied by the ionization cross section. We can easily extend this expression to two edges from elements A and B, in which case the I_T drops out and we can write

$$\frac{N_A}{N_B} = \frac{I_K^A \sigma_K^B}{I_K^B \sigma_K^A} \tag{39.5}$$

Similar expressions apply to L, M edges, etc., and combinations of edges can be used. So you see that if you are quantifying more than one element, then you don't need to gather the ZLP, which saves hitting the PDA or CCD with this high-intensity signal.

In both equations 39.4 and 39.5 we assumed that we could accurately subtract the background under the edge and that we know σ. Unfortunately, as you'll see, both background subtraction and determination of σ are non-trivial. We will discuss these points later, but first we must take account of the practical realities of spectrum acquisition, and modify the equations accordingly.

- First, you can't gather the whole of the spectrum out to the beam energy, E_0, because above ~2 keV the intensity decreases to a level close to the system noise.
- Second, while ionization-loss electrons can theoretically have any energy between E_C and E_0, in practice the intensity in the edge falls to the background level within about 100 eV of the threshold, E_C.
- Third, the background-extrapolation process becomes increasingly inaccurate beyond ~100 eV.

For all these reasons, it is imperative to restrict integration of the edge intensities to some window, Δ, usually in the range 20–100 eV. So we modify equation 39.4 to give

$$I_K(\Delta) = N\sigma_K(\Delta)I_T(\Delta) \tag{39.6}$$

The term $I_T(\Delta)$ is more correctly written as $I_l(\Delta)$ where I_l is the intensity of the zero-loss (direct beam) electrons combined with the low-loss electrons over an energy-loss window Δ. Only if we have true single scattering can we use I_T and we'll discuss the conditions for this later.

As we discussed, EELS has the tremendous advantage that the energy-loss electrons are predominantly forward scattered and so you can easily gather most of the signal. So because we never manage to collect the full angular range of energy-loss electrons, we must further modify the equation by including the collection angle β and write

$$I_K(\beta\Delta) = N\sigma_K(\beta\Delta)I_l(\beta\Delta) \tag{39.7}$$

This factor $\sigma_K(\beta\Delta)$ is the partial ionization cross section.

From this equation therefore, the absolute quantification for N is given by

$$N = \frac{I_K(\beta\Delta)}{I_l(\beta\Delta)\sigma_K(\beta\Delta)} \tag{39.8}$$

For a ratio of two elements A and B, the low-loss intensity drops out again as in equation 39.5

$$\frac{N_A}{N_B} = \frac{I_K^A(\beta\Delta)\sigma_K^B(\beta\Delta)}{I_K^B(\beta\Delta)\sigma_K^A(\beta\Delta)} \tag{39.9}$$

We can draw a direct analogy between this equation and the Cliff-Lorimer expression (equation 35.2) used in thin-film XEDS. In both cases, the composition ratio C_A/C_B or N_A/N_B is related to the intensity ratio I_A/I_B through a sensitivity factor which we call the k_{AB} factor in XEDS and which in EELS is the ratio of two partial cross sections, σ^B/σ^A.

Remember that the major assumption in this whole approach is that the electrons undergo *a single-scattering event*. In practice, it's difficult to avoid some plural scattering, although in very thin specimens the approximation remains valid, if errors of ±10–20% are acceptable. If plural scattering is significant, then it must be removed by deconvolution, which we'll discuss in Section 39.6. You should also note when using the ratio equation your analysis is a lot better if the two edges are similar in shape, i.e., both K edges, or both L edges, otherwise the approximations inherent in equation 39.9 will be less accurate.

In summary, equations 39.8 and 39.9 give us, respectively, an absolute value of the number of atoms/unit area of the specimen or a ratio of the number of atoms of the elements A and B either at a given analysis point or within a filtered image. To get this information experimentally, you have to carry out two essential steps

- Background subtraction to obtain I_K (and hence, N) for each element A, B, etc.
- Determination of the partial ionization cross section $\sigma_K(\beta\Delta)$ to get the ratio N_A/N_B.

So again, you can see why it is important to know β.

39.4.B Background Subtraction

The background is a rapidly changing continuum decreasing from a maximum intensity just after the plasmon peak at about 15–25 eV, down to a minimum at which it is indistinguishable from the system noise, typically at $E > \sim 2$ keV. In addition to plural scattering, there is also the possibility of single-scattering contributions to the background from the tails of preceding ionization edges and perhaps contributions from the spectrometer itself. Because of the complexity of these contributions, it has not been possible to model the background from first principles, as is possible in XEDS using Kramers' law.

Despite the complexity of the various contributions to the background, the methods for subtraction are relatively simple. There are two ways commonly used to remove the background

- Curve fitting.
- Using difference spectra.

Curve Fitting: You select a window δ in the background before the edge onset and fit a curve to the intensity in the window. Then you extrapolate the curve over another window Δ under the edge. This process is shown schematically in Figure 39.9, and experimentally in Figure 39.10.

We assume that the energy dependence of the background has the form

$$I = A E^{-r} \quad (39.10)$$

where I is the intensity in the channel of energy loss E, and A and r are constants. The fitting parameters are only valid over a limited energy range because they

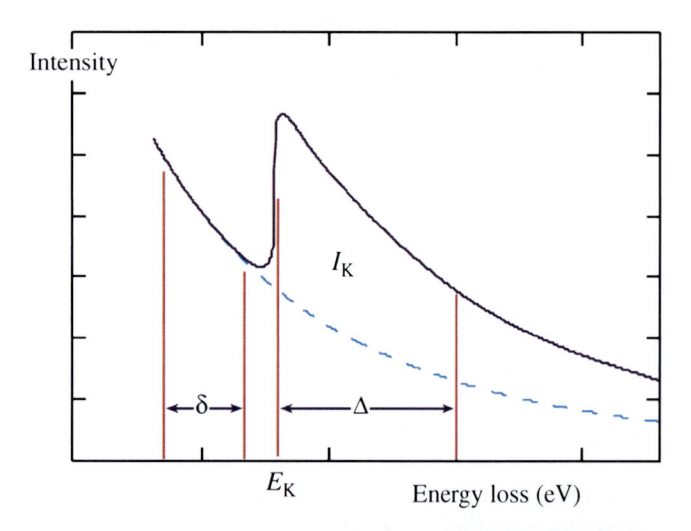

FIGURE 39.9. The parameters required for background extrapolation and subtraction under an ionization edge. The pre-edge fitting window δ is extrapolated over a post-edge window Δ then the intensity under the extrapolated line is subtracted from the total intensity in the window Δ to give the desired edge intensity I_K.

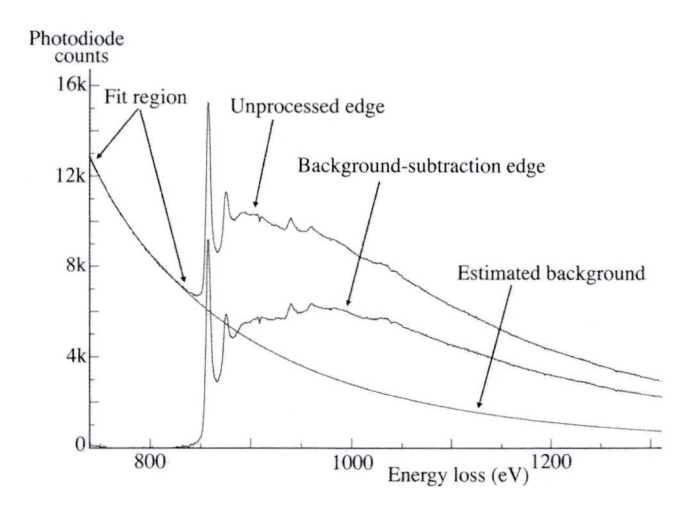

FIGURE 39.10. Comparison of an experimental Ni $L_{2,3}$ edge before and after background subtraction. The fitted region before the unprocessed edge is extrapolated to give the estimated background which is then subtracted leaving the (total) edge intensity (note: there is no edge window Δ shown here).

depend on E. The exponent r is typically in the range 2–5, but A can vary tremendously. We can see some trends in how r varies. The value of r decreases as

- The specimen thickness, t, increases.
- The collection angle, β, increases.
- The electron energy loss, E, increases.

The fit of the curve to the tail of a preceding edge shows a similar power-law dependence to the background, and may be fitted similarly, i.e., $I = BE^{-s}$. The fitting window δ should not be <10 channels and should not be >30% of E_K. In practice, however, you might not be able to fit the background over such a wide window if another edge is present within that range, which limits the goodness of fit of the curve.

You should choose the extrapolation window, Δ, such that the ratio of the finish to the start energies, $E(\text{finish})/E(\text{start})$, is < 1.5; so Δ is smaller for lower edge energies. Using larger windows, although improving the statistics of the edge intensity, eventually reduces the accuracy of the quantification because the fitting parameters A and r are only valid over ~100 eV. If there's a lot of ELNES, either use a larger Δ to minimize its effect or avoid it in the extrapolation window unless the quantification routine can handle it.

Instead of the simple power-law fit, you can in fact use any expression, such as an exponential, polynomial, or log-polynomial, so long as it provides a good fit to the background and gives acceptable answers for known specimens. Polynomial expressions can behave erratically if you extrapolate them over a large Δ, so use them cautiously. Generally, the power law seems adequate for most purposes except close to the plasmon peaks ($E < \sim 100$ eV). Clearly, the background channels closest to the edge onset will influence the extrapolation most strongly, and various weighting schemes have been proposed. A noisy spectrum will be particularly susceptible to poor fitting, unless some type of weighting is used.

> **FITTING WINDOWS**
> There are two fitting windows: δ before the edge and Δ after the edge. Each has constraints for good background fitting.

We can judge the goodness of fit of a particular power-law expression qualitatively by looking at the extrapolation to ensure that it is heading toward the post-edge background and not substantially under- or over-cutting the spectrum. More quantitatively, we can assign a χ^2 value based on a linear least-squares fit to the experimental spectrum. The least squares fit can be conveniently tied in with a weighting scheme using the expression

$$\chi^2 = \sum_i \frac{(y - y_i)^2}{y^2} \quad (39.11)$$

where y_i is the number of counts in the ith channel and $y = \ln_e I$. (Look back at Section 35.3.B.) The squared term in the denominator ensures suitable weighting of the channels close to the edge. Alternatively, the Gatan software includes 'smart' feedback which forces the background extrapolation to merge with the experimental background intensity well after the edge.

Difference Spectra: You can also remove the background using a first-difference approach (which is equivalent to differentiating the spectra). This method is particularly suited to PEELS since it simply involves taking two spectra, offset in energy by a few eV, and subtracting one from the other. As shown in Figure 39.11, the difference process results in the slowly varying background being reduced to zero and the rapidly varying ionization edge intensity showing up as classic difference peaks, similar to what you may have seen in Auger spectra. This is the *only* way to remove the background if your specimen thickness changes over the area of analysis and it also has the advantage that it suppresses spectral artifacts common to PEELS, particularly the channel-to-channel gain variation.

> **DIFFERENTIATE A SPECTRUM**
> The difference spectrum is a numerical method of differentiating the spectrum. It emphasizes changes in the spectrum.
> The top-hat filter gives a second-difference spectrum. Difference and/or division are all carried out digitally.

Another kind of difference method involves convoluting the experimental spectrum with a top-hat or similar filter function, as we described in Section 35.3 for XEDS (Michel and Bonnet). Top-hat filtering effectively gives a second-difference spectrum which also removes the background but exacerbates some artifacts.

Background subtraction for energy-filtered imaging can be achieved in several ways.

The first and most usual method of background subtraction while filtering, is called the three-window method (Jeanguillaume et al. 1978). Two pre-edge windows are used to calculate the background fit and one post-edge window in which the extrapolated background is subtracted from the total intensity to leave the edge intensity (equivalent to adding another pre-edge window in Figure 39.9). Egerton has shown that the intensity in the background window under the edge (I_b) is related to the intensity in the two pre-edge windows (I_1 and I_2) by

$$I_b = [A/(1-r)][E_h^{1-r} - E_l^{1-r}] \quad (39.12)$$

where A and r are the usual factors in the background-fitting equation (39.10) determined from I_1 and I_2 and E_h and E_l are the high- and low-energy values defining the extrapolation window under the edge. To factor out the thickness effects, a low-loss image has to be acquired and divided into the K edge image, as in equation 39.8. Alternatively, two edges can be quantified and divided to give relative quantitative images, as in the quantification equation 39.9. Selection of the energy windows and choice of their width are subject to all the limitations we discussed for background subtraction and peak integration in that same section. Because the specimen thickness will often vary over the area being imaged, this method must be applied at every pixel in the image.

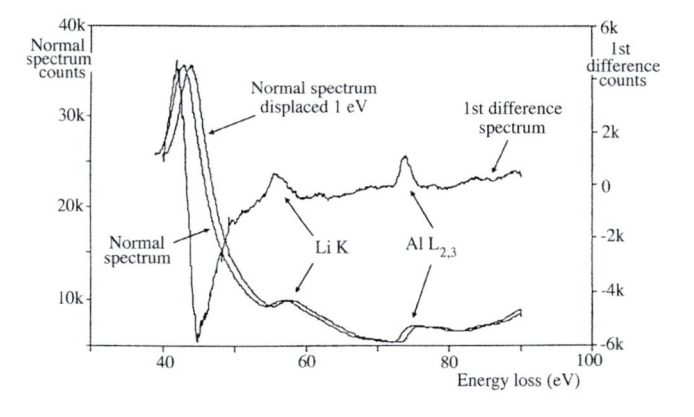

FIGURE 39.11. First-difference method of background subtraction, showing two PEELS spectra from a specimen of Al Li displaced by 1 eV and subtracted from one another to give a (first-difference) spectrum in which the background intensity falls to a straight line of close to zero counts and the small Li K and Al L$_{2,3}$ edges are clearly revealed.

> **METHODS FOR BACKGROUND SUBTRACTION**
> The three-window method: use two pre-edge windows to calculate a background.
> The jump ratio: calculate by dividing two signals.
> Maximum-likelihood: when you only have a few channels to use for the subtraction.

The second method that is the commonly used method is to simply divide the signal in the edge by the signal in a background window just preceding the edge. These so-called jump-ratio images are only qualitative but give useful information, as we'll see in Section 39.9.

The third background-subtraction approach often used in ETEM imaging is the maximum-likelihood method (Unser et al.), which is useful when only a few channels are available for estimating the background and the peak intensity.

Kothleiner and Hofer describe the various parameters that control your choice of the best three windows. In general, many of the considerations for the selection of δ and Δ for spectral acquisition also apply during imaging.

39.4.C Edge Integration

The edge integration procedure you use depends on how you removed the background. If you used a power-law approach, then remember that there is a limit over which the edge integration window Δ is valid. The value of Δ should be large enough to maximize the integrated intensity, but not so large that the errors in your background extrapolation dominate. Often the presence of another edge limits the upper end of the integration window. The lower end is usually defined from the edge onset, E_K, but if there is strong (well defined) near-edge structure, such as in the B K edge or the Ca L_{23} edge, then your integration window should start at an energy above these, unless the quantification schemes can handle ELNES (see below). If you subtracted the background using a first-difference approach, then you determine the peak intensity by fitting the experimental spectrum to a reference spectrum from a known standard using multiple least-squares fitting. We'll talk more about this when we discuss deconvolution of spectra.

39.4.D The Partial Ionization Cross Section

There are several ways we can determine the partial ionization cross section, $\sigma(\beta\Delta)$. We either use a theoretical approach or compare the experimental spectra with known standard spectra.

Theoretical calculation: The most common approach is that due to Egerton (the 1979 and 1981 papers) who produced two short computer programs to model the K and L shell partial cross sections. The programs are called SIGMAK and SIGMAL, respectively. They are public-domain software, have been updated regularly, and are stored in your EELS computer system. They are standard parts of the Gatan software. The codes are given in the appendices of Egerton's book. The cross sections are modeled by approximating the atom in question to an isolated hydrogen atom with a charge on the nucleus equal to the atomic number Z of the atom, but with no outer-shell electrons.

At first sight, this so-called hydrogenic cross section is an absurd approximation! The approach is actually tractable because the hydrogen-atom wave function can be expressed analytically by Schrödinger's wave equation, which can be modified to account for the increased charge on atoms above hydrogen. Because this treatment neglects the outer-shell electrons, it is best suited to K shell electrons. Figure 39.12 shows comparison between the measured nitrogen-K intensity and that computed using SIGMAK. As you can see, the SIGMAK hydrogenic model essentially ignores the near-edge and post-edge fine structure (which would be absent in the spectrum from a hydrogen atom), but still gives a reasonably good fit to the experimental edge. Figure 39.13 compares the Cr L edge with the SIGMAK model. The L shell fit is almost as good as the K fit, although the white lines are imperfectly modeled. These programs are very widely used since they are simple to understand and easy and quick to apply. An alternative approach uses empirical parameterized equations to modify σ for the effects of β and Δ; Egerton gives the appropriate codes which you can also download from URL #2.

There are more complex methods which calculate the cross section in a more realistic way than the

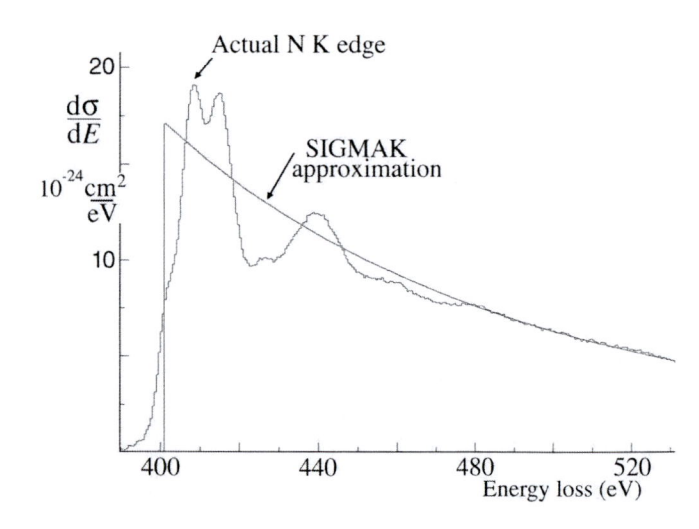

FIGURE 39.12. Comparison between an experimental N-K edge and a hydrogenic fit to the edge obtained using the SIGMAK program. The fit makes no attempt to model the near-edge fine structure but the total area under the fit is still a close approximation to the area under the experimental edge.

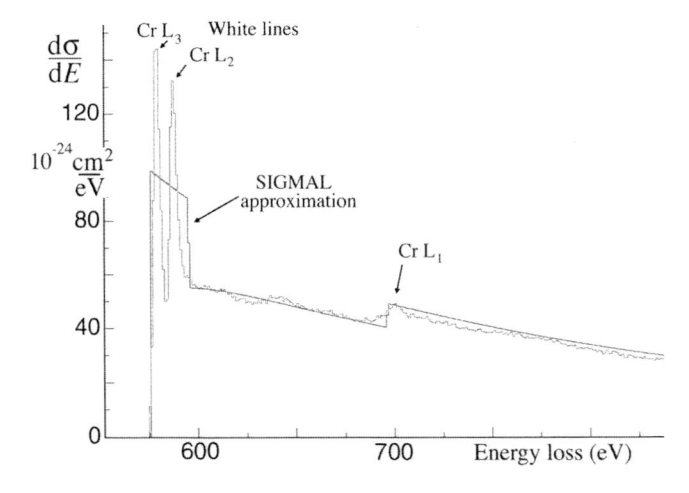

FIGURE 39.13. Comparison between an experimental Cr $L_{2,3}$ edge and a hydrogenic fit obtained using the SIGMAL program. The fit makes no attempt to model the intense white lines, but only makes a rough estimate of their average intensity.

SIGMAK/L hydrogenic models, e.g., using the Hartree-Slater model which is available in the Gatan software or atomic-physics approaches which are better for the more complex L and M (and even N) edges (see papers by Rez and Hofer et al.). Egerton has compared experimental and theoretical cross sections; the M shell data (which are the worst case) are shown in Figure 39.14. The data are actually plotted in terms of the oscillator strength γ (which is a measure of the response of the atom to the incident electron). This term is the integral of the generalized oscillator strength (GOS), which is proportional to the differential cross section, so just think of γ as being proportional to σ. There is still relatively poor agreement between experiment and theory for the M shell. Note that the models in Figure 39.14 are all atomic rather than hydrogenic. Similar data in Egerton's paper show better agreement for K and L shells. These models, while more precise, require substantially longer computing time but this is fast becoming less of a problem. Given the other sources of error in EELS analysis, you rarely need to go to such lengths to obtain a better value of $\sigma(\beta\Delta)$ and, unless you are well versed in the physics of ionization cross sections, you should probably stick with the SIGMAK/L methods, particularly for routine quantification.

Experimental Determination: Rather than calculating σ theoretically, you can generate a value experimentally using known standards. This approach is of course exactly analogous to the experimental *k*-factor approach for XEDS quantification in which the cross section is automatically included (along with the fluorescence yield and other factors). It is surprising at first sight that the classic XEDS approach of using standards has not been widely used in EELS, but the reason is obvious when you remember the large number of variables that affect the EELS data. The standard and unknown must have the same thickness, the same bonding characteristic, and the spectra must be gathered under identical conditions; in particular β, Δ, E_0, and t must be the same. So your standard would basically be your unknown!

Again, it is the problem of thickness measurement that appears to be the main limitation to improving the accuracy of analysis.

In summary, there are two approaches to the determination of $\sigma(\beta\Delta)$: theoretical calculation and experimental measurement. In contrast to XEDS, the theoretical approaches dominate. There is good evidence that, particularly for the lighter elements, for which EELS is best suited, the simple and quick hydrogenic model is usually adequate. However, for the heavier elements, where the M shell is used for analysis, the calculations are getting better although for such elements, it's still probably better to revert to X-ray analysis. Leapman has given a lucid and much more detailed description of the various methods of background subtraction and peak integration necessary for quantification.

One last point worth noting is that, before carrying out an experiment or gathering a long spectrum image for quantification, it is worth simulating the spectra to see if the experiment will produce useful data. We

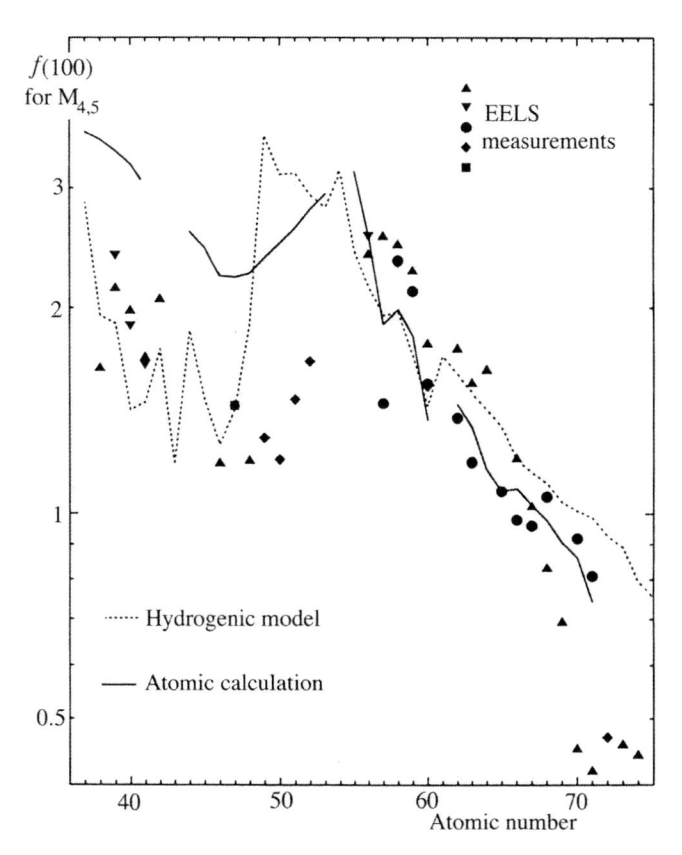

FIGURE 39.14. Comparison of the experimental and theoretical approaches to determination of the $M_{4,5}$ ionization cross section shown in terms of the variation in the dipole oscillator strength (*f*) as a function of atomic number.

described the advantages of DTSA for similar aspects of XEDS data in Chapter 33 and the companion text chapter. Gatan offers the EELS Advisor software (URL #3) to help take the uncertainty out of your planned experiment. Like DTSA, the EELS advisor allows you to simulate both spectra and images and will let you know if the element you seek will be detectable under your planned experimental conditions. It may tell you that the specimen is too thick or the amount of the element is below the detection or spatial resolution limits and that you need to change one or more of the many experimental variables that you have at your disposal.

39.5 MEASURING THICKNESS FROM THE CORE-LOSS SPECTRUM

While you may think we're now in a position where we have all the data needed to solve the quantification equations 39.8 and 39.9, our assumption all along has been that the spectra were the result of single scattering and we neglected plural scattering. In practice, there will *always* be some plural-scattering contribution to the ionization edges.

> **COMBINATION**
> The combination of a plasmon interaction and an ionization will show up as a bump about 15–25 eV past the onset of the edge.

This effect is shown schematically back in Figure 39.2E. So how do we go about correcting this? We can either make our specimens so thin that plural scattering is negligible or we can deconvolute the spectra. The former approach is better but sometimes unrealistic. The latter approach is mathematically simple but, as we've taken pains to point out on several occasions, deconvolution can be misleading and create spectral artifacts if not done properly; so we need to examine deconvolution in more detail. First, let's look at how we determine t because EELS offers us a simple method for this.

We saw back in Section 38.3.C that the plasmon-peak intensity is a measure of the specimen thickness but there is also thickness information in any energy-loss spectrum since the *total* amount of inelastic scatter increases with specimen thickness. So we can write a parallel expression to equation 38.7

$$t = \lambda \ln \frac{I_t}{I_0} \quad (39.13)$$

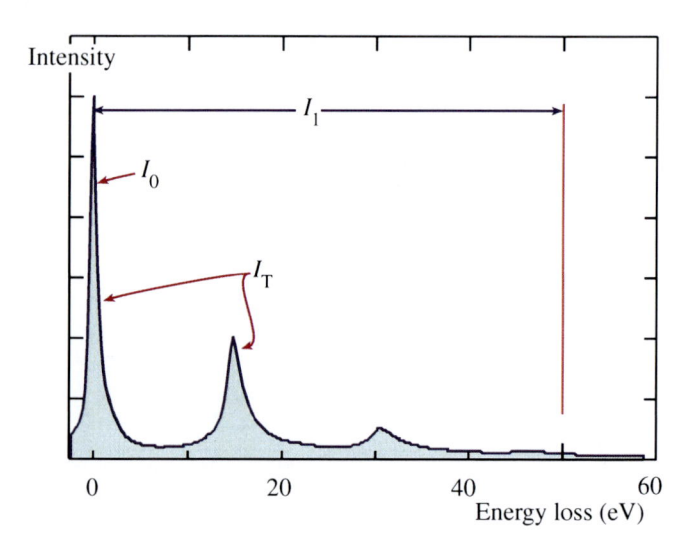

FIGURE 39.15. Definition of the zero-loss counts (I_0) and the total counts (I_T) required for thickness determination. I_T is effectively equivalent to the low-loss counts (I_l) out to ~50 eV, including I_0.

where I_0 is the ZLP intensity, I_t is the total intensity in the low-loss spectrum out to 50 eV, including I_0 (as shown in Figure 39.15) and λ is the average mean free path for these low-energy losses. We ignore any intensity above ~50 eV because, even though that's where all the interesting ionization edges are, it is a negligible fraction of I_t (as is apparent if you go back and look at Figure 37.1). To determine λ in equation 39.13, we use a parameterization based on many experimental measurements (Malis et al.)

$$\lambda = \frac{106 F (E_0 / E_m)}{\ln(2\beta E_0 / E_m)} \quad (39.14)$$

where λ is in nm, E_0 in keV, β in mrad, F is a relativistic correction factor, and E_m is the average energy loss in eV which, for a material of average atomic number Z, is given by

$$E_m = 7.6 Z^{0.36} \quad (39.15)$$

The relativistic factor (F) is given by

$$F = \frac{1 + E_0/1022}{(1 + E_0/511)^2} \quad (3.16)$$

You can easily store these equations in the TEM computer or in your calculator and they give t with an accuracy of ~±20 % so long as the β is < ~15 mrads at 100 keV. In addition to this and the plasmon-peak-intensity approach, there are other methods (in Egerton's book) for determining thickness from various aspects of the EEL spectrum, but Malis' parameterization method is by far the most widely used. From what you've learned so far it should be obvious that if you can determine a thickness by comparing the intensity in two regions of the spectrum, we can just as easily form an image from each of the two intensities in equation 39.13, and thus extract a thickness image as we did for XEDS in Figure 36.13C.

You may find that, when you measure t, Murphy's law is operating and the area you're interested in is too thick for quantitative core-loss analysis. Then you'll have to deconvolute the spectra to make the single-scattering assumption valid.

39.6 DECONVOLUTION

We saw back in Figure 39.2 that the plural scattering adds intensity to the ionization edge, mainly as a result of combined inner (ionization) and outer-shell (plasmon) losses.

We can approximate the experimental ionization edge as a true single-scattering (hydrogenic) edge convoluted with the plasmon, or low-loss, spectrum.

The aim of deconvolution therefore, as shown schematically in Figure 39.16, is to extract the single-scattering contribution from the plural-scattering intensity in the spectrum. We'll describe two methods, the Fourier-Log and the Fourier-Ratio, both developed in Egerton's book and available for download at URL #1. Both methods are also incorporated in the Gatan software. Using a small β increases the deconvolution error since the plural-scattered electrons have a wide angular distribution and so more of them are excluded as β decreases.

The Fourier-Log method removes the effects of plural scattering from the whole spectrum. The technique describes the spectrum in terms of the sum of individual scattering components, i.e., the zero loss (elastic contribution) plus the single-scattering spectrum plus the double-scattering spectrum, etc. Each term is convoluted with the instrument response function, which is a measure of how much the spectrometer degrades the generated spectrum; in the case of a PEELS, this is the point-spread function we described in Section 37.5. The Fourier transform of the whole spectrum (F) is then given by

$$F = F(0) \exp\left(\frac{F(\mathcal{E})}{I_0}\right) \quad (39.17)$$

where $F(0)$ is the transform of the elastic contribution, $F(\mathcal{E})$ is the single-scattering transform and I_0 is the zero-loss intensity. So to get the single-scattering transform you take logarithms of both sides, hence, the name of the technique.

Extracting the single-scattering spectrum would ideally involve an inverse transformation of $F(\mathcal{E})$, but this results in too much noise in the spectrum. There are various ways around this problem, the simplest of which is to approximate the zero-loss peak to a delta function. After deconvolution, you can subtract the background in the usual way, prior to quantification.

> **DECONVOLUTION METHODS**
> The Fourier-Log method: deconvolute then subtract background.
> The Fourier-Ratio technique: subtract background then deconvolute.
> Multiple least-squares fitting: when the specimen is not uniformly thin.
> All three methods are approximations.

The danger of any deconvolution is that you may introduce artifacts into the single-scattering spectrum, e.g., from artifacts in the original spectrum. Despite the

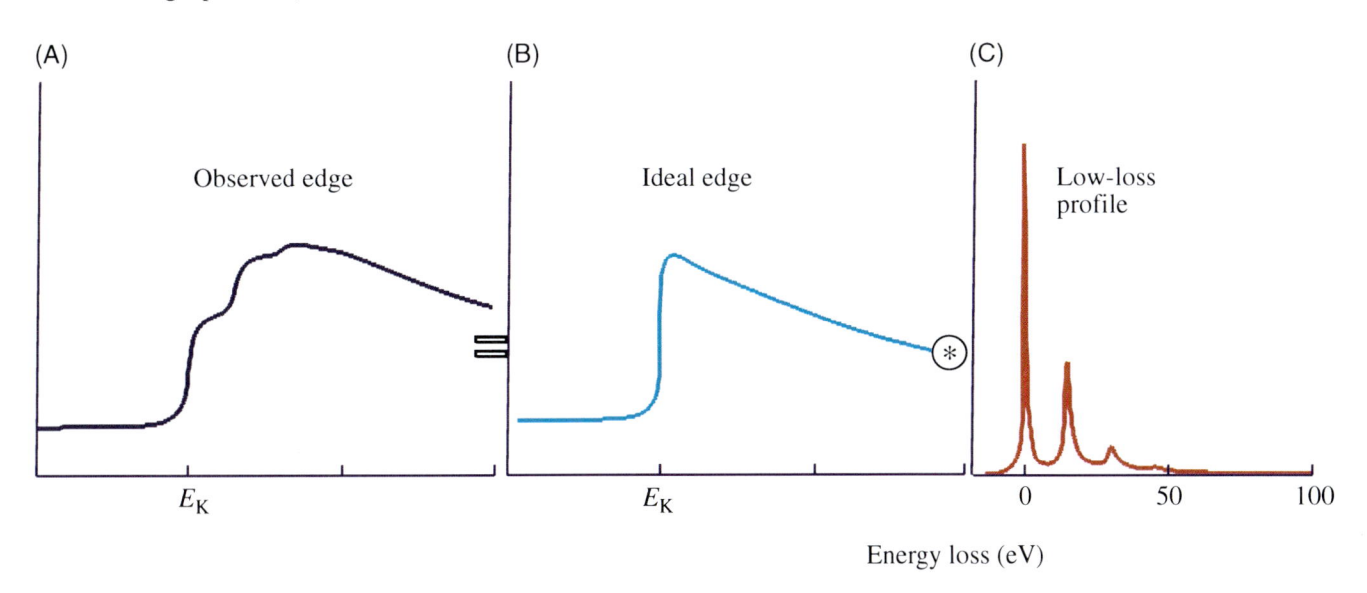

FIGURE 39.16. The experimentally observed ionization-edge intensity (A) consists of the convolution of the hydrogenic single-scattering ionization edge intensity (B) with the low-loss plasmon intensity profile (C).

assumptions and approximations, the net result of deconvolution is often an increase in the ionization-edge jump ratio. This improvement is important when you are attempting to detect small ionization edges from trace elements, or the presence of edges in spectra from thick specimens. An example of Fourier-Log deconvolution is shown in Figure 39.17.

The Fourier-Ratio technique: This approach approximates the experimental spectrum to the ideal single-scattering spectrum $F(E)$, convoluted with the low-loss spectrum. We define the low-loss portion of the spectrum as the region up to ~50 eV, including the ZLP, but before the appearance of any ionization edges. So we can now write

$$F' = F(E) \cdot F(P) \quad (39.18)$$

where F' is the Fourier transform of the experimental intensity distribution around the ionization edge and $F(P)$ is the Fourier transform of the low-loss (mainly plasmon) spectrum. In this equation, therefore, the instrument response is approximated by the low-loss spectrum rather than the ZLP. If we rearrange equation 39.18 to give a ratio (hence the name of the technique)

$$F(E) = \frac{F'}{F(P)} \quad (39.19)$$

We now obtain the single-scattering distribution by carrying out an inverse transformation. In contrast to the Fourier-Log technique, you must subtract the background intensity before deconvolution. Again, to avoid the problem of increased noise, it is necessary to multiply equation 39.19 by the transform of the ZLP.

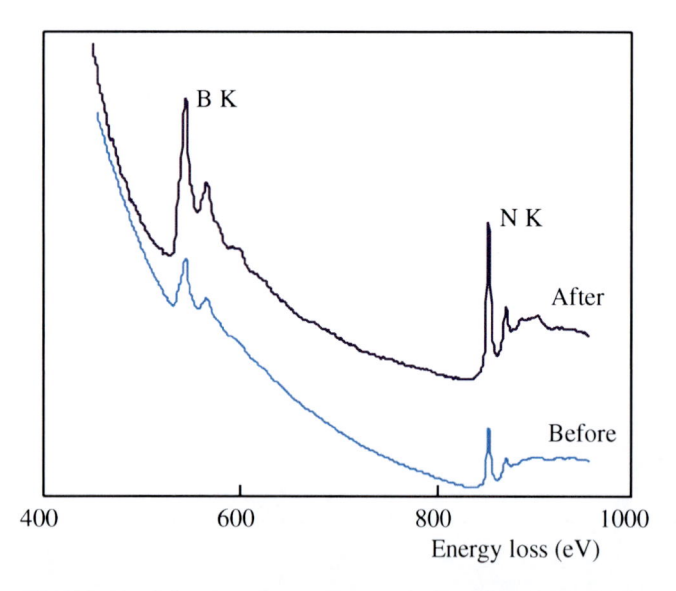

FIGURE 39.17. Spectrum from a thick crystal of BN before and after Fourier-Log deconvolution. The jump ratio in the deconvoluted spectrum (which is displaced vertically for clarity) is clearly increased by the process.

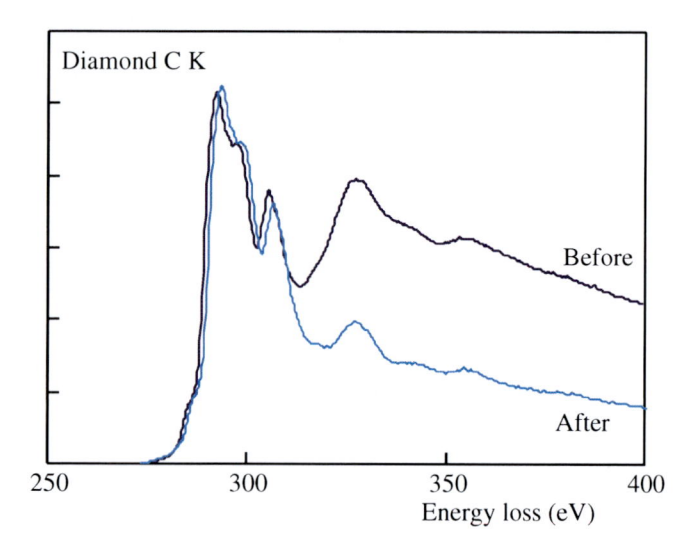

FIGURE 39.18. A carbon-K edge from a thick specimen of diamond before and after Fourier-Ratio deconvolution. You can see how the plural-scattering contribution to the post-edge structure is removed.

Figure 39.18 shows a carbon K edge after Fourier-Ratio deconvolution.

Multiple least-squares fitting: If your specimen is not uniformly thin, Fourier techniques won't work. Then you should use multiple least-squares (MLS) fitting of convoluted standard reference spectra (Leapman 2004). A single-scattering reference spectrum $R0(E)$ in the region of the edge to be quantified is convoluted with the first plasmon-loss portion of the unknown spectrum (P) and the resultant spectrum $R1(E) = P*R0(E)$ is used to generate several reference spectra ($R2(E) = P*R1(E)$, etc.). These reference spectra are then fitted to the experimental spectrum using MLS routines and specific fitting parameters are obtained. An experimental set of Fe, Co, and Cu reference spectra is shown in Figure 39.19A and the actual fit to part of the experimental spectrum from an intermetallic in a Cu-Be-Co alloy is shown in Figure 39.19B.

In summary, to quantify ionization-loss spectra you need a single-scattering spectrum, which can be approximated if you have very thin specimens or generated by deconvolution of your experimental spectrum. It is arguable that, given the stringency of the single approximation, it might be wise to deconvolute all core-loss spectra prior to quantification, but the uncertain effects of the possible errors introduced by deconvolution mean that you should do this cautiously. Often you'll find it useful to deconvolute the point-spread junction from all PEELS spectra, since this sharpens the edge onset and any ELNES intensity variations.

DECONVOLUTION CAUTION
Always check the validity of the deconvolution routine by applying it to spectra from a known specimen obtained over a range of thickness.

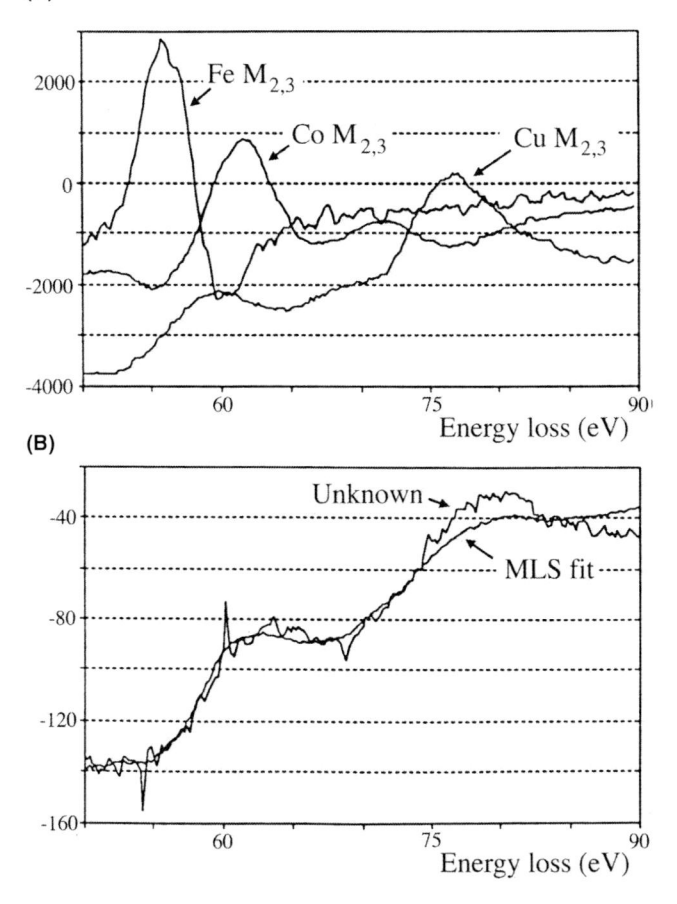

FIGURE 39.19. (A) Three first-difference low-loss, M-edge reference spectra from Fe, Co, and Cu superimposed on a low-energy portion of an experimental (first-difference) spectrum from an intermetallic particle in a Cu-Be-Co alloy. (B) MLS fit of the combined reference spectra to the experimental spectrum showing the good fit that can be obtained.

39.7 CORRECTION FOR CONVERGENCE OF THE INCIDENT BEAM

If you're working in STEM mode to get high spatial resolution, then it is possible that the beam-convergence angle α may introduce an error into your quantification. When α is equal to or greater than β, convergence effects can limit the accuracy because the experimental angular distribution of scattered electrons will be wider than expected (yet again, a good reason to know β). Therefore, you have to convolute the angular distribution of the ionization-loss electrons with the beam convergence angle. Joy (1986b) proposed handling this through a simple equation which calculates the effective reduction (R) in the partial cross section $\sigma(\beta\Delta)$ when $\alpha > \beta$.

$$R = \frac{\left[\ln\left(1 + \frac{\alpha^2}{\theta_E^2}\right)\beta^2\right]}{\left[\ln\left(1 + \frac{\beta^2}{\theta_E^2}\right)\alpha^2\right]} \quad (39.20)$$

where θ_E is the characteristic scattering angle. A similar reduction factor is incorporated in the Gatan software quantification routines. So you can see that if α is small (particularly if it is smaller than β) then R is $<<1$ and the effect of beam convergence is negligible. Generally, with the typical range of probe-limiting apertures in a STEM, convergence angles should not be larger than 5–10 mrads so it should always be feasible to make sure that β is large enough. However, a note of caution is worthwhile because, with C_s correction, it is now possible to use much larger convergence angles to increase the probe current without degrading the probe size.

39.8 THE EFFECT OF THE SPECIMEN ORIENTATION

In crystalline specimens, diffraction may influence the intensity of the ionization edge. This effect may be particularly large if your specimen is oriented close to strong two-beam conditions and, as we saw back in Chapter 35, this can be used to good effect in ALCHEMI. Both X-ray emission and ionization-loss intensity can change because of electron channeling effects close to the Bragg condition. At the Bragg condition, the degree of beam-specimen interaction increases, compared with zone-axis illumination where no strong scatter occurs and the energy-loss processes behave similarly. This phenomenon (known as the Borrmann effect in XEDS) is not important for low-energy edges, but intensity changes of a factor of 2 have been reported for Al and Mg K edges (Taftø and Krivanek). The use of large α minimizes the problem in XEDS, but beam-convergence effects are themselves a problem in EELS as we just described. Unless you're an alchemist, the easiest way to avoid orientation effects is simply to operate under kinematical conditions and stay well away from any bend centers or bend contours, just as in XEDS.

39.9 EFTEM IMAGING WITH IONIZATION EDGES

There are countless examples of EELS analyses using ionization-loss edges in the general references that we have given at the end of this and the other EELS chapters. As we described back in Section 37.8, you have several experimental options such as point analyses, spectrum-line profiles, and various forms of energy-filtered TEM imaging. By far the most powerful method of analysis, as with XEDS, is to form EFTEM images. You can either select a given edge from which to form a single image or you can gather a spectrum image and select the specific energy later. The former method is the norm for in-column filters while the latter is more common for post-column GIFs. We'll look at both but refer you to the 1995 text on the subject by Reimer for more theoretical and practical details and the review by Hofer and Warbichler for many illustrative examples. You can also image portions of the fine structure in the edge as we'll show in the next chapter, but here we will emphasize the power of ionization-loss analysis with reference to elemental images only. It is good to draw comparisons with what can be achieved with elemental imaging with XEDS, as we described back in Chapters 32–35.

39.9.A Qualitative Imaging

EFTEM images using the intensity in ionization edges will obviously correspond to elemental maps. The simplest way to get this information is to *subtract* a pre-edge background image from a post-edge image; this two-window subtraction method gives a passable qualitative elemental distribution. Alternatively, you can just *ratio* images of pre- and post-edge windows then what you get is called a jump-ratio image (see Figure 39.20B and C) which, again, is not quantitative and the intensity just reflects the edge to background ratio. Obviously, both of these qualitative methods will work best if (i) the jump ratio is high, (ii) the edge intensity is clearly visible above the background, and (iii) both thickness and diffraction conditions remain reasonably constant across the mapped region. Otherwise, the interpretation of any intensity changes is fraught with danger and artifacts abound. In fact, unless your specimen is very thin ($t \leq 0.1 \lambda$) it's probably not worth bothering with this qualitative approach and you may as well spend your time forming quantitative EFTEM images.

39.9.B Quantitative Imaging

If, at each pixel, we carry out a background subtraction, an edge integration, and multiply the resulting intensity by the partial ionization cross section ratio, we should get quantitative images of the distribution of specific elements. There isn't too much difference, in principle, between quantifying a spectrum as we've just described in equations 39.8 and 39.9 and forming an image. The main difference is in fact in the method of background subtraction, which we've also described already in Section 39.4.B.

The most common method for quantitative EFTEM imaging involves acquiring three images from electrons in selected energy windows: two from the background preceding the edge and one from under the edge as we've already described. This approach works equally well for an in-column filter or for a post-column GIF; you can either operate in TEM mode and acquire three images or operate in STEM mode and choose whether you select just the three energy windows or gather a full spectrum image. The only difference is that the TEM images can be acquired in a few seconds while the STEM approach may require minutes or even hours and is probably only worth doing if you are acquiring a full spectrum image. Figure 39.20 shows ratio images and fully quantitative images pulled out of a spectrum-image data cube.

In both instruments, to form a specific filtered image, you shift the energy spectrum until the desired energy window passes through the energy-selecting slit (see the schematic diagrams in Figures 37.13 and 37.15). The energy shift is actually achieved by changing the accelerating voltage of the TEM so that electrons of different energies stay on-axis and thus in focus through the spectrometer. The quality of your images is governed by the same factors that control the spectrum: good jump ratios to get good signals, well-separated peaks to ensure good background-fitting statistics, and a good thin specimen to permit valid quantification under single-scattering conditions. Once acquired, your filtered images can be subject to advanced post-specimen processing, such as pixel-clustering methods (Cutrona et al.) or the standard methods, such as color assignments for different elements, as shown back in Figure 37.17D. If you're at all uncertain about whether the experiment will work, remember to simulate it first, using the EELS Advisor software (URL #2).

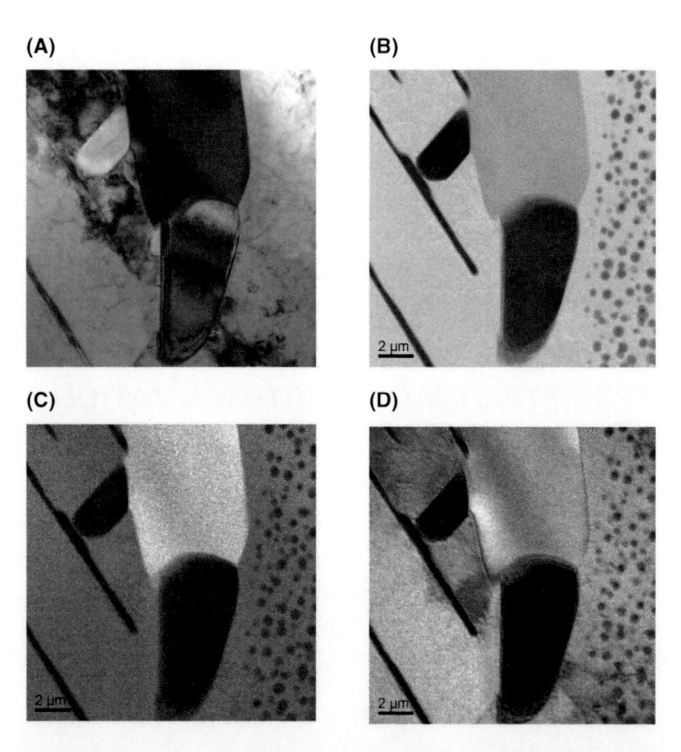

FIGURE 39.20. (A) BF image of precipitates in a stainless-steel foil. The other images were obtained from specific energy-loss electrons and illustrate both jump ratio and fully quantitative images. (B) Fe M jump ratio image; (C) Cr L jump ratio image; (D) quantitative Cr map.

EFTEM imaging will only continue to improve as all the instrumental factors that we have already discussed become common. We will have C_s correction, monochromation, and better spectrometers with more uniform transmissivity achieved via higher-order aberration correction.

39.10 SPATIAL RESOLUTION: ATOMIC-COLUMN EELS

In contrast to the situation in XEDS, beam spreading is not a major factor in determining the source of the EELS signal and so the many factors that influence beam spreading are mainly irrelevant. The spectrometer only collects those electrons emanating from the specimen in a narrow cone, as shown in Figure 39.21 Therefore, energy-loss electrons that are elastically scattered through large angles are excluded from contributing to your spectrum. Remember that for XEDS these same high-angle electrons would still generate X-rays some distance from the incident-probe position, and these X-rays would be detected by XEDS. In the absence of a contribution from beam spreading, the spatial resolution of ionization-loss spectrometry depends on the mode of analysis

- The factor controlling the resolution in STEM mode, or in a probe-forming (diffraction) mode on a TEM, is mainly the size of the probe; because of the strong forward-scattered signal, we can easily get data with probe sizes < 0.2 nm and with aberration correction we can break the Ångstrom barrier.
- When we operate in TEM mode, the spatial resolution is a function of the selecting aperture, i.e., the spectrometer entrance aperture and its effective size at the plane of the specimen. Lens aberrations usually limit the spatial resolution, as we showed back in Section 37.4.C.

In addition to the usual factors affecting the probe size, such as the diffraction limit and lens aberrations (go back and read Chapter 5), another factor that we have to consider in EELS, but ignore in XEDS (although it occurs in X-ray generation also) is the phenomenon of delocalization.

> **DELOCALIZATION IN EELS**
> Delocalization is the ejection of an inner-shell electron by the passage of a high-energy electron some distance from the atom. It's as if the beam electron scares the core-shell electron sufficiently to eject it without actually laying a finger on it!

The scale of this wave-mechanical effect is inversely proportional to the energy loss over dimensions of a few nm (which, of course, is very large if you're worried about atomic-level spatial resolution). Egerton gives the following simplified expression for the diameter (d_{50}) containing 50% of the inelastic intensity

$$(d_{50})^2 = \left(\frac{0.5\lambda}{\theta_E^{3/4}}\right)^2 + \left(\frac{0.6\lambda}{\beta}\right)^2 \quad (39.21)$$

where you should recognize all the terms and yet again appreciate the importance of knowing your collection angle. This expression gives a localization of ~1 nm for $E = 50$ eV and 0.4 nm for $E \simeq 300$ V (C edge K) and might therefore appear to prevent atomic-resolution EELS. Fortunately, delocalization does not appear to be a factor in STEM images of single, isolated atoms and resolution appears to be primarily determined by the width of the probe, even for light atoms. So it appears that the usual factors, such as probe aberrations, signal to background in the EELS signal, and damage are much more important in terms of EELS spatial resolution. All the experimental evidence seems to agree with the secondary role for delocalization in that HAADF imaging of single atoms has been long established (see Section 22.4 and the companion text); atomic-level changes in chemistry were first demonstrated by ionization loss in the early 1990s (Browning et al. and Batson). The same process that confines the electron beam to channel along the atomic columns in HAADF imaging also means that the spectroscopic

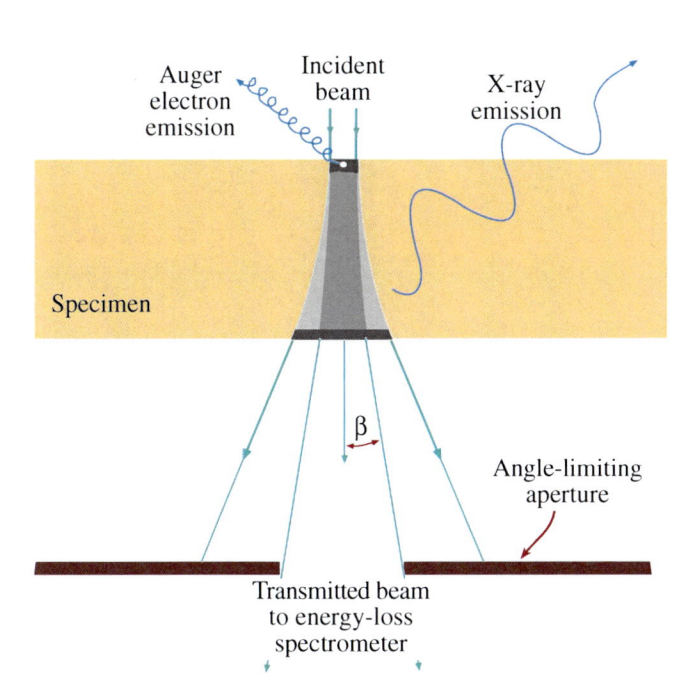

FIGURE 39.21. The effect of the spectrometer collection angle is to limit the contribution to the spectrum from high-angle scattered electrons, thus ensuring high spatial resolution. In contrast, X-rays can be detected from the whole beam-specimen interaction volume.

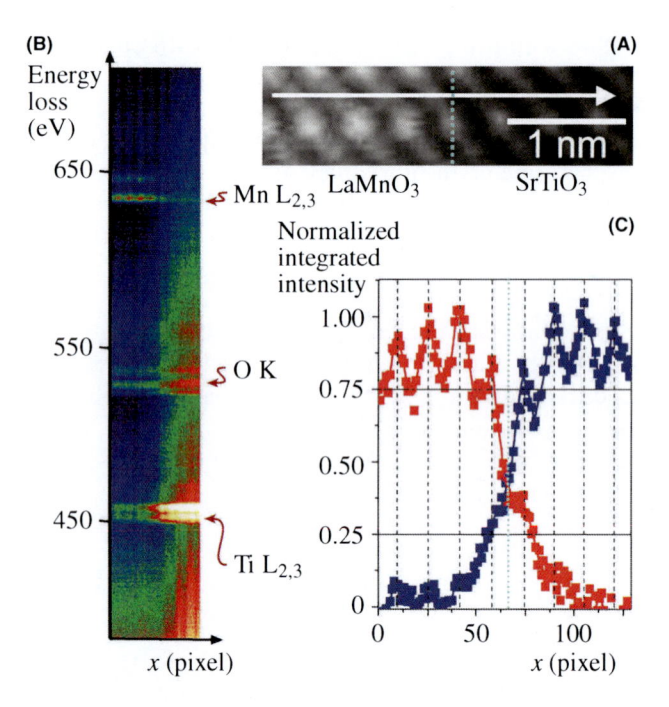

FIGURE 39.22. (A) HAADF image of a LaMnO$_3$/SrTiO$_3$ interface (blue dashed line). (B) EELS linescan along the arrowed direction in (A). Approximate positions of the Ti L$_{2,3}$, O K and Mn L$_{2,3}$ absorption edges are highlighted. The ripple corresponds to atomic-plane positions along the scan. (C) Normalized integrated intensities (40-eV window) under the Ti L$_{2,3}$ (blue) and the Mn L$_{2,3}$ (red) edges. Black dotted lines show the estimated positions of the respective MnO$_2$ and TiO$_2$ atomic planes.

As is obvious in Figure 39.21, EELS has an inherently higher signal collection efficiency than XEDS. Conversely, as we've seen, it also has a correspondingly poorer signal to background because of the higher plural scattering. But, as for the best spatial resolution, more signal wins in the end. Leapman and Hunt argued in 1991 that, in most situations, PEELS is more sensitive to the presence of small amounts of material than XEDS. This has been borne out in experimental studies over many years and the latest improvements in FEG sources, C_s correction, and spectrometer hardware have brought us to the point where combined imaging and spectroscopic analysis (including fine structure and associated electronic effects) of single atoms is achievable. Figure 39.23 shows single-atom analytical sensitivity with atomic-level spatial resolution. So while, in principle, the inverse relationship between analytical sensitivity and spatial resolution that we described for XEDS in Figure 36.11 applies, when you're able to detect single atoms with atomic resolution you've just about reached the fundamental limit of any analytical technique.

Conclusion: analysis using ionization edges offers both atomic-resolution spatial resolution and single-atom analytical sensitivity.

signal is similarly localized. Because the EELS intensities are so much higher than XEDS, we are able to extract EELS spectra which contain information from a single atomic column, as shown in Figure 39.22). Both the Ti and Mn integrated signals in Figure 39.22C show an approximate decrease of 50% (or higher) within an atomic plane of the interface indicating true atomic-level resolution.

It is possible even to detect the presence of single atoms of high-Z elements on individual atomic columns of lower-Z elements (see next section) and C_s correction now makes atomic-resolution EELS almost straightforward (Varela et al.)

39.11 DETECTION LIMITS

The detection limits for ionization-loss spectrometry are governed by the same factors as we discussed for XEDS, so we have to optimize several factors

- The edge intensity.
- The signal to background ratio (jump ratio).
- The efficiency of signal detection.
- The time of analysis.

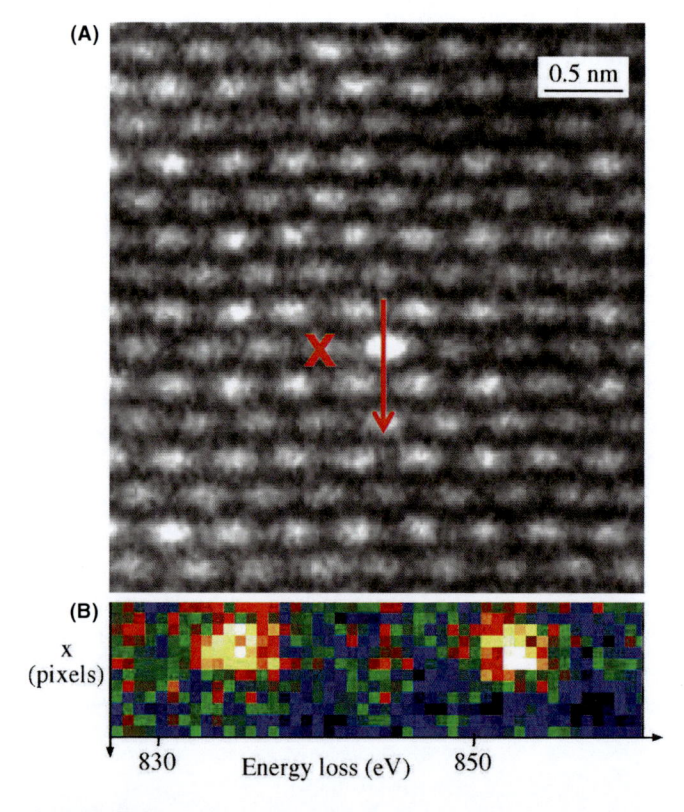

FIGURE 39.23. (A) HAADF image of a La impurity atom (X) in single atomic column in CaTiO$_3$. (B) Spectrum line-intensity profiles showing the white lines in the La M$_{4,5}$ edge as the beam scans across the La atom (along the red arrow). The white lines appear only when the La atom is scanned.

CHAPTER SUMMARY

The ionization edges can be used to give quantitative elemental analyses and quantitative images from all the elements in the periodic table, using a simple ratio equation. Beware, however, of the many experimental variables you have to define for your TEM, PEELS, energy filter, and (most importantly) your very thin specimen. Compared to XEDS there have been relatively few quantitative analyses or composition profiles measured using EELS, but quantitative imaging is becoming much more common.

To use Egerton's ratio equation you have to

- Subtract the background using a power law, or MLS approach. The former is easier. The latter is better for complex spectra.
- Integrate the edge intensity. That's usually straightforward.
- Determine the partial ionization cross section $\sigma_K(\beta\Delta)$. Calculate $\sigma_K(\beta\Delta)$ with SIGMAK and SIGMAL for most K and L edges.
- For M edges use a known standard or, better still, use XEDS.
- For the lightest elements (e.g., Li) use a known standard.

The biggest limitation to quantification is that your specimens have to be much thinner than one mean free path (typically $<<$ 50 nm) otherwise deconvolution routines are needed, which can introduce artifacts of their own.

Ionization-loss imaging is becoming widespread because of the increased availability of in-column and post-column filters and is the recommended method for quantitative analysis.

Spatial resolution and minimum detection are better than for XEDS. Combined atomic-column resolution and single-atom detection has been demonstrated.

BOOKS AND REVIEWS

Ahn, CC Ed. 2004 *Transmission Electron Energy-Loss Spectrometry in Materials Science and the EELS Atlas* 2nd Ed. Wiley-VCH Weinheim Germany. Must-read chapters on quantification and imaging.

Brydson, R 2001 *Electron Energy-Loss Spectroscopy* Bios (Royal Microsc. Soc.) Oxford UK. Great basic introduction; lots of practical tips.

Egerton, RF 1996 *Electron Energy-Loss Spectroscopy in the Electron Microscope* 2nd Ed. Plenum Press New York. Still the EELS bible. Read more here about the Bethe ridge.

Hofer, F and Warbichler, P 2004 *Elemental Mapping Using Energy-Filtered Imaging* in *Transmission Electron Energy-Loss Spectrometry in Materials Science and the EELS Atlas* 2nd Ed. **159**–233 Ed. CC Ahn Wiley-VCH Weinheim Germany. A thorough review of EFTEM/ESI, full of outstanding examples.

Joy, DC 1986a *The Basic Principles of EELS*, 1986b *Quantitative Microanalysis using EELS* in *Principles of Analytical Electron Microscopy* 249–276 and 277–299 Eds. DC Joy, A. Romig Jr. and JI Goldstein Plenum Press New York. Introduction to the principles and in-depth discussion of the experimental details necessary for quantification

Kohler-Redlich, P and Mayer, J 2003 *Quantitative Analytical Transmission Electron Microscopy* in *High-Resolution Imaging and Spectrometry of Materials* 119–187 Eds. F Ernst and M Rühle Springer New York. Integrated review of quantitative EELS and other TEM techniques with an emphasis on interfacial studies.

Reimer, L Ed. 1995 *Energy-Filtering Transmission Electron Microscopy* Springer New York. The first book on EFTEM/ESI with all you need to know about the theory and practice, but very few applications.

APPLICATIONS

Batson, PE 1993 *Simultaneous STEM Imaging and Electron Energy-Loss Spectroscopy with Atomic-Column Sensitivity* Nature **366** 727–728.

Browning, ND Chisholm, MF and Pennycook, SJ 1993 *Atomic-Resolution Chemical Analysis Using a Scanning Transmission Electron Microscope* Nature **366** 143–146.

Cutrona, J, Bonnet, N, Herbin, M and Hofer, F 2005 *Advances in the Segmentation of Multi-Component Microanalytical Images* Ultramicrosc. **103** 141–152.

Egerton, RF 1979 *K-Shell Ionization Cross-Sections For Use in Microanalysis* Ultramicrosc. **4** 169–179.

Egerton, RF 1981 *SIGMAL; A Program For Calculating L-shell Ionization Cross-Sections* in Proc. 39th EMSA Meeting 198–199 Ed. G.W. Bailey Claitors Baton Rouge LA.

Egerton, RF 1993 *Oscillator-Strength Parameterization of Inner-Shell Cross Sections* Ultramicrosc. **50** 13–28.

Hofer, F, Golob, P and Brunegger, A 1988 *EELS Quantification of the Elements Sr to W by Means of M_{45} Edges* Ultramicrosc. **25** 81–84.

Jeanguillaume, C, Trebbia, P and Colliex, C 1978 *About the Use of Electron Energy-Loss Spectroscopy for Chemical Mapping of Thin Foils with High Spatial Resolution* Ultramicrosc. **3** 237–249.

Leapman, RD and Hunt, JA 1991 *Comparison of Detection Limits for EELS and EDXS* Microsc. Microanal. Microstruct. **2** 231–244.

Leapman RD (2004) *EELS Quantitative Analysis* in Transmission Electron Energy Loss Spectrometry in Materials Science and the EELS Atlas 2nd Ed. 49–96 Ed CC Ahn Wiley-VCH Weinheim Germany. A great reference source.

Malis, T, Cheng, S and Egerton, RF 1988 *EELS Log-Ratio Technique for Specimen-Thickness Measurement in the TEM* J. Electron Microsc. Tech. **8** 193–200.

Michel, J. and Bonnet, N 2001 *Optimization of Digital Filters for the Detection of Trace Elements in EELS. III – Gaussian, Homomorphic and Adaptive Filters* Ultramicrosc. **88** 231–242.

Muller, DA and Silcox, J 1995 *Delocalization in Inelastic Scattering* Ultramicrosc. **59** 195–213. If you're physics-oriented and want to read more about delocalization in EELS.

Rez, P 2003 *Electron Ionization Cross Sections for Atomic Subshells* Microscopy and Microanalysis **9** 42–53.

Taftø, J and Krivanek, OL 1982 *Site-Specific Valence Determination by Electron Energy-Loss Spectroscopy* Phys. Rev. Lett. **48** 560–563.

Unser, M, Ellis, JR, Pun, T and Eden, M 1986 *Optimal Background Estimation in EELS* J. Microsc. **145** 245–256.

Varela, M, Lupini, AR, van Bentham, K, Borisevich, AY, Chisholm, MF, Shibata, N, Abe, E and Pennycook, SJ 2005 *Materials Characterization in the Aberration–Corrected Scanning Transmission Electron Microscope* Annu. Rev. Mater. Res. **35** 539–569.

URLs
1) http://www.cemes.fr/&7Eeelsdb/
2) http://laser.phys.ualberta.ca/~egerton/programs/programs.htm3.
3) http://www.gatan.com/software/eels_advisor.php

SELF-ASSESSMENT QUESTIONS

Q39.1 Is it better to do ionization–loss analysis at 100 keV or 200 keV? Justify your answer.

Q39.2 Why should you not operate with too small a collection angle if attempting to do quantitative analysis?

Q39.3 Why should you not operate with too large a collection angle if attempting to do quantitative analysis?

Q39.4 What is the compromise you make when choosing a large or small collection angle?

Q39.5 What kind of energy resolution do you need for quantitative analysis. Does this resolution requirement permit you to improve other aspects of the spectral acquisition?

Q39.6 Define the jump ratio. How can you increase this ratio for a given ionization edge?

Q39.7 Why do we not define the ionization edge energy as the peak-intensity channel of the edge?

Q39.8 What are white lines and why are they called this?

Q39.9 Define N, I, σ, and δ, including the appropriate units.

Q39.10 What relationship exists (if any) between the background-fitting window and the background-extrapolation window?

Q39.11 Under what circumstances might you use a 'goodness of fit' criterion for the background-extrapolation procedure?

Q39.12 Why do we use the term 'partial' to describe the ionization cross section (σ) used in the Egerton equation?

Q39.13 Why do the simplistic SIGMAK and SIGMAL models still give a reasonable approximation to fitting the edge intensity?

Q39.14 Why isn't there a SIGMAM model and in its absence, how do you quantify M edges?

Q39.15 Why would you deconvolute your spectrum before attempting to quantify it?

Q39.16 Why is the background subtraction procedure in EELS far more important and difficult than in XEDS?

Q39.17 What are the units of quantification of the specimen composition obtained via the ratio method?

Q39.18 What is the greatest uncertainty in the quantification procedure?

Q39.19 Why might you need to correct the partial ionization cross section for too large a convergence angle instead of simply decreasing the C2 aperture?

Q39.20 Why is the spatial resolution of EELS fundamentally better than that of XEDS?

Q39.21 What is delocalization and how might it affect your EELS analysis?
Q39.22 Why do we call an ideal edge 'hydrogenic'?
Q39.23 Why does a hydrogenic ionization edge look like a triangle?
Q39.24 Why don't we see ionization edges from core levels much deeper than about 2 keV?

TEXT-SPECIFIC QUESTIONS

T39.1 Explain the nomenclature in Figure 39.1.

T39.2 Examine Figure 38.1 and compare with Figure 39.3. Why are both the zero loss and plasmon loss peaks effectively Gaussian in shape when all the ionization edges have a much broader intensity distribution extending asymmetrically over several tens of eV?

T39.3 Is the background intensity as drawn in Figure 39.2 realistic and if not, why not?

T39.4 If the energy resolution of the EELS spectrometer is on the order of a few eV at worst, why is it not possible to discriminate ionization edges that are several tens of eV apart without resorting to deconvolution?

T39.5 Given that TiN and TiC have equal numbers of atoms of each elemental constituent, why are the intensities so very different in the EELS spectra in Figure 39.8? What other spectroscopic method might you use to distinguish TiC and TiN, and why can't this method be used in an AEM?

T39.6 Does the first difference approach (Figures 39.10 and 39.19) actually remove the background intensity? If not, why not? How does this method compare with the top-hat filter approach used in XEDS?

T39.7 Why do we recommend using experimental standards for the best k-factor determination in XEDS but generally prefer calculation of partial ionization cross section values for EELS quantification?

T39.8 Calculate the thickness of an Fe specimen for which the low-loss intensity is 10% of the zero-loss intensity in a spectrum gathered with a collection angle of 100 mrads at 100 keV.

T39.9 Compare and contrast the two principal methods of deconvolution in Figures 39.17 and 39.18.

T39.10 Why can't we simply expand the integration windows δ and Δ in Figure 39.9 in order to increase the goodness of fit of the background subtraction and increase the total number of counts in the edge, respectively?

T39.11 Compare and contrast the expression for quantification of an ionization-loss spectrum (equation 39.5) with the Cliff-Lorimer expression for quantification of a characteristic X-ray spectrum (equation 35.2).

T39.12 Justify using the gross simplifications of the hydrogenic approach to model ionization edges shown in Figures 39.12 and 39.13.

T39.13 Look at Figure 39.6 and explain the relationship between the penetration of the electron into the potential well and the value of the energy loss.

T39.14 List the pros and cons of deconvoluting out the plural-scattering contributions and the point-spread function from a spectrum.

T39.15 Calculate the maximum convergence angle (α) you should use, such that the cross section for carbon K shell ionization (100 kV, β = 20 mrad) needs no correction. State any assumptions.

T39.16 Why has the spatial resolution of EELS analysis received so little study compared to that of X-ray analysis?

T39.17 Compare and contrast two different background-subtraction methods used for EFTEM with two different methods used for spectroscopy.

T39.18 Compare and contrast the experimental factors that limit spatial resolution in XEDS and EELS.

T39.19 Compare and contrast the experimental factors that limit analytical detection limits in XEDS and EELS.

T39.20 How do you know if your specimen is too thick for EELS analysis and how do you minimize the problem (apart from making a thinner specimen)?

T39.21 Why is EELS so much more efficient at collecting electrons compared to XEDS collecting X-rays, as summarized in Figures 39.4 and 39.5, given that the actual collection angles are about the same?

T39.22 Why is the B K edge so much more intense than the N K edge in Figure 39.3 when the atomic ratio of B:N in boring nitride is 1:1?

T39.23 What is a typical collection angle for X-rays in an XEDS spectrometer and how does it compare with the collection angle for electrons in an EELS?

40
Fine Structure and Finer Details

CHAPTER PREVIEW

In the previous chapter, we described elemental analysis using ionization edges, but there is much more than just elemental information in the ionization edges and this distinguishes EELS from XEDS. There are detailed intensity variations in the core-loss spectra called energy-loss near-edge structure (ELNES) and extended energy-loss fine structure (EXELFS). From this fine structure, which we can resolve because of the high-energy resolution inherent in EELS, we can obtain data on how the ionized atom is bonded, the coordination of that specific atom, and its density of states. As always, we can use any intensity changes to create filtered images which show the distribution of, e.g., regions of different bonding states. Furthermore, we can probe the distribution of other atoms around the ionized atom (i.e., determine the radial-distribution function (RDF) which is very useful for the study of amorphous materials) and we can study momentum-resolved EELS, observe the anisotropy of chemical bonds, combine EELS with tomography, inter alia. Understanding these phenomena often requires that we use certain concepts from atomic and quantum physics. The non-physicist can skip some sections at this time and just concentrate on the results. The rewards of working through this topic will be an appreciation of some of the more powerful aspects of EELS.

> **WHY LOOK AT FINE STRUCTURE ?**
> If high spatial resolution is important, you can't obtain this additional information by any other spectroscopic technique.

This fine structure is all the more useful because we now have the ability to simulate the spectra using atomic-structure calculations, which help us understand the details in the spectra. A full appreciation of the calculations is beyond the scope of the book but this is a growing field that will only assume more significance.

As a wrap-up to EELS and the book as a whole, we'll finish by saying a few words on some of the more esoteric aspects of TEM, such as angular-resolved spectrometry, radial-distribution-function determination, Compton scattering, core-level shifts, and tomographic EELS that are not yet in the mainstream but, with continuing advances in instrumentation and computation, will surely grow in importance.

40.1 WHY DOES FINE STRUCTURE OCCUR?

We saw in Section 39.1 that the ionization edges have intensity variations superimposed on the ideal hydrogenic sawtooth shape. The stronger oscillations occur within about 30–50 eV of the onset of the edge (ELNES) and the weaker ones extend out for several hundred eV as the edge intensity diminishes (EXELFS). This fine structure contains a wealth of useful information, but to understand its origins you have to use some ideas from quantum physics.

One way we can look at this process is to switch from a particle to a wave model of the electron, as we've done before, e.g., when we talked about diffraction in Part 2. Then we can imagine that any excess energy ($>E_c$) that the ejected electron possesses is a wave emanating from the ionized atom. Now, if this wave has only a few eV of excess energy, it undergoes plural, elastic scattering from the surrounding atoms, as shown schematically

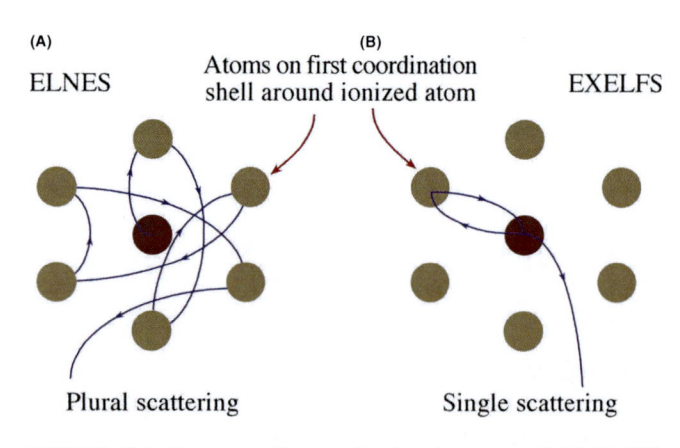

FIGURE 40.1. Schematic diagram showing the source of (A) ELNES and (B) EXELFS. The excess energy retained by the electron escaping above the Fermi level creates a wave radiating from the ionized atom and is scattered by surrounding atoms. The low-energy ELNES arises from plural scatter and is affected by the bonding between the atoms. The higher-energy EXELFS approximates to single scatter and is affected by the local atomic arrangement.

in Figure 40.1A, and this scattering is responsible for the ELNES, as we'll show. If the wave has even more excess energy, then, because of the smaller interaction cross section for higher-energy electrons (as we've already seen many times) it is less likely to be scattered by the surrounding atoms. In fact, we can approximate the cause of the ELNES to a single-scattering event, as shown in Figure 40.1B. Thus, EXELFS and ELNES can be viewed as a continuum of electron-scattering phenomena, with the arbitrary distinction that ELNES is confined to a few tens of eV past the edge onset while EXELFS extends for several hundred eV past the edge onset. There are other ways to explain fine structure and we'll mention some of these later when we talk about modeling the phenomena.

> **ELNES AND EXELFS I**
> Both arise because the ionization process can impart more than the critical ionization energy (E_c) needed by the core electron to be ejected from its inner shell.
> Energy-Loss Near-Edge Structure
> EXtended Energy-Loss Fine Structure

You should know that similar fine-structure effects can occur in X-ray spectra, but are usually not resolvable in the TEM because of the poor resolution of the semiconductor XEDS detector. However, we did note that experimental high-resolution X-ray detectors can resolve bonding effects in terms of shifts in X-ray peaks (see Figure 32.9C). In fact, there is a whole field of X-ray spectrometry that is used for studying atomic bonding (X-ray absorption near-edge structure or XANES) and atom positions and structure (extended X-ray absorption fine structure or EXAFS). These techniques are analogous to ELNES and EXELFS, but require a synchrotron to generate sufficient signal. This is one of the few examples where TEM is the cheaper characterization technique.

Most of this chapter deals with the experimental measurement and basic theoretical simulation of ELNES and EXELFS; the information we give is augmented in the companion text. While ELNES arises from plural scattering and is thus a more complex process than EXELFS, it is much more widely used, because it gives a more intense signal and the information it reveals has been used to study a very wide range of materials. So we'll discuss ELNES first.

40.2 ELNES PHYSICS

40.2.A Principles

As you know well, when an atom is ionized, it is raised from its ground state to an excited state leaving a hole in an inner shell. The core electron must receive enough energy from the beam electron to be ejected from its shell, but it may not receive enough to escape to the vacuum level. So it is still not completely free of the Coulomb attraction to the nucleus. In such circumstances, the final state of the core electron will be in one of a range of possible energy levels above the Fermi energy (E_F). You may recall that the Fermi level (or the Fermi surface in three dimensions) is the boundary between the filled states and the unfilled states in the weakly bound conduction/valence bands (although, strictly speaking, this statement is only true when $T = 0$ K). In a metal, there is no separate valence band and E_F sits somewhere in the conduction band, as shown schematically in the classic energy level diagram of an atom in Figure 40.2. In an insulator or a semiconductor, E_F is between the valence band (in which all the states are filled) and the conduction band (which has no filled states). The possible energy values that can be imparted to the ejected electron are controlled by the energy distribution of these unfilled states and, therefore, the energy lost by the incident electron similarly reflects this distribution of the unfilled states. One philosophical point of quantum uncertainty is that these states don't exist until an electron appears in them, but we'll conveniently ignore this.

So, the excited electron can reside in any of the unfilled states, but what's crucial here is that there is not an equal probability of the electron ending up in each possible unfilled state. Some empty states are more likely to be filled than others because there are more states within certain energy ranges than in others. This uneven distribution of empty energy levels is termed the density of (unfilled) states (DOS) and this is also shown

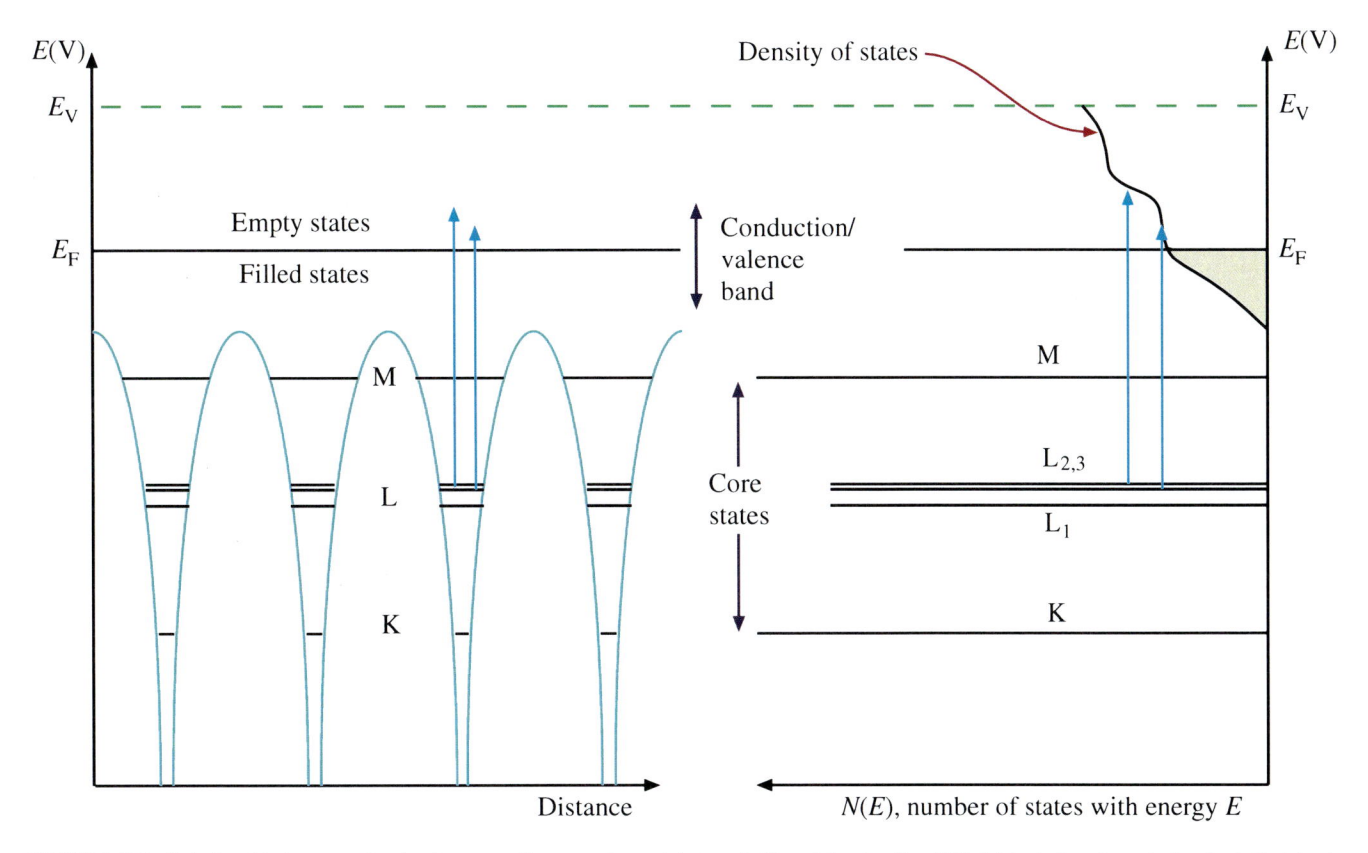

FIGURE 40.2. Relationship between the classic energy diagram of a metal atom (left) and the density of filled (shaded) and empty (unshaded) states in the conduction/valence band (right). The DOS is approximately a quadratic function on which small variations are superimposed. Ionization results in electrons being ejected from the core states into empty states above the Fermi level E_F.

in Figure 40.2. Because of the greater probability of electrons filling certain unoccupied states above E_F, the ELNES intensity is greater at the energy losses corresponding to these high DOS regions above the Fermi energy (which can be thought of as equivalent to the critical ionization energy E_C), as shown in Figure 40.3.

> **ELNES**
> This variation in intensity, extending several tens of eV above the ionization edge onset, E_C, is the ELNES, and it effectively mirrors the unfilled DOS above E_F.

The importance of ELNES is that the DOS is extremely sensitive to changes in the bonding or the valence state of the atom. For example, if you look ahead to Figure 40.5, the carbon K ELNES is different for graphite, diamond, and buckyballs and the Cu L ELNES changes when Cu is oxidized to CuO. On an even more detailed level, we can even deduce the coordination of the ionized atom from the shape of the ELNES.

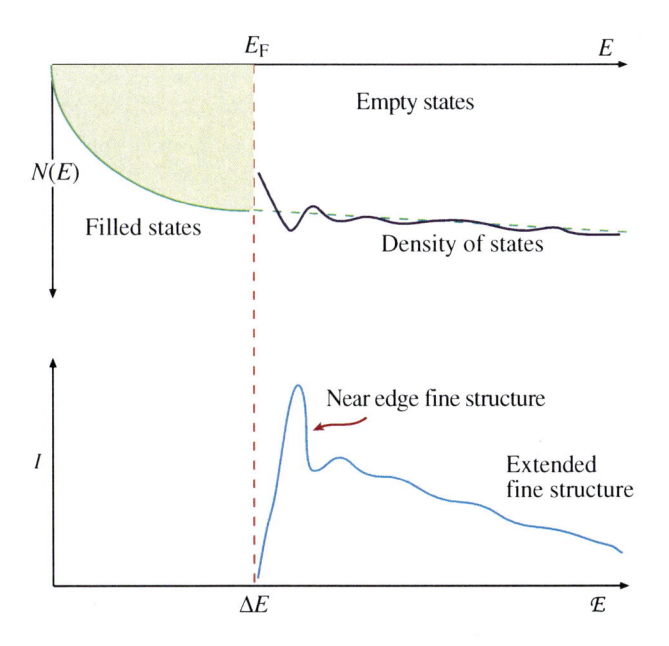

FIGURE 40.3. Relationship between the empty DOS and the ELNES intensity in the ionization edge fine structure. Note the equivalence between the Fermi energy E_F and the ionization edge onset E_C. Electrons ejected from the inner shells reside preferentially in regions of the DOS that have the greatest density of unfilled states. The filled states below E_F are drawn as a quadratic function, but this is an approximation.

40.2 ELNES PHYSICS

> **DOS AND FERMI SURFACES**
> Even if you don't understand the intricacies of the DOS and Fermi surfaces, you can still deduce bonding information by comparing your experimental ELNES with that from standard specimens of known valence state or coordination.

We discuss this fingerprinting approach in Section 40.2.D below and you can check it in the EELS Atlas or at URL #1. Remember, we did exactly the same type of fingerprinting of different phases with the low-loss spectra in Section 38.3.A.

40.2.B White Lines

Perhaps the most startling example of ELNES is the presence of the white lines, which we first saw in Chapter 39; these lines are intense sharp peaks on certain ionization edges. These sharp peaks arise because in certain elements the core electrons are excited into well-defined empty states, not a broad continuum, as in Figure 40.3. The $L_{2,3}$ edges of the transition metals and the $M_{4,5}$ edges of the rare-earth elements show such lines. The white lines in the Fe L edge are the L_3 and L_2 edges, respectively, as shown in Figure 40.4, and these specific lines arise because the d shell has unfilled states. (We'll explain what happened to L_1 later.) To explain these lines we need a little more quantum physics, which you can skip if you wish and go to the last paragraph of this section. You should also be aware that there is disagreement as to whether white lines are truly fine structure or strictly ionization edge (atomic) intensity; but we'll leave this somewhat arcane discussion to those who know better (another cause of fracas in bars at M&M meetings).

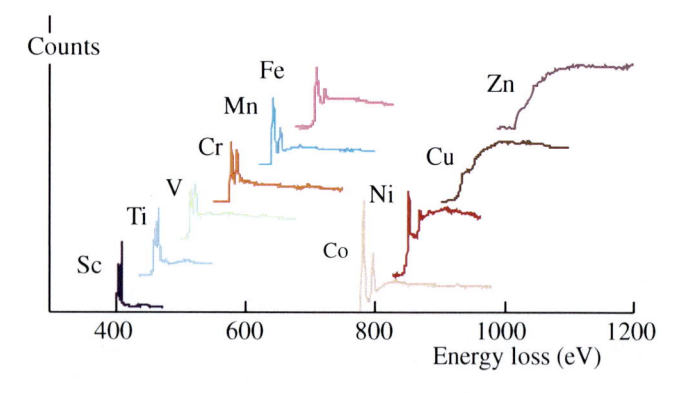

FIGURE 40.4. Spectra from the transition metals show a variation in the L_3 and L_2 white-line intensity ratios reflecting the variation in the number of core L-shell electrons ejected into unfilled d states. Note that Cu and Zn show no white lines because their d shells are full. The L_3 and L_2 white lines in the Fe L edge are the only ones that show the expected L_3:L_2 of ~2:1.

40.2.C Quantum Aspects

First, remember that the various electron energy levels, K, L, M, etc., correspond to principal quantum numbers (n) equal to 1, 2, 3, etc. Within those energy levels, the electrons may have s, p, d, or f states, for which the angular-momentum quantum number (l) equals 0, 1, 2, 3, respectively. The notation s, p, d, f comes from the original description of the atomic-spectral lines arising from these electron states, namely, sharp, principal, diffuse, and fine, although these terms have no counterpart in the EELS spectra we obtain.

As we noted in Section 39.1, the nomenclature $L_{2,3}$ arises from the fact that the L shell, from which the electron was ejected, has different energy levels. Such separation of the energies of the core states is called *spin-orbit splitting*.

Because the L electrons in levels 2 and 3 are in the p state, quantum theory demands that the sum (j) of their spin quantum number (s) and angular-momentum quantum numbers (l) is governed by the Pauli exclusion principle such that $j(=s+l)$ can only be equal to 1/2, 3/2, 5/2, etc. The spin quantum number, s (not to be confused with the s state), can only equal ±1/2. Taking all this into account, along with other quantum-number restrictions, it turns out that in the higher energy (i.e., more tightly bound) L_2 shell, we can have two p electrons with $j = ±1/2$ while in the L_3 shell, we can have four p electrons with $j = ±1/2, ±3/2$. Therefore, we might expect twice as many electrons to be excited from the L_3 shell as from the L_2 shell giving an L_3/L_2 intensity ratio (above the edge intensity, not above background) of 2. While this rule is approximately obeyed in the Fe spectrum shown in Figure 40.4, in practice, the ratio is seen to increase along the transition metal series from 0.8 for Ti to 3 for Ni, as is also seen in the spectral sequence in Figure 40.4.

Now these p-state electrons in the L shell cannot be excited to just any unoccupied state.

So for the p state ($l = 1$) the only permitted final states are either an s state ($l = 0$) or a d state ($l = 2$). Consequently, the core electrons are ejected primarily into the unoccupied d states in the conduction band, since there are few available s states there.

> **DIPOLE-SELECTION RULE**
> The change Δl in the angular momentum quantum number between the initial and final states must equal ±1.

It is because of the dipole-selection rule that we don't see a strong L_1 edge in the spectrum. The L_1 edge sits closer to the nucleus than the L_2 and L_3 edges and its electrons are in the s state ($l=0$) so they can only be excited to a p state ($l = 1$), but not to a d state ($l = 2$), or to another s state. Since there are few unfilled p states in the conduction band of transition metals and they are much more spread out in energy than the d states, the L_1 intensity is very low and the peak is broad and may even be invisible in the $L_{2,3}$ post-edge structure.

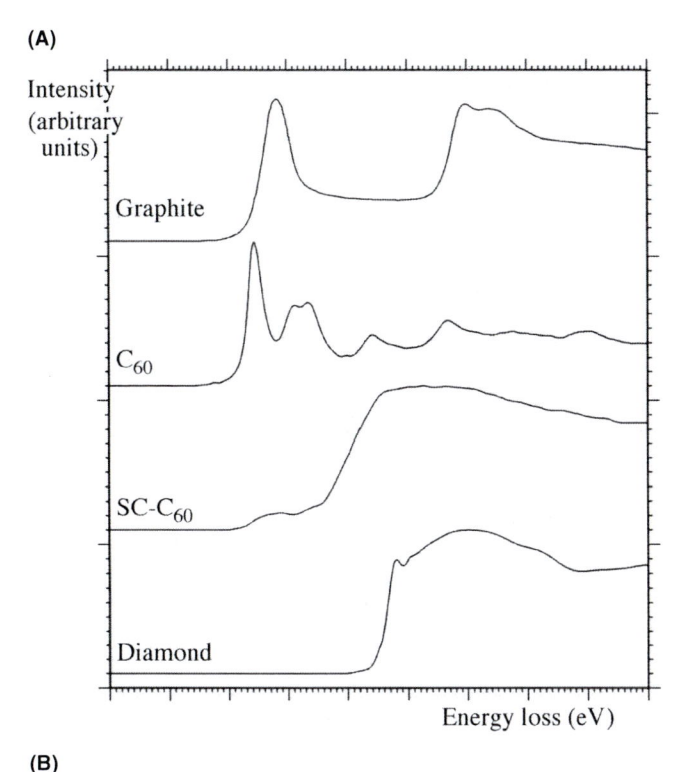

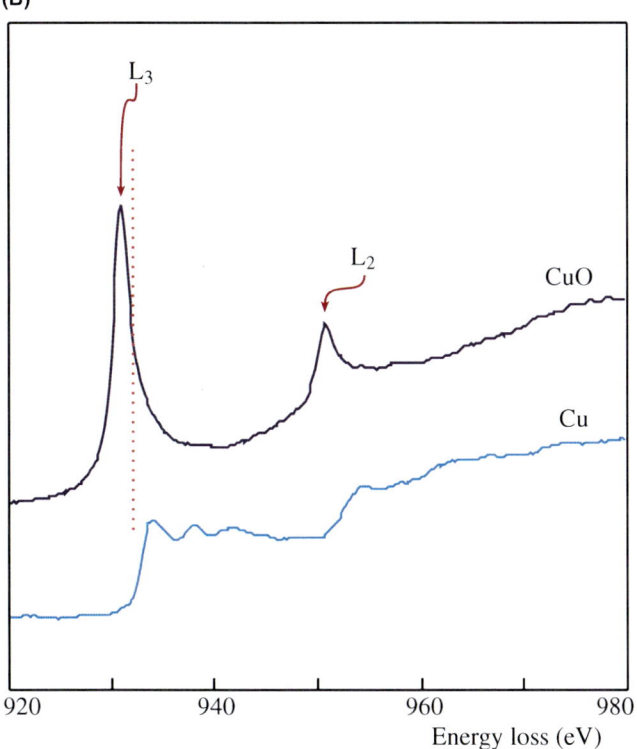

FIGURE 40.5. (A) Differences between the ELNES of the carbon-K edge from various forms of carbon. (B) Change in the Cu $L_{2,3}$ edge ELNES as Cu metal is oxidized and the filled d states lose electrons, thus permitting the appearance of white lines.

The energy width of the white lines is also affected by the time it takes for the ionized state to decay. One form of Heisenberg's uncertainty principle states that $\Delta E \Delta t = h/4\pi$, so a rapid decay gives a wide peak. For example, the Fe L_2 ionization can be rapidly compensated by an electron from the L_3 shell filling the hole and ejecting an Auger electron from the d shell. (This is called a Coster-Kronig transition.) A conduction band electron could also fill the L_2 core hole but the L_3 core hole can *only* be filled from the conduction band. Therefore, because there are two possible ways to fill the L_2 core hole, the L_2 line has a shorter Δt and a larger E than the L_3 line, which is much sharper.

In elements that don't have strong white lines, the ELNES is still present but appears just as weaker oscillations in intensity, which still reflect the DOS, and which, like the white lines, we can calculate and predict (much more of this later) (e.g., look at the pure-Cu ELNES in Figure 40.5C compared with the Fe ELNES in Figure 40.4).

40.3 APPLICATIONS OF ELNES

So let's see how all of this physics can be useful. (Is this an oxymoron?) The ELNES has been found to be dependent on details of the local atomic environment, such as coordination, valence state, and the type of bonding. Measurement of the fine structure, understanding how it is related to the electronic structure and ultimately to materials properties, can answer some hitherto-unsolved problems, particularly those where changes in bonding occur over small distances in your specimen. If you look at Figure 40.5, you'll see the carbon K edges for graphite and diamond. The carbon atom has hybridized s and p orbitals (termed σ and π in molecular-orbital theory). Graphite contains sp2 bonds in the basal plane with Van der Waals bonding between the planes. In contrast, the diamond structure has four directional, hybridized, sp3 covalent bonds and the atoms are tetrahedrally coordinated rather than arranged in graphitic sheets. The strong peak K edge at 284 eV identifies the empty π* states into which the K shell electrons are transferred in graphite, while the diamond K edge has no π* peak but shows a strong σ* peak at about 290 eV. This kind of information is also extremely useful in the study of thin diamond and diamond-like carbon films, which are of great interest to both semiconductor manufacturers and the coatings industry (sunglasses in particular). Carbon films can be made with a continuous range of graphitic and diamond-like character and it is possible to deduce the relative fraction of sp3 (diamond) and sp2 (graphite) bonding from the K edge ELNES (Bruley et al.). In today's world of carbon nanotubes, buckyballs, and graphene, all these newer forms of carbon can easily be distinguished by their ELNES. For example, carbon K-edge spectra from C$_{60}$ (Buckminsterfullerene of Buckyballs) are also shown in Figure 40.5 in the standard and shock-compressed form. Another useful example is given in Figure 40.5B, where the changes in the Cu L$_{2,3}$ edge with oxidation are shown. This is a classic example. Since Cu metal has all its 3d states filled, there are no white lines in spectra from the metal. Upon oxidation, some 3d electrons are transferred to the oxygen, leaving unfilled states, and the white lines appear in the oxide spectrum. Note also that the onset of the oxide edge is different from that of the metal, because this electron transfer changes the value of E_C.

ELNES changes often occur at interfaces where the bonding changes locally over less than 1 nm. In Figure 40.6 the Si-K edge ELNES is seen to change across a Si-SiO$_2$ interface because the Si bonding changes. In this example, you can see the extraordinary power of an FEG STEM to provide simultaneous atomic-level images and spectra localized to individual atomic columns (even though this work (from Batson) is now more than 15 years old). The combination of Z-contrast imaging (see Section 22.4) and PEELS is arguably the most powerful analytical technique for atomic characterization, as we showed in Figures 39.22 and 39.23.

> **THE CHEMICAL SHIFT**
> This difference in edge-onset energies is called a chemical shift and also helps to fingerprint the specimen. (More on this in Section 40.6.)

Bonding may also be changed by local segregation and one of the more powerful examples of ELNES is the detection of bonding changes associated with elemental segregation to interfaces, which can cause extraordinary changes in the mechanical properties of metals and alloys. For example, Ni$_3$Al has great potential as a high-temperature intermetallic, but is limited by its inherent brittle behavior resulting in intergranular fracture. It has been known for many years that this brittle behavior can be countered by the addition of a fraction of a percent of B which is known to segregate to the boundaries. Why this segregation results in a major ductility improvement was unknown until it was shown that, at B-containing boundaries, the Ni L$_{2,3}$ edge exhibits slight ELNES changes consistent with the more metallic-like bonding of pure Ni (see Muller et al.). In a complementary study, Keast et al. measured Bi segregation to Cu grain boundaries, and observed ELNES changes in the Cu L$_{2,3}$ edge, consistent with the Cu atoms in the boundary taking on a less-metallic bonding state (see Figure 40.7). This ELNES change, which is equivalent to the transfer of less than 0.3 electrons (whatever that means) from each Cu atom at the boundary to an adjacent Bi atom, may account for the brittle behavior of Cu doped with as little as 20 ppm Bi; an extraordinary change in mechanical behavior, noted first in 1874. Understanding the role of slight electronic bonding effects in such macroscopic behavior as brittle behavior may help to transform the power-generation industry for example, which spends billions of dollars removing impurity elements that cause catastrophic failure of pressure vessels if left to segregate to grain boundaries.

Such studies of ELNES are probably the most widely used aspects of EELS and the literature abounds with ELNES studies of valency determination and atomic coordination. Some examples include bonding changes at oxide interfaces on Si (Botton et al.) and probing the structure of potential next-generation Hf-based gate oxides in Si semiconductors (McComb et al.). Reviews of the potential and practical applications of ELNES have been given by Keast et al. and Brydson et al.

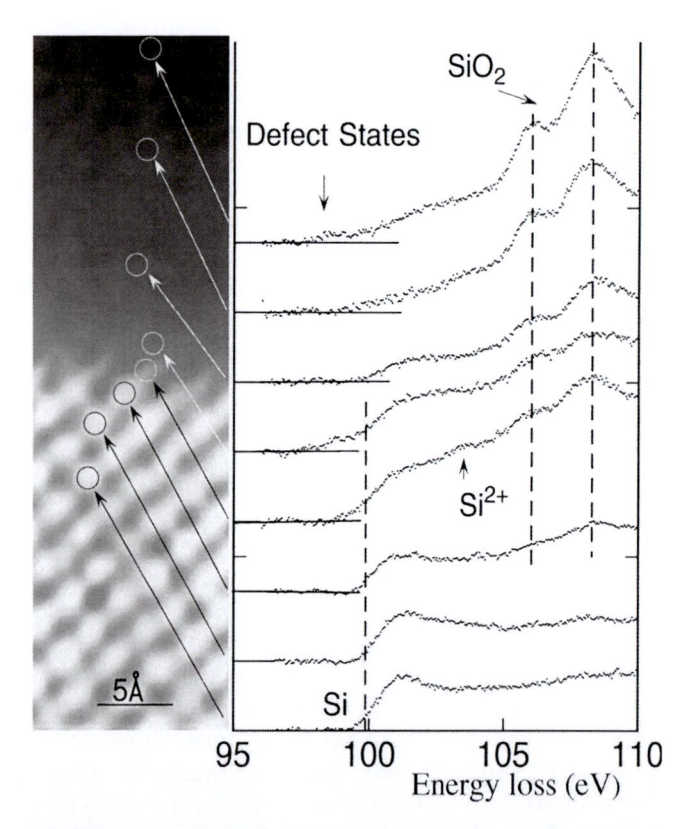

FIGURE 40.6. The change in the ELNES of the Si L edge across an interface between crystalline Si and amorphous SiO$_2$. Local electronic changes at the atomic level are easily discerned.

40.4 ELNES FINGERPRINTING

Although the ELNES is directly related to the details of the electronic structure, interpretation of particular features in an experimental spectrum is not always

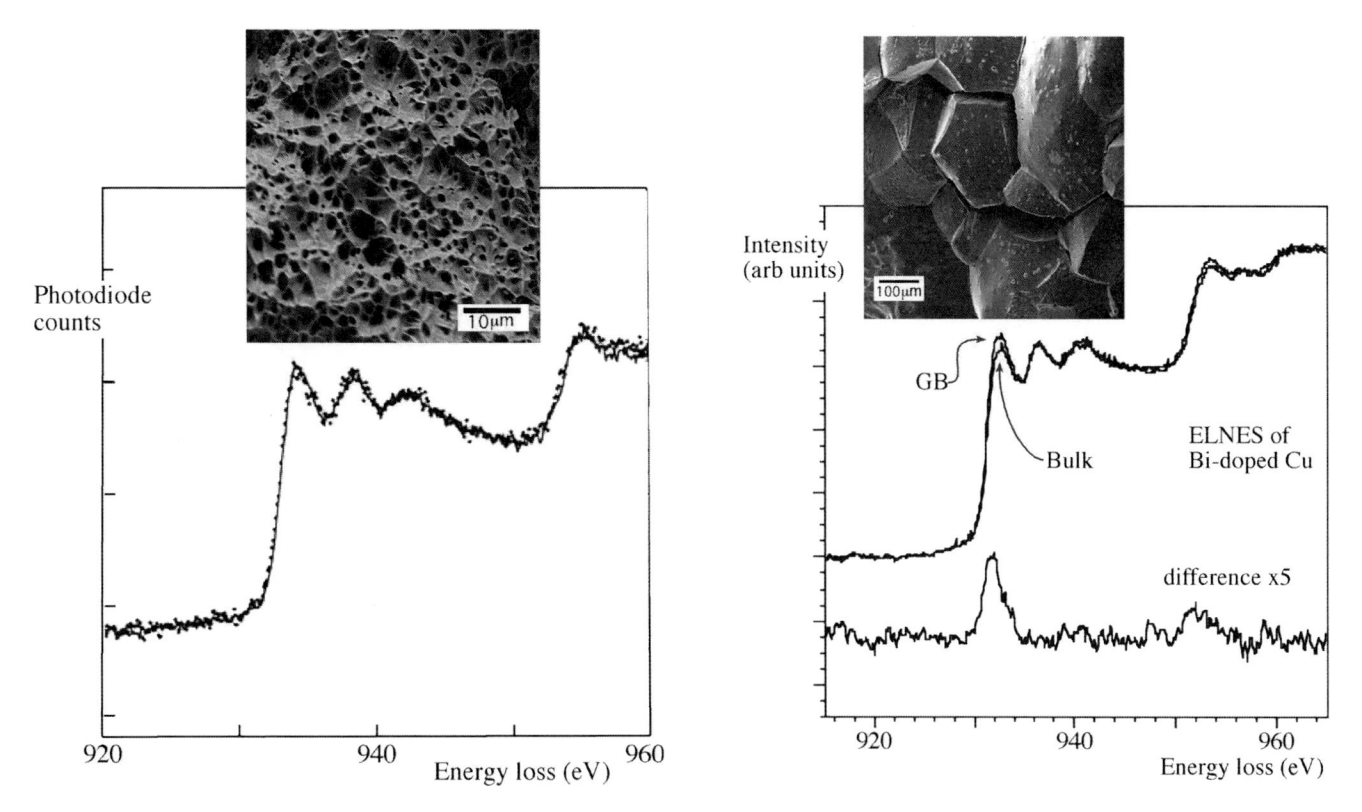

FIGURE 40.7. Change in ELNES due to impurity segregation. (A) Cu L$_{2,3}$ ELNES in pure Cu. (B) Slight Cu L$_{2,3}$ ELNES change between the bulk (grain interior) and a grain boundary to which Bi is segregated. The effect is magnified 5× in the difference plot. The two SE images show the extraordinary change in fracture behavior of ductile, pure Cu and brittle, Bi-doped Cu.

straightforward and you may not have the capability to carry out the atomic-structure calculations that we'll describe below. If this is the case, don't despair because you can still use a fingerprinting approach without fully knowing the details of the electronic structure. The idea behind such fingerprinting, as we've seen for the low-loss spectra, is that the general form of the ELNES is predominantly sensitive to the nearest-neighbor coordination and so it changes with changes in the structure. An example is provided by the Al-L$_{2,3}$ ELNES and Al-K ELNES of aluminum-oxygen materials which are sensitive to the local coordination of Al (i.e., whether octahedral or tetrahedral). Likewise, Figure 40.8 shows the experimental Mn-L$_{2,3}$ ELNES for different minerals in which the valence state of the Mn varies from 2+ to 4+. Again, we use the word fingerprint to emphasize that it is not necessary to understand the details of the DOS of complex materials in order to be able to interpret the ELNES spectra. Direct comparison with spectra from known standards is often all that is required for probable identification of the bonding state of a specific atom in your specimen. But because the matches are rarely perfect (given all the experimental and specimen variables that may affect the detailed intensity in the fine structure) go back and read our caution about low-loss fingerprinting in Section 38.3.A and apply it to your ELNES fingerprinting. Note that you don't need the very best energy resolution to carry out fingerprinting: the data in Figure 40.8 were taken many years ago from a standard PEELS system and, for many cases, a LaB$_6$ source is fine.

A theoretical calculation of the unoccupied DOS will always be useful in understanding or predicting features in the ELNES. In the next section, we'll show some examples where modeling the ELNES has helped our interpretation.

40.5 ELNES CALCULATIONS

Many attempts have been made to compare the experimental ELNES with calculations of the DOS in simple materials, such as metals and oxides. Great strides have been made in the last few years, mainly in improvements in models of the atomic potentials and in the computing power needed to pursue the calculations. This aspect is transforming the study of ELNES from an esoteric field to one with broad applications in materials science. This topic is also addressed in substantial depth in the companion text.

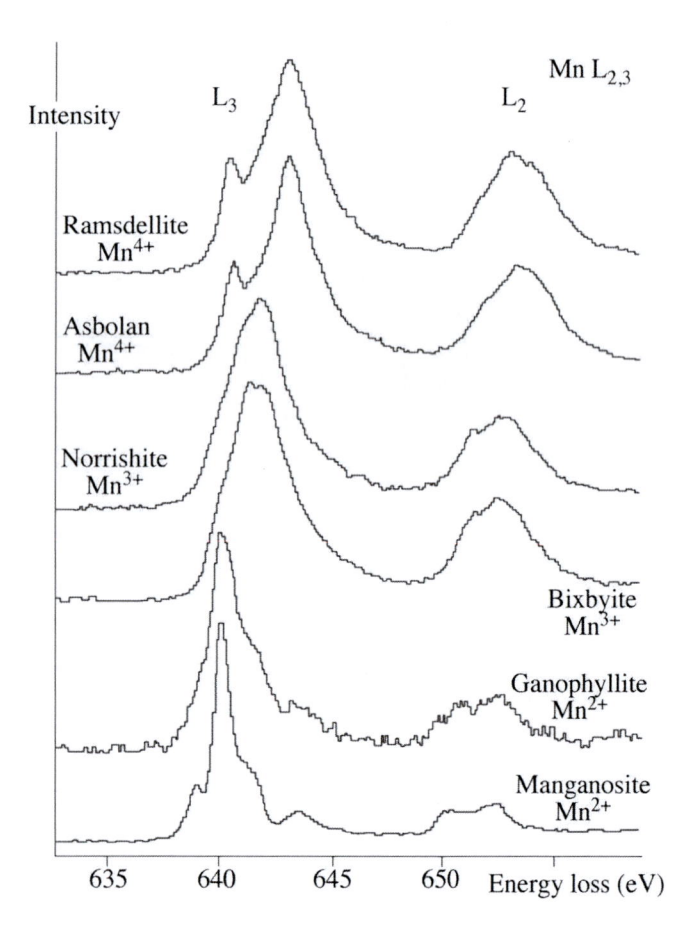

FIGURE 40.8. Comparison of Mn $L_{2,3}$ ELNES from a range of minerals in which the Mn coordination and hence valence state changes. The L_2 and L_3 white lines broaden as the oxidation state increases from +2 to +4 and in some cases the L_3 peak splits into two peaks. Understanding why such changes happen from an electronic standpoint is not necessary for identifying the different minerals or valence states.

40.5.A The Potential Choice

Calculating the electronic structure in solids involves solving the Schrödinger equation for each electron in the potential of the solid, including the Coulomb potential of both the nuclei and all the other electrons. (Now might be a good time to move on to the next section if you haven't recognized too many words in the previous sentence!) We also have to include terms due to the fact that the electrons are affected by the presence of other electrons and their behavior is correlated (i.e., they aren't isolated particles). Given the large number of electrons involved in any calculation, we often use an approach called density-functional theory (DFT) (if you're into physics and want to go to the next level (or two) then you need to read Finnis' (2003) book on atomic-modeling which includes DFT). Out of DFT comes a simplifying assumption, which we call the local-density approximation (LDA). Within the LDA method, we choose one of three different approaches to perform our calculations, which basically come down to a choice of atomic potential

- We can calculate the band structure directly in reciprocal space. This is usually described as the band-theory approach and the electron states are formed in a repeating crystal lattice. If this reminds you of Bloch states, back in Chapter 14, you're right.
- We can describe the electron states in terms of molecular orbitals (MO).
- We can calculate the effect of multiple scattering (MS) of the electron wave in real space based on the model shown in Figure 40.1A.

A range of band-structure methods are used and they go by rather strange names, such as augmented plane wave (APW), full-potential linearized APW (FLAPW) (URL #2), augmented spherical wave (ASW) (URL #3), CASTEP (URL #4), Layer Korringa-Kohn-Rostoker (LKKR), pseudopotentials, and other methods. The URLs will lead you to the sites for the various public-domain or commercial versions of the software. The best name by far is the muffin-tin (MT) potential (URL #5), which is spherically symmetric at the atomic positions and flat between them. (Apparently, to some physicists, this shape looks like a cross section of a tin used to bake muffins.) This model modifies the classic energy diagram, as shown in Figure 40.9. The MT form is useful because it generates wave functions that we can break down into the various angular-momentum components (which describe the partial DOS which is reflected in the ELNES). However, most MT approaches assume the crystal lattice is infinite and you need Bloch's theorem for the wave-function calculations which give the DOS. You'll probably find these techniques computationally challenging and not very flexible. But with the advent of easily available, high-performance, parallel computing, we are no longer so constrained (see the next section). For example, only recently have MO theorists been able model large unit cells, planar interfaces, and those (now-ubiquitous) amorphous materials.

> **THE TERMINOLOGY**
>
> Pseudopotential
> MT: Muffin-tin potential
> DOS: density of states
> DFT: density functional theory
> LDA: local density approximation
> APW: the augmented plane wave
> FLAPW: the full-potential linearized APW

MO theory is just an extension of using molecular orbitals to describe solids. (Not surprisingly, this approach is often used by chemists!) To use this approach, we have to divide our specimen into separate molecular units. If we calculate the MOs for each unit,

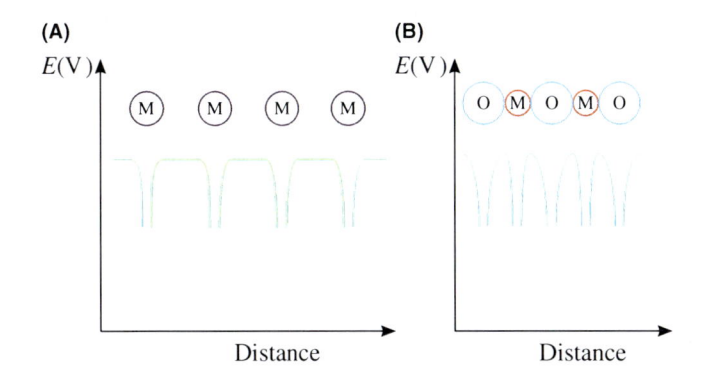

FIGURE 40.9. The muffin-tin potential energy diagram for (A) a metal and (B) an oxide. Note the symmetry of the potential wells for the metal and the asymmetry for the oxide.

we can then interpret the ELNES spectra in terms of core-shell electrons being ejected into unoccupied MOs. These unoccupied MOs arise because the excited-atom orbitals interact (i.e., bond) with nearest-neighbor atoms. (This is the π/π^* and σ/σ^* bonding/antibonding orbital notation that we used to describe the C-K shell ELNES back in Figure 40.5A). We can extend this idea and imagine the various MOs as simply linear combinations of atomic orbitals (which is then called the LCAO approach). LCAO works well if the orbitals are occupied but, for unoccupied orbitals, we have to use the self-consistent field (SCF) method, which basically assumes that the atoms are organized in a localized molecular cluster which then uses a version of the MS method which we'll now describe.

MS (not Word!) calculations are based on the interpretation of the ELNES as scattering of the electron wave that emerges from the ionized atom by atomic shells around that excited atom (which we started with back in Figure 40.1). Still the most elegant MS method is that due to Durham et al. The Durham method first divides the cluster of atoms into shells, each approximately equidistant from the ionized atom. We then solve the scattering within each atomic shell in turn. Finally, we consider scattering between different atomic shells. Since we have to calculate *all* possible scattering paths it's much easier in crystals because we can use their symmetry to facilitate our calculations. A self-consistent version of the MS method called FEFF (now in version 8) is commercially available (URL #6) and highly recommended. You can extend this shell-by-shell MS approach to model much more complex amorphous systems, incommensurate structures, and non-periodic atomic arrangements at planar interfaces and defects. You'll find that the MS calculations predict modulations in the near-edge intensity, which correspond directly to the DOS of the ionized atom. So you should be aware that these calculations are only an *interpretation* of what actually happens to the electron after it emerges above the Fermi level. Also, many calculations of ELNES only show reasonable agreement with experiment when the effect of the core hole is included, so we now need to explain this terminology.

Figure 40.10 gives a comparison of the experimental C-K edge ELNES from TiC with the results of calculations using several different potentials, none of which reproduce the experimental spectrum precisely but all of which capture some aspects of the general shape.

40.5.B Core Holes and Excitons

Having chosen an atomic potential, we actually determine the ELNES (using the MS approach) by calculating all possible inter- and intra-shell scattering events suffered by the electron after it emerges above the Fermi level. One of the problems that confuses this issue is that the ionization event results in a hole in the core shell which, of course, changes the atomic potential.

A bit more physics: The ionization process occurs in the time taken for the beam electron to traverse the diameter of the particular inner shell. We know that a 200-keV electron has a velocity, $v = 2.7 \times 10^8$ ms^{-1}, and the K-shell of oxygen, for example, has a diameter of ~ 0.01 nm, so the ionization process occurs over $\sim 10^{-19} - 10^{-20}$ s. By comparison, the atom stays in its excited state (which is a

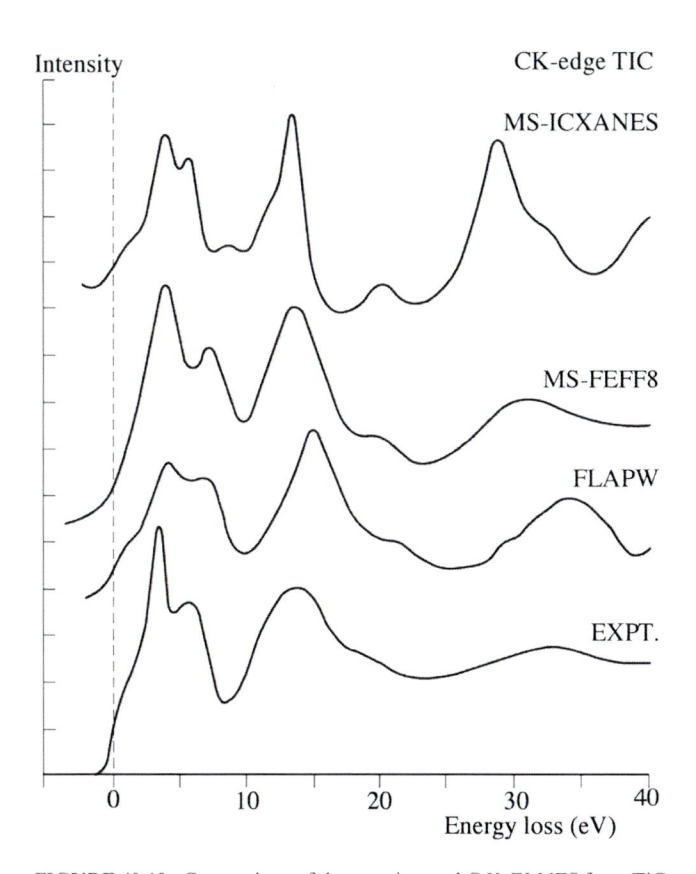

FIGURE 40.10. Comparison of the experimental C K-ELNES from TiC (*Expt.*) with the results of theoretical modeling calculations using both a band-structure code (*FLAPW*), and two different MS codes (*FEFF8* and *ICXANES*).

> **REAL MATERIALS**
> In ceramics and semiconductors, the ionized electron remains localized to the ionized atom. It may interact with the hole creating an electron-core hole bound state termed an exciton. Creation of an exciton may influence the ELNES; this remains a matter of some debate.

combination of the lifetimes of the excited electron and the lifetime of the hole in the inner-shell) for much longer, because the hole decays in $\sim 10^{-14}$–10^{-15} s. Because the lifetime of a hole is $10^5 \times$ longer than the excitation process, the outermost electron states, including the final state of the excitation process, will experience an attractive potential because of the core hole, which behaves like an extra nuclear charge on the atom. So, in fact, all we do to compensate for this is assume that the ionized atom now has a nuclear charge of $Z+1$, rather than Z, because the missing electron lowers the shielding affect of the core electrons. This extra positive charge may be shielded by other electrons (e.g., weakly bound valence electrons) which will move in response to the existence of the hole and reduce its effect. Despite this screening (go back and check Section 3.5 to remind yourself what this term means), the core-hole potential will tend to attract the outer electron states more strongly. So the available final states for the ejected electron, in the presence of the long-lived hole, will tend to be more sensitive to the short-range environment of the excited atom and, of course, this will be reflected in the ELNES.

40.5.C Comparison of ELNES Calculations and Experiments

The 1982 seminal paper in the field of ELNES experiments on transition metals and oxides is by Leapman et al. For further examples, you should read the review articles which we mentioned at the end of Section 40.3. We'll just show a couple of examples here but the literature contains many. The difference due to different coordinations is obvious. The sharp peak at the Al L edge onset is thought to be an exciton. This effect is not well modeled by the theory, which otherwise makes a good match with the experimental data.

The electron energy-loss near-edge structure (ELNES) at the O K edge has been studied in yttria-stabilized zirconia (YSZ) (Ostanin et al.). The electronic structure of YSZ for compositions between 3 and 15 mol.% Y_2O_3 has been computed using a pseudopotential-based technique to calculate the local relaxations near the O vacancies. The results showed phase transition from the tetragonal to cubic YSZ at 10 mol.% of Y_2O_3, reproducing experimental observations. Using the relaxed defect geometry, calculation of the ELNES was carried out using the full-potential linear muffin-tin orbital method. The results show very good agreement with the experimental O K-edge signal, demonstrating the power of using ELNES to probe the stabilization mechanism in doped metal oxides.

If an atom exists in two different environments in a structure, then we can make the approximation that the ELNES is simply a linear superposition of the contributions from the two environments and experiments tend to support this simple approach.

It's perhaps best to conclude with the conclusion of Duscher et al.

> "We have reached a level of agreement between theory and experiment not achieved previously in such a range of different materials by including a localized core-hole effect. There was no significant difference between both the methods used, a core hole in an all-electron method and the $Z+1$ approximation. This approach is sufficiently robust to proceed to interface structures."

In other words, theoretical calculations are a well-established and useful field of ELNES research and this is explored much further in the companion text. One last and obvious point to note is that, because the ELNES signals are often quite strong, it is straightforward to map out different portions of the ELNES, thus imaging, in effect, changes in the DOS and localized variations in atomic bonding, with the usual high resolution expected of EFTEM images. An example is shown in Figure 40.11.

40.6 CHEMICAL SHIFTS IN THE EDGE ONSET

We can think of the atoms in our specimen as having different charges with respect to one another (which we otherwise call electronegativity). So any changes in the charge in different systems will lead to changes in the binding energies of the various (occupied and unoccupied) electron states and it's reasonable to ask if we could detect this binding-energy change in EELS. We already know that (for a hypothetical, single, isolated, hydrogen atom) the ionization-edge threshold energy is effectively the critical ionization energy, E_C. However, in a real material, the experimental-edge onset corresponds to the difference in energy between the initial state and the lowest unoccupied final state *in the presence of the core-hole*. More often than not, it is extremely difficult to determine accurately the threshold energy, which often lies above the experimental-edge onset. Changes in the effective charge on the atom affect the energies of both the initial and final states. Unlike the deep-lying core orbitals, the outer orbitals are easily influenced by factors, such as bonding. If we consider changing from a metal to an insulator, the presence of a band gap in the insulator will result in a shift of the edge

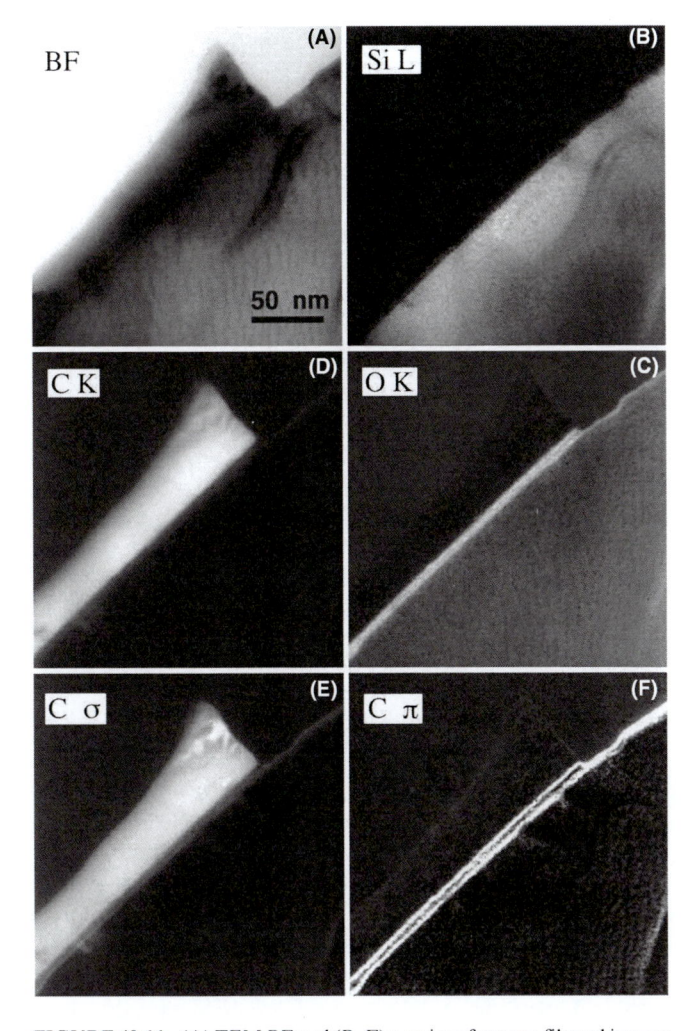

FIGURE 40.11. (A) TEM BF and (B–F) a series of energy-filtered images revealing the Si, C, and O elemental distributions and the carbon bonding maps at the interface between a diamond-like carbon film and a Si substrate. In the oxygen-rich amorphous layer at the interface, there is a double layer of carbon atoms that is primarily π-bonded (and possibly arises from carbon contamination in surface grooves at the interface) (F). The carbon film is predominantly σ-bonded (E), indicating a high degree of diamond-like character.

spectra and comparing them with reference materials. While we can correlate the edge-onset energies with such variables as oxidation states, atomic charge, and coordination, there is room for considerable improvement in calculation of the true edge-onset energies. It's worth noting that the possibility of a chemical shift in the edge may make it difficult to interpret ELNES intensity changes detected by difference techniques (such as the example shown back in Figure 40.7B). However, careful experimentation should minimize this danger.

40.7 EXELFS

If, after an atom is ionized, the ejected electron does not fill an empty state but escapes outside the atom, then it acts like a free electron (typically with energy >50 eV). We can interpret this excess energy as an electron wave, which can be diffracted by the atoms in the structure around the original ionization site. Because the electron has higher energy than those which gave rise to multiple-scattering ELNES, the diffraction is assumed to be a single-scattering event, as shown back in Figure 40.1B, and this diffraction causes oscillations on the otherwise smooth DOS. We call these oscillations extended energy-loss fine structure or EXELFS. As with any diffraction event, there is information in these EXELFS ripples about atomic positions and the atomic information comes from a relatively short range (the first few nearest neighbors) since this weak electron doesn't scatter from more distant atoms.

> **ELNES AND EXELFS II**
> ELNES is multiple scattering and EXELFS is single scattering, although the two phenomena overlap since, e.g., the L_1 ELNES peak is often far enough past the edge onset to be included in the EXELFS.

The EXELFS modulations start about 50 eV above the ionization-edge energy, are each 20–50 eV wide, as shown in Figure 40.12A, and occur over several hundred eV. EXELFS is closely analogous to the oscillations seen in the extended X-ray absorption (edge) fine structure (EXAFS) in synchrotron X-ray spectra and is one reason why EELS has long been described as a synchrotron in a TEM (albeit much cheaper than your typical synchrotron). One significant difference is that EXAFS results in complete photoabsorption of the incident X-ray while EXELFS involves absorption of only a small fraction of the energy of the beam electron. We can carry this analogy a little further. Both EXAFS and EXELFS give us structural information from materials in which there are strong, local, atomic correlations. Both techniques are atom specific so, in principle, we can solve even the most complex multi-component

onset to higher energy loss. For example, the Al-L_{23} edge shifts from 73 eV in the metal to 77 eV in Al_2O_3, and we saw similar behavior for the Cu/CuO spectra in Figure 40.5B. Another example is provided by the C-K edges in Figure 40.5A, in which shifts in the π* peak position are easier to see than changes in the rather ill-defined edge onsets.

Similar edge-onset shifts are well known in XPS and are called chemical shifts. They are reasonably well understood and we can often predict them theoretically. However, the electron-excitation process in EELS is more complicated than X-ray induced ionization (detected in XPS), particularly the unavoidable presence of the core hole and the variable extent to which it is screened by the remaining electrons. Consequently, in comparison with X-rays, little systematic work on EELS chemical shifts has been done, apart from fingerprinting experimental

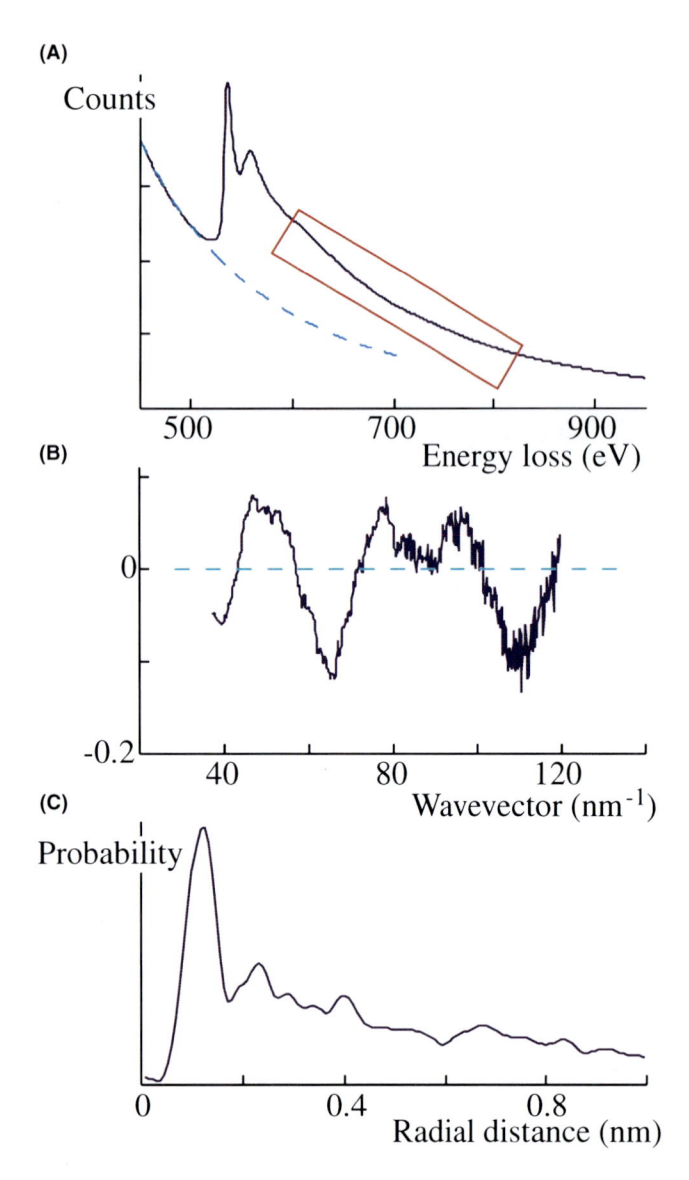

FIGURE 40.12. (A) EXELFS modulations on an ionization edge. (B) From the oscillations in the post-edge spectrum it is necessary to transfer the data to *k*-space before (C) Fourier transforming the data to produce a radial-distribution function.

structures, if the information around all the atoms can be accessed.

There are, however, limitations to conventional EXAFS.

- We can't easily access the K-edges below 3 keV since X-ray absorption at these low energies requires thin specimens for transmission-EXAFS and a low absorption atmosphere over the entire X-ray beam path (source-specimen-detector).
- As you know from the XEDS chapters, X-rays cannot easily be focused to a sub-micrometer spot, so EXAFS has a relatively low spatial resolution, although this is constantly improving. EXELFS offers us the unique ability to obtain atomic and electronic structure with nanometer-scale spatial resolution.
- Since TEMs operate in high vacuum and use thin specimens, EELS is more naturally suited to K-edge analysis of low-Z elements (as well as L-edge analysis of higher-Z elements) than low-energy EXAFS.

There is the usual price to pay for high spatial resolution in that the EXELFS signal is noisy and you'll find that extracting high-quality atomic information is much more challenging than for EXAFS where there is no shortage of signal.

40.7.A RDF via EXELFS

With EXELFS we can determine the partial radial distribution function (RDF) around a specific atom, and we are not restricted to the heavier atoms ($Z>18$) needed for EXAFS. So there is great potential for studying materials, such as low-Z glasses, amorphous Si, bulk metallic glasses, and quasicrystalline structures (both of the latter two often contain relatively low-Z elements such as Be, Mg, Al, P). In particular, since glasses lack any long-range periodic structure, we are limited in the techniques to determine their atomic structure. As you've already seen back in Section 18.7, diffraction of electrons (or X-rays or neutrons) from glass provides only diffuse information. To get atomic-structure information from glasses, you have to employ resonance signals from the Å-level and EXELFS can do that. The high spatial resolution of EXELFS is obviously advantageous and all your data can be compared with your images and the rest of the TEM-based information that you acquire from the analyzed volume. However, you can't get good EXELFS unless your specimen is very thin and you'll also have to consider phase effects, which are averaged out in EXAFS. Despite these apparent advantages of EXELFS, RDF work continues to be dominated by synchrotron X-ray sources because of the intensity of the signal. If you're interested in pursuing this (and EXELFS for that matter), a good place to start is the text by Koningsberger and Prins.

> **DECONVOLUTION AGAIN**
> Deconvolution is always the first step if your specimen isn't thin enough, i.e., if the plasmon peak is greater than 10% of the ZLP.

Experimentally, it's not easy to see the EXELFS modulations because they are only ~5% of the edge intensity, and so you need good counting statistics. This is one of the rare cases where you might find a thermionic source useful because it can deliver more total current than a FEG and, for this application, energy resolution is not so important. TEM diffraction mode will also increase your total signal intensity. Either way, however, you pay a price in terms of a loss of spatial resolution and an increased chance of specimen damage. If you need the best spatial resolution, a FEG STEM should (as usual) be your instrument of choice.

So, we're interested in EXELFS because of the structural information contained in the weak intensity oscillations. To extract this information, you can use the commercial Gatan software (see Section 1.6) or the EXELFS version of Rehr's FEFF code (URL #6). Public-domain EXELFS software is also available at URL #7.

You first have to ensure that your spectrum contains only single-scattering information, otherwise the plural-scattering intensity may mask the small EXELFS peaks. Deconvoluting the point-spread function may also help sharpen the faint modulations.

Next, you have to remove the background if it wasn't done prior to deconvolution. Then your spectrum intensity has to be converted to an electron-wave function in k-space (reciprocal space) where

$$k = \frac{2\pi}{\lambda} = \frac{[2m_0(E-E_K)]^{\frac{1}{2}}}{h} \qquad (40.1)$$

where E_K is the edge onset energy, E is the energy of the ejected electron, of wavelength λ, and the rest of the terms have their usual meaning. The electron-wave interference gives periodic intensity maxima in k-space when

$$\left(\frac{2a}{\lambda}\right)2\pi + \Phi = 2\pi n \qquad (40.2)$$

Here a is the distance from the ionized atom to the first scattering atom, and Φ is the phase shift that accompanies the scattering. Therefore, we expect to see periodic maxima occurring for $n = 1, 2$, etc., and for different interatomic spacings. We obtain the atomic spacing by Fourier transforming the k-space modulations to give the RDF, originating at the ionized atom. Obviously, these two equations don't tell the full story and there is more about data analysis in Egerton's book.

When we have the RDF, we ought to be able to determine the local atomic environment, if the various interferences can be discriminated and identified. Peaks in the RDF indicate the probability of an atom occurring a certain distance from the ionized-atom site. Figure 40.12 shows a summary of the EXELFS data extraction technique. Despite the low signal problems, EXELFS studies appeared at the earliest stages of EELS research (see, e.g., Leapman and Cosslett) and references to the technique have continued to surface in the literature over the intervening decades. For example, Sikora et al. have compared EXELFS and EXAFS applied to crystalline materials. Alamgir et al. have made comparisons of the two techniques for the study of slow-cooled, bulk metallic glasses (a fascinating new range of materials). Figure 40.13 shows the extraction of EXELFS data from the P-K edge (inaccessible in a synchrotron spectrum) from Pd-Ni-P, a model amorphous metal. The various steps involve first the isolation of the fine structure (Figure 40.13A) and expressing it as a function of momentum transfer (k), $\chi(k)$, from beyond the P K-edge (Figure 40.13B). The Fourier transform of $\chi(k)$, FT[$\chi(k)$], is proportional to the partial radial distribution function of atoms around P (Figure 40.13C). Upon back-Fourier transformation of the first peak FT[$\chi(k)$], the contribution to $\chi(k)$ from the first coordination shell around P in these glasses is determined (Figure 40.13D). Similar exercises could be performed from the second and higher shells although the quality of the signal degrades rapidly. These data are then fitted with calculated $\chi(k)$ functions of various model structures using the multiple-scattering ab-initio code FEFF7 and a possible model of the coordination of Pd and Ni atoms around the P atoms is created.

40.7.B RDF via Energy-Filtered Diffraction

RDF data acquired through EXELFS complement another TEM method of acquiring RDF information. This involves energy filtering of SADPs by scanning the pattern across the entrance aperture to the PEELS using post-specimen scan coils. (See, e.g., McBride and Cockayne; see also Section 18.7.) Effectively, a full spectrum is available at each scattering angle, but, in fact, only the zero-loss (ideally only the elastic) electrons are required. The plot of the ZLP intensity as a function of scattering angle constitutes a line profile across a filtered DP from which the RDF can be extracted. This process does not have the spatial resolution of EXELFS, since typical SADPs are integrated over ~0.2–1 μm^2, but the signal is much stronger. Accuracies of ±0.001 nm in nearest-neighbor distances can be obtained, and the process is rapid enough to be performed on-line.

40.7.C A Final Thought Experiment

ELNES and EXELFS are really quite remarkable demonstrations of quantum theory and the wave-particle duality. Consider that the EXELFS part of the spectrum only contains electrons that have been scattered by electrons in the specimen atoms, and yet we are able to deduce information about what happened to those atoms *long after the beam-specimen interaction occurred* and also deduce where the scattering atom sits in the structure!

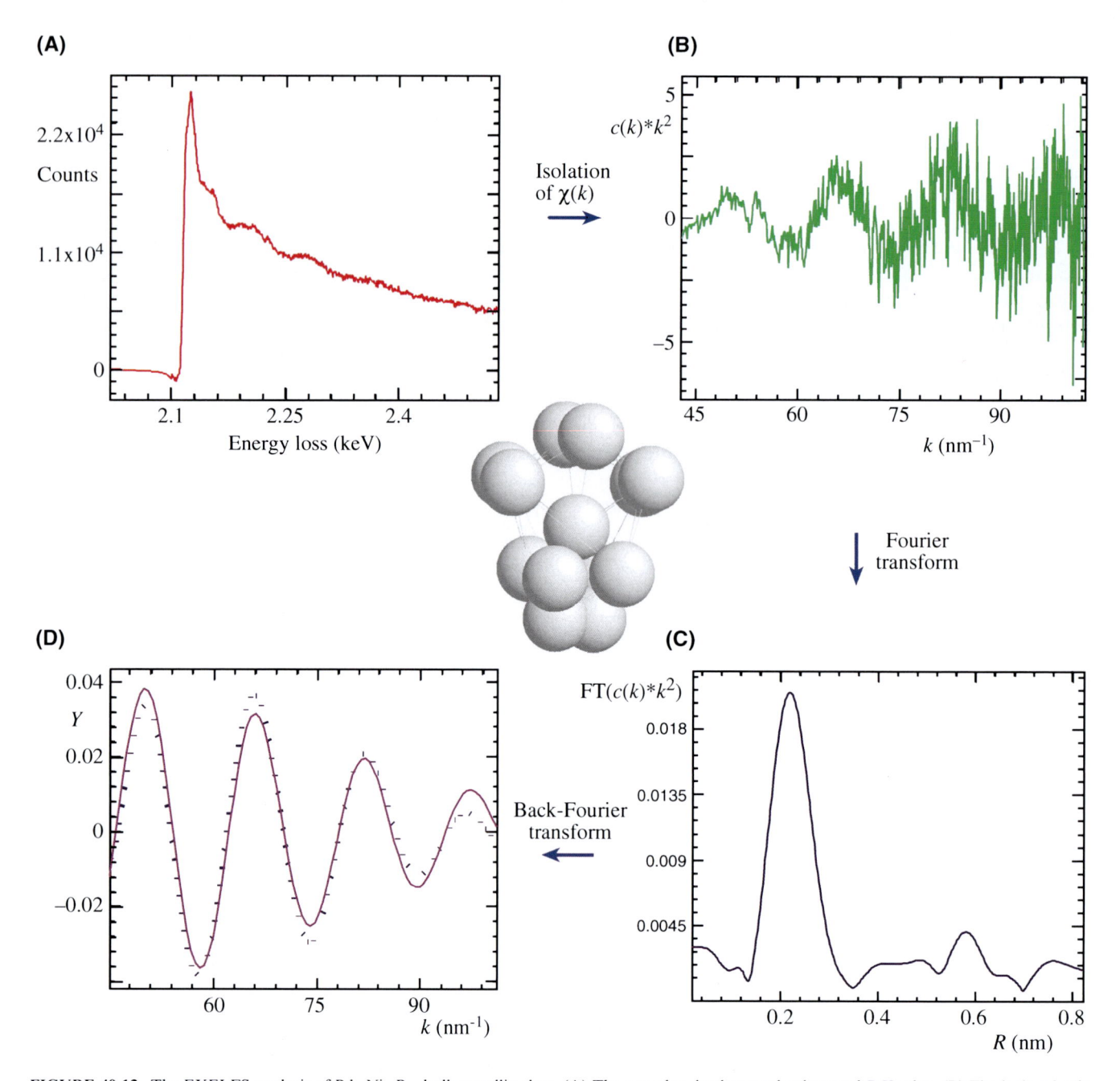

FIGURE 40.13. The EXELFS analysis of $Pd_{30}Ni_{50}P_{20}$ bulk metallic glass. (A) The pre-edge, background-subtracted P K-edge. (B) The isolated $\chi(k)$ data. (C) The Fourier transform of $\chi(k)$ to radial space, $FT[\chi(k)]$, and (D) the back-Fourier transform of the first peak of $FT[\chi(k)]$ back to k-space (dots) and fitting with a calculated function for a tetragonal dodecahedron (dashed line). The model in the center is a tetragonal dodecahedron with a P atom at the center surrounded by a first nearest-neighbor shell of Pd and Ni atoms deduced from the EXELFS.

An (approximately wrong) particle-based analogy would be to imagine that we are catching bowling balls that have been thrown at pins, arranged in a certain pattern. (Although instructive, this exercise is best carried out as a thought experiment.) From the velocity (energy) of the balls that we catch, we are able not only to identify the weight of the pin that was hit (i.e., identify the characteristic ionization edge), but we can also deduce how the pin fell down and where it rolled (the ELNES). Furthermore, we can also work out the spatial arrangement of the surrounding pins that didn't fall down (the EXELFS).

So how does the beam electron know where the core electron went after it left the inner shell? The answer lies in the fact that the bowling ball (particle) analogy is totally inadequate. In fact, only certain electron transitions are allowed and the beam electron can therefore only transfer certain quantized energies to the core electron, not a continuum of possible energies. So the beam electron does, in effect, 'know' the possible final state of the core electron since it reflects that state in its energy loss. (If you really understand this, then you should pat yourself on the back.)

40.8 ANGLE-RESOLVED EELS

Most of the time so far, we've been talking about gathering spectra and images by sending the direct beam into the spectrometer and splitting it up into its component energies. This is often called *spatially resolved* EELS since we map out a specific region or gather individual spectra from different spatial locations on the specimen. However, we have occasionally mentioned that the *angle* of scatter of the energy-loss electrons is important, and there is a whole field of EELS research that studies angle-resolved spectra. To do this, we just scan the DP across the PEELS entrance aperture and gather spectra at different angles as for the RDF measurements that we just described. However, rather than studying the energy of electrons primarily, this technique emphasizes the determination of the *momentum* of the energy-loss electrons. Momentum-transfer studies were pioneered by Silcox and co-workers. Now with FEG STEMs you can get even more information about the symmetry of electronic states, which complements spatially resolved ELNES (e.g., Wang et al.).

Because of such angular effects, the size of the spectrometer entrance aperture and/or the collection angle β may influence the details of ELNES. If the final state of the ejected electron has a definite directionality, as it will in an anisotropic crystal, the ELNES for such a specimen will depend on both the scattering angle θ and the crystal orientation. The classic paper on orientation or momentum dependence is the study of graphite and BN (Leapman et al.).

There are various ways of performing angle-resolved EELS and Botton and co-workers describe many methods and applications. For a given energy loss and a given specimen orientation, the momentum transfer at zero scattering angle, θ, is parallel to the electron beam ($q_\parallel$). As θ is increased, a component perpendicular to the beam is introduced ($q_\perp$) and, at approximately θ_E, $q_\perp$ becomes dominant. To obtain angle-resolved ELNES, we have to measure the spectra as a function of the angle between a crystal direction and the direction of momentum transfer.

First, you can keep the orientation fixed and change the collection aperture. Figure 40.14 shows angle-resolved ELNES of a very thin (~ 30 nm) graphite flake obtained by changing the size of the collection aperture in a STEM. The π* states are parallel to the *c*-axis, which in this case is parallel to the electron beam. Therefore, the corresponding π* peak intensity is larger, relative to the σ* states, when we have a small collection aperture and $q_\parallel$ dominates. Similar effects would be obtained if the orientation of the graphite planes to the electron beam was changed.

Second, you can keep your collection angle fixed at a small angle $< \theta_E$ and measure the ELNES as a function of θ. This is particularly easy with energy filtering which can display angular scattering distributions of specific energy-loss electrons.

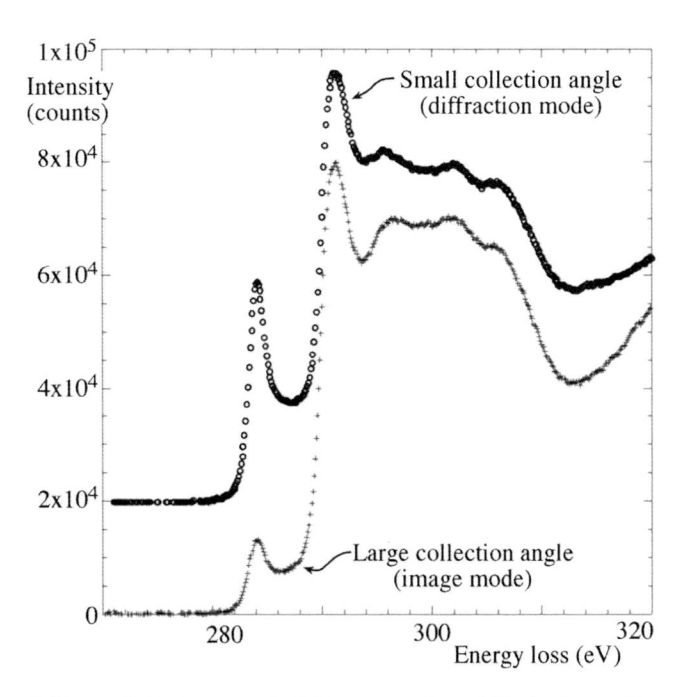

FIGURE 40.14. Image mode EEL spectra of the C-K edge in graphite showing changes in the relative intensities of the π* and σ* peaks under two different collection angles due to the directional scattering variation from the sp2 and sp3 bonds in the graphite.

You can also use the '45° method' (Botton) which combines tilting the specimen at 45° to the principal axis and measuring the ELNES at ±θ, where θ is chosen to select both $q_\perp$ and $q_\parallel$. If you make the beam convergent (e.g., in STEM mode for high spatial resolution) then the angular resolution of this technique is reduced.

One practical aspect of angle-resolved EELS is the study of Compton scattering, which is the ejection of outer-shell electrons by high-energy photons or electrons. We can detect these Compton-scattered electrons by observing the EELS spectrum at a high scattering angle (θ ~100 mrad), either by displacing the objective aperture to select an off-axis portion of the diffraction pattern or by tilting the incident beam. This process has been used to analyze the angular and energy distribution of Compton-scattered electrons and determine bonding information, since the Compton-scattering process is influenced by the binding energy.

40.9 EELS TOMOGRAPHY

We've talked about tomography in various parts of the book and how taking a series of images at different tilts allows you to reconstruct 3D information about features within your specimen. You should also read the chapter in the companion text which gives an in-depth review of the various experimental challenges for this kind of imaging.

Just as for X-ray images, it is possible to use a full-tilt series of EFTEM images to build up a full 3D image

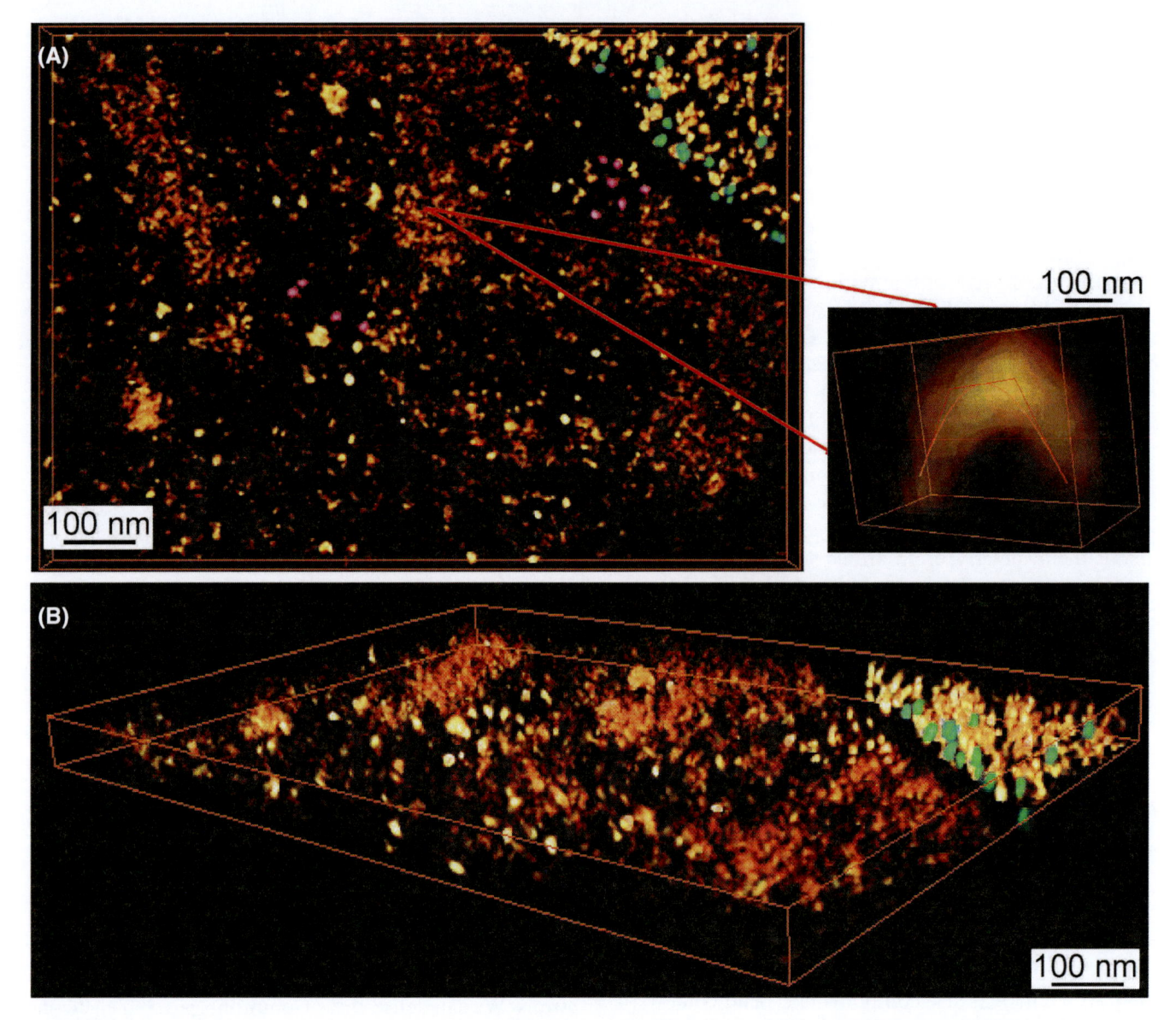

FIGURE 40.15. (A) 2D projection (top view) and (B) energy-filtered, 3D-tomographic image reconstructed from multiple P-$L_{2,3}$ edge filtered images showing the phosphorus distribution in an unstained plastic section of freeze-substituted *Drosophila* larva. The principal region is part of a cell nucleus and the top right is a region of cytoplasm containing ribosomes. The phosphorus distribution reflects the distribution of nucleic acid. Ribosomes (colored green) are known to contain about 7000 P atoms in their RNA. Another series of phosphorus-containing dense particles of unknown origin are present in the nucleus (see enlarged inset).

from a set of 2D projections. (See the paper by Midgley and Weyland.) Because ionization-edge composition images are not susceptible to significant contrast changes as a function of tilt, they are ideal for tomographic reconstruction which reveals the surface features, growth angles/facets, and other features that are not easily obtained from the usual 2D projection image. EFTEM tomography is similar to XEDS tomography which we didn't discuss: EFTEM tomography has the advantage of much quicker generation because XEDS tomography is only feasible via a series of tilted STEM images which take much longer to produce.

Figure 40.15A shows one of the P-L ionization-edge images from a tilt series and Figure 40.15B shows the reconstructed tomographic image of the distribution of P in a cell from a fruit fly. In this area, the biologists are well ahead of the materials scientists and Leapman et al., in 2004, gave a fine example of the application of EELS tomography to discerning the 3-D shape of ribosomes (not quite EELS of eels, but of nematode worms, which look a bit like eels!). In contrast to ionization-edge images, plasmon images retain significant diffraction contrast and so are less useful for tomography.

EFTEM tomography is an area that will see increasing applications, particularly as nanotechnologists pursue their dream of manufacturing devices in a controlled manner from the atom/molecule upward. The ability of EFTEM (and to a lesser extent, XEDS) to reveal the actual shape (combined with the quantitative local chemistry) of quantum dots, gate oxides, and other sub-nanometer fabrications will seriously enhance TEM's role in this growing field.

The ability to extract 3D information by tomographic methods is just one of several examples of the extraordinary advances that have taken place in the TEM field since the first edition of this book was published more than 10 years ago.

It is worth concluding now, as we did then, that we encourage you to experiment with your TEM at all opportunities and never think that there is nothing new to discover. The current generation of students growing up with an expectation of, and familiarity with, full computer control of everything and access to information immediately from anywhere should be able to combine these skills to make TEMs perform in ways that the former generation of more manually oriented TEM operators could never dream of. For example, we haven't even mentioned time-resolved EELS although the strong low-loss signal and efficient collection means that gathering spectra with millisecond or even microsecond resolution is not out of the question and ultra-fast (nanosecond) TEM imaging using laser-excited sources is fast becoming a reality. No doubt there will be a serious need for yet another edition of 'TEM: a text for nanotechnologists' a decade hence, which we sincerely hope will be written by some of those who started their careers by reading this book, rather than by those who wrote it.

CHAPTER SUMMARY

You can appreciate now that there is a wealth of fine detail in the EEL spectrum beyond the relatively strong plasmon peaks and the ionization edges. To extract this information, you need a single-scattering (deconvoluted) spectrum and occasionally some sophisticated mathematical analysis. Interpretation of the data is still limited by our lack of knowledge of the physics of the electron-specimen interactions. However, considerable research is going on into EELS fine-structure studies which are the future of the technique. Both the experimental methods and the theoretical calculations are still developing. We have introduced several specialized topics

- Energy-loss near-edge structure.
- Extended energy-loss fine structure.
- RDF determination.
- Angle-resolved (momentum-transfer) EELS.
- EELS tomography.

However, we have really only given you a suspicion of the potential of these topics. If EELS becomes a technique you use in your research, we recommend watching its development and that of related techniques in the reference sources we have given you in the last four chapters, particularly the quadrennial EELS workshops (see Chapter 37), conferences such as the biannual Frontiers of Electron Microscopy in Materials Science (FEMMS), and the proceedings of the various national and international microscopy and analysis societies which, if you haven't joined by now, you should do so immediately!

BOOKS AND REVIEWS

Brydson, RMD, Sauer, H and Engel, W 2004 *Probing Materials Chemistry using ELNES* in *Transmission Electron Energy Loss Spectrometry in Materials Science and the EELS Atlas* 2nd Ed. 223–270 Ed. CC Ahn Wiley-VCH Weinheim Germany. Comprehensive review of chemical information via the ELNES spectrum; exceptional bibliography.

Egerton, RF 1996 *Electron Energy-Loss Spectroscopy in the Electron Microscope* 2nd edition Plenum Press New York. Everything you need to know on fine structure.

Finnis, MW 2003 *Interatomic Forces in Condensed Matter* Oxford University Press, New York. Essential reading if you are modeling spectra.

Keast, VJ, Scott, AJ, Brydson, R, Williams, DB and Bruley, J 2001 *Electron Energy-Loss Near-Edge Structure – a Tool for the Investigation of Electronic Structure on the Nanometre Scale* J. Microsc. **20** 135–175. Broad-based review with lots of examples.

Koningsberger, DC and Prins, R 1988 *X-Ray Absorption: Principles, Applications, Techniques of EXAFS, SEXAFS and XANES* Wiley New York. Probably more than you'll ever want to know about the X-ray analogs of EELS.

Raether, H 1965 *Electron Energy-Loss Spectroscopy* in Springer Tracts in Modern Physics Springer-Verlag New York. *The* source if you really want to know the physics of EELS.

CALCULATIONS AND TECHNIQUE

Durham, PJ, Pendry, JB and Hodges, CH 1982 *Calculation of X-ray Absorption Near Edge Structure, XANES* Comp. Phys. Comm. **25** 193–205.

Duscher, G, Buczko, R, Pennycook, SJ and Pantelides, ST 2001 *Core-Hole Effects on Energy-Loss Near-Edge Structure* Ultramicrosc. **86** 355–362.

Leapman, RD and Cosslett, VE 1976 *Extended Fine Structure Above the X-ray Edge in Electron Energy Loss Spectra* J. Phys. D: Appl. Phys. **9** L29–L32.

Midgley, PA and Weyland, M 2003 *3D Electron Microscopy in the Physical Sciences: the Development of Z-Contrast and EFTEM Tomography* Ultramicrosc, **96** 413–431.

McBride, W, and Cockayne, DJH 2003 *The Structure of Nanovolumes of Amorphous Materials* J. Non-Cryst. Sol. **318** 233–238.

MOMENTUM TRANSFER STUDIES

Botton, GA, Boothroyd, CB and Stobbs, WM 1995 *Momentum Dependent Energy Loss Near Edge Structures Using a CTEM: the Reliability of the Methods Available* Ultramicrosc. **59** 93–107.

Leapman, RD, Grunes, LA and Fejes, PL 1982 *Study of the L_{23} Edges in the 3d Transition Metals and Their Oxides by Electron-Energy Loss Spectroscopy with Comparisons to Theory*. Phys. Rev. **25**(12) 7157–73.

Leapman, RD and Silcox, J 1979, *Orientation Dependence of Core Edges in Electron-Energy-Loss Spectra from Anisotropic Materials* Phys. Rev. Lett. **42** 1361–1364.

Wang, YY, Cheng, SC, Dravid, VP and Zhang, FC 1995, *Momentum-Transfer Resolved Electron Energy Loss Spectroscopy of Solids: Problems, Solutions and Applications* Ultramicrosc. **59** 109–119.

APPLICATIONS

Alamgir, FM, Jain, H, Williams, DB and Schwarz, R 2003 *The Structure of a Metallic Glass System Using EELFS and EXAFS as Complementary Probes* Micron **34** 433–439.

Batson, PE 1993 *Carbon 1s Near-Edge-Absorption Fine Structure in Graphite* Phys. Rev. B **48** 2608–2610.

Botton, GA 2005 *A New Approach to Study Bond Anisotropy With EELS* J. Electr. Spectr. Rel. Phen. **143** 129-137.

Botton, GA, Gupta, JA, Landheer, D, McCaffrey, JP. Sproule, GI and Graham, MJ 2002 *Electron Energy Loss Spectroscopy of Interfacial Layer Formation in Gd_2O_3 Films Deposited Directly on Si (001)* J. Appl. Phys. **91** 2921–2924. Bond changes at oxide interfaces.

Bruley, J, Williams, DB, Cuomo, JJ and Pappas, DP 1995 *Quantitative Near-Edge Structure Analysis of Diamond-like Carbon in the Electron Microscope Using a Two-Window Method* J. Microsc. **180** 22–32.

Keast, VJ, Bruley, J, Rez, P, Maclaren, JM and Williams, DB 1998 *Chemistry and Bonding Changes Associated with the Segregation of Bi to Grain Boundaries in Cu* Acta Mater. **46** 481–490.

Leapman, RD, Kocsis, E, Zhang, G, Talbot, TL and Laquerriere, P 2004 *Three-Dimensional Distributions of Elements in Biological Samples by Energy-Filtered Electron Tomography* Ultramicrosc. **100** 115–125.

McComb, DW, Craven, AJ, Hamilton, DA and MacKenzie, M 2004 *Probing Local Coordination Environments in High-k Materials for Gate Stack Applications* Appl. Phys. Lett. **84** 4523–4525.

Muller, DA, Subramanian, S, Batson, PE, Silcox, J and Sass, SL 1996 *Structure, Chemistry and Bonding at Grain Boundaries in Ni_3Al-I. The Role of Boron in Ductilizing Grain Boundaries* Acta Mater. **44** 1637–1645.

Ostanin, S, Craven, AJ, McComb, DW, Vlachos, D, Alavi, A, Paxton, AT and Finnis, MW 2002 *Electron Energy-Loss Near-Edge Shape as a Probe to Investigate the Stabilization of Yttria-Stabilized Zirconia* Phys. Rev. B **65** 224109–117.

Sikora, T, Hug, G, Jaouen, M and Rehr, JJ 2000 *Multiple-Scattering EXAFS and EXELFS of Titanium Aluminum Alloys* Phys. Rev. B **62** 1723–1732.

URLs

1) www.cemes.fr/~eelsdb
2) www.flapw.de
3) www.physik.uni-augsburg.de/~eyert/aswhome.shtml
4) www.castep.org
5) http://hermes.phys.uwm.edu/projects/elecstruct/mufpot/MufPot.TOC.html
6) http://feff.phys.washington.edu
7) www.cemes.fr/epsilon/home/main.php

SELF-ASSESSMENT QUESTIONS

Q40.1 Why does the ionization edge extend beyond the critical ionization energy (the edge onset) to give ELNES and EXELFS, rather than exist simply as a peak at the critical ionization energy?

Q40.2 What is the Fermi level/Fermi surface and why is it crucial to our understanding of the energy-loss process?

Q40.3 What is the density of states (DOS) and why are there both filled and unfilled DOS?

Q40.4 Relate the K, L, etc., core shells to the principal quantum numbers (n).

Q40.5 State the Pauli exclusion principle and explain why this is relevant to ELNES.

Q40.6 What is spin-orbit splitting and why is this relevant to ELNES?

Q40.7 What is the dipole-selection rule and why is this relevant to ELNES?

Q40.8 Why does the ionization edge onset for a specific elemental core loss sometimes shift when that element is bonded differently?

Q40.9 What is XANES, how is it detected, and what is its relation to ELNES?

Q40.10 Why do bonding changes change the ELNES?

Q40.11 What useful information is contained in the EXELFS spectrum?

Q40.12 Why is EXELFS such a challenging technique to apply?

Q40.13 What is an exciton?

Q40.14 What is a core hole?

Q40.15 Why is there bonding information in both the low-loss and high-loss spectrum?

Q40.16 Distinguish angle-resolved and spatial-resolved EELS.

Q40.17 Why is angle-resolved EELS linked to the concept of electron momentum transfer?

Q40.18 What is the RDF, why is it useful, and how can you measure it?

Q40.19 Why would you want to calculate the ELNES intensity?

Q40.20 What is Compton scattering and how can we study this in EELS?

Q40.21 Under which circumstance would you choose to use an ELNES spectrum as a fingerprint and what precautions should you take when drawing conclusions from a potential match?

TEXT-SPECIFIC QUESTIONS

T40.1 Distinguish single, multiple, and plural scattering for EELS. How do these definitions compare with scattering terms used in high-resolution imaging?

T40.2 Figure 40.1 gives an electron-wave description of the generation of ELNES and EXELFS. Can you use a particle analogy to describe the process?

T40.3 Is Figure 40.2 drawn for a crystalline metal or an amorphous semiconductor? Explain your answer and thus indicate how the figure would change if the other kind of material were being illustrated.

T40.4 In Figure 40.3 there appears to be no intensity in the ionization edge corresponding to the filled states. Why is this? In a real spectrum there would indeed be intensity before the ionization edge. What would cause this?

T40.5 Why does the Cu L edge in Figures 40.4 and 40.5B exhibit no intense white lines at the edge onset like the rest of the transition metal series in Figure 40.4?

T40.6 In old specimens and older TEMs, the diamond K edge (like in Figure 40.5A) sometimes shows residual intensity preceding the ionization-edge onset, at roughly the same energy as the π^* sp2 peak in the graphite and C_{60} edges shown above. Since diamond has no sp2-bonded carbon, can you speculate what might be giving rise to this intensity?

T40.7 Why are the muffin-tin potential wells in Figure 40.9 symmetric for the metal but asymmetric for the oxide?

T40.8 Look at the comparison of calculated and experimental spectra in Figure 40.10. These calculations were done more than a decade ago. Go on the Web and see if you can find better examples of calculated edge shapes that show a better fit to experimental spectra. If you can't, what conclusions can you draw about calculating ELNES. If you can, what different conclusions can you draw?

T40.9 How do you think that correcting the spherical aberration in the objective lens will improve the study of energy-loss fine structure? Do you think the addition of electron gun monochromators will affect the study of this same phenomenon?

T40.10 What crucial information can be gained about the behavior of semiconductor interfaces and gate oxides via ELNES? (Hint: Google PE Batson and read his papers.)

T40.11 Under what circumstances would you choose an MO rather than an MS approach to calculating the near-edge spectrum?

T40.12 List the principal differences between FLAPW, ASW, CASTEP, and LKKR.

T40.13 ELNES fingerprinting can distinguish different mineral species as in Figure 40.8. Why should we ever bother to use XEDS to study the same problem? Does the beam-sensitivity of many minerals have a role to play in deciding what technique to use? If so, explain what.

T40.14 Why does the signal in Figure 40.12B become noisier at larger wavevectors?

T40.15 Given what you know about the crystal structures of graphite and diamond, would you expect either of their energy-loss spectra to be sensitive to crystallographic orientation? If so, how do you think the fine scale features of the relevant spectrum in Figure 40.5A might change with orientation? (Hint: look at Figure 40.14.)

T40.16 Compare and contrast EXAFS and EXELFS for studying short-range atomic structures. Why would you use EXELFS when TEM diffraction patterns give similar short-range atomic structural information?

T40.17 In addition to ELNES and energy-filtered diffraction for RDF determination, can you think of other ways to explore the structure of glasses using TEM?

T40.18 For both momentum-resolved and tomographic EELS, we have to tilt the specimen considerably. What are the experimental challenges to doing this and how might they be overcome?

T40.19 If the low-loss spectrum reveals the valence states of the atoms in the specimen why do we not use this part of the spectrum more often for bonding studies but instead use ELNES which only explores the unfilled DOS (i.e., the electrons that aren't there)?

T40.20 Explain why K-shell ionization results in a hydrogenic edge.

T40.21 Explain why L shell ionization gives L_1, L_2, and L_3 edges.

T40.22 Why is the L_1 edge rarely visible, thus leaving the usual L edge as the $L_{2,3}$ in spectra from transition metals?

T40.23 Similarly, why is the $M_{4,5}$ edge the expected M edge in the rare earths?

T40.24 Explain why EELS edges and X-ray absorption edges are effectively the same phenomenon.

Index

A

A₃B ordered fcc, 262–263
Aberration, 6–7, 54, 61, 64, 81, 82, 91, 92, 98, 99, 103, 105, 106, 107, 108, 109, 110–111, 133, 141, 148, 150, 155, 158, 161, 162, 295, 372, 373, 485, 488, 493, 513, 522, 525, 590, 636, 663, 682, 683, 684, 691, 692, 696, 707, 735
  aberration-free focus, 492
  coma, 494, 498
  function, 485, 495
  See also Chromatic aberration; Spherical aberration
Absorption, 241, 413–414, 654–656
  absorption-free intensity ratio, 190
  anomalous, 319, 413, 414, 416, 449
  contrast, 374
  correction, 601, 653, 654–656, 671
  distance, 435
  edge, 588, 607, 613, 645, 717, 736
  of electrons, 118, 374
  extrapolation techniques for correction, 671, 725, 726, 727, 728
  parameters, 434, 458
  path length, 655
  of X-rays, 28, 294, 599, 600, 634, 641, 653, 656, 660, 671, 694, 717, 742, 751, 752
Accelerating voltage
  calibration of, 149–150, 168–169
  continuous kV control for CBED, 217–218, 341
  effect on Bloch waves, 252–253
  effect on EELS, 190, 681–682, 684
  effect on Ewald sphere, 355
  effect on X-rays, 615
Adaptive filter, 572, 573
Airy disk, 32, 33, 107, 109, 110
ALCHEMI, 517, 657–658, 669
Allowed reflections, 258, 259, 265, 287, 348, 349, 350
Amorphous
  carbon, 35, 163, 183, 185, 295, 374, 551, 554, 612, 719, 722
  See also Holey carbon film
  germanium, 587–588
  layer, 502, 571, 670, 751
  materials, 197, 293–295, 373, 415, 502, 504, 528, 703, 741, 748
  specimen, 377–378

Amplitude contrast, 106, 371–386, 411, 458, 504, 505
  See also Contrast
Amplitude of diffracted beam, 223
Amplitude-phase diagrams, 31–32
Analog
  collection, 102
  to digital converter, 607
  images, 115, 125
  pulse processing, 591
Analog dot mapping, 617–618
Analytical electron microscopy (AEM), 7, 25, 53, 54, 62, 66, 75, 76, 80, 81, 82, 83, 97, 99, 103, 111, 121, 132, 133, 138, 143, 144, 150, 184, 185, 186, 352, 581, 584, 586, 588, 589, 590, 592, 593, 594, 595, 597, 598–600, 605, 606, 607, 608, 609, 611, 612, 613, 614, 615, 616, 617, 618, 625, 626, 627, 628, 630, 633, 634, 639, 647, 648, 651, 655, 663, 672, 674, 682, 690, 693, 694, 696, 706, 717
Angle, 26–27, 83, 685–688
  See also Bragg; Collection semiangle; Convergence semiangle; Incidence semiangle
Angle-resolved EELS, 755
Angular-momentum quantum number, 744
Annular condenser aperture, 157
Annular dark field (ADF), 122, 160, 161, 162, 329, 373, 376, 377, 379, 380, 384, 385, 635, 659
  image, 161
  See also Dark field (DF), detector
Anodic dissolution, 178
Anomalous X-ray generation, 641
  See also Absorption
Anticontaminator, 130
Anti-phase (domain) boundaries (APB), 229, 263, 420, 426, 427, 428, 429, 434, 503
Aperture
  alignment of C2, 147
  condenser (C2), 111, 157, 493
  differential pumping, 131, 683, 687
  function, 485, 486, 488
  objective, 101, 109, 111, 152, 154–158, 161, 162, 165, 167, 207, 278, 332, 372–373, 375, 376–379, 381, 382, 385–386, 389, 390, 396, 408, 411, 413, 448, 466, 489, 500, 511, 512–514, 516, 517, 519, 520, 521, 528, 529, 535, 539, 540, 551, 552, 573, 600, 670, 686–688, 691, 706, 721, 755
  virtual, 152, 154, 491, 492
  virtual C2, 152, 154
  See also Diaphragm
Artifact
  in EELS, 730
  in image, 9, 10, 542
  of specimen preparation, 190
  X-ray peak, 605, 606–607, 613, 625, 628, 672
Artificial color, 124, 555
Artificial superlattice, 264, 265, 415
Ashby-Brown contrast, 456
Astigmatism
  condenser, 162
  intermediate, 163
  objective, 162, 163, 164, 169, 466
Atomic
  basis, 259, 260, 262
  correction factor, 640, 650
  number, 11, 16, 24, 26, 29, 30, 39, 41, 42, 57, 58, 59, 60, 122, 224, 237, 258, 284, 373, 378, 497, 635, 639, 640, 643, 650, 665, 672, 728, 729, 730
  scattering amplitude, 40, 45, 49, 258, 261, 294, 336, 517, 669
  scattering factor, 44–45, 223, 257, 378
  structure, 48, 55, 380, 389, 493, 679, 741, 747, 752
Atomic-column EELS, 567, 735–736
Auger electron spectrometer (AES), 53, 55, 61, 62
Augmented plane wave, 748
Automated crystallography, 305
Automated orientation determination, 305
Automatic beam alignment, 560
Automatic peak identification, 627–630
Averaging images, 554–556
Axis-angle pair, 303

B

Back focal plane, 94, 95, 111, 152, 162, 204, 205, 373, 491, 573, 683, 685–688, 691
See also Lens
Background
  extrapolation, 725, 726, 727, 728
  modeling, 643
  subtraction, 342, 550, 555, 641–644, 646, 650, 659, 725, 726–728, 729, 734
  See also Bremsstrahlung
Backscattered electron (BSE) detection, 115, 230
Baking, Band gap (semiconductor), 709–710
  image, 710
Bandwidth, 118, 119
Bar, 128, 157, 188, 202, 517, 588, 594, 611
Barn, 27
Basal plane, 261, 449, 745
Basis vectors, 563–565
Beam-defining aperture, 122, 329, 380, 466, 599, 610, 615
Beam (electron)
  blanking, 101
  broadening, 84, 87, 666, 673
  coherence, 533
  convergence, 538–540
  See also Convergence semiangle
  current, 13, 65, 76, 78, 81, 82–83, 84, 87, 107, 110, 121, 149, 182, 592, 593, 596–597, 599, 601, 617, 640, 641, 653, 667, 668, 670, 672, 674, 694, 721
  damage, 10–11, 30, 53–68, 76, 86, 123, 164, 263, 556, 557, 632, 652, 673
  deflection, 398
  diameter, 82, 83–85, 107, 326, 329, 636, 664, 665, 666, 667
  diffracted, 17, 24, 40, 41, 47–49, 53, 155, 156, 157, 166, 198, 199, 201, 202, 204, 215, 216, 221–231, 235, 240, 257, 265, 271, 272, 273, 274, 280, 285, 296, 298, 305, 313, 337, 361, 371, 372, 381–382, 383, 385, 407, 408, 421, 444, 463, 469, 470, 471, 472, 488, 513, 517, 519, 525, 534, 535, 536, 537, 641, 701
  See also Diffracted beam
  diffracted amplitude, 272, 413
  direct, 24, 25, 34–36, 48, 53, 116, 117, 152, 155–157, 159, 160, 161, 162, 166, 168, 169, 198, 202, 203, 204, 205, 215, 225, 229, 294, 317, 337, 349, 350
  direction, 26, 44, 45, 204, 216, 217, 248, 283, 288, 299, 300, 301, 302, 304, 312, 317, 318, 348, 349, 350, 543, 600, 614
  energy, *See* Accelerating voltage
  incident, 325, 384
  many-beam conditions, *See* Many-beam
  parallel, 92, 94, 141, 142–143, 145, 146, 147, 148, 152, 158, 163, 283–305, 311, 324, 325, 326, 339, 340, 352, 385, 386, 511, 721
  shape, 667
  splitter, 397, 525
  tilting, 147, 199, 285, 382, 467, 669, 755
  translation, 147
  two-beam conditions, *See* Two-beam approximation
Beam-sensitive materials709
  See also Beam (electron), damage
Beam-specimen interaction volume, 323, 598, 663, 664, 665, 672, 735
Bend contour, 352, 385, 386, 407, 411–412, 413, 415, 441, 493, 521, 641, 658, 733
Beryllium
  grid, 609, 612
  oxide, 612
  specimen holder, 613
  window, 586, 587, 598, 607, 628, 633, 650, 651, 654
Bethe cross section, 58
Bethe ridge, 718
Biprism, 77, 397, 398, 525
Black level, 377, 528
Black/white contrast, 449
Bloch theorem, 237
Bloch wall, 517
Bloch wave
  absorption, 241
  amplitude, 252–253
  coefficient, 238
  kinematical condition, 221
Body-centered lattice, 357
Boersch effect, 683
Bohr radius, 42, 63, 378, 705
Bohr theory, 55
Bolometer, 590, 591, 592, 594, 663
Borrmann effect, 657, 658, 733
Boundary, *See* Grain, boundary; Interface; Phase boundary
Bragg
  angle, 34, 49, 83, 169, 200, 201, 202, 205, 222, 223, 229, 230, 312, 327, 339, 340, 353, 381, 408, 451, 680, 687, 721
  beam, 221, 230, 239, 245, 246, 247, 250, 254, 431, 492, 534, 536
  See also Diffracted beam
  condition, 204, 207, 213–216, 227, 228, 241, 253, 272, 274, 294
  diffraction, 26, 201, 202, 211, 214, 222, 231, 311, 312, 320, 339, 373, 381, 699
  law, 199, 200–202, 211, 213–214, 217, 218, 319, 411, 488, 519, 590
  plane, 249, 319, 416
  reflection, 49, 157, 202, 208, 262, 305, 448, 492, 537
Bravais lattice, 267, 347
Bremsstrahlung, 40, 58, 60, 135, 168, 582, 598, 605, 606, 608–614, 618, 626, 632, 635, 641–643, 673
  coherent, 613–614, 626, 642
  See also Background
Bright field (BF)
  detector, 122, 159, 160, 161, 162, 326, 366, 373, 380, 384, 385, 521
  high-order BF, 521
  image, 155–159, 161, 163, 165, 166, 168, 169, 182, 230, 295, 304, 330, 331, 352, 362, 365, 372, 374–376, 379–381, 386, 407–409, 411, 412, 414, 415, 424, 425, 427, 428, 434, 444, 452, 455, 458, 468, 472, 473, 475–477, 512, 516, 521, 573, 599, 617, 669, 670, 687, 734
  in STEM, 304, 373, 376, 385, 600, 617, 670
  symmetry, 361, 366
Brightness (gun), 79, 116, 150, 327, 375, 626
Brillouin-zone boundary (BZBs), 245, 253
Buckyballs, 743, 745
Bulk holder, 135, 175
  See also Specimen, holder
Bulk modulus, 457
Burgers vector, 278, 320, 339, 396, 402, 420, 441–444, 446–449, 456, 458, 469, 473–474, 476
  See also Dislocation

C

Calibration
  of accelerating voltage, 168–169
  of camera length, 165–166
  of focal increment, 169
  of illumination system, 149–150
  of image rotation, 100–101
  of magnification, 164–165
Camera constant, 166, 355
Camera length, 154, 155, 161, 162, 165–167, 197, 198, 217, 218, 284, 302, 317, 318, 326, 327, 328, 329, 336, 361, 373, 379, 398, 686, 687, 688, 721
c/a ratio, 259, 314
Carbon
  amorphous, 35, 163, 183, 185, 295, 374, 551, 554, 612, 719, 722
  contamination, 586
  film, 86, 162, 163, 164, 173, 183–185, 374, 375, 397, 521, 554, 612, 722, 745, 751
  nanotube, 73, 81, 365, 366, 695, 745
  See also Holey carbon film
Cartesian-vector notation, 260
Cathode-ray tube (CRT), 115
Cathodoluminescence, 53, 62–63, 116, 122, 523–524
Cauliflower structure, 583
CCD-based WDS, 590

Center of symmetry, 230, 236, 240, 358, 361, 435
Centrosymmetric point group, 358
Channel-to-channel gain variation, 689, 727
Channeling, 230, 339, 366, 379, 646, 657–658, 733
Characteristic length, 221, 222, 223–224, 225, 237, 478
 See also Extinction distance
Characteristic scattering angle, 63, 700, 705, 717, 719, 733
Charge-collection microscopy, 62–63, 523–524
Charge-coupled device (CCD) camera, 10, 116, 120–121
Charge-density determination, 366
Chemically sensitive images, 517
Chemically sensitive reflections, 261, 262, 263, 519, 567
Chemical resolution, 626
Chemical shift, 746, 750–751
Chemical wire/string saw, 176
Chi-squared, 645
Chromatic aberration, 6, 8, 9, 91, 103, 104–106, 108, 109, 148–149, 159, 334, 377, 491, 495–497, 533, 680, 681, 687–688, 690, 702, 703, 706, 711
 See also Aberration
Chromium-film standard (C film), 160, 189, 717
Clamping ring, 132, 133, 135, 181
Cleavage, 176, 187, 191, 414
Cliff-Lorimer
 equation, 640, 641, 646, 647, 648, 652, 656, 657, 673
 k factor, 646
Coherence, *See* Beam (electron), coherence; Bremsstrahlung, coherent; Spatial coherence
 Foucault imaging, 516–517
 Fresnel imaging, 516
 interference, 31
 particles, 456, 457
 processing, 526
 scattering, 39, 202, 319
Coincident-site lattice (CSL), 500
Cold FEG, 498
Cold trap, *See* Anticontaminator
Collection semiangle, 34
Collimator, 598–599, 611
Column approximation, 223, 229–230, 421, 423, 426, 433, 434–436, 443, 457, 478
Coma-free alignment, 493, 498
Comis, 433–436, 458
Composition
 measurement, 7, 667–668
 profile, 616, 659, 667
Compton scattering, 741, 755
Computer simulation, 311, 351, 362, 365, 392, 414, 456, 492, 493, 539, 665

 See also Image, simulation of
Condenser 1 lens, 83, 84, 85, 142, 144, 145, 148, 149, 150, 158, 167, 326, 327, 334, 584, 721
Condenser 2 lens, 83, 111, 116, 142–150, 158, 167, 207, 326, 327, 328, 330, 332, 584, 610, 615, 632, 721
 aberration, 81, 82, 91, 92, 108, 111, 150, 373, 525, 725
 alignment, 161–162
 aperture, 101–102
 calibration, 154
 defocusing, 95–96
 diaphragm, 101–102
 See also Lens
Condenser objective condition, 326
Condenser-objective lens, 325, 326
Conduction band, 59, 62, 64, 66, 117, 245, 585, 605, 709, 720, 742, 744–745
Confidence limit, 632, 648
Confocal microscopy, 4, 8
Conical diffraction, 157, 291
Conical scanning, 157
Conjugate plane, 94, 152, 160, 161
Constructive interference, *See* Interference
Contamination, 62, 81, 99, 102, 106, 118, 124, 127, 130, 132, 135, 137, 138, 149, 161, 162, 181, 189, 190, 277, 324, 341, 494, 519, 523, 584–586, 588, 595, 609, 612, 615, 620, 626, 632, 636, 641, 647, 658, 670–671, 673, 751
Continuum, 55, 238, 293, 443, 605, 606, 613, 642, 726, 742, 744, 754
 See also Background; Bremsstrahlung
Contrast
 difference, 371, 372–373, 374, 376
 Fresnel, 374, 389, 397, 399–402, 540–541
 inside-outside, 448, 450, 476
 minimum, 378, 492, 493, 497, 498, 561
 minimum defocus, 498
 topographic, 519, 521
 transfer function, 17, 485, 487, 494, 506
 See also Amplitude contrast; Diffraction, contrast; Phase contrast
Convergence semiangle, 84
 correction (Cs correction), 6–8, 10, 62, 68, 76, 82, 84, 104, 108, 124, 355, 597, 647, 659, 663, 668, 672, 674, 733, 735, 736
Convergent beam
 diffraction, 671
 See also Higher-order Laue zone (HOLZ)
 energy-filtered, 77
 imaging (CBIM), 332, 334
Cooling holder, 132, 134, 135–136, 324, 336
 See also Specimen, holder
Core-hole effect, 750
Core-loss image, 715, 717, 719, 730, 731, 732, 741

Coster-Kronig transition, 59, 745
Coulomb force, 24, 36, 39, 40, 199, 221
Count rate, 588–594, 596–599, 601, 607, 608, 616, 618, 625, 627, 630, 632, 636, 641, 658, 659, 667, 673, 717
Coupled harmonic oscillator, 231
Coupled pendulum, 477
Critical energy, *See* Ionization
Cross-correlating image, 561
Cross-correlation function, 568
Cross section
 differential, 23, 27, 39, 41, 42, 44, 45, 63, 729
 elastic, 27, 28, 44, 54, 58
 experimental, 27–28
 generalized, 433–434
 modified Bethe-Heitler, 58
 partial ionization, 723, 725, 728–730, 734
 phonon, 63–64
 plasmon differential, 63–64
 relativistic Hartree-Fock, 545
 Rutherford, 41–43
 screened-relativistic Rutherford, 42–43
Cross-section specimen preparation, 182
Cross-tie wall, 517
Cryogenic pump, 130
Cryo-transfer holder, 136
 See also Specimen, holder
Crystal
 A-face centered, 357
 B-face centered, 357
 cubic, 198, 241, 258, 260, 261, 273, 277, 289, 340, 356, 357, 394
 diamond cubic, 289, 314, 420
 high-symmetry pole, 341
 I-centered, 357
 imperfect, 386, 443, 534
 low-index, 289
 low-symmetry zone axis, 338, 356
 non-centrosymmetric, 230, 236, 358, 435
 orientation of, 304, 317–318, 435, 493, 755
 orthorhombic, 213, 354, 357
 perfect, 197, 224, 235, 236, 259, 279, 386, 392, 396, 397, 421–423, 426, 435, 441, 443, 445, 447, 459, 463, 471, 473, 478, 492, 505, 533, 534, 536, 537, 540, 542
 plane, 9, 48, 49, 138, 204, 208, 286, 311, 336, 339, 657
 pole, 287, 288
 potential of, 230, 236, 238, 239, 240
 primitive, 257–258, 267, 356
 projected potential of, 534
 simple cubic, 241, 258, 261, 273
 tetragonal distortion of, 415
 zone axis, 17, 204, 213
 See also Lattice
Crystal Kit, 267
Crystallographic convention, 204, 230
Crystallographic shear, 503

Curie temperature, 429, 517
Current, 74–79, 82–86
   See also Beam (electron)
Current centering
   dark, 118, 689–690, 709, 711
   density, 162
Curve fitting, 645, 726
Cut-off angle, 700, 706
Cyanide solution, 174
Cyclotron radius, 99, 100

D
Dark field (DF)
   annular, 85, 122, 161
   centered, 155–156
   detector, 160
   diffuse, 295
   displaced-aperture, 155, 156, 412
   external control for, 515
   focus of, 492
   high-angle annular, 122, 144, 379–381, 386, 710, 735, 736
   image, 206, 332
   multiple, 206, 207
   STEM, 376–377
   through focus (2½D), 513
   tilt control, 382
   See also Weak-beam dark field
Dead layer, 118, 585, 586, 595, 596, 607, 651
Dead time, 592–593, 596–598, 601, 607–608, 611, 627, 630
Debye-Waller factor, 336, 348, 435, 533
Decision limit, 674
Deconvolution, 630–632, 667, 688–689, 694, 700, 701, 705, 706, 707, 715, 721, 725, 728, 730, 731–733, 752, 753
Defect
   computer modeling of, 432–433
   core, 421
   unit cell, 361–362
   See also Dislocation; Grain, boundary; Stacking fault; Twin boundary
Defocus condition, 492, 513
Defocused CBED patterns, 329–330
Defocus image, 557
Deformable-ion approximation, 435
Delocalization, 390, 497, 498, 500, 735
Delta (δ) fringe, 427–429
Delta function, 731
Density, 16, 27, 28, 29, 63, 65, 74, 75, 76, 79, 81, 82, 115, 120, 124, 164, 174, 176, 180, 190, 293, 362, 364, 365, 371, 373, 378, 401, 442, 444, 448, 450, 487, 495, 503, 514, 541, 563, 590, 651, 653, 654, 655, 656, 665, 676, 679, 706, 707, 709, 710, 721, 741, 742, 743, 748
Density-functional theory, 711, 748
Density of states, 590, 721, 742, 743, 744, 745, 747, 748, 749, 750, 751

Depth
   distribution of X-ray production, 655, 665
   of field, 8, 91, 92, 101, 103, 110–111, 513
   of focus, 8, 91, 92, 101, 103, 110–111
   fringes, 426, 471
Desktop Microscopist, 685
Desk-Top Spectrum Analyzer (DTSA), 16, 608, 628, 641, 652, 674, 730
Detectability limits, 54, 475, 581, 589, 590, 626, 631, 632, 663, 672, 673, 674, 675, 715, 736
Detection quantum efficiency (DQE), 116, 118, 119, 121, 123
Detector (electron), 10, 24, 27, 117–122, 159, 372, 373, 389, 511, 523, 586
   depletion region of, 117, 118
   envelope function, 485, 491, 492, 494, 495, 496, 497
   gain of, 118, 119
   STEM, 326, 366, 372, 385, 386, 511, 528, 687
   See also Spectrometer (EELS)
Detector (X-ray), *See* Spectrometer (X-ray energy-dispersive); Spectrometer (X-ray wavelength-dispersive)
Determination limit, 674
Deviation parameter, 216, 273, 297, 298, 353, 371, 382–384, 415, 435, 496
   See also Excitation error
Diamond-cubic structure, 420
Diamond window, 187
Diaphragm, 101–102, 122, 132, 144, 145, 147, 148, 149, 153, 154, 155, 156, 324, 326, 332, 333, 373, 375, 385, 485, 502, 515, 521, 528, 594, 601, 608, 609–611, 613, 621, 627
   self-cleaning, 102
   top-hat C2, 611, 621
   See also Aperture
Dielectric constant, 42, 694, 701, 706, 710, 711
   determination of, 705
   image, 705
Dielectric response, 679, 699, 705
Difference spectrum, 643, 690, 727, 733
Differential hysteresis imaging, 165
Differential pumping aperture, 131, 683, 687
Differentiating the image, 556
Diffracted beam, 47–49, 156, 166, 204, 221–231, 265, 337, 383, 408, 519, 534, 701
   amplitude of, 31–33, 371, 444
   intensity of, 47, 215, 273, 274
Diffracting plane, 199, 201, 202, 204, 208, 213, 246, 287, 289, 312, 318, 319, 320, 332, 340, 371, 390, 396, 407, 411, 412, 416, 441, 442, 444, 445, 449, 452, 454, 463, 469
   See also Bragg, plane
Diffraction
   camera, 165, 198–199
   center, 162

   contrast, 197–207, 313, 371–386
   convergent beam, *See* Convergent beam, diffraction
   from dislocations, 278–279
   double, 222, 296–298, 304, 394
   extra reflection, 264
   Fraunhofer, 30–31
   Fresnel, 30–31
   grating, 31–32, 164, 165, 273, 573, 590
   group, 365–366
   indexing, 213
   180° inversion of, 31
   mode, 116, 152, 153, 154, 161, 166, 167, 206, 326, 330, 334, 466, 685, 687, 688, 721, 735, 753
   multiple, 296, 364
   nanodiffraction, 283, 291, 323, 347, 365, 366
   oblique-textured, 291
   pattern, 17, 49, 198, 204, 207, 372, 382, 383, 384, 391, 394, 755
   ring, 155, 287, 293
   rocking-beam, 230
   rotation, 167–168
   scanning-beam, 365
   selected area, *See* Selected area diffraction (SAD)
   shell scattering, 749
   single-crystal, 168
   split spot in, 516
   spot spacing in, 336
   streak in, 254
   systematic absence in, 304
   systematic row in, 332
   vector (g), 201
Diffraction coupling, 685
Diffractogram, 17, 493, 551, 552–554, 555, 560, 561, 574
Diffuse scattering, 329, 711
   See also Scanning transmission electron microscope (STEM)
Diffusion pump, 129, 130, 131, 180, 191
Diffusion coefficient, 104
Digital
   filtering, 643, 644
   image, 16, 117, 124, 155, 528, 556, 619
   mapping, 618–620
   pulse processing, 598, 635
   recording, 131, 434
Digital Micrograph, 552, 560, 570, 573
Dimpling, 177, 178, 191
Diode array, 683
   saturation of, 691, 723
Dipole selection rule, 744
Direct beam, *See* Beam (electron), direct
Discommensurate structure, 278
Discommensuration wall, 503
Disk of least confusion, 103
Dislocation
   array, 278–279, 394, 450–451, 452, 455
   contrast from, 444–448
   core of, 402, 443, 448, 469

density, 442
dipole, 448–450, 476
dissociated, 451, 463, 473–477
edge, 442, 444, 451, 452, 459, 469, 471, 475
end-on, 400, 401, 402
faulted dipole, 441, 476
faulted loop, 448
inclined, 458
interfacial, 453
intersecting, 458
line direction, 449
loop, 448–450
misfit, 448, 453, 455, 456
network, 448, 453
node, 448
ordered array of, 278
pair, 450
partial, 445–447, 450, 451, 463, 473–476
screw, 441, 442, 444, 447, 452, 454, 457, 458, 459, 470, 475
strain field of, 279, 371, 447, 451, 457
superlattice, 447
transformation, 453, 456
See also Burgers vector; Strain
Disordered/ordered region, 263, 517–518
Dispersion diagram
branches of, 247, 249, 250, 251, 253, 254
plane (of spectrometer), 246, 247, 249
relation, 250–251
surface, 247–250
Displacement
damage, 67, 68, 86, 721
energy, 67
field, 433, 435, 441, 442, 443, 444, 447–448, 456, 458
vector, 348, 349, 350, 458
Display resolution, 592, 614, 627, 630, 722
Double-period image, 564
Double-tilt holder, 134, 324
See also Specimen, holder
Drift
correction, 617, 636
rate, 464, 498
tube, 681, 682, 684, 722
Dwell time, 124, 636, 710, 722
Dynamical
absence, 258
calculation of intensity, 364–365
condition, 362–363
contrast in CBED, 347
coupling, 239
diffraction, 203, 221, 222, 225, 274, 296, 297, 298, 329, 331, 358, 364, 641, 669
scattering, 30, 203, 221, 258, 265, 272, 298, 311, 319, 323, 342, 358, 558
See also Diffraction
Dynamic experiments, 526–528

E

Edge, *See* Ionization
EELS advisor software, 730, 734

EELS, *See* Electron energy-loss spectrometry (EELS)
EELS tomography 755–757
Effective EELS aperture diameter, 688
EFTEM imaging, 685, 703, 733–735
Elastic
coherent, 25
constant, 454, 457, 458
cross section, 41–43
mean-free path, 23
scattering, 25–26, 27, 28, 30, 39–49, 64, 199, 324, 329, 332, 339, 340, 373, 374, 378, 380, 381, 664, 665, 699, 702, 741
See also Scanning transmission electron microscope (STEM)
Elasticity theory, 421, 448, 458, 474, 475
ELD software, 559
Electric-field potential, 236
Electro-discharge machining, 176
Electron
backscatter pattern (EBSP), beam, *See* Beam (electron)
beam-induced current (EBIC), 62, 136, 523
channeling, 366, 379, 733
charge, 523–524
crystallography, 47, 324, 558, 559
detector, *See* Detector (electron)
diffraction, 8–9, 34–36
dose, 65
See also Beam (electron), damage
source, 73–87
Electron-electron interaction, 40, 41, 42, 66, 683, 699
Electron-hole pair, 62–63, 117, 118, 524, 585, 586, 587, 591, 592, 593, 595, 611
Electronic-structure tools, 711
Electron-spectroscopic imaging, 690
Electropolishing, 178, 179, 183
Electrostatic lens, 78, 80, 96
See also Lens
ELP (energy-loss program), 16
Empty state, 751
EMS (electron microscope simulation program), 17, 127, 267, 287, 314, 491, 534, 571
Enantiomorphism, 347, 363–364
Energy
EELS, 715–736
spectrometry (EELS) resolution, 76
spread, 73, 74, 76–77, 79, 81, 82, 85–86, 105, 148–149, 496, 498, 693, 701
window, 635, 703, 734
X-ray, 590, 591–593
See also Electron energy-loss spectrometry (EELS)
Energy-dispersive spectrometry, 7
See also Spectrometer (X-ray energy-dispersive)
Energy filtering, 7, 8, 106, 334, 336, 342, 352, 366, 478, 696, 702, 703, 753, 755
Energy-loss, *See* Electron energy-loss spectrometry (EELS)

Electron energy-loss spectrometry (EELS)
angle-resolved, 755–756
collection efficiency of, 735–736
collection mode, 735
detectability limit of, 736
diffraction mode, 735
image mode, 755
imaging, 735–736
microanalysis by, 589
parallel collection, 681
serial collection, 685
spatial resolution of, 735–736
See also Spectrometer (EELS)
Energy-loss spectrum
artifacts in, 688–689
atlas of, 720, 723
channeling effect in, 733
deconvolution of, 731–733
extended fine structure in, 743
extrapolation window in, 726, 727
families of edges in, 723
fine structure in, 736
gun, 693
See also Gun, holography
interferometer, 398
lens, *See* Lens
microscope microanalyzer, 646
momentum, 741
near-edge structure in, 723
parameterization of, 730
phase, *See* Phase boundary
potential energy, 749
power-law fit, 726
rest mass, 700
scattering, *See* Diffraction, shell scattering
source, 735–736
See also Gun
structure factor, 567
See also Structure factor
velocity of, 700
wavelength of, 705
wave vector, 752
Envelope function, 485, 491, 492, 494, 495, 496, 497
Epitaxy, 138, 168, 296
Errors in peak identification, 634
Errors in quantification, 647–648
Escape peak, 606, 612, 614, 628, 630, 656
Eucentric
height, 151, 164, 165, 166, 169, 330, 515
plane, 100–101, 151, 167, 326, 327, 330, 333, 334, 599
specimen, 295
See also Goniometer
Ewald Sphere, 214–218, 235, 241, 248, 249, 252, 271, 273–275, 279–281, 290–292, 298, 312, 318, 324, 336–338, 351, 355, 411, 430, 431, 447, 449, 464, 466, 467, 470, 478, 537, 559

Excitation error, 216–217, 224, 227, 252, 318, 353, 463, 702
   effective, 228–229, 407, 464
   See also Deviation parameter
Exciton, 66, 750
Exposure time, 123, 158, 207, 305, 337, 464, 466
Extended energy-loss fine structure (EXELFS), 294, 717, 718, 741, 742, 751–754, 757
Extended X-ray absorption fine structure (EXAFS), 294, 717, 751–753
Extinction distance
   apparent, 435
   determination of, 435
   effective, 245, 252, 408, 467, 471, 669
Extraction replica, 185, 191, 377, 378, 616, 724
Extraction voltage, 80–81
Extrinsic stacking fault, 473

F

Face-centered cubic, 259
   See also Crystal; Lattice
Fano factor, 593
Faraday cup, 82, 85, 101, 121–122, 596, 614, 653, 670
Fast Fourier transform (FFT), 534, 536
Fe^{55} source, 593
Fermi
   energy, 720, 742, 743
   level, 55, 720, 742, 743, 749
   surface, 63, 277, 278, 742
FIB (Focused-ion beam), 11, 157, 186, 188–189, 373, 634
Field-effect transistor (FET), 585, 586, 589, 591
Field emission, 61, 73, 74–75, 80–81, 498
   See also Gun
Filtered image, 120, 434, 550, 572, 681, 683, 692, 694–696, 699, 702, 703, 709, 717, 722, 725, 734, 741, 751, 756
Filter mask, 573
Fine structure, *See* Electron energy-loss spectrometry (EELS)
Fingerprinting, 704, 708, 744, 746–747, 751
Fiori definition, 614, 673
   See also Peak-to-background ratio
First-difference spectrum, 727, 733
First-order Laue zone, 351
   See also Higher-order Laue zone (HOLZ)
Fitting parameter, 645
Fixed-pattern readout noise, 689
Flat-field correction, 570
Fluctuation microscopy, 294, 366, 528
Fluorescence (light), 116
Fluorescence (X-ray), 607, 609, 612, 613, 630, 632, 639, 640, 647, 653, 654
   correction, 656–657
   yield, 55, 59, 587, 605–606, 628, 650, 715, 721, 729

Flux lines, 517, 526, 527
   See also Magnetic correction, flux lines
Focus, 110–111, 148, 151, 329–332, 399, 490–491, 682–683, 895–896
   See also Lens; Overfocus; Underfocus
Focused-ion beam, *See* FIB (Focused-ion beam)
Focusing circle (WDS), 681
Forbidden electron energies, 247
Forbidden reflection, 258, 263, 265, 288, 296, 299, 300, 301, 349, 350, 366
   See also Diffraction, pattern; Systematic absence
Foucault image, 516
Fourier
   analysis, 572
   coefficient, 17, 237, 240, 246
   component, 435
   deconvolution (logarithmic, ratio), 705
   fast (Fourier) transform (FFT), 534, 536
   filtering, 551, 571
   inverse transform, 536, 731, 732
   reconstruction, 551–552
   series, 237, 238
   transform, 485, 487, 536, 551, 568, 572, 573, 731, 732, 752–754
Frame averaging, 464, 466, 478, 550, 555, 557
Frame grabber, 551
Frame time, 120, 121
Free electron, 59, 61, 705, 708, 751
Free-electron density, 63, 706, 707, 710
Fresnel
   biprism, 77, 397–398
   contrast, 374, 389, 397–402, 540–541
   diffraction, 30–31, 229, 535
   fringe, 86, 87, 163, 389, 397, 400, 401, 402–403, 540–541, 575
   image, 403
   zone construction, 229
Friedel's law, 358
Full-potential linearized augmented plane wave (FLAPW), 748
Full Width at Half Maximum (FWHM), 83, 84, 85, 149, 150, 593–595, 605, 628, 631, 632, 643, 644, 667, 683, 684, 693, 694
Full Width at Tenth Maximum (FWTM), 84, 85, 86, 326, 594, 595, 643, 659, 664, 667, 668, 674
FWTM/FWHM ratio, 595, 601

G

Gas bubble, 399–400
Gas-flow proportional counter, 590
Gatan image filter (GIF), 681
Gaussian
   curve fitting, 645
   diameter, 84
   image plane, 103–104, 106–110, 488, 664
   intensity, 83, 84, 85
   statistics, 647, 673

Generalized-oscillator strength (GOS), 729
Generated X ray emission, 600, 612, 628, 634
Germanium detector, 587–588
   See also Spectrometer (X-ray energy-dispersive)
Ghost peak, 690, 691, 701, 723
Glaser, 491, 494
Glass layer, 278
Glide plane (dislocation), 443, 444
Glide plane (symmetry), 444
Goniometer, 132, 133, 151, 285
   See also Eucentric
GP zone, 277
Grain
   boundary, 6, 17, 216, 276, 278, 286, 302, 366, 380, 392, 395, 397, 400, 402, 409, 419, 420, 500, 502, 503, 517, 616, 659, 694, 747
   See also Stacking fault; Twin boundary
   coincident-site lattice, 500
   high-angle, 402
   low-angle, 400, 402
   rotation, 430
   small-angle, 454
   size, 116, 123, 197, 283, 284, 290, 291, 293
   texture, 292
   tilt, 402
   twist, 455
Gray level, 371, 555, 563
Gray scale, 123, 570, 635, 636
g·R contrast, 443, 450
Great circle, 286–287, 302
   See also Stereographic projection
Grid, 75, 77, 133, 134, 173, 174, 175, 182–185, 187, 188, 274, 458, 588, 594, 598, 599, 600, 609, 610, 611, 612, 621, 633
Gun
   alignment of, 498
   brightness of, 498
   crossover, 80
   emission current, 79, 82
   field-emission, 80–81
   filament, 81, 493
   flashing of, 81
   holography, 81
   lanthanum hexaboride, 81–82
   saturation of, 81
   self-biasing, 78
   tungsten, 81
   undersaturated image of, 493
   Wehnelt cylinder, 77, 78
   See also Electron, source; Field emission
G vector, 204, 224–225, 238, 253, 314, 315, 351, 384, 393, 425, 433, 442, 453, 458, 669
   See also Diffraction, vector (g)

H

Handedness, 363, 364
Hartree-Slater model, 729

Heisenberg's uncertainty principle, 745
Hexagonal close-packed crystal, 259, 267
  See also Crystal
Higher-order Laue zone (HOLZ)
  indexing, 348–352
  line, 351–352
  plane, 339, 340
  ring, 336, 337, 338, 341, 348, 351, 354, 355–356, 361, 362, 363, 364
  scattering, 334, 336
  shift vector t, 357
  simulation of, 351, 362
  See also Convergent beam, diffraction; Kikuchi diffraction, line
Higher-order reflection, 202, 207, 289
Higher-order waves, 48
Higher-order X-ray lines, 57
High-resolution TEM, 483–506
High voltage, 6, 13, 75, 76, 77, 96, 104–105, 108, 119, 179, 450, 451, 490, 495, 522, 588, 608, 681, 691, 722
High-voltage electron microscope, 6
History of the TEM, 5
Holder, See Specimen, holder
Hole-count, 608, 609
Holey carbon film, 86, 162, 163, 183, 184
Hollow-cone diffraction, 157, 291–293, 295
Hollow-cone illumination, 157, 291, 293, 381
Hollow-cone image, 158, 295, 528
Holography, 81, 397, 496, 524–526
Howie-Whelan equations, 224–226, 407, 413, 414, 421, 433, 434, 436, 442–443, 457
Hydrocarbon contamination, 138, 612
  See also Contamination
Hydrogenic edge, 717, 718, 719, 724, 731

I
Ice, 595
Illumination system, 75, 77, 78, 83, 93, 101, 141, 142–150, 161, 163, 311, 334, 584, 605, 608
  See also Condenser 1 lens; Condenser 2 lens
Image
  analysis of, 453, 491
  calculation of, 17, 537
  contrast in, 24, 77, 109, 222, 384, 469, 475
  coupling, 685
  of defects, 235, 409, 443
  delocalization in, 500, 735
  distance, 94, 95
    See also Lens
  drift, 76
  of flux lines, 526
  formation, 92, 93
  lattice-fringe, 392, 502, 559
  matching, 436, 540
  plane, 94, 95, 96

  See also Lens
  processing of, 16, 493, 549–556
  rotation of, 99, 167, 384
  simulation of, 15, 389, 429, 433, 458, 533–545, 563
  of sublattice, 500
Imaging system of TEM, 164–168
Incidence semiangle, 34
  See also Angle; Scanning transmission electron microscope (STEM), incoherent
Incommensurate structure, 503
Incomplete charge collection, 593, 595, 645
Incomplete read-out, 690
Inelastic, See Scanning transmission electron microscope (STEM), inelastic
Information limit, 492, 495
Information theory, 495
Infrared sensors, 362
In-hole spectrum, 610
In-line holography, 524
Inner potential, 235, 236, 238, 541, 545
  mean, 237
  scaled mean, 238
Inner-shell ionization, See Ionization; Instrument response function
In-situ holders, 135
In situ TEM, 138, 526, 531
Instrument response function, 731
Instrument spectrum, 613
Integration, 42, 273, 458, 470, 644
  total, 691
  window, 728
Integration approach, 273
Intensity, 371
  See also Spectrum, electron energy-loss, X-ray
Interaction constant, 486
Interband scattering, 254, 466, 478
Interband transition, 63, 66, 705, 709, 710, 711
Interface, 44, 275, 295, 430
  contrast, 667–668
  dislocation, 456
  interphase, 286, 302, 616, 617, 667
  semicoherent, 456
  strain at, 396
  See also Grain, boundary; Phase boundary
Interference, 13, 31, 40, 47, 48, 77, 185, 200, 202, 203, 241, 394, 398, 526
  constructive, 33, 34, 45, 213, 214
  destructive, 48, 200, 203
  fringe, 77, 396, 397, 398
Intergranular film, 402, 541
Intermediate-voltage electron microscope, 76, 323, 589, 600, 608, 616, 618, 626, 635, 636, 668, 673, 675
Internal-fluorescence peak, 612, 613, 630
International Tables, 265–266
Internet, 15–17, 117

Intersecting chord construction, 465
Inter-shell scattering, 749
Interstitial atom, 277
Intraband transition, 700, 709
Intra-shell scattering, 749
Intrinsic Ge detector, 585, 587–588
  See also Spectrometer (X-ray energy-dispersive)
Inversion domain boundary, 420, 503
Invisibility criterion, 424, 445
Ion beam blocker, 181
Ionic crystal, 66
Ionization, 57–59, 667, 715–721
  critical energy for, 55, 57, 58, 628, 682, 717, 723, 742, 743, 750
  cross section for, 55, 57, 588, 599, 605, 615, 650, 652, 717, 723, 725, 728, 734
  edge, 717, 723, 733–735
  integration of, 717
  intensity of, 715, 733
  jump ratio of, 722, 732
  onset of, 730
  shape of, 717
Ion milling, 174, 178–181, 182, 186
Ion pump, 130, 131, 137

J
Jump-ratio image, 728, 734

K
Kernel, 556
Kikuchi diffraction, 311–320
  3g, 464, 465
  band, 313, 314, 318, 319, 325, 327, 339, 340, 341, 348, 361, 512
  deficient, 313, 319, 340
  excess, 314, 318
  line, 311–313, 318, 327, 339–340, 366, 382, 413, 464, 466
  map, 303, 313–318, 319
  pair, 313, 314, 317, 318, 340
  pattern, 303, 311, 312, 314, 315, 317, 319, 325, 464–465
Kinematical diffraction, 235, 280, 358, 362, 521
  approximation, 221, 463, 470, 472
  crystallography, 559
  equation, 464
  integral, 469, 470
  intensity, 294
Kinematically forbidden reflection, 258, 265, 296
  See also Forbidden reflection; Systematic absence
Kinetic energy, 14, 68, 230, 235, 236, 238, 253, 477
K(k_{AB}) factor, 640, 646
  calculation of, 648–652
  error in, 647–648
  experimental values of, 646–647
Knock-on damage, 65, 66, 67, 68, 626, 646
  See also Beam (electron), damage
Kossel, 8, 323, 327, 336, 339, 348, 352, 354
  cone, 312, 313

Kossel (cont.)
  pattern, 327, 328, 331, 332, 341, 355, 356, 361
Kossel-Möllenstedt (K-M), 327, 328, 330, 336, 337, 342, 348, 352, 353, 354, 356, 361, 671
  conditions, 327, 330, 336, 342, 348, 354
  fringe, 352, 353, 361, 671
  pattern, 327, 328, 356, 361
Kramers' cross section, 60
Kramers-Kronig analysis, 705
Kramers' Law, 606, 643, 645, 726
k space, 534, 536, 752, 753, 754
Kurdjumov–Sachs, 303
k vector, 199, 200, 239, 246, 248, 249, 251, 313, 325, 432, 537

L

L_{12} structure, 262, 427
Laplacian filtering, 556
Large-angle convergent-beam electron-diffraction patterns, 330–332, 412
Lattice, 211–212, 257–258, 356–357, 361–363, 389–392, 400–402, 567–568
  centering, 338, 354, 356–357
  defect, 400–402, 407, 408, 422, 442, 454
    See also Dislocation
  fringe, 389–392, 502, 525, 559, 573
  imaging, 393, 500, 517, 567–568
  misfit, 415, 454
  parameter, 363, 394, 396, 415, 420, 424, 427, 454, 456, 540
  point, 215, 216, 237, 257, 259, 262, 278, 336, 357, 470
  strain, 347, 361–363, 456
  vector, 200, 211, 212, 239, 297, 302, 420, 421, 424, 427, 428, 429, 485, 489
    See also Crystal
Leak detection, 131–132
Least-squares refinement, 568
Lens, 91–112, 145–146, 148–149, 150–152, 161–164, 681–684
  aberration of, 485, 488
    See also Chromatic aberration; Spherical aberration
  astigmatism, 106, 162, 163
    See also Astigmatism
  asymmetric, 682, 683
  auto-focusing, 151
  auxiliary, 116
  bore of, 97, 145
  condenser, See Condenser 1 lens; Condenser 2 lens
  condenser-objective, 142, 143, 145–146
  current, 97, 151, 164, 330, 496, 513, 560
  defects, 99, 103, 107, 148, 331, 494, 553
  demagnification, 95, 144, 145, 154
  focal length of, 104
  focal plane of, 91, 92, 94
  focus of, 163

    See also Overfocus; Underfocus
  gap, 145
  hysteresis, 165
  immersion, 98
  intermediate, 152, 153, 154, 155, 162, 163, 164, 166, 167, 205, 206, 207, 330, 517, 683, 685, 687, 691
  low-field, 517
  mini-, 145
  Newton's equation, 95
  objective, See Objective lens
  octupole, 99, 104, 105, 106, 692
  optic axis of, 92, 93, 99, 104, 147
  pincushion distortion, 106, 165
  polepiece of, 143, 144, 599
  post-spectrometer, 683
  projector, 111, 131, 141, 162, 166, 167, 682, 683, 685, 686, 687, 691
  projector crossover, 682, 684, 686, 687
  ray diagram, 103, 325, 333
  rotation center of, 161–162, 163
  sextupole, 99, 683, 692
  snorkel, 98
  superconducting, 98, 99
  symmetric plane of, 151, 152
  thin, 92, 94, 95
  wobbling of, 148, 162
Library standard, 645, 704
Light element, 59, 61, 587, 590, 594, 595, 626, 679
Line analysis, 651, 695
Linear combination of atomic orbitals, 749
Linear elasticity, 442, 448, 495
Line of no contrast, 456, 457
Liquid N_2, 66, 129, 130, 132, 133, 180, 324, 363, 584, 585, 586, 595, 626
  dewar, 133, 584, 594
  holder, 132, 324, 336
Lithography, 187
Local-density approximation, 748
Long-period superlattice, 264–265
Long-range ordering, 279
Lorentz force, 99, 100, 398, 516
  microscopy, 81, 398, 515–517
Low-dose microscopy, 377, 556
Low-loss, 680, 690, 693, 699–711, 722, 727, 732, 733, 744
  intensity, 725
  spectrum, 694, 701, 703–711, 730, 731, 732
    See also Electron energy-loss spectrometry (EELS); Plasmon, fingerprinting

M

Magnetic correction, 514–515
  domain wall, 514, 516
  flux lines, 517, 526
  induction, 516, 517
  prism spectrometer, 679, 681, 682, 716
  recording media, 514
  specimen, 514–517

Magnification, 5, 7, 8, 11, 53, 76, 82, 86, 91, 95–96, 104, 106, 109, 110, 142, 143, 145, 147, 151, 153, 155, 161, 162, 164–165, 167, 168, 169, 206, 230, 264, 284, 327, 328, 330, 384, 389, 458, 466, 475, 493, 498, 512, 513, 518, 560, 562, 589, 599, 638, 670, 683, 686, 687, 688, 692
Many-beam, 240, 245, 390, 391, 408, 409, 433, 435, 436, 473, 478, 534, 565
  calculation, 436
  conditions, 478, 565
  images, 390, 391, 470, 473, 534
Mask, 174, 187, 324, 334, 336, 338, 342, 397, 476, 521, 523, 551, 556, 572, 573, 586, 631, 705, 723, 753
Mass-absorption coefficient, 654
  See also Absorption, of X-rays
Mass-thickness contrast, 185, 371, 373–379, 381, 382, 384, 407, 511, 696, 702
  See also Contrast
Materials examples in text
  Ag, 59, 66, 258, 374, 420, 424, 527, 588, 608, 609, 611, 649, 651
  Ag_2Al, 314, 316
  Ag_2Se, 291
  Al, 35, 45, 54, 64, 67, 68, 82, 117, 118, 122, 166, 174, 205, 224, 258, 262, 263, 303, 352, 353, 372, 374, 385, 397, 420, 427, 429, 543, 583, 611, 618, 633, 651, 656, 704, 706, 708, 727, 733, 750, 752
  Al-Ag, 649
  $Al_xGa_{1-x}As$, 263, 264, 420, 517, 519, 568
  Al_2O_3, 265, 266, 279, 296, 297, 298, 349, 379, 395, 410, 412, 413, 429, 519, 751
  Al_3Li, 262, 303, 383
  AlAs, 567, 568
  Al-Cu, 506
  Al-Li-Cu, 9
  Al-Mn-Pd, 504, 505
  Al-Zn, 618
  Au, 29, 35, 43, 45, 64, 68, 102, 117, 160, 166, 173, 224, 263, 374, 397, 400, 424, 452, 453, 492, 523, 553, 585, 610, 635, 651, 665
  Au_4Mn, 500, 501, 502
  B, 187, 587, 613, 689, 721
  $BaTiO_3$, 429
  Be, 59, 67, 173, 237, 258, 587, 588, 612, 633, 702, 732, 733, 752
  biotite, 644
  Bi-Sr-Ca-Cu-O, 504
  BN, 585, 590, 689, 705, 717, 719, 723, 732, 755
  Ca, 59, 258, 504, 505, 633, 634, 638, 646, 671, 728
  carbon, 35, 43, 59, 86, 163, 173, 183, 185, 207, 264, 295, 324, 365, 374, 375, 378, 397, 505, 521, 523, 528, 541, 551, 587, 612, 615, 633, 636, 670, 688, 695, 719, 722, 745, 751

See also Amorphous; Holey carbon film
carbon nanotube, 73, 81, 365, 366, 695, 745
catalyst particles, 4, 10, 617
CdTe, 181, 182
Co, 68, 258, 429, 457, 526, 630, 649, 651, 653, 732, 733
CoGa, 262, 395
Cr film, 518, 593, 612, 614, 615
CsCl, 262, 420
Cu, 29, 42, 43, 44, 59, 68, 173, 174, 187, 201, 237, 258, 263, 304, 305, 364, 385, 424, 429, 457, 466, 470, 471, 473, 475, 596, 611, 612, 618, 629, 633, 648, 665, 674, 675, 732, 743, 745, 746, 747, 751
Cu-Al, 649
Cu$_3$Au, 262, 263
CuAu, 420
CuCl$_{16}$PC, 557, 558
Cu-Co, 457
CuZn, 262
diamond, 173, 176, 178, 184, 224, 262, 289, 314, 401, 420, 585, 732, 743, 745, 751
Fe, 67, 68, 224, 258, 420, 502, 517, 630, 631, 632, 646, 652, 674, 732, 734, 744, 745
See also Materials examples in text, stainless steel, steel
FeAl, 262
Fe$_3$Al, 262, 426
Fe$_2$O$_3$, 295, 296, 297, 298, 395, 503
Fe$_3$O$_4$, 521
Fe-Cr-Ni, 633
Fe-Cr-O, 631
Fe-Mo, 649
Fe-Ni, 515
ferrite, 456
ferroelectric, 429, 514
GaAs, 8, 175, 176, 177, 186, 187, 236, 258, 263, 289, 351, 414, 420, 428, 517, 519, 521, 567, 568
Ge, 59, 67, 168, 224, 237, 245, 262, 380, 381, 392, 424, 454, 503, 527, 538, 552, 553, 560, 566, 583, 588, 594, 607, 632, 649
glass
  metallic, 293, 752, 753, 754
  oxide, 633
  silica, 583
  silicate, 517
graphite, 164, 186, 342, 410, 448, 528, 554, 583, 743, 745, 755
See also Carbon
hematite, 296, 297, 298, 456
high-T$_c$ superconductor, 99, 514
hydrofluoric acid, 173
icosahedral quasicrystal, 504
InAs, 392
K$_2$O·7Nb$_2$O$_5$, 558
latex particle, 374, 375, 376, 377

Mg, 57, 67, 258, 465, 607, 608, 629, 633, 646, 651, 654, 655, 672, 706, 733, 752
MgO, 68, 86, 176, 224, 261, 290, 317, 397, 409, 467, 468, 545, 655
Mo, 68, 101, 102, 264, 265, 277, 596, 608, 609, 610, 611, 617, 649, 659, 672
MoO$_3$, 167, 168
Na, 67, 258, 261, 293, 589, 627, 651, 653, 706, 725
NaCl, 176, 186, 261, 264
nanocrystals, 157, 283, 284, 290, 291, 293, 295
Nb, 67, 68, 506, 571, 572, 619, 620, 628, 649, 651, 653
Nb-Al, 649
Nb$_{12}$O$_{29}$, 539
Ni, 68, 173, 258, 262, 353, 394, 427, 449, 451, 502, 515, 586, 589, 593, 595, 596, 599, 619, 628, 633, 653, 655, 656, 657, 672, 680, 720, 726, 744, 746, 753, 754
NiAl, 262, 263, 420, 427, 503, 656
Ni$_3$Al, 258, 262, 263, 420, 427, 446, 473, 656, 746
Ni-Cr-Mo, 649
NiFe$_2$O$_4$, 420, 500
NiO, 175, 261, 394, 400, 402, 420, 426, 429, 453, 454, 455, 500, 501, 502, 513, 552, 593, 596, 599, 611, 614, 654, 720
nitric acid, 173, 174
ordered intermetallic alloy, 500
Pb, 66, 67, 375, 588, 598, 630, 649, 651
perchloric acid, 173, 174, 634
perovskite, 504
polymer, 3, 10, 30, 41, 60, 65, 66, 67, 86, 99, 109, 123, 124, 135, 138, 181, 184, 373, 375, 376, 377, 585, 587, 598, 702, 706, 709, 710
polystyrene, 710
polytype, 504
polytypoid, 504
Pt, 101, 102, 116, 173, 185, 188, 379, 506, 609
quantum-well heterostructure, 182
quartz, 10, 66
quasicrystal, 198, 504–505, 506, 543–544, 752
Sb, 629
Si, 120, 129, 181, 197, 528, 587, 588–589
SiC, 420, 429, 504, 695
Sigma (σ) phase, 729
Si/Mo superlattice, 264–265
Si$_3$N$_4$, 187, 188, 503, 695, 710
SiO$_2$, 65, 174, 177, 293, 376, 380, 381, 402, 566, 693, 710, 720, 746
SnSe, 452
SnTe, 396
spinel, 6, 217, 392, 401, 402, 420, 428, 454, 455, 456, 472, 493, 500, 501, 502, 513, 552
SrTiO$_3$, 6, 381, 706, 736

stainless steel, 174, 326, 341, 420, 424, 616, 633, 634, 657, 674, 724, 734
steel, 11, 378, 427, 514, 633, 659, 723
superconductor, 99, 173, 504, 505, 526
Ta, 68, 136, 137, 598, 627, 629
Ti$_3$Al, 420
Ti, 7, 67, 68, 130, 314, 629, 630, 631, 635, 651, 716, 723, 736, 744
TiAl, 262
TiC, 723, 724, 749
TiN, 723, 724
TiO$_2$, 428, 562, 563, 631, 635, 736
U, 487, 505, 506, 715
vanadium carbide, 264, 278, 500
wurtzite, 262, 420, 590
Y, 118
YBCO, 396, 397, 565
yttrium-aluminum garnet, 118
Zn, 68, 258, 447, 448, 449, 618, 627, 649, 651, 653, 708, 744
ZnO, 262, 348, 420, 541
ZnS, 116, 118, 237
Materials safety data sheet, 173
Mean-free path, 23
  elastic, 39–50
  inelastic, 53–69
  plasmon, 63–64
Mean-square vibrational amplitude, 435
Mechanical punch, 176
Microanalysis, 132, 133, 589, 657
  qualitative, 581
  quantitative, 76, 364, 433, 434, 478
  See also Spectrometer (EELS); Spectrometer (X-ray energy-dispersive); Spectrometer (X-ray wavelength-dispersive)
Microcalorimeter, 590–591
Microdensitometer, 85, 371, 550, 551
Microdiffraction, 528
  See also Convergent beam, diffraction
Microdomain, 517
Miller-Bravais notation, 260
Miller indices, 46, 49, 204, 212
Mini lens, 145, 146
Minimum contrast, 378, 492, 493, 497, 498, 561
  detectability, 379, 497
  detectable mass, 663, 674
  detectable signal-to-noise ratio, 497
  mass fraction, 663, 674
  resolvable distance, 107
MINIPACK-1, 571
Mirror plane, 358, 360, 361
  See also Point group; Symmetry
Mirror prism, 691
Modulated structure, 503, 504
Moiré fringes, 284, 298, 392, 393–397, 456
  complex, 396–397
  general, 393
  rotational, 393, 394
  translational, 393, 394
Molecular-orbital theory, 745

Möllenstedt spectrometer, 352
Momentum transfer, 753, 755
Monochromator, 76, 86, 105, 106, 319, 681, 693–694, 701, 705, 707, 722
Monte-Carlo simulation, 523
Moore's Law, 362
Moseley's Law, 58
Muffin-tin potential, 748, 749
Multi-channel analyzer (MCA), 591
Multi-element spectrum, 644, 647, 654
Multi-phase specimen, 665
Multi-photon microscopy, 4
Multiple domains, 304
Multiple least-squares fitting, 645, 731, 732
Multiple scattering, See Scanning transmission electron microscope (STEM), multiple
Multislice calculation, 533, 534, 536, 543, 544, 571
Multivariate statistical analysis, 619, 620, 659
Multi-walled nanotube, 365, 366
Murphy's law, 123, 707, 731

N

Nanocharacterization, 4
Nanodiffraction, 283, 291, 323, 324, 347, 365, 366
Nanomaterials, 4, 174, 483, 694, 704
Nanoparticles, 4, 189, 271, 276, 302, 366, 400, 632, 635, 695
Nanostructured electronics, 362
Nanotechnology, 3, 4, 154, 323, 324
Nanotubes, 4, 73, 77, 81, 365, 366, 689, 694, 695, 705, 745
Nanowires, 4
Near-field calculation, 535
Near-field microscopy, 4, 30, 535
Near-field regime, 229, 397
Néel wall, 517
Nematode worms, 756
NIST, 14, 29, 44, 58, 304, 608, 631, 633, 646, 647, 652, 653, 654
 multi-element glass, 647
 oxide glass, 633
 Sandia/ICPD electron diffraction database, 304
 thin-film standard (SRM 2063), 631, 641, 654
Noise, 81, 115, 116, 118, 119, 121, 122, 124, 376, 380, 386, 464, 466, 478, 492, 495, 497, 498, 522, 528, 541, 549, 556, 557, 561, 562, 563, 565, 567, 570, 572, 586, 588, 591, 593, 594, 598, 608, 619, 620, 631, 659, 675, 688, 689, 709, 725, 732
 reduction, 556, 557, 565, 675
 See also Signal-to-noise ratio

O

Objective lens, 101, 111, 152, 154–158, 161, 162, 167, 207, 278, 332, 372–373, 375, 376–379, 382, 385–386, 389, 390, 396, 411, 413, 448, 466, 489, 500, 511, 512–514, 516, 519, 520, 521, 528, 529, 539, 540, 552, 573, 600, 670, 686–688, 691, 706, 721, 755
 aperture, 91, 101–102
 astigmatism, 106, 162, 163
 collection semiangle of, 34
 defocus, 162, 163, 331, 553, 565
 diaphragm, 156
 focal increment of, 169
 instability of, 466
 polepiece, 143, 144, 599
 rotation alignment of, 162
 transfer function of, 485, 486, 487–488, 490, 491, 492, 494, 495, 552, 560
 See also Lens
Oblique-textured electron DP, 291
Omega (Ω) filter, 681, 691–692
On-axis image, 297, 391, 559, 560
Optical bench, 92, 198, 549, 573–574
Optical system, 82, 386, 389, 483–484, 490, 495
Optic axis, 34, 75, 79, 92, 96, 99, 100, 101, 104, 106, 110, 131, 143, 144, 147, 149, 151, 155, 158, 161, 162, 198, 205, 248, 285, 305, 313, 317, 382, 384, 390, 397, 463, 466, 485, 493, 514, 515, 519, 533, 599, 688, 691, 701
 See also Lens
Ordering, 263, 264, 272, 277, 278, 279, 366, 427, 517
 long-range, 279
 short-range, 277, 278, 517
Orientation imaging, 319
Orientation mapping, 305
Orientation relationship, 204, 283, 289, 302–303, 305, 339, 500
 cube/cube, 303
 Kurdjumov–Sachs, 303
 Nishiyama–Wasserman, 303
 precipitate-matrix, 302
O-ring, 131, 132, 133
Overfocus, 96, 116, 143, 145, 147, 148, 149, 162, 163, 164, 165, 207, 325, 330, 331, 400, 485, 500, 514, 515, 517, 626
 See also Lens; Underfocus
Overvoltage, 57, 626
Oxide layer, 504, 519, 710

P

Parallax shift, 511, 512, 513
Paraxial ray condition, 100, 104
Particle on a substrate, 81, 396, 413
Pascal, 128
Passband, 492–493
Path difference, 31, 33, 48, 49, 200, 201, 202, 488
Path length, 33, 100, 599, 600, 609, 655, 656, 682
 See also Absorption, of X-rays
Pathological overlap, 628, 630
Pattern recognition, 561, 562–563, 569

Pauli exclusion principle, 744
Peak-to-background ratio, 614
Peak (X-ray characteristic), 469–470, 605–606, 614, 627–634, 644–646, 701–702
 deconvolution of, 630–632
 integration of, 644–646, 650, 727, 729
 overlap of, 589, 590, 597, 627, 630
 visibility of, 632–634
Periodic continuation method, 542
Phase boundary, 191, 419, 420, 429, 447, 502, 503
 distortion function, 488
 of electron wave, 31, 47
 factor, 191, 419, 420, 429, 447, 502, 503
 grating, 534, 535, 536
 negative, 487
 object approximation, 486
 reconstructed, 557
 shift, 46, 471, 486, 488, 491, 526, 753
 transformation, 136, 456, 526
 See also Contrast, difference; Interface
Phase contrast, 77, 86, 106, 163, 164, 169, 371, 373, 380, 381, 389–403, 411, 487, 488, 490, 492, 493, 494, 495, 505, 511, 515, 543, 545, 557, 668, 696, 703
Phasor diagram, 31, 32, 421, 426, 470–473
Phonon, 59, 63–64, 336, 680, 700, 701, 702
Phosphorescence, 116
Photo-diode array, 683
 See also Diode array
Photographic dodging, 550
Photographic emulsion, 66, 122–124, 197, 464, 561
Photomultiplier, 117, 118–120
 See also Scintillator-photomultiplier detector
p-i-n device, 586
Pixel, 120, 121, 123, 124, 159, 305, 496, 497, 556, 562, 563, 564, 567, 568, 569, 571, 590, 618, 619, 620, 636, 658, 675, 685, 694, 727, 734, 736
Pixel-clustering, 734
Planar defect, 250, 254, 263, 275–277, 286, 302, 347, 419–436, 452, 472, 503, 504, 600, 616, 669, 670
 inclined, 250, 431, 669
 See also Grain, boundary; Stacking fault; Twin boundary
Planar interface, 275, 503, 600, 694, 748, 749
Plane normal, 213, 287, 288, 299, 300, 301, 302, 303, 317, 419, 458, 519, 538, 670, 682
Plane wave, 31, 32, 33, 40, 45, 46, 48, 49, 200, 237, 245, 249, 250, 748
 amplitude, 239–241
Plasma cleaner/cleaning, 132, 137, 138, 189, 626
Plasmon, 63–64, 109, 680, 682, 693, 700, 703, 705–708, 710, 717, 719, 720, 722, 726, 731, 732, 752, 756

energy, 63, 64, 109, 706, 709
excitation, 54, 63, 700, 717
fingerprinting, 704, 708
frequency, 63
loss, 682, 702, 706, 707, 708, 709, 711, 717, 718, 719, 731, 732
  See also Low-loss, spectrum
mean-free path, 63–64
peak, 63, 680, 701, 703, 704, 705, 706, 707, 708, 709, 710, 717, 720, 722, 723, 726, 730, 752
Plural elastic scattering, 741
  See also Elastic, scattering; Scanning transmission electron microscope (STEM), elastic
p-n junction, 62, 117, 118, 523
Point analysis, 354
Point defect, 65, 277, 278, 448, 502
  See also Interstitial atom; Vacancy
Point group, 9, 332, 347, 354, 358, 359, 361, 364, 366
  determination of, 358
  symmetry of, 237, 358, 361
  two-dimensional, 354, 486, 511
Point-to-point resolution, 493, 506
Point-spread function, 483, 593, 630, 688–689, 701, 753
Poisson's ratio, 444, 457
Poisson statistics, 593
Polepiece, See Lens
Polycrystalline material, 290–291, 293, 319, 452, 613
Polymer, 3, 10, 30, 41, 60, 64, 65, 66, 67, 86, 99, 109, 123, 124, 132, 135, 138, 181, 183, 184, 373, 375, 376, 377, 585, 587, 588, 598, 702, 706, 709, 710
Polytype, 504
Polytypoid, 504
Position-tagged spectrometry, 620, 659
Post-specimen lens, 85, 110, 161, 326, 329, 342, 379, 683, 686
  See also Lens
Potential
  inner, 235, 236, 237, 238, 239, 240, 397, 399, 400, 402, 540, 541, 545
  periodic, 236, 237, 254
  projected, 486, 487, 511, 534, 537, 538, 541, 557, 563, 565
  well, 120, 402, 541, 542, 720, 721, 749
Precession CBED, 342
Precession diffraction, 147, 158, 284, 285, 293, 295
Precision ion milling, 181
Precision ion polishing, 181
Primitive great circle, 286, 287
  See also Stereogram
Primitive lattice, 257–258, 356, 357
  See also Crystal; Lattice
Principal quantum number, 744
Probability map, 536
Probe, 8, 9, 48, 81, 82, 95, 98, 103, 111, 124, 135, 144, 146, 148, 149, 150, 158,
161, 189, 291, 325, 339, 340, 362, 377, 402, 496, 522, 584, 589, 590, 610, 611, 612, 614, 615, 616, 618, 625, 635, 636, 647, 659, 668, 672, 673, 680, 694, 733, 735, 750
  current, 84, 85, 110, 124, 143, 149, 150, 610, 612, 614, 616, 617, 618, 626, 627, 641, 653, 668, 683, 733
  size, 82, 84, 85, 86, 144, 145, 148, 149, 150, 326, 523, 590, 614, 626, 627, 636, 640, 641, 659, 668, 673, 733, 735
  See also Beam (electron)
Processing HRTEM image, See Image, processing of
Propagator matrix, 423
Pulse processing, 591, 592, 596, 598, 635, 636
Pump, vacuum, 127–138, 178, 180, 181, 184, 189, 586, 683, 687
  cryogenic, 130
  diffusion, 129, 130, 131, 180, 191
  dry, 189
  ion, 130, 131, 137
  roughing, 128–129, 130
  turbomolecular, 129–130

Q
Quadrupole, 98, 99, 104, 105, 683, 692
  See also Lens
Qualitative mapping, 635
Qualitative microanalysis, 581
Quantifying HRTEM images, 549–575
Quantitative chemical lattice imaging, 517, 567–568
  defect contrast imaging, 411, 422, 470
  HRTEM, 567
  image analysis, 561–562
  mass-thickness contrast, 373–379, 381, 382, 384, 696, 702
  microanalysis, 132, 133, 589, 657
Quantitative mapping, 658, 659, 671
QUANTITEM, 563–567
Quantum-mechanical convention, 230
Quantum number, 744
Quasicrystal structure, 198, 504–505, 506, 543–544, 752

R
Racemic mixture, 636
Radial-distribution function (RDF), 293, 294, 373, 703, 741, 752, 753, 755
Radiation damage, 6, 10, 53, 64, 65, 66, 68, 119, 448
  See also Beam (electron), damage
Radiolysis, 64, 65, 66, 68, 646
  See also Beam (electron), damage
Ray diagram, 91, 92–94, 100, 101, 102, 103, 104, 105, 111, 143, 147, 152, 154, 156, 157, 198, 295, 325, 326, 327, 328, 330, 333, 610, 692
Rayleigh criterion, 5, 84, 107, 108, 490

Rayleigh disk, 108, 484
Real space, 211, 212, 213, 226, 236, 258, 262, 264, 265, 271, 279, 286, 302, 319, 332, 355, 356, 410, 412–413, 427, 485, 492, 534, 536, 542, 572, 748
  approach, 534, 536, 563, 572
  crystallography, 412–413
  patching method, 542
  unit cell, 262
  vector, 236
Reciprocal lattice, 200, 202, 211–212, 213–216, 235, 254, 258, 259, 260, 262, 271, 273, 278, 280, 289, 290, 292, 297, 336, 337, 348, 356, 357, 430, 431, 435, 470, 489, 538, 552
  formulation of, 535
  origin of, 215
  point, 202, 215, 216, 253, 258, 262, 278, 290, 336, 351, 357, 470, 552
  rod, 214, 273, 337
  See also Relrod
  spacing, 338, 538
  vector, 200, 211, 212, 213, 235, 239, 290, 297, 302, 485, 489
  See also Diffraction, vector (g)
Reciprocity theorem, 94, 381, 386, 521
Recombination center, 524
Reference spectra, 701, 723, 732, 733
Reflection electron microscopy, 420, 519–520
Reflection high-energy electron diffraction, 519
Refractive index, 5, 225, 230, 238, 239, 358
Relative-transition probability, 650
Relative transmission, 670
Relativistic effect, 6, 14, 41, 42, 700
Relrod, 214, 215, 216, 249, 271–277, 279, 280, 281, 289, 336, 337, 338, 410, 430, 431, 441
  See also Reciprocal lattice, rod
Replica, 164, 165, 185, 377, 378, 616, 724
Resolution, 5–7, 91–112, 483–507, 589, 663–676, 735
  atomic level, 381
  limit, 4, 6, 103, 104, 109, 323, 490, 492, 493, 494, 594, 701, 703, 730
  theoretical, 107–108, 594
Resolving power, 5, 33, 106, 107, 124
Reverse-bias detector, 585
Richardson's Law, 74
Right-hand rule, 99
Rigid-body translation, 420
Rose corrector, 494

S
Safety, 10, 102, 173–174, 175, 176, 178, 189
Scan coil, 147, 149, 157, 158, 159, 165, 293, 295, 305, 326, 366, 753
Scanning image, 101, 115, 116, 118, 122, 124, 159, 161, 376

Scanning transmission electron microscope (STEM), 8, 158–161, 326, 372–373, 376–377, 384–386, 528
  annular dark-field image, 161
  bright-field image, 122, 159–161
  coherent, 373, 379
  cross-section, 158, 168
  dark-field image, 161
  detectors in, 326, 366, 372–373, 385, 386, 511, 528, 687
  diffraction contrast in, 384–386
  digital imaging, 618
  elastic, 373
  factor, *See* Atomic, scattering factor
  forward, 122, 735
  image magnification in, 165
  incoherent, 39, 43, 106, 107, 378, 379, 381
  inelastic, 319
  inter-shell, 749
  intra-shell, 749
  mass-thickness contrast in, 377
  matrix, 185, 374
  mode, 124, 151, 159, 166, 326, 584, 626, 632, 685, 721, 733, 734, 735, 755
  multiple, 493, 699, 748, 753
  multiple-scattering calculations, 753
  nuclear, 42, 45
  plural, 493, 699, 700, 721
  post-specimen, 110, 161, 326, 342
  Rutherford, 41–44, 161, 373, 374, 378
  semiangle of, 84, 378
  single, 29, 43
  strength, 83, 135, 326, 354, 584
  thermal-diffuse, 64, 336, 338
  Z contrast, 44, 379–381, 506, 543, 545, 567, 746
  See also Angle; Coherent; Elastic; Z contrast, scattering
Scherzer, 490–491, 492, 494, 495, 498, 500, 553
  defocus, 490–491
Schottky, 62, 73, 74, 75, 81, 82, 117, 497, 498, 617, 683, 693
  diode, 117
  See also Detector (electron)
  emitter, 75, 498
Schrödinger equation, 46, 222, 230, 235, 236, 237, 238, 239, 748
Scintillation, 116
Scintillator-photomultiplier detector, 118–120
Screw axis, 543, 544
  See also Space Group; Symmetry, screw axis
Secondary electron, 24, 53, 54, 60–62, 115, 118, 188, 373, 522, 709
  detector, 373
  fast, 705
  imaging of, 522–523
  slow, 605, 606
  types of, 522

Segregation to boundaries, 659
Selected area diffraction (SAD), 141, 152–155, 156, 157, 160, 166, 167, 204–207, 283, 284–285, 289, 295, 311, 313, 323, 324, 325, 326, 330, 332, 333, 334, 339, 342, 347, 348, 352, 365, 376, 411, 412, 413, 493, 498, 505, 525, 551, 680, 688
  aperture, 152, 154, 155, 158, 166, 167, 205, 206, 332, 339, 376, 411, 493, 525, 551, 688
  error, 498
  pattern exposure, 154, 157, 160, 205, 206, 283, 284, 412
Selection rules, 257, 258, 267, 289, 302, 744
Semiangle, *See* Angle; Bragg; Collection semiangle; Convergence semiangle; Incidence semiangle
Semiconductor detector, 117–118, 119, 122, 161, 523, 581, 585–589, 594, 607
Semi-quantitative analysis, 723
Shadowing, 185, 374, 375, 377, 378, 541
Shape effect, 271, 273, 290, 559, 572
Short-range ordering, 277, 278, 517
Side-entry holder, 132, 133, 134, 135, 150, 285
  See also Specimen, holder
SIGMAK(L) program, 729
Signal-to-background ratio (jump ratio), 703, 722, 728, 732, 734, 736
Signal-to-noise ratio, 116, 118, 124, 497, 498, 556, 570, 688
Signal processing, 376, 377, 572, 589, 590, 594, 606
Silicon-drift detector, 588–589
Silicon dumbbells, 391, 575
Si(Li) detector, 585, 586, 587, 588, 591, 594, 595, 606, 607, 626, 628, 630, 638, 651
  See also Spectrometer (X-ray energy-dispersive)
Simulated probe image, 668
Single-atom detection, 663, 715
Single-atom imaging, 54, 378, 663, 674, 679, 715, 736
Single-electron counting, 700, 709
Single-electron interaction, 709
Single-period image, 564
Single scattering, *See* Scanning transmission electron microscope (STEM), single
Single-sideband holography, 524
SI units, 14, 65, 128, 654, 665
Slow-scan CCD, 478, 553, 559, 570, 692
  See also Charge-coupled device (CCD) camera
Small-angle cleaving, 186
Small circle, 262, 287, 428
  See also Stereogram
Smearing function, 483, 487
  See also Point-spread function

Space group, 9, 17, 47, 266, 267, 296, 347, 354, 358, 361, 540, 558
Spatial coherence, 77, 398, 491, 498
Spatial resolution, 8, 29, 54, 62, 76, 77, 85, 87, 148, 323, 324, 325, 329, 347, 348, 352, 362, 490, 523, 581, 589, 619, 625, 626, 640, 647, 658, 659, 663–676, 688, 705, 721, 730, 733, 735–736, 752, 753, 755
Specimen
  90°-wedge, 186, 187, 414
  artifacts in, 541
  bulk, 25, 44, 135, 136, 284, 520, 589, 599, 607, 608, 617, 635, 639, 640, 643, 646, 647, 650, 655, 674
  cooling of, 10
  damage to, 10, 24, 64, 636, 673, 680, 753
  See also Beam (electron), damage
  density of, 656
  double-tilt, 134, 135
  drift of, 136, 376, 466, 495, 496, 584, 616, 620, 641, 647, 658, 668, 673
  EBIC, 136, 523
  heating, 64, 65–66
  height of, 101, 150, 151, 327, 512, 683
  See also z control
  holder, 10, 11, 82, 97, 121, 127, 132–133, 134, 135, 150, 151, 169, 175, 187, 207, 512, 514, 612, 613, 653, 655
  low-background, 135, 324, 584, 600, 626
  multiple, 134, 135
  orientation of, 238, 323, 494, 614, 630, 721, 733, 755
  preparation of, 11, 134, 173–192, 416, 503, 616, 626, 633, 669, 671
  quick change, 134
  rotation of, 285
  self-supporting, 173, 174, 175
  single-tilt, 134, 135, 324
  single-tilt rotation, 324
  spring clips for, 133
  straining, 136, 137
  surface of, 61, 62, 65, 136, 158, 179, 275, 395, 411, 433, 448, 455, 522, 523, 599, 633, 707
  thickness of, 11, 29, 63, 109, 110, 111, 164, 197, 323, 329, 352, 402, 466, 487, 565, 595, 627, 654, 655, 656, 669, 671, 675, 679, 702, 705, 708, 721, 726, 727, 730
  See also Thickness of specimen
  tilt axis, 169
  tilting of, 134, 181, 187, 228, 274, 285, 289, 382, 394, 430, 447, 511, 515, 671, 755
  top-entry, 133, 134, 136, 169
  transmission function, 485
  vibration, 495
  wedge-shaped, 274, 408, 410, 564, 565
Spectrometer (EELS)
  aberrations of, 682, 683, 684
  artifacts in, 689–690

calibration of, 684
collection semiangle of, 685–688
dispersion of, 683
entrance aperture of, 681, 682, 684, 686, 687, 688
entrance slit of, 294
focusing of, 682–683
object plane of, 683, 685
post-spectrometer slit, 683
resolution of, 683–684, 722
See also Electron energy-loss spectrometry (EELS)
Spectrometer (X-ray energy-dispersive), 7, 581
artifacts of, 689–690
Au absorption edge, 613
Au contact layer, 651
automatic shutter, 611
clock time, 593, 596, 597
collection angle of, 101, 598–599
contamination of, 132
dead layer, 585, 586, 595, 596, 651
dead time, 592, 593, 596, 597, 598
efficiency of, 715, 721
escape peak, 606, 607, 612, 614, 628, 630, 656
incomplete-charge collection, 59
internal-fluorescence peak, 612, 613, 630
leakage current of, 593, 689, 690, 691
live time, 593, 596, 614
performance criteria for, 688
residuals in, 645
shutter, 594, 595, 600, 611
sum peak, 607, 608, 628, 630, 632
system peaks, 613, 627, 628, 632
take-off angle of, 614, 654, 655
time constant, 593, 596, 597, 598, 627, 630
window, 702, 703, 725
Spectrometer (X-ray wavelength-dispersive), 589–591
Spectrum
electron energy-loss, 680–681
X-ray, 26, 57, 60, 175, 591, 605–621, 643, 699
Spectrum imaging, 605, 619–620, 659, 680
Spectrum-line profile, 616, 617, 620, 695, 733
Sphere of projection, 286
See also Stereographic projection
Spherical aberration, 6, 84, 100, 103–104, 108, 148, 205, 331, 488, 490, 494, 610, 635, 703
broadening, 84
coefficient, 84, 104
error, 108
See also Aberration
Spinodal decomposition, 504
Spinorbit splitting, 744
Spin quantum number, 744
Spot mode, 144, 149, 326, 584, 615–616
Spurious peak, 628

Spurious X-ray, 608, 609, 611, 635, 655
Sputtering, 65, 66, 68, 178, 180, 181, 646, 721
Stacking fault, 67, 250, 275, 279, 419, 420, 421, 422, 424–427, 441, 446, 447, 448, 458, 471, 472, 473, 475, 476, 503, 524
contrast, 424, 425, 426
energy, 446, 448, 458, 473, 476
fringes, 402
inclined, 472
intrinsic, 475
overlapping, 426–427
See also Planar defect
Stage, 24, 34, 62, 68, 81, 98, 101, 115, 118, 119, 121, 122, 123, 127, 130, 131, 132–133, 136, 137, 138, 150–152, 154, 169, 177, 188, 197, 214, 223, 224, 327, 330, 347, 374, 466, 512, 515, 522, 523, 540, 584, 586, 587, 589, 598, 599, 605, 611, 612, 614, 616, 650, 655, 673, 753
Staining, 66
Standard Cr film, 518, 593, 612, 614, 615
Standard specimen, 164, 168, 305, 362, 599, 652, 744
Stationary-phase method, 470
Statistical criterion, 632, 673, 674
Statistically significant peak, 632
Stereogram, 286, 287, 303, 359
Stereographic projection, 286–287, 288, 302, 311, 315, 317, 348, 358
Stereology, 512
Stereomicroscopy, 285, 442, 450
Stigmators, 106, 149, 162, 163, 164
See also Astigmatism
Strain, 127, 135, 136, 278, 279, 339, 347, 361–363, 394, 396, 399, 400, 415, 441–442, 443, 444, 447, 448, 451, 452, 456, 457, 468–469, 476, 526
analysis of, 362
contrast, 399, 456, 457
field, 279, 371, 394, 399, 400, 441–459, 468–469, 476
lattice, 347, 361–363, 456
measurement of, 347, 352, 361–363
Strain engineering, 362
Straining holder, 136, 137, 551
See also Specimen, holder
Strain-layer superlattice, 362
Stress, 74, 81, 133, 134, 136, 182, 189, 324, 341, 375, 441, 477, 486, 495, 504, 527, 625
field, 136, 441
Strong-beam image, 382, 421, 424, 425, 467, 474, 476
Structure correlation, 294
Structure factor, 46–47, 223, 224, 257, 258–259, 260, 261, 262, 267, 272, 274, 284, 290, 332, 336, 347, 351, 356, 364–365
Structure-factor determination, 261, 364–365

Structurefactor-modulus restoration, 559
Student *t* value, 648
Substitutional atom, 277, 657
Substitutional site, 658
Substrate, 4, 81, 186, 187, 296, 302, 379, 393, 394, 395, 396, 397, 413, 452, 518, 519, 583, 751
Summation, 84, 108, 223, 239, 272–273, 569, 654
Sum peak, 592, 606, 607, 608, 628, 630, 632, 633, 638
Superlattice, 262–265, 277, 362, 415, 420, 427, 447, 504, 517
dislocation, 447
reflection, 262–265, 427, 517
See also Ordering
Surface barrier detector, 117, 118
diffusion, 527
dislocation, 452, 453
faceting, 392
groove, 540–542, 751
imaging, 519–521
layer, 81, 117, 182, 187, 380, 451, 452, 453, 519
plasmon, 707
reconstruction, 519
relaxation, 363, 458, 517, 527
of specimen, 61, 62, 65, 136, 179, 275, 395, 411, 433, 448, 455, 511, 522, 523, 599, 633, 707
Symmetry
bright-field projection, 155, 159–161, 361
determination of, 295, 340, 354, 357–361, 366
elements of, 266, 287, 305, 357, 358, 359, 361
glide plane, 442, 443, 444, 449
inversion, 359, 360, 365
projection-diffraction, 361
rotational, 360, 361
rotation axis, 358, 360, 412
screw axis, 543, 544
three dimensional, 364
translational, 504, 544
whole-pattern (WP), 361
zone-axis, 348, 352, 380, 383, 412, 458
See also Mirror plane; Point group; Space group
Systematic absence, 259, 260, 290, 304, 348, 354, 356, 357
See also Forbidden reflection
Systematic row, 202, 205, 230, 249, 332, 411, 435, 457, 458, 464, 465, 466, 657
See also Diffraction, pattern
System peaks, 613, 627, 628, 632

T

Template, 305, 396, 397, 562, 563, 564, 567, 568, 569, 570, 572, 573, 659, 693
Temporal coherence, 76–77, 85
See also Spatial coherence

Texture, 283, 284, 290, 291, 292, 319, 557
  See also Grain
Thalidomide, 363
Thermal-contraction coefficient, 363
Thermal-diffuse scattering, 64, 338
Thermionic gun, 74, 77–80, 82, 144
  See also Gun
Thickness of specimen, 703
  dependence, 538, 567
  determination of, 352–354, 361, 415, 670, 671, 730
  effect on contrast, 373–379
  effective, 337, 415
  fringe, 389, 407, 408–411, 413, 414–415, 424, 426, 430, 432, 446, 452, 456, 467, 468, 472, 521, 669
  image, 731
  simulation of, 414
Thin-foil criterion, 640, 654
  See also Cliff-Lorimer
Thin-foil effect, 217–218, 273–274
  See also Shape effect
Threshold energy, 67, 750
  See also Beam (electron), damage
Through-focus dark-field, 513
  See also Dark field (DF)
Through-focus image, 402, 493, 506
Tie line, 248, 249, 250, 251, 252, 254, 413, 431, 432
  See also Dispersion diagram, surface
Tilted-beam condition, 390
Tomography, 10, 135, 442, 511–512, 741, 755–757
Top bottom effect, 297, 298
Top-hat filter function, 643, 644
Topotaxy, 296
Torr, 128, 129, 130
Trace element, 625, 632, 634, 659, 672, 732
Transfer function, 485, 486, 487–488, 490, 491, 492, 494, 495, 552, 560
  effective, 491
  See also Contrast, transfer function
Translation boundary, 419
Transmission electron microscope, 3–18
Tripod polisher, 177, 178
TV camera, 111, 118–120, 141, 326
  See also Charge-coupled device (CCD) camera
Twin boundary, 6, 169, 207, 217, 276, 402, 419, 420, 427, 454, 455, 472, 503
  See also Grain, boundary; Stacking fault
Twin-jet apparatus, 178
Twin lens, 145, 146
Two-beam approximation, 224, 227, 390, 424, 435, 478
  calculation, 457
  condition, 224, 230, 296, 352, 381–382, 383, 384, 385, 422, 444, 658, 669, 733
Two-photon microscopy, 4
Two-window method, 643

U
Ultrahigh vacuum, 13, 55, 128, 129–130
Ultramicrotomy, 180, 183–184, 634
Ultra-soft X-rays, 590
Ultrasonic cleaning, 176, 182, 184
Ultra-thin window, 586, 587, 588
  See also Spectrometer (X-ray energy-dispersive)
Uncollimated electrons, 610
Underfocus, 96, 142, 143, 145, 147, 148, 149, 154, 162, 163, 164, 165, 198, 205, 207, 229, 330, 381, 382, 400, 466, 492, 500, 514, 515, 516, 517, 553
  See also Lens; Overfocus
Unfilled states, 55, 742, 743, 744, 746
Unit cell, 46, 47, 212, 223, 257, 258, 261, 262, 264, 272, 273, 274, 324, 338, 347, 354–357, 421, 427, 435, 447, 500, 540, 542, 544, 558, 563, 567, 571, 655, 665, 669, 748
  determination of, 354–357
  image, 567
  scattering amplitude from, 258
  volume of, 28, 65
  See also Crystal; Lattice

V
Vacancy, 66, 67, 277, 449, 528
  loop, 449
  See also Dislocation, loop
  ordered array, 264, 278, 542
Vacuum
  backing valve, 131
  high, 128, 129–131, 523, 752
  low, 128
  pumps, 127
  rough, 128
  tweezers, 175, 189
  ultrahigh, 129–131
  wave vector in, 249
Vacuum level, 742
Valence-electron density, 365, 709
Valence state image, 679, 743, 744, 745, 747, 748
Valence-ultraviolet spectrum, 705
Végard's law, 363
Video image, 550
  See also Charge-coupled device (CCD) camera
Viewing screen, 5, 11, 26, 75, 76, 82, 85, 97, 115, 116–117, 119, 124, 141, 143, 144, 146, 148, 149, 151, 152, 153, 154, 155, 156, 162, 166, 199, 224, 371, 372, 681, 682, 685, 686, 691, 694, 696
Visible-light microscope, 5, 86, 91, 198
Void, 374
Voltage centering, 162, 493
Von Laue, 48, 49, 358
  condition, 199, 214, 216, 271
  equations, 48, 49, 201, 203, 213–214
  groups, 49
  X-ray pattern, 218

zone, 216, 217, 289, 290, 311, 323, 335–338, 349, 351, 355, 356, 357, 458, 614
  See also Higher-order Laue zone (HOLZ); Zero-order Laue zone

W
Wafering saw, 176
Water vapor, 124, 127, 132, 136, 189, 586, 587, 595
Wave
  diffracted, 48, 199, 200, 259, 536
  equation, 235–236, 247, 495, 728
  function, 33, 222, 224, 227, 236, 237, 241, 252, 399, 486, 487, 535, 728, 748, 753
  incident, 34, 36, 48, 199, 215, 216, 486
  matching construction, 248
  total amplitude of, 31
  total function, 222, 224, 227–228, 236, 252
  vector, 45, 199, 200, 215, 216, 222, 224, 227, 238, 239, 241, 245, 246, 248, 249, 251, 399, 432, 435
Wavefront, 30, 31, 32, 45, 103, 199, 200, 223
Wavelength-dispersive spectrometer (WDS), 589–591
  See also Spectrometer (X-ray wavelength dispersive)
Weak-beam dark field, 463–479
  3g reflection, 464, 466
  condition, 453
  of dislocation, 469–470
  thickness fringes in, 467
  See also Dark field (DF)
Weak phase-object approximation, 486
WEBeMAPS, 350, 351
Wedge specimen, 186, 187, 249, 250, 275, 279, 408, 414, 430, 431, 446, 472, 641
Wehnelt, 75, 77, 78, 79, 132
  bias, 75, 78, 79
  cylinder, 77, 78
  See also Gun; Thermionic gun
Weighting factor, 258
Weiss zone law, 204, 289, 335, 348
White line, 463, 723, 728, 729, 736, 744, 745, 746, 748
White noise, 497, 572
White radiation, 198, 215
Wien2k code, 711
Wien Filter, 693
Windowless (X-ray energy-dispersive spectrometer) detector, 587, 630, 723
Window polishing, 183
Window (X-ray energy dispersive spectrometer), 586–587
Wobbling (lens), 148, 162
Work function, 74
World Wide Web, 15
Wulff net, 286, 287, 302

X

X-ray
- absorption of, 28, 294, 599, 600, 634, 641, 653, 656, 671, 694, 717, 742, 751, 752
  See also Absorption, of X-ray
- absorption near-edge structure, 742
- atomic-number correction, 640, 650
- bremsstrahlung, 40, 55, 60, 135, 598, 605, 606, 608, 609, 610, 611, 613, 635, 643
- characteristic, 653
- count rate, 589, 598, 599, 616, 625, 641, 658, 673
- depth distribution of, 654, 655, 665
- detector, *See* Spectrometer (X-ray energy-dispersive)
- diffraction, 5, 30, 197, 198, 215, 290, 545
- emission of, 55–60, 61, 63, 650, 654, 657, 658, 669, 715, 733, 735
- energy of, 7, 55, 58, 60, 581, 582, 585, 587, 588, 591, 593, 594, 595, 605, 606, 626, 629, 641, 643, 652, 653
- energy-dispersive spectrometry (XEDS), *See* Microanalysis; Spectrometer (X-ray energy-dispersive)
- families of lines, 605, 628, 629
- fluorescence, 612, 721
- image, 605, 616–620, 634, 636, 658, 755
- interaction volume, 663, 664, 665, 672, 674, 735
- map, 381, 550, 567, 584, 588, 616, 617, 618, 619, 658–659, 672
- peak-to-background ratio, 673
- scattering factor, 435
- spectra, 87, 581, 590, 605–621, 645, 647, 653, 688, 717, 723, 742, 751
  See also Microanalysis; Spectrometer (X-ray energy-dispersive); Spectrometer (X-ray wavelength-dispersive)

Y

YAG scintillator, 685, 688, 691
Young's modulus, 457
Young's slits, 13, 30, 31, 279

Z

ZAF correction, 639
Z contrast, 44, 161, 379–381, 506, 543, 545, 567, 746
- scattering, 379, 380
z control, 151, 327, 330, 332, 512
Zero-energy strobe peak, 631
Zero-loss, 630, 671, 680, 683, 690, 691, 699, 701–703, 707, 725, 730, 731, 753
- integral, 729
- peak, 701–703
Zero-loss filter, 703
Zero-order Laue zone, 216, 289
Zero-order wave, 48
Zeta-factor, 652–654, 656
z-factor, 641, 646, 653, 654, 656, 659, 671, 672, 675
Zone axis, 348, 352, 380, 383, 412, 458
- high-symmetry, 338, 356
- image, 537
- low-symmetry, 338, 356
- orientation, 348, 380, 614
- pattern, 216, 303, 319, 412
- symmetry, 338, 356

Printed by Books on Demand, Germany